COMBINATORIAL OPTIMIZATION IN COMMUNICATION NETWORKS

Combinatorial Optimization

VOLUME 18

Through monographs and contributed works the objective of the series is to publish state of the art expository research covering all topics in the field of combinatorial optimization. In addition, the series will include books, which are suitable for graduate level courses in computer science, engineering, business, applied mathematics, and operations research.

Combinatorial (or discrete) optimization problems arise in various applications, including communications network design, VLSI design, machine vision, airline crew scheduling, corporate planning, computer-aided design and manufacturing, database query design, cellular telephone frequency assignment, constraint directed reasoning, and computational biology. The topics of the books will cover complexity analysis and algorithm design (parallel and serial), computational experiments and application in science and engineering.

COMBINATORIAL OPTIMIZATION IN COMMUNICATION NETWORKS

Edited by

MAGGIE XIAOYAN CHENG
University of Missouri, Rolla, Missouri

YINGSHU LI
Georgia State University, Atlanta, Georgia

DING-ZHU DU
University of Texas at Dallas, Richardson, Texas

e-ISBN: 0-387-29026-5

ISBN-13: 978-1-4419-3968-5

e-ISBN-13: 978-0-387-29026-3

Printed on acid-free paper.

AMS Subject Classifications: 90C27, 05C85, 8R10

9 8 7 6 5 4 3 2 1

springer.com

Contents

Part II: Combinatorial Optimization in Optical and Interconnection Networks

Part III: Combinatorial Optimization in Other Network Applications

Preface

Combinatorial optimization problems arise in all areas of technology, and they certainly have implications for many network problems, for instance, infrastructure deployment, resource management, routing, and QoS provisioning. To advance and promote the theory and applications of combinatorial optimization in communication networks, this volume addresses the unique intersection of the two areas.

There are many combinatorial optimization books on the market and numerous research papers in the literature addressing specific network problems, such as routing optimization, scheduling, and resource allocation. This book is the first to bridge the optimization and networking research communities. To improve the quality and coherence of the book, we carefully selected papers that represent state-of-the-art research, interacted with authors to make every chapter harmonically fit in with the same central theme—combinatorial optimization in communication networks—and finally the book took its current shape.

The book is mainly concerned with network problems that involve one or more combinatorial optimization solution techniques. Having in mind a list of combinatorial optimization methods and a list of network problems, we see a high-degree bipartite graph between them. Two approaches were considered: optimization method oriented (starting from combinatorial optimization methods and finding appropriate network problems as examples) and network problem oriented (focusing on specific network problems and seeking appropriate combinatorial optimization methods to solve them). We finally decided to use the problem-oriented approach, mainly because of the availability of papers: most papers in the recent literature appear to address very specific network problems, and combinatorial optimization comes as a convenient problem solver.

The three editors each bring a different perspective to this book: one is a world-renowned expert in operations research and complexity theory, and has been active in wireless networks, optical networks, and switching networks in recent years, the other two are active in wireless network research, with each having a different focus area. As a result, combinatorial optimization methods in all network technologies are collected in this book, with most papers focused on the wireless networking area.

The book covers a collection of network problems that need combinatorial optimization. Most chapters start from the problem to be addressed, introduce the required background information, describe the combinatorial optimization approach, and some even provide follow-up references for interested readers. It can be used as a handy reference book for research scientists in communication networks and operations research. When used as a textbook, it can be used for a graduate-level network course for specific network problems and their solutions resulting from some optimization techniques, or for a course on combinatorial optimization, using the network problems as real-life examples to enhance student understanding.

Maggie Xiaoyan Cheng

Yingshu Li

Ding-Zhu Du

Introduction

Aims

Combinatorial optimization is concerned with the arrangement, grouping, ordering, or selection of discrete objects from a finite set. Many problems involve seeking a best configuration of a set of parameters to achieve desired objectives. Combinatorial optimization exists everywhere. In communication networks, combinatorial optimization is used in network design and management to meet various operational needs. Classical applications of combinatorial optimization in communication networks trace back to the use of shortest paths in the Internet routing and spanning trees in the bridged LAN configuration. A typical application of combinatorial optimization in networks usually involves mathematical modeling of the problem, with an objective to reduce network deployment cost or operation cost, to provide better quality of service, or to improve the network performance. The recent boom in network research created many new problems that require new insights into their mathematical structures, novel approaches, and efficient solution techniques. The objective of this book is to expose such new findings and advance the theory and applications of combinatorial optimization in communication networks.

Scope

The scope of this book includes some important combinatorial optimization problems arising in optical networks, wireless ad hoc networks, sensor networks, mobile communication systems, and satellite networks.

Example network problems covered include media access control, routing optimization, topology control, resource allocation, and management, QoS provisioning in various wireless networks, light-path establishment in optical networks, etc. The specific combinatorial optimization techniques adopted by the authors are quite diverse, on the other hand. Instead of focusing on the combinatorial optimization techniques and extending them to network problems, most authors adopted an approach of starting from a network problem, analyzing its intrinsic structure, studying its computational complexity, and then developing an efficient solution for it.

Overview

The book includes combinatorial optimization problems in wireless networks, optical networks, and interconnection networks, as well as other network applications. Wireless network research is the major part, partly due to its high availability in recent literature, and partly because all three editors are currently active in this research community. Among the 22 chapters, 12 chapters are dedicated to wireless networks, 4 chapters are dedicated to specific problems in optical networks and interconnection networks, and the remaining 6 chapters are for other more general network applications that are not restricted to one particular network technology. Consequently, the book is divided into three logical parts: part I, wireless networks, part II, optical and interconnection networks, and part III, other network applications.

Part I consists of 12 chapters, all in wireless networks. Chapter 1 introduces topology control algorithms to achieve better energy efficiency and network capacity for both homogeneous and heterogeneous wireless networks. Chapter 2 includes several lines of research in cellular systems, including channel assignment, location management, base station placement, and code division multiple access, and it addresses combinatorial optimization problems in these topics. Chapter 3 discusses the problem of optimally sharing a single server/channel among multiple users/queues in wireless communication systems. Chapter 4 introduces strategies to improve network performance by using multipath routing in ad hoc networks, and specifically addresses the end-to-end multipath routing and N-to-1 multipath routing, and presents the protocols to find multiple paths and policies on the usage of multiple paths. Chapter 5 addresses several optimization problems in ad hoc networks, including the minimum size virtual backbone problem, transmission power control problem, and sensor node localization problem. Chapter 6 introduces stochastic linear programming and its application in proactive resource allocation in heterogeneous wireless networks. Chapter 7 addresses approaches to select working sensors in wireless sensor networks in order to maximize the total lifetime of the network, and meanwhile, satisfying the coverage and connectivity requirements. Chapter 8 addresses QoS provisioning for adaptive multimedia in mobile wireless networks. It introduces an abstract general traffic model and an optimal call admission control scheme that guarantees the QoS requirements and maximizes the utilization. Chapter 9 provides a summary of optimal power assignment algorithms in DS-CDMA networks, introduces a new collision model for DS-CDMA networks, and presents the collision probability and throughput analysis under this model. Chapter 10 introduces information-directed routing that jointly optimizes for maximal information gain and minimal communication cost in sensor networks. Chapter 11 includes call admission and handoff management strategies for multimedia LEO satellite networks. Chapter 12 introduces the time slot allocation problem in MFTDMA (Multi-Frequency Time-Division Multiple Access) satellite networks, in which the throughput is optimized through a linear and integer programming approach.

Part II consists of 4 chapters. Chapter 13 presents a new optimization technique for the lightpath establishment problem in optical networks that considers routing and wavelength assignment jointly. Chapter 14 introduces complexity models for WDM switching networks and provides complexity bounds under different request models. Chapter 15 describes interconnection network models and summarizes the topological properties of the most widely used networks. Chapter 16 includes a brief survey of some bounded degree Cayley networks on their routing, diameter, and fault tolerance properties, and presents an anonymous leader election algorithm in interconnection networks with bounded degree.

Part III consists of 6 chapters. Chapter 17 addresses several optimization techniques including dynamic programming, integer linear programming, Steiner tree construction, clustering, and their applications in routing problems in packet switching networks, WDM optical networks, and wireless ad hoc networks. Chapter 18 introduces an optimal scheduling algorithm that minimizes the ratio of the response time to its service time for on-demand data broadcasts. Chapter 19 includes optimal stream replication and bandwidth allocation problems in simulcasting systems for real-time video distribution. Chapter 20 presents a fast failure recovery scheme in high-speed networks that implements the minimum cost source-based rerouting. Chapter 21 introduces a primal-dual algorithm for the dynamic facility location problem, which has applications in many network problems such as network design, information flow routing, and cache distribution on the Internet. Finally, Chapter 22 presents preliminary work exploring more efficient approaches for hard combinatorial optimization problems that have significant implications for communication networks.

Part I

Combinatorial Optimization in *Wireless Networks*

Chapter 1

Topology Control in Wireless Multihop Networks

Ning Li and Jennifer C. Hou
Department of Computer Science
University of Illinois at Urbana-Champaign Urbana, IL 61801
E-mail: {nli,jhou}@cs.uiuc.edu

1 Introduction

With the rapid growth of wireless communication infrastructures over the recent years, new challenges have been posed on the system and analysis of wireless mobile ad hoc networks. Energy efficiency [1] and network capacity [2] are among the most important performance metrics, and topology control and management — how to determine the transmission power of each node so as to maintain network connectivity while consuming the minimum possible power and improving the network capacity with spatial reuse — has emerged to be one of the most important issues [1].

Specifically, in a wireless network where every node transmits with its maximal transmission power, the network topology is implicitly built by the routing protocol. In particular, each node keeps a list of neighbor nodes that are within the transmission range. On the other hand, in a topology controlled wireless network, instead of transmitting using the maximum possible power, nodes collaboratively determine their transmission power and define the topology of the wireless network by the neighbor relation under certain criteria. That is, each node has the opportunity of choosing the set of neighbors it would like to communicate with, by adjusting its transmission power. The network topology is thus defined by having each node form its own proper neighbor relation, subject to maintaining network connectivity.

The importance of topology control lies in the fact that it critically affects the system performance in several ways. In addition to reducing energy consumption and improving network capacity, topology control also has an impact on contention for the medium. Collisions can be mitigated as much as possible by choosing the smallest transmission power subject to maintaining network connectivity [3], [4].

Many centralized or localized geometric structures have been used for topology control in wireless networks, for instance, Minimum Spanning Tree (MST) [5], Relative Neighborhood Graph (RNG) [6], Gabriel Graph (GG), Delaunay Triangulation [7], and Yao structure [8], just to name a few. In this chapter, we present several sparse geometric structures that are based on the MST and minimum spanning graph for topology control. We first introduce the *Local Minimum Spanning Tree* (LMST) [9], a localized topology control algorithm for homogeneous wireless networks where the maximum transmission range of each node is the same. Then we present two localized topology control algorithms [10], *Directed Relative Neighborhood Graph* (DRNG) and *Directed Local Spanning Subgraph* (DLSS), for heterogeneous wireless networks where the maximum transmission range of each node may be different. We also consider the fault tolerance issue under topology control. Note that by reducing the number of wireless links in the network, topology control actually decreases the degree of routing redundancy. As a result, the topology thus derived is more susceptible to node failures/departures. To deal with the fault tolerance issue, we introduce a centralized algorithm, the *Fault-Tolerant Global Spanning Subgraph* (FGSS), and a localized algorithm, the *Fault-Tolerant Local Spanning Subgraph* (FLSS) [11]. Both FGSS and FLSS preserve network k-connectivity. Simulation results show that the algorithms discussed in this chapter are not only more energy efficient than existing approaches, but also significantly improve the network capacity.

The rest of this chapter is organized as follows. We define the network model in Section 2. After setting the stage for discussion, we present LMST for homogeneous networks in Section 3, and DRNG and DLSS for heterogeneous networks in Section 4. We then discuss in Section 5 the issue of fault tolerance and introduce FGSS and FLSS. Finally, we present in Section 6 a performance study of all the topology control algorithms discussed in the chapter, and conclude the chapter in Section 7 with several research avenues for future work.

2 Network Model

Let the topology of a multihop wireless network be represented by a simple directed graph $G = (V(G), E(G))$ in the 2-D plane, where $V(G) = \{v_1, v_2, \dots, v_n\}$ is the set of nodes (vertices) and $E(G)$ is the set of links (edges) in the network. Each node has a unique *id* (such as an IP/MAC address). Here we assume $id(v_i) = i$ for simplicity. Although G is usually assumed to be geometric in the literature, here we only assume that G is a general graph, i.e., $E(G) = \{(u, v) : v \text{ can receive } u\text{'s transmission correctly}\}$. We also assume that the wireless channel is symmetric, and each node is able to gather its own location information via, for example, several lightweight localization techniques for wireless networks [12], [13], [14]. In what follows, we first define the following terms and notations, and then outline the design requirements that one should meet to devise effective topology control algorithms.

Definition 2.1 (Visible Neighborhood.) *The visible neighborhood N_u^V is the set of nodes that node u can reach by using the maximum transmission power; i.e., $N_u^V = \{v \in V(G) : (u, v) \in E(G)\}$. For each node $u \in V(G)$, let $G_u^V = (V(G_u^V), E(G_u^V))$ be the induced subgraph of G such that $V(G_u^V) = N_u^V$.*

Definition 2.2 (Weight Function.) *Given two edges $(u_1, v_1), (u_2, v_2) \in E(G)$ and the Euclidean distance function $d(\cdot, \cdot)$, the weight function $w : E \mapsto R$ satisfies:*

$$\begin{aligned} & w(u_1, v_1) > w(u_2, v_2) \\ \Leftrightarrow \quad & d(u_1, v_1) > d(u_2, v_2) \\ \textit{or} \quad & (d(u_1, v_1) = d(u_2, v_2) \ \&\& \ \max\{id(u_1), id(v_1)\} > \max\{id(u_2), id(v_2)\}) \\ \textit{or} \quad & (d(u_1, v_1) = d(u_2, v_2) \ \&\& \ \max\{id(u_1), id(v_1)\} = \max\{id(u_2), id(v_2)\} \\ & \quad \&\& \ \min\{id(u_1), id(v_1)\} > \min\{id(u_2), id(v_2)\}). \end{aligned}$$

The weight function w ensures that two edges with different end-nodes have different weights. As most of the topology control algorithms introduced below are executed by each node in a decentralized manner, the weight function defined above is used to guarantee a unique outcome of the topology control algorithms. Also note that $w(u, v) = w(v, u)$.

Definition 2.3 (Neighbor Set.) *Node v is an* out-neighbor *of node u (and u is an* in-neighbor *of v) under an algorithm ALG, denoted $u \xrightarrow{ALG} v$, if and only if there exists an edge (u, v) in the topology generated by the algorithm.*

In particular, we use $u \rightarrow v$ *to denote the neighbor relation in* G. $u \overset{ALG}{\longleftrightarrow} v$ *if and only if* $u \xrightarrow{ALG} v$ *and* $v \xrightarrow{ALG} u$. *The* out-neighbor set *of node* u *is* $N^{out}_{ALG}(u) = \{v \in V(G) : u \xrightarrow{ALG} v\}$, *and the* in-neighbor set *of* u *is* $N^{in}_{ALG}(u) = \{v \in V(G) : v \xrightarrow{ALG} u\}$.

Definition 2.4 (Degree.) *The* out-degree *of a node* u *under an algorithm ALG, denoted* $deg^{out}_{ALG}(u)$, *is the number of out-neighbors of* u; *i.e.,* $deg^{out}_{ALG}(u) = |N^{out}_{ALG}(u)|$. *Similarly, the* in-degree *of a node* u, *denoted* $deg^{in}_{ALG}(u)$, *is the number of in-neighbors; i.e.,* $deg^{in}_{ALG}(u) = |N^{in}_{ALG}(u)|$.

Definition 2.5 (Topology.) *The topology generated by an algorithm ALG is a directed graph* $G_{ALG} = (E(G_{ALG}), V(G_{ALG}))$, *where* $V(G_{ALG}) = V(G)$, *and* $E(G_{ALG}) = \{(u, v) : u \xrightarrow{ALG} v, u, v \in V(G_{ALG})\}$.

Definition 2.6 (Radius.) *The radius,* R_u, *of node* u *is defined as the distance between node* u *and its farthest neighbor (in terms of Euclidean distance), i.e,* $R_u = \max_{v \in N^{out}_{ALG}(u)}\{d(u, v)\}$.

Definition 2.7 (Connectivity.) *For any topology generated by an algorithm ALG, node* u *is said to be* connected *to node* v *(denoted* $u \Rightarrow v$*) if there exists a path* $(p_0 = u, p_1, \ldots, p_{m-1}, p_m = v)$ *such that* $p_i \xrightarrow{ALG} p_{i+1}, i = 0, 1, \ldots, m-1$, *where* $p_k \in V(G_{ALG}), k = 0, 1, \ldots, m$. *It follows that* $u \Rightarrow v$ *if* $u \Rightarrow p$ *and* $p \Rightarrow v$ *for some* $p \in V(G_{ALG})$.

Definition 2.8 (BiDirectionality.) *A topology generated by an algorithm ALG is* bidirectional, *if for any two nodes* $u, v \in V(G_{ALG})$, $u \in N^{out}_{ALG}(v)$ *implies* $v \in N^{out}_{ALG}(u)$.

Definition 2.9 (Bidirectional Connectivity.) *For any topology generated by an algorithm ALG, node* u *is said to be* bidirectionally connected to *node* v *(denoted* $u \Leftrightarrow v$*) if there exists a path* $(p_0 = u, p_1, \ldots, p_{m-1}, p_m = v)$ *such that* $p_i \overset{ALG}{\longleftrightarrow} p_{i+1}, i = 0, 1, \ldots, m-1$, *where* $p_k \in V(G_{ALG}), k = 0, 1, \ldots, m$. *It follows that* $u \Leftrightarrow v$ *if* $u \Leftrightarrow p$ *and* $p \Leftrightarrow v$ *for some* $p \in V(G_{ALG})$.

Definition 2.10 (Addition and Removal.) *The* Addition *operation adds an extra edge* (v, u) *into* G_{ALG} *if* $(u, v) \in E(G_{ALG})$ *and* $(v, u) \notin E(G_{ALG})$. *The* Removal *operation deletes any edge* $(u, v) \in E(G_{ALG})$ *if* $(v, u) \notin E(G_{ALG})$.

Both *Addition* and *Removal* operations attempt to create a bidirectional topology by either converting unidirectional edges into bidirectional ones or removing all unidirectional edges. The resulting topology after *Addition* or *Removal* is alway bidirectional, if the transmission range for each node is the same. If the transmission range for each node is not the same, the result of *Removal* is still bidirectional, but the result of *Addition* may not be bidirectional.

Requirements for effective topology control algorithms. An effective topology control algorithm should meet several requirements [9]

(1) The algorithm should preserve network connectivity (or k-connectivity for the purpose of fault tolerance). This is the fundamental requirement of topology control.

(2) Because there is usually no central authority in a multihop wireless network, each node has to make its own decision based on the information collected from the network. That is, the topology control algorithm should be distributed.

(3) As the network topology may change as a result of mobility and/or node failure, a topology control algorithm may have to be executed multiple times in response to mobility or network dynamics. It is thus desirable that the algorithm depend only on the information collected *locally* so as to reduce the control message overhead and the delay incurred in topology management.

(4) It is desirable if all the links in the network topology induced by the topology control algorithm are bidirectional. Bidirectional links guarantee the existence of reverse paths, and facilitate link-level acknowledgment [3] and handshaking mechanisms, such as the floor acquisition mechanism *request-to-send/clear-to-send* (RTS/CTS).

3 Topology Control in Homogeneous Networks

In this section, we consider a homogeneous wireless network where every node has the same transmission range d_{max}. As a result, the network topology G becomes an undirected graph. The homogeneity assumption has been commonly used in most of the topology control algorithms in the literature, perhaps except for [10, 15]. We now introduce a localized topology control algorithm, *Local Minimum Spanning Tree* (LMST) [9], for homogeneous networks.

3.1 LMST: Local Minimum Spanning Tree

The proposed algorithm consists of three steps: information collection, topology construction, and construction of topology with only bidirectional links.

The information needed by each node u in the topology construction process is the information of all nodes in its visible neighborhood, N_u^V. This can be obtained by having each node broadcast periodically a Hello message using its maximal transmission power. The information contained in a Hello message should at least include the *id* and the position of the node. The time interval between two broadcasts of Hello messages depends on the level of node mobility, and can be determined using a probabilistic model in [9].

After obtaining the neighborhood information, each node u builds its local minimum spanning tree T_u that spans all the nodes within its visible neighborhood N_u^V. The time it takes to build the MST varies from $O(m \log n)$ (the original Prim's algorithm [16]) to almost linear of m (the optimal algorithm [17]), where n is the number of vertices and m is the number of edges. Node v is a neighbor of node u if and only if (u, v) is a link on the local MST built by u. The network topology under LMST is all the nodes in V and their individually perceived neighbor relations. Note that the topology is *not* a simple superposition of all local MSTs.

Definition 3.1 (LMST.) *In the Local Minimum Spanning Tree (LMST), node v is a neighbor of node u, denoted $u \xrightarrow{LMST} v$, if and only if $(u, v) \in E(T_u)$. That is, v is a neighbor of u if and only if v is on u's local MST T_u, and is one hop away from u.*

Because the neighbor relation is determined locally by each node, some links in the final topology may be unidirectional; i.e., $u \xrightarrow{LMST} v$ does not necessarily imply $v \xrightarrow{LMST} u$. An example is given in Figure 1, where $d(u, v) = d < d_{\max}$, $d(u, w_4) < d_{\max}$, $d(u, w_i) > d_{\max}$, $i = 1, 2, 3$, and $d(v, w_j) < d_{\max}$, $j = 1, 2, 3, 4$. Because $N_u^V = \{u, v, w_4\}$, $u \xrightarrow{LMST} v$ and $u \xrightarrow{LMST} w_4$. Also $N_v^V = \{u, v, w_1, w_2, w_3, w_4\}$; hence $v \xrightarrow{LMST} w_1$. Here link (u, v) is unidirectional. We can apply either *Addition* or *Removal* to obtain a bidirectionally connected topology.

Definition 3.2 (Topology G^+_{ALG}.) *The topology G^+_{ALG} generated by an algorithm ALG is an undirected graph $G^+_{ALG} = (V(G^+_{ALG}), E(G^+_{ALG}))$, where $V(G^+_{FLSS}) = V(G_{ALG})$, and $E^+_{ALG} = \{(u, v) : (u, v) \in E(G_{ALG})$ or $(v, u) \in E(G_{ALG})\}$.*

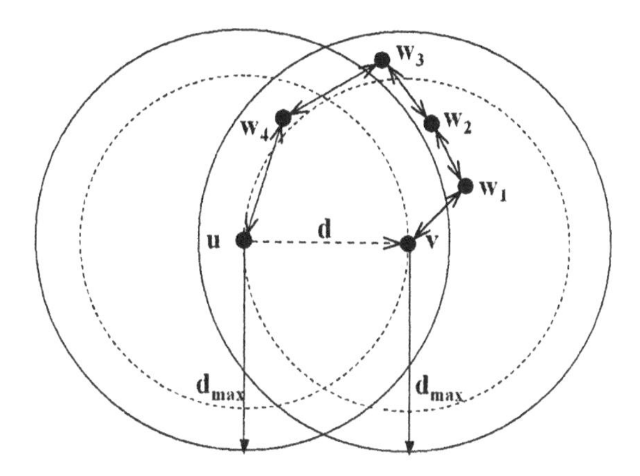

Figure 1: Links in the topology derived by LMST may be unidirectional.

Definition 3.3 (Topology G^-_{ALG}.) *The topology G^-_{FLSS} generated by an algorithm ALG is an undirected graph $G^-_{ALG} = (V(G^-_{ALG}), E(G^-_{ALG}))$, where $V(G^-_{ALG}) = V(G_{ALG})$, and $E^-_{ALG} = \{(u,v) : (u,v) \in E(G_{ALG})$ and $(v,u) \in E(G_{ALG})\}$.*

3.2 Properties of LMST

In this section, we prove several desirable properties of LMST, including its connectivity and degree bound. The following lemma is important to the proof of connectivity.

Lemma 3.4 *For any edge $(u,v) \in E(G)$, $u \Leftrightarrow v$ in G_{LMST}.*

Proof. Let all the edges $(u,v) \in E(G)$ be sorted in the ascending order of the weight; i.e., $w(u_1,v_1) < w(u_2,v_2) < \cdots < w(u_l,v_l)$, where l is the total number of edges in G. We prove by induction.

1. *Basis*: The first edge (u_1,v_1) satisfies $w(u_1,v_1) = \min_{(u,v)\in E(G)} \{w(u,v)\}$. Because the shortest edge is always on the local MST, we have $u_1 \stackrel{LMST}{\longleftrightarrow} v_1$, which means $u_1 \Leftrightarrow v_1$.

2. *Induction*: Assume the hypothesis holds for all edges $(u_i, v_i), 1 \le i < k$, we prove $u_k \Leftrightarrow v_k$ in G_{LMST}. If $u_k \stackrel{LMST}{\longleftrightarrow} v_k$, then $u_k \Leftrightarrow v_k$. Otherwise, without loss of generality, assume $u_k \nrightarrow v_k$. In the local topology construction of u_k, before edge (u_k, v_k) is inspected, there must already exist a path $p = (w_0 = u_k, w_1, w_2, \ldots, w_{m-1}, w_m = v_k)$

from u_k to v_k, where $(w_i, w_{i+1}) \in E(T_{u_k}), i = 0, 1, \ldots, m-1$. Because edges are inserted in the ascending order of the weight, we have $w(w_i, w_{i+1}) < w(u_k, v_k)$. Applying the induction hypothesis to each pair $(w_i, w_{i+1}), i = 0, 1, \ldots, m-1$, we have $w_i \Leftrightarrow w_{i+1}$. Therefore, $u_k \Leftrightarrow v_k$. □

Lemma 3.4 shows that, for any edge $(u, v) \in E(G)$, either $u \overset{LMST}{\longleftrightarrow} v$, or u and v are bidirectionally connected to each other in G_{LMST} via links of smaller weight.

Theorem 3.5 (Connectivity of LMST.) *If G is connected, then G_{LMST}, G^+_{LMST} and G^-_{LMST} are all connected.*

Proof. We only need to prove that G^-_{LMST} preserves the connectivity of G, for $E(G^-_{LMST}) \subseteq E(G_{LMST}) \subseteq E(G^+_{LMST})$. Suppose G is connected. For any two nodes $u, v \in V(G)$, there exists at least one path $p = (w_0 = u, w_1, w_2, \ldots, w_{m-1}, w_m = v)$ from u to v, where $(w_i, w_{i+1}) \in E(G), i = 0, 1, \ldots, m-1$. Because $w_i \Leftrightarrow w_{i+1}$ by Lemma 3.4, we have $u \Leftrightarrow v$ in G_{LMST}. Because p is bidirectional in G_{LMST}, the removal of unidirectional links does not affect the existence of p. Therefore, $u \Leftrightarrow v$ in G^-_{LMST}; i.e., G^-_{LMST} preserves the connectivity of G. □

It has been observed that any minimum spanning tree of a finite set of points in the plane has a maximum node degree of six [18]. We prove this property independently for LMST. We only need to prove the degree bound for G^+_{LMST} because the degree bound of G_{LMST} or G^-_{LMST} can only be lower.

Lemma 3.6 *Given three nodes $u, v, p \in V(G)$ satisfying $w(u, p) < w(u, v)$ and $w(p, v) < w(u, v)$, then $u \nrightarrow v$ and $v \nrightarrow u$ in G^+_{LMST}.*

Proof. We only need to consider the case where $(u, v) \in E(G)$ because $(u, v) \notin E(G)$ would imply $u \nrightarrow v$ and $v \nrightarrow u$. Consider the local topology construction of u and v. Before we insert (u, v) into T_u or T_v, the two edges (u, p) and (p, v) have already been processed because $w(u, p) < w(u, v)$ and $w(p, v) < w(u, v)$. Thus $u \Leftrightarrow p$ and $p \Leftrightarrow v$ by Lemma 3.4, which means $u \Leftrightarrow v$. Therefore, (u, v) should be inserted into neither T_u nor T_v; i.e., $u \nrightarrow v$ and $v \nrightarrow u$ in G^+_{LMST}. □

Before stating the next corollary, we give the definition of the *Relative Neighborhood Graph* (RNG).

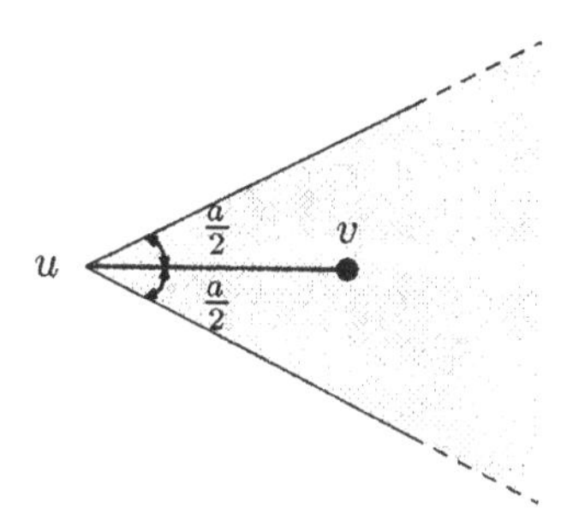

Figure 2: The definition of $cone(u, \alpha, v)$.

Definition 3.7 (Neighbor Relation in RNG.) *For RNG [19], [20],* $u \overset{RNG}{\longleftrightarrow} v$ *if and only if there does not exist a third node p such that* $w(u,p) < w(u,v)$ *and* $w(p,v) < w(u,v)$. *Or equivalently, there is no node inside the shaded area in Figure 3(a).*

The following corollary is a byproduct of Lemma 3.6.

Corollary 3.8 *The topology by LMST is a subgraph of the topology by RNG; i.e.,* $G_{LMST} \subseteq G_{RNG}$.

Definition 3.9 *A* $cone(u, \alpha, v)$ *is the unbounded shaded region shown in Figure 2.*

Theorem 3.10 (Degree Bound) *The out-degree of a node in* G^+_{LMST} *is bounded by six; i.e.,* $deg^{out}_{LMST+}(u) \leq 6, \forall u \in V(G^+_{LMST})$.

Proof. First we prove by contradiction that if $v \in N^{out}_{LMST+}(u)$, then there cannot exist any other node $w \in N^{out}_{LMST+}(u)$ that lies inside $Cone(u, 2\pi/3, v)$. Assume that such a node w exists; then $\angle wuv < \pi/3$. If $w(u,w) > w(u,v)$, then $\angle uvw > \pi/3 > \angle wuv$. We have $w(u,w) > w(v,w)$, which implies $u \nrightarrow w$ by Lemma 3.6. If $w(u,w) < w(u,v)$, then $\angle uwv > \pi/3 > \angle wuv$. We have $w(u,v) > w(v,w)$, which implies $u \nrightarrow v$ by Lemma 3.6. Both scenarios contradict the assumption that $v, w \in N^{out}_{LMST+}(u)$.

Consider any node $u \in V(G^+_{LMST})$. Put the nodes in $N^{out}_{LMST+}(u)$ in order such that for the ith node w_i and the jth node w_j $(j > i)$, $w(u, w_j) > w(u, w_i)$. We have proved that $\angle w_i u w_j \geq \pi/3$; i.e., node w_j cannot reside inside $Cone(u, 2\pi/3, w_i)$. Therefore, node u cannot have any neighbor other than node w_i inside $Cone(u, 2\pi/3, w_i)$. By induction on the rank of nodes in $N^{out}_{LMST+}(u)$, the maximal number of neighbors that u can have is no greater than six; i.e., $deg^{out}_{LMST+}(u) \leq 6$. □

4 Topology Control in Heterogeneous Networks

As mentioned in Section 3, most of the topology control algorithms in the literature assume homogeneous networks. However, this may not always hold in practice due to various reasons. First, even devices of the same type may have slightly different transmission ranges. Second, there exist devices of dramatically different capabilities in the same network. Third, the transmission range may be time-varying or affected by environmental stimuli.

In this section, we consider a heterogeneous wireless network where the maximum transmission range of each node may be different. In this case the network topology G becomes a directed graph. Let $r_{\min}$ and $r_{\max}$ be the smallest and the largest transmission ranges among all nodes in the network, respectively. We first show that most of the topology control algorithms that are devised under the homogeneity assumption cannot be directly applied to heterogeneous networks. Then we introduce two localized topology control algorithms, *Directed Relative Neighborhood Graph* (DRNG) and *Directed Local Spanning Subgraph* (DLSS).

4.1 Motivations

In this section, we give several examples that show topology control algorithms devised under the homogeneity assumption, e.g., CBTC [21], RNG [6], and LMST, may render disconnectivity in heterogeneous networks [10], thus motivating the need for new topology control algorithms for heterogeneous networks.

CBTC and RNG. Two of the other well-known topology control algorithms for homogeneous networks are *Cone-Based Topology Control* (CBTC) [21] and *Relative Neighborhood Graph RNG* [6]. CBTC(α) is a two-phase algorithm in which each node finds the minimum power p such that transmitting with p ensures that it can reach some node in every cone of degree α. The algorithm has been analytically shown to preserve network connectivity if $\alpha < 5\pi/6$. It has also ensured that every link between nodes is bidirectional. Several optimizations to the basic algorithm are also discussed, which include: (i) a *shrink-back* operation can be added at the end to allow a boundary node to broadcast with less power, if doing so does not reduce the cone coverage; (ii) if $\alpha < 2\pi/3$, asymmetric edges can be removed while maintaining the network connectivity; and (iii) if there exists an edge from u to v_1 and from u to v_2, respectively, the longer edge can be removed while preserving connectivity, as long as $d(v_1, v_2) < \max\{d(u, v_1), d(u, v_2)\}$.

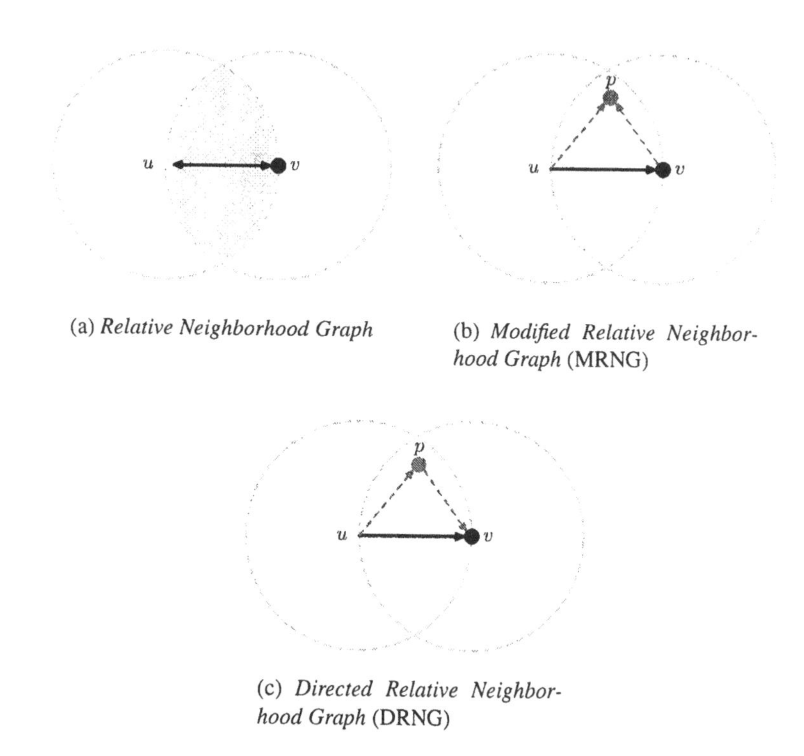

(a) *Relative Neighborhood Graph*

(b) *Modified Relative Neighborhood Graph* (MRNG)

(c) *Directed Relative Neighborhood Graph* (DRNG)

Figure 3: The definitions of RNG, MRNG, and DRNG.

To facilitate the introduction of RNG, we give the definition of the *Relative Neighborhood Graph* (RNG) below.

Definition 4.1 (Neighbor Relation in RNG.) *For RNG [19],[20],* $u \overset{RNG}{\longleftrightarrow} v$ *if and only if there does not exist a third node p such that* $w(u,p) < w(u,v)$ *and* $w(p,v) < w(u,v)$. *Or equivalently, there is no node inside the shaded area in Figure 3(a).*

The notion of RNG is proposed in [6] to facilitate topology initialization of wireless networks. Based on local knowledge, each node makes decisions to derive the network topology based on RNG. The network topology thus derived has been reported to exhibit good overall performance in terms of power usage, interference, and reliability.

Counterexamples. As shown in Figure 4, the network topology derived under $CBTC(\frac{2}{3}\pi)$ [21] (without optimization) may not preserve connectivity in

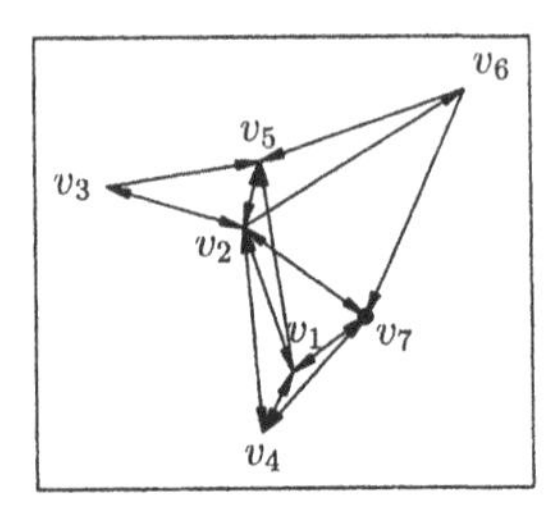

(a) Original topology (without topology control) is strongly connected

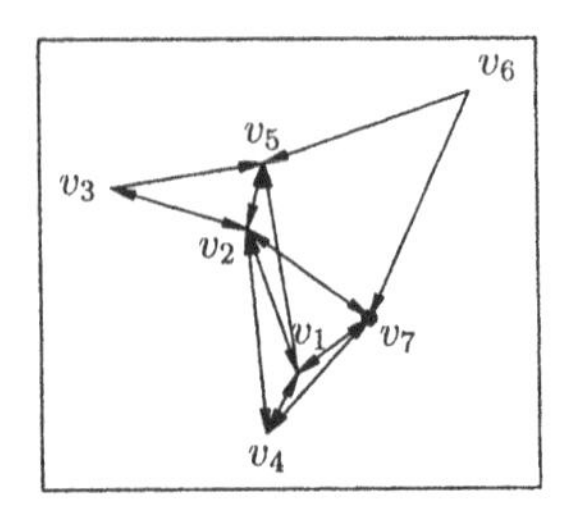

(b) Topology by $CBTC(\frac{2\pi}{3})$ without optimization is not strongly connected: there is no path from v_2 to v_6

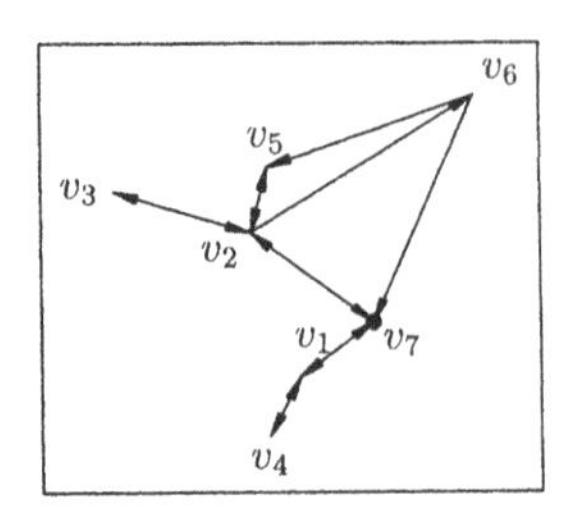

(c) Topology by DLSS is strongly connected

Figure 4: An example that shows $CBTC(\frac{2\pi}{3})$ may render disconnectivity in heterogeneous networks. There is no path from v_2 to v_6 due to the loss of edge (v_2, v_6), which is discarded by v_2 because v_5 and v_7 have already provided the necessary coverage.

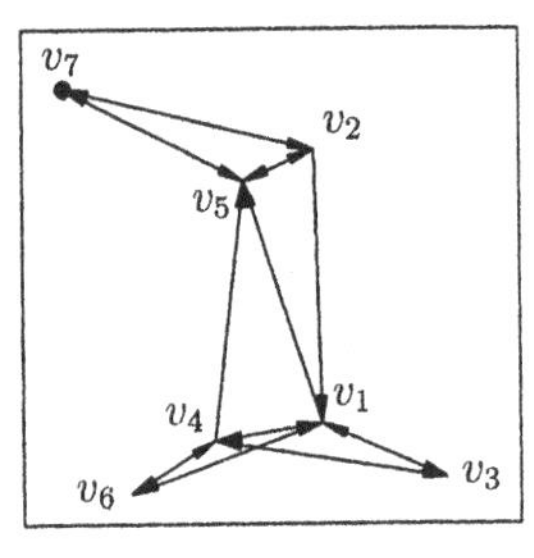

(a) Original topology (without topology control) is strongly connected

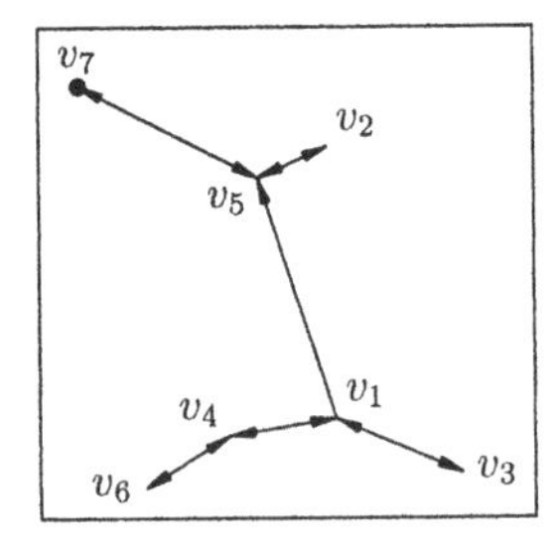

(b) Topology by RNG is not strongly connected: there is no path from v_5 to v_1

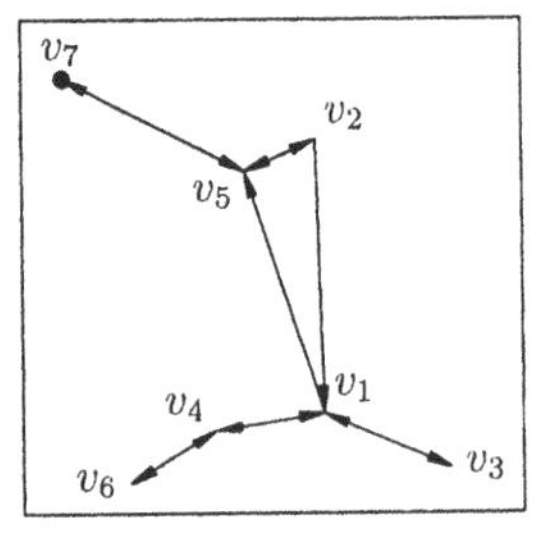

(c) Topology by DLSS is strongly connected

Figure 5: An example that shows RNG may render disconnectivity in heterogeneous networks. There is no path from v_5 to v_1 due to the loss of edge (v_2, v_1), which is discarded because $|(v_5, v_1)| < |(v_2, v_1)|$, and $|(v_5, v_2)| < |(v_2, v_1)|$.

a heterogeneous network. (The arrows in the figure indicate the direction of the links.) Similarly, as shown in Figure 5, the network topology derived under RNG may be disconnected in a heterogeneous network. As RNG is defined for undirected graphs only, we may modify its definition for directed graphs.

Definition 4.2 (MRNG.) *For* Modified Relative Neighborhood Graph *(MRNG), $u \xrightarrow{MRNG} v$ if and only if there does not exist a third node p such that $w(u,p) < w(u,v), d(u,p) \leq r_u$ and $w(p,v) < w(u,v), d(v,p) \leq r_v$ (Figure 3(b)).*

In spite of the modification, as shown in Figure 6, the topology derived under MRNG may still be disconnected in a heterogeneous network.

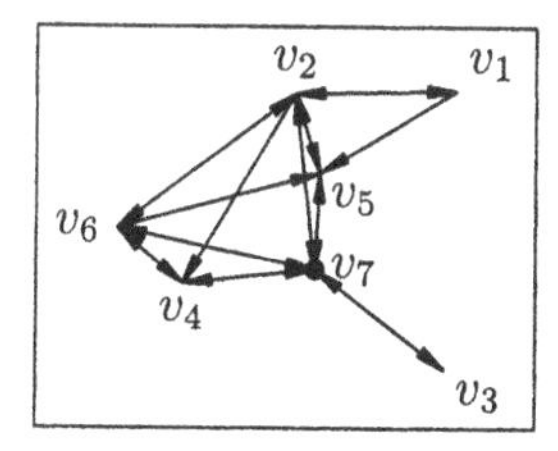

(a) Original topology (without topology control) is strongly connected

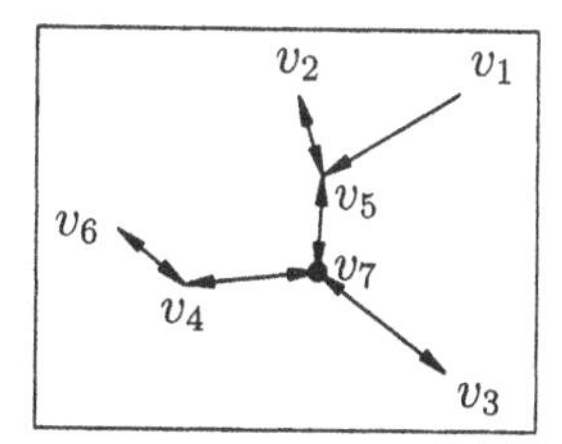

(b) Topology by MRNG is not strongly connected: there is no path from v_5 to v_1

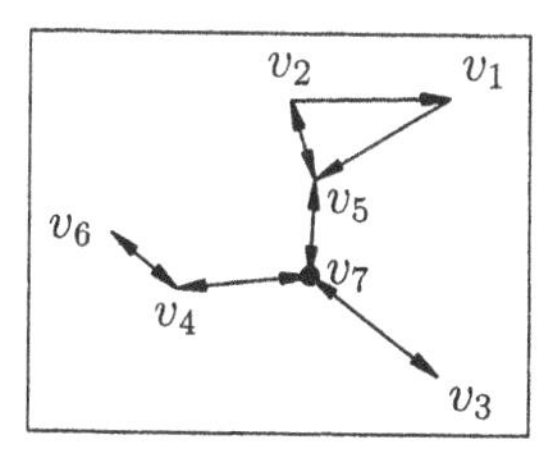

(c) Topology by DLSS is strongly connected

Figure 6: An example that shows MRNG may render disconnectivity in heterogeneous networks. There is no path from v_5 to v_1 due to the loss of edge (v_2, v_1), which is discarded because $|(v_2, v_5)| < |(v_2, v_1)|$, and $|(v_1, v_5)| < |(v_2, v_1)|$.

One possible extension to LMST is for each node to build a local *directed* minimum spanning tree [22], [23], [24] and keep only neighbors within one hop. Unfortunately, as shown in Figure 7, the resulting topology does not preserve strong connectivity. In the next subsection, we elaborate on how to revise LMST and RNG so that strong connectivity is preserved in heterogeneous networks.

4.2 Localized Algorithms: DRNG and DLSS

In this section, we present two localized topology control algorithms, *Directed Relative Neighborhood Graph* (DRNG) and *Directed Local Spanning Subgraph* (DLSS), for heterogeneous networks [10]. Both algorithms are composed of three steps: information collection, topology construction, and construction of topology with only bidirectional links. The first and the last steps are essentially the same as those described in Section 3.1. Therefore, we only

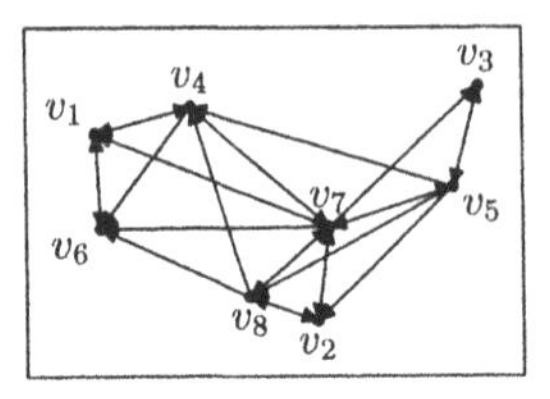

(a) Original topology (without topology control) is strongly connected

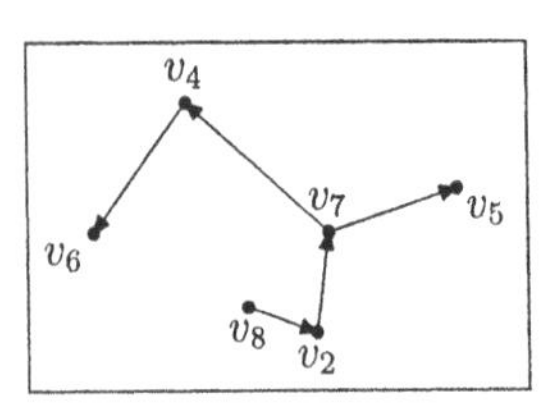

(b) The local directed MST rooted at v_8

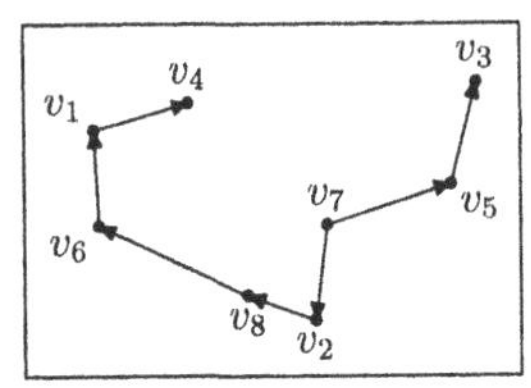

(c) The local directed MST rooted at v_7

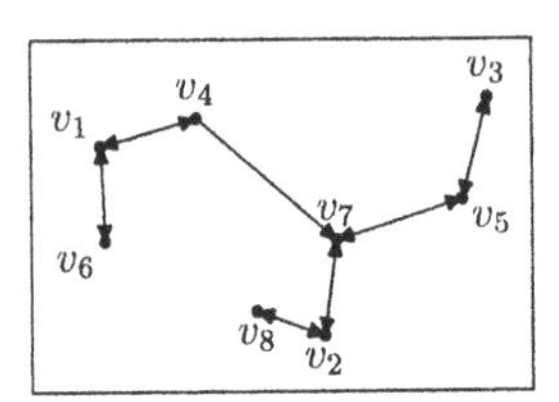

(d) The resulting topology is not strongly connected: there is no path from v_7 to v_4

Figure 7: An example that shows the algorithm in which each node builds a local directed minimum spanning tree and only keeps the one-hop neighbors may result in disconnectivity.

elaborate on the step of topology construction here. Essentially, instead of building a directed local minimum spanning tree (as in LMST) or using MRNG (as in RNG-based topology control) to define the neighbor relation, a node will use the following definition.

Definition 4.3 (DRNG.) *For* Directed Relative Neighborhood Graph *(DRNG),* $v \xrightarrow{DRNG} u$ *if and only if* $v \in N_u^V$ *and there does not exist another node* $p \in N_u^V$ *such that* $w(u,p) < w(u,v)$ *and* $w(p,v) < w(u,v), d(p,v) \leq r_p$ *(see Figure 3(b)).*

Definition 4.4 (DLSS.) *For* Directed Local Spanning Subgraph *(DLSS),* $v \xrightarrow{DLSS} u$ *if and only if* $(u,v) \in E(S_u)$*, where* S_u *is the output of DLSS(*u*) (Figure 8). Hence node* v *is a neighbor of node* u *if and only if node* v *is on node* u*'s directed local spanning graph* S_u*, and is one hop away from node* u*.*

Procedure: DLSS(u)
Input: G_u^V, the induced subgraph of G that spans the visible neighborhood of u;
Output: $S_u = (V(S_u), E(S_u))$, the local spanning subgraph of G_u^V;
begin
1: $V(S_u) := V$, $E(S_u) := \emptyset$;
2: Sort all edges in $E(G_u^V)$ in the ascending order of weight
3: **for** each edge (u_0, v_0) in the order
4: **if** u_0 is not connected to v_0 in S_u
5: $E(S_u) := E(S_u) \cup \{(u_0, v_0)\}$;
6: **endif**
7: **end**
end

Figure 8: DLSS Algorithm.

DRNG and DLSS are natural extensions of RNG and LMST for heterogeneous networks, respectively. Conceptually, in DLSS instead of computing a directed local MST that minimizes the *total* cost of all the edges in the subgraph (Section 4.1), each node computes a directed local subgraph (Figure 10) that minimizes the *maximum* cost among all edges in the subgraph.

4.3 Properties of DRNG and DLSS

In this section, we discuss several desirable properties of DRNG and DLSS by presenting a sequence of lemmas and theorems. In particular, Lemma 4.5, Theorem 4.6, and Lemma 4.7 can be proved by keeping in mind that G is a directed graph and following the same line of argument in Section 3.1. Note that we can only prove that $u \Rightarrow v$ in Lemma 4.5, because $u_k \rightarrow v_k$ does not guarantee that $v_k \rightarrow u_k$. Theorem 4.6 can be proved with the use of Lemma 4.5, in the same fashion Theorem 3.5 was proved with the use of Lemma 3.4.

Lemma 4.5 *For any edge $(u, v) \in E(G)$, we have $u \Rightarrow v$ in G_{DLSS}.*

Theorem 4.6 (Connectivity of DLSS.) *G_{DLSS} preserves the connectivity of G, i.e., G_{DLSS} is strongly connected if G is strongly connected.*

Lemma 4.7 *Given three nodes $u, v, p \in V(G_{DLSS})$ satisfying $w(u,p) < w(u,v)$ and $w(p,v) < w(u,v)$, $d(p,v) \leq r_p$, then $u \nrightarrow v$ in G_{DLSS}.*

By leveraging Theorem 4.6 to prove that DRNG preserves strong connectivity, we first prove the following lemma.

Lemma 4.8 *The edge set of G_{DLSS} is a subset of the edge set of G_{DRNG}, i.e., $E(G_{DLSS}) \subseteq E(G_{DRNG})$.*

Proof. We prove by contradiction. For any edge $(u, v) \in E(G_{DLSS})$, assume $(u, v) \notin E(G_{DRNG})$. From the definition of $DRNG$, there must exist a third node p such that $w(u, p) < w(u, v), d(u, p) \leq r_u$ and $w(p, v) < w(u, v), d(p, v) \leq r_p$. By Lemma 4.7, $u \nrightarrow v$ in G_{DLSS}; i.e., $(u, v) \notin E(G_{DLSS})$. □

The following theorem that proves DRNG preserves strong connectivity is a direct result of Theorem 4.6 and Lemma 4.8.

Theorem 4.9 (Connectivity of DRNG) *If G is strongly connected, then G_{DRNG} is also strongly connected.*

Let $Disk(u, r)$ denote the disk of radius r, centered at node u. Then the following lemma is a direct result of the definition of DRNG.

Lemma 4.10 *Given three nodes $u, v, p \in V(G_{DRNG})$ satisfying $w(u, p) < w(u, v)$ and $w(p, v) < w(u, v)$, $d(p, v) \leq r_p$, then $u \nrightarrow v$ in G_{DRNG}.*

Now to derive the in-degree bound, we state the following corollary (which is a direct result of Lemma 4.7 and Lemma 4.10).

Corollary 4.11 *If v is an out-neighbor of u in G_{DLSS} or G_{DRNG}, and $d(u, v) \geq r_{min}$, then u can not have any other out-neighbor inside $Disk(v, r_{min})$.*

Based on the above corollary, the following two theorems that give the in-degree bound of G_{DLSS} and G_{DRNG} can be proved by following the same line of arguments found in Theorem 3.10.

Theorem 4.12 *For any node u in G_{DLSS} or G_{DRNG}, the number of out-neighbors that are inside $Disk(u, r_{min})$ is at most 6.*

Theorem 4.13 (In Degree Bound.) *The in-degree of any node in G_{DLSS} or G_{DRNG} is bounded by 6.*

To derive the out-degree bound, we prove the following theorem.

Theorem 4.14 (Out-Degree Bound.) *The out-degree of any node in G_{DLSS} or G_{DRNG} is bounded by a constant that depends only on r_{max} and r_{min}.*

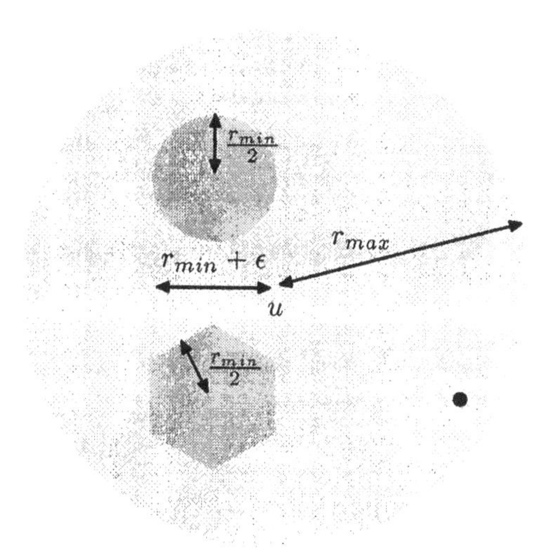

Figure 9: An example that shows the out-degree in a heterogeneous network can be very large. The transmission range of u is $r_{\max}$ and the transmission range for all other nodes is $r_{\min}$, where $r_{\max} = 2(r_{\min} + \epsilon)$, $\epsilon > 0$. All nodes are so arranged that the distance between any node and its closest neighbor is $r_{\min} + \epsilon$. Therefore, the only links that exist are those from u to the other nodes. Because it is impossible to relay packets, u has to use its maximal transmission power and keeps all 18 neighbors.

Proof. By Theorem 4.12, for any node u in G_{DLSS} or G_{DRNG}, there are at most 6 out-neighbors inside $Disk(u, r_{\min})$. Also by Corollary 4.11, the set of disks $\{Disk(v, \frac{r_{\min}}{2}) : v \in N^{out}(u), v \notin Disk(u, r_{\min})\}$ are disjoint. Therefore, the total number of out-neighbors of u is bounded by:

$$c_1 = 6 + \left\lceil \frac{\pi[(r_{\max} + \frac{r_{\min}}{2})^2 - (\frac{r_{\min}}{2})^2]}{\pi(\frac{r_{\min}}{2})^2} \right\rceil = 4\lceil \beta(\beta + 1) \rceil + 6,$$

where $\beta = \frac{r_{\max}}{r_{\min}}$. Figure 9 shows a scenario where the maximum out-degree of u is achieved if $\epsilon \to 0$. Based on the scenario, we further tighten the bound. Because the hexagonal area (as shown in Figure 9) centered at every neighbor of u is disjoint with each other, the total number of neighbors of u is bounded by:

$$c_2 = \left\lceil \frac{\pi(r_{\max} + \frac{r_{\min}}{\sqrt{3}})^2}{\frac{\sqrt{3}}{2} r_{\min}^2} \right\rceil - 1 = \left\lceil \frac{2\pi}{\sqrt{3}} (\beta + \frac{1}{\sqrt{3}})^2 \right\rceil - 1.$$

□

The bound given in Theorem 4.12 is applicable to both DLSS and DRNG, but the out-degree of the same node in G_{DLSS} is always smaller than that in G_{DRNG} because $E(G_{DLSS}) \subseteq E(G_{DRNG})$. Although this given bound is quite large, the average out-degree of nodes is actually not as large. In particular, because $\sum_{v \in V} deg^{in}(v) = \sum_{v \in V} deg^{out}(v)$, we have $E[deg^{out}(v)] = \frac{1}{n} \sum_{v \in V} deg^{in}(v) \leq \frac{1}{n} \cdot 6n = 6$.

5 Fault-Tolerant Topology Control

In this section, we discuss the issue of fault tolerance for topology control in wireless networks. By using the smallest possible transmission power that is connectivity-preserving, topology control algorithms usually render topologies with a smaller number of links. This implies that there exist less redundant routing paths in the network, and the topologies thus derived are more susceptible to unpredictable system events such as node failures.

To deal with this issue, one should figure in an adequate level of routing redundancy into topology control. This can be realized by considering k-connectivity in the process of topology construction. In particular, a k-vertex connected network is $k - 1$ fault-tolerant; i.e., it can survive failure of at most $k - 1$ nodes.

Definition 5.1 (k-vertex connectivity.) *A graph G is k-vertex connected if for any two vertices $v_1, v_2 \in V(G)$, there are k pairwise-vertex-disjoint paths from v_1 to v_2. Or equivalently, a graph is k-vertex connected if the removal of any $k - 1$ nodes (and all the related links) does not partition the network.*

In the following discussion, we concentrate on k-vertex connectivity, and for the simplicity of notation, use k-connectivity to refer to k-vertex connectivity. We assume the network is heterogeneous, and consider directed network topologies. We first propose in Section 5.1 a centralized algorithm, FGSS$_k$ [11], that will serve as the theoretical base, and then present in Section 5.2, a localized algorithm, FLSS$_k$ [11], that can be practically implemented in real environments.

5.1 FGSS$_k$: Fault-Tolerant Global Spanning Subgraph

In this subsection, we present a centralized greedy algorithm, FGSS$_k$, that builds k-connected spanning subgraphs. Conceptually, FGSS$_k$ is a generalized version of Kruskal's algorithm [25] for $k \geq 2$, where Kruskal's algorithm is a well-known algorithm that constructs the minimum spanning tree (1-connected spanning subgraph) of a given graph. The algorithm is given in Figure 10.

Procedure: FGSS$_k$
Input: $G(V, E)$, a k-connected simple graph;
Output: $G_k(V_k, E_k)$, a k-connected spanning subgraph of G;
begin
1: $V_k := V, E_k := \emptyset$;
2: Sort all edges in E in the ascending order of weight
3: **for** each edge (u_0, v_0) in the order
4: **if** u_0 is not k-connected to v_0 in G_k
5: $E_k := E_k \cup \{(u_0, v_0)\}$;
6: **endif**
7: **end**
end

Figure 10: FGSS$_k$ algorithm.

Time complexity of FGSS$_k$. By using network flow techniques [26], a query on whether two vertices are k-connected can be answered in $O(n + m)$ time for any fixed k, where n is the number of vertices and m is the number of edges in the graph. For $k \leq 3$, there also exist $O(1)$ time algorithms [27]. Therefore, the time complexity of FGSS$_k$ is $O(m(n+m))$, and can be improved to $O(m)$ for $k \leq 3$.

Properties of FGSS$_k$. Although FGSS$_k$ is a generalized version of Kruskal's algorithm, the techniques used to prove k-connectivity are completely different. To prepare the proof that FGSS$_k$ preserves k-connectivity, we first provide two lemmas. These two lemmas are also critical to the proofs in Section 5.2.

Let the path from node u to node v in G be represented by a set p of vertices on the path, i.e., $p = \{u, w_1, w_2, \ldots, w_l, v\}$. Let $S_{uv}(F)$ be a maximal set of pairwise-vertex-disjoint paths from u to v in F. Thus for $\forall p_1, p_2 \in S_{uv}(F)$, we have $p_1 \cap p_2 = \{u, v\}$.

Lemma 5.2 *Let u_1 and u_2 be two vertices in a k-connected undirected graph F. If u_1 and u_2 are k-connected after the removal of edge (u_1, u_2), then $F - (u_1, u_2)$ is still k-connected.*

Proof. Equivalently, we prove that $F' = F - (u_1, u_2)$ is connected after the removal of any $k-1$ vertices from F'. Consider any two vertices v_1 and v_2 in F'. Without loss of generality, we assume $\{u_1, u_2\} \cap \{v_1, v_2\} = \emptyset$. (The other cases can be proved following a similar line of argument.) We now prove that v_1 is still connected to v_2 after the removal of the set of any $k-1$ vertices $W = \{w_1, w_2, \ldots, w_{k-1}\}$, where $w_i \in V(F') - \{v_1, v_2\}$. This is obviously true if (v_1, v_2) is an edge in F. Therefore, we only consider the case where there is no edge from v_1 to v_2 in F. Since F is k-connected, $|S_{v_1 v_2}(F)| \geq k$.

Let F'' be the resulting graph after (u_1, u_2) and W (and related edges) are removed from F, and let s_1 be the number of paths in $S_{v_1 v_2}(F')$ that are broken due to the removal of vertices in W; i.e., $s_1 = |\{p \in S_{v_1 v_2}(F') : \exists w \in W, w \in p\}|$. Because the paths in $S_{v_1 v_2}(F')$ are pairwise-vertex-disjoint, the removal of any one vertex in W breaks at most one path in the set. Given $|W| = k-1$, we have $s_1 \leq k-1$.

If $|S_{v_1 v_2}(F')| \geq k$, then $|S_{v_1 v_2}(F'')| \geq |S_{v_1 v_2}(F')| - s_1 \geq 1$; i.e., v_1 is still connected to v_2 in F''. Now we consider the case where $|S_{v_1 v_2}(F')| < k$. This occurs only when the removal of (u_1, u_2) breaks one path $p^0 \in S_{v_1 v_2}(F)$. Without loss of generality, let the order of vertices on the path be v_1, u_1, u_2, v_2. Because the removal of (u_1, u_2) reduces the number of pairwise-vertex-disjoint paths between v_1 and v_2 by at most one, $|S_{v_1 v_2}(F) - \{p^0\}| \geq k-1$. Hence $|S_{v_1 v_2}(F')| = k-1$. Now we consider two cases.

1. $s_1 < k-1$: $|S_{v_1 v_2}(F'')| \geq |S_{v_1 v_2}(F')| - s_1 \geq 1$; i.e., v_1 is still connected to v_2 in F''.

2. $s_1 = k - 1$: in this case every vertex in W belongs to some path in $S_{v_1v_2}(F')$. Because p^0 is internally disjoint with all paths in $S_{v_1v_2}(F')$, we have $p^0 \cap W = \emptyset$. Thus v_1 is connected to u_1 and u_2 is connected to v_2 in F''. Let s_2 be the number of paths in $S_{u_1u_2}(F')$ that are broken due to the removal of vertices in W; i.e., $s_2 = |\{p \in S_{u_1u_2}(F') : \exists w \in W, w \in p\}|$. Because $|S_{u_1u_2}(F')| \geq k$ and $s_2 \leq k-1$, $|S_{u_1u_2}(F'')| \geq 1$; i.e., u_1 is still connected to u_2 in F''. Therefore, v_1 is still connected to v_2 in F''.

We have proved that for any two vertices $v_1, v_2 \in F'$, v_1 is connected to v_2 after the removal of any $k-1$ vertices from $F' - \{v_1, v_2\}$. Therefore, F' is k-connected. □

Lemma 5.3 *Let G and G' be two undirected simple graphs such that $V(G) = V(G')$. If G is k-connected, and every edge $(u,v) \in E(G) - E(G')$ satisfies that u is k-connected to v in $G - \{(u_0,v_0) \in E(G) : w(u_0,v_0) \geq w(u,v)\}$, then G' is also k-connected.*

Proof. Let $E = E(G) - E(G') = \{(u_1,v_1),(u_2,v_2),\ldots,(u_m,v_m)\}$ be a set of edges in descending order of weight, i.e., $w(u_1,v_1) \geq w(u_2,v_2) \geq \cdots \geq w(u_m,v_m)$. We define a series of graphs that are subgraphs of G: $G^0 = G$, and $G^i = G^{i-1} - (u_i,v_i)$, $i = 1,2,\ldots,m$. Now we prove by induction.

1. *Base*: $G^0 = G$ is k-connected.

2. *Induction*: If G^{i-1} is k-connected, we prove that G^i is k-connected, where $i = 1,2,\ldots,m$. Because $G - \{(u_0,v_0) \in E(G) : w(u_0,v_0) \geq w(u_i,v_i)\} \subseteq G^{i-1} - (u_i,v_i)$, u_i is k-connected to v_i in $G^{i-1} - (u_i,v_i)$. Applying Lemma 5.2 to G^{i-1}, we can prove that $G^i = G^{i-1} - (u_i,v_i)$ is still k-connected.

Now we have proved that G^m is k-connected. Because $E(G^m) \subseteq E(G')$, G' is also k-connected. □

Now we are in a position to prove that FGSS_k preserves k-connectivity.

Theorem 5.4 *FGSS_k can preserve k-connectivity of G; i.e., G_k is k-connected if G is k-connected.*

Proof. Because edges are inserted into G_k in an ascending order, whether u is k-connected to v at the moment before (u,v) is inserted depends only on the edges of smaller weights. Therefore, every edge $(u, v) \in E_0 = E(G) - E(G_k)$ satisfies that u is k-connected to v in $G - \{(u, v) \in E(G) : w(u, v) \geq w(u_0, v_0)\}$. We can prove that G_k preserves k-connectivity of G by applying Lemma 5.3 to G_k. □

Let $\rho(F)$ be the largest radius of all nodes in F; i.e., $\rho(F) = \max_{u \in V(F)}\{R_u\}$. Now we prove that FGSS$_k$ achieves the min-max optimality. That is, let $SS_k(G)$ be the set of all k-connected spanning subgraphs of G; then $\rho(G_k) = \min\{\rho(F) : F \in SS_k(G)\}$. This optimality is proved in [5] for the case $k = 2$, and we extend the result to arbitrary k.

Theorem 5.5 *The maximum transmission radius (or power) among all nodes is minimized by FGSS$_k$, i.e.,* $\rho(G_k) = \min\{\rho(F) : F \in SS_k(G)\}$.

Proof. Suppose G is k-connected. By Theorem 5.4, G_k is also k-connected. Let (u, v) be the last edge that is inserted into G_k; then we have $w(u, v) = \max_{(u_0, v_0) \in E(G_k)} \{w(u_0, v_0)\}$ and $R_u = R_v = \rho(G_k)$. Let $G'_k = G_k - (u, v)$, we have $|S_{uv}(G'_k)| < k$; otherwise according to the algorithm in Figure 10, (u, v) should not be included in G_k. Now consider a graph $H = (V(H), E(H))$, where $V(H) = V(G)$ and $E(H) = \{(u_0, v_0) \in E(G) : w(u_0, v_0) < w(u, v)\}$. If we can prove that H is not k-connected, we can then conclude that any $F \in SS_k(G)$ must have at least one edge equal to or longer than (u, v), which means $\rho(G_k) = \min\{\rho(F) : F \in SS_k(G)\}$.

Now we prove by contradiction that H is not k-connected. Assume H is k-connected and hence $|S_{uv}(H)| \geq k$. We have $E(H) \not\subseteq E(G'_k)$; otherwise, $|S_{uv}(G'_k)| \geq |S_{uv}(H)| \geq k$. Therefore, $E_0 = E(H) - E(G'_k) \neq \emptyset$. Because edges are inserted into G'_k in an ascending order, $\forall (u_1, v_1) \in E_0$ satisfies that u_1 is k-connected to v_1 in $H - \{(u_0, v_0) \in E(H) : w(u_0, v_0) \geq w(u_1, v_1)\}$. By Lemma 5.3, we can prove that G'_k is k-connected. This means $|S_{uv}(G'_k)| \geq k$, which is a contradiction. □

The min-max optimality of FGSS$_k$ is an important feature, and helps improve the energy-related performance. For example, let the network lifetime be defined as the time it takes for the first node to deplete its energy. If we assume a static network in which each node has the same energy and may send data to any other node with equal probability, then the network lifetime is approximately the same as the lifetime of the node that uses the maximum radius among all nodes. By minimizing the maximum radius (and transmission power), FGSS$_k$ achieves the maximum network lifetime.

Procedure: FLSS$_k$
Input: G_u^V, u's visible neighborhood;
Output: S_u, a k-connected spanning subgraph of G_u^V;
begin
1: $V(S_u) := V(G_u^V)$, $E(S_u) := \emptyset$;
2: Sort all edges in $E(G_u^V)$ in ascending order of weight
3: **for** each edge (u_0, v_0) in the order
4: **if** u_0 is not k-connected to v_0 in S_u
5: $E(S_u) := E(S_u) \cup \{(u_0, v_0)\}$;
6: **endif**
7: **end**
end

Figure 11: FLSS$_k$ algorithm.

5.2 FLSS$_k$: Fault-Tolerant Local Spanning Subgraph

FGSS$_k$ is a centralized algorithm that requires knowledge of global information, and hence cannot be applied directly to wireless networks in which collecting/updating global information may consume significant power. In this section, we present a localized algorithm, *Fault-Tolerant Local Spanning Subgraph* (FLSS), that is derived based on FGSS$_k$. FLSS is also composed of three steps: information collection, topology construction, and construction of topology with only bidirectional links. Again, as the first and the last steps are essentially the same as those described in Section 3.1, we only elaborate on the topology construction step here.

In $FLSS_k$, each node independently executes the algorithm described in Figure 11. Then the topology can be defined as follows.

Definition 5.6 (FLSS$_k$.) *In the Fault-tolerant Local Spanning Subgraph (FLSS$_k$), node v is a neighbor of node u, denoted $u \xrightarrow{FLSS} v$, if and only if $(u, v) \in E(S_u)$. That is, v is a neighbor of u if and only if v is on u's local spanning subgraph S_u, and is one hop away from u.*

Theorem 5.7 (Connectivity of FLSS$_k$.) *If G is k-connected, then G_{FLSS}, G^+_{FLSS} and G^-_{FLSS} are all k-connected.*

Proof. We only need to prove that G^-_{FLSS} preserves k-connectivity of G, for $E(G^-_{FLSS}) \subseteq E(G_{FLSS}) \subseteq E(G^+_{FLSS})$. Because G^-_{FLSS} is bidirectional, we can treat it as an undirected graph. Let $E = E(G) - E(G^-_{FLSS})$. For any edge $e = (u, v) \in E$, at least one of (u, v) and (v, u) was not in G_{FLSS}, because $e \notin E(G^-_{FLSS})$. Without loss of generality assume (u, v) was not in G_{FLSS}. Thus in the process of local topology construction of node u, u was already k-connected to v before (u, v) was inspected. Because edges are inserted in ascending order, whether u is k-connected to v at the moment before (u, v) is inspected depends only on the edges of smaller weights. Therefore, u is k-connected to v in $G - \{(u_0, v_0) \in E(G) : w(u_0, v_0) > w(u, v)\}$. Let $G' = G^-_{FLSS}$; then we can conclude that G^-_{FLSS} is k-connected by Lemma 5.3. □

Definition 5.8 (Strictly Localized Algorithms.) *An algorithm is strictly localized if its operation on any node u is based only on the information that is originated from the nodes in N_u^V.*

Let $LSS_k(G)$ be the set of all k-connected spanning subgraphs of G that are constructed by strictly localized algorithms. Now we prove that FLSS achieves the min-max optimality among all strictly localized algorithms; i.e., $\rho(G_{FLSS}) = \min\{\rho(F) : F \in LSS_k(G)\}$.

Theorem 5.9 *Among all strictly localized algorithms, $FLSS_k$ minimizes the maximum transmission radius (or power) of nodes in the network; i.e., $\rho(G_{FLSS})$ $= \min\{\rho(F) : F \in LSS_k(G)\}$.*

Proof. Suppose G is k-connected. Let (u, v) be the last edge inserted into G_{FLSS}. We have $w(u, v) = \max_{(u_0,v_0)\in E(G_{FLSS})} \{w(u_0, v_0)\}$ and $R_u = R_v = \rho(G_{FLSS})$. Let G_0 be the induced subgraph of G_{FLSS} where $V(G_0) = N_u^V$, and let $G'_0 = G_0 - \{(u, v)\}$. We have $|S_{uv}(G'_0)| < k$; otherwise (u, v) should not be included in G_0. Also, we define $H_0 = (V(H_0), E(H_0))$, where $V(H_0) = V(G_u^V)$ and $E(H_0) = \{(u_0, v_0) \in E(G_u^V) : w(u_0, v_0) < w(u, v)\}$.

To prove that H_0 is not k-connected, we replace G, G_k, G'_k, and H with G_u^V, G_0, G'_0, and H_0, respectively, and follow essentially the proof in Theorem 5.5. After proving H_0 is not k-connected, we consider the following cases.

1. u is k-connected to v in G_u^V: Because H_0 is not k-connected, any $F \in LSS_k(G)$ should have had at least one edge equal to or longer than (u, v),
2. u is not k-connected to v in G_u^V: In this case to preserve the connectedness as much as possible, any $F \in LSS_k(G)$ should have included (u, v);

In both cases, $\rho(F) \geq \rho(G_u^V) = \rho(G_{FLSS})$, which means $\rho(G_{FLSS}) = \min\{\rho(F) : F \in LSS_k(G)\}$. □

6 Simulation Study

In this section, we evaluate the performance of all the algorithms presented in Sections 3–5. Because FLSS is a generalized version of LMST and DLSS, we only compare FLSS_k against other fault-tolerant algorithms such as $\text{CBTC}(\frac{2\pi}{3k})$ [28] and $\text{Yao}_{p,k}$ [8]. In compliance with the guideline given in [8], the parameter p in $\text{Yao}_{p,k}$ is set to 6 in order to minimize the average power. We also give the performance of the centralized algorithm FGSS_k as a baseline.

In the first set of simulations, we use *J-Sim* [29] and carry out a simulation study to evaluate the performance with respect to node degree, maximum radius, and energy saving. In the simulation scenario considered, n nodes are uniformly distributed in a 1000 m × 1000 m region. The transmission range of all nodes is 261.195 m, which corresponds to a transmission power of 0.28183815 watt under the free space propagation model. We vary the number of nodes in the region from 70 to 300. Each data point is the average of 50 simulation runs.

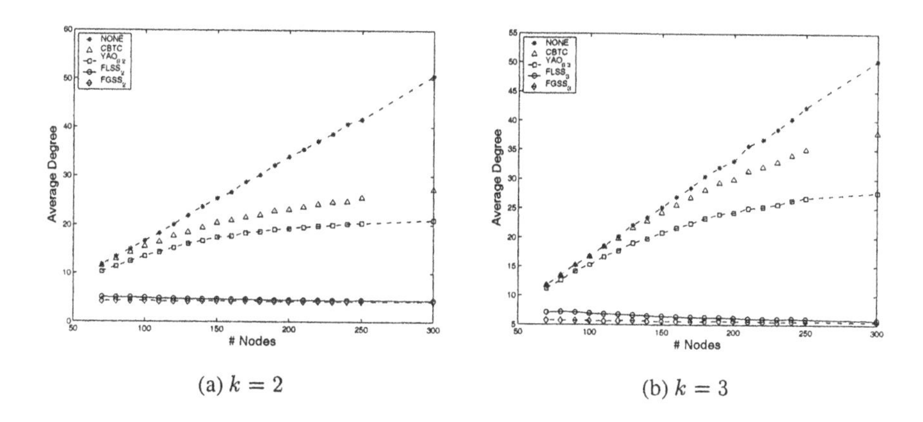

(a) $k = 2$ (b) $k = 3$

Figure 12: Comparison of $\text{CBTC}(\frac{2\pi}{3k})$, $\text{YAO}_{6,k}$, FLSS_k, and FGSS_k with respect to the average node degree.

Node Degree. First we compare the average node degree of the topologies derived under different algorithms, where the node degree is defined as the number of nodes within the transmission radius of a node. Node degree is a good indication of the level of MAC interference (and hence the extent of spatial reuse), i.e., the smaller the degree of a node, the fewer nodes its transmission may interfere with, and potentially affect. Figure 12 shows the average node degree of the topologies derived under NONE (with no topology control), CBTC($\frac{2\pi}{3k}$), $\text{YAO}_{6,k}$, FLSS_k, and FGSS_k, for $k = 2$ and $k = 3$. The average degree under NONE increases almost linearly with the number of nodes. The average degree under CBTC($\frac{2\pi}{3k}$) and $\text{YAO}_{6,k}$ also increases as the number of nodes increases. In contrast, the average degree under FGSS_k and FLSS_k actually decreases. The average degrees of both CBTC($\frac{2\pi}{3k}$) and $\text{YAO}_{6,k}$ are much higher than that of FGSS/FLSS.

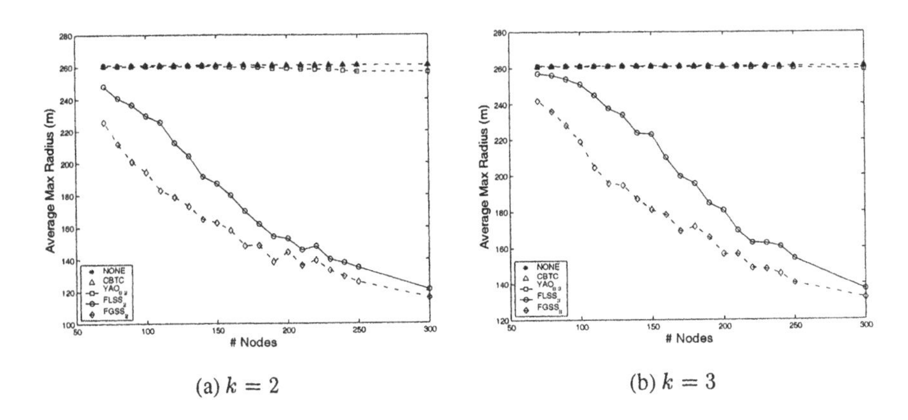

(a) $k = 2$

(b) $k = 3$

Figure 13: Comparison of CBTC($\frac{2\pi}{3k}$), $\text{YAO}_{6,k}$, FLSS_k, and FGSS_k with respect to the average maximum radius.

Maximum Radius. The average maximum radius for the topologies derived under CBTC($\frac{2\pi}{3k}$), $\text{YAO}_{6,k}$, FLSS_k, and FGSS_k are shown in Figure 13, for $k = 2$ and $k = 3$. The average maximum radius of CBTC($\frac{2\pi}{3k}$) or $\text{YAO}_{6,k}$ comes very close to that of NONE, which implies that CBTC($\frac{2\pi}{3k}$) and $\text{YAO}_{6,k}$ cannot really improve the network lifetime. In contrast, the average maximum radius of FLSS_k is significantly smaller, which implies that it can greatly prolong the network lifetime. Moreover, its performance is very close to that of the centralized algorithm FGSS_k.

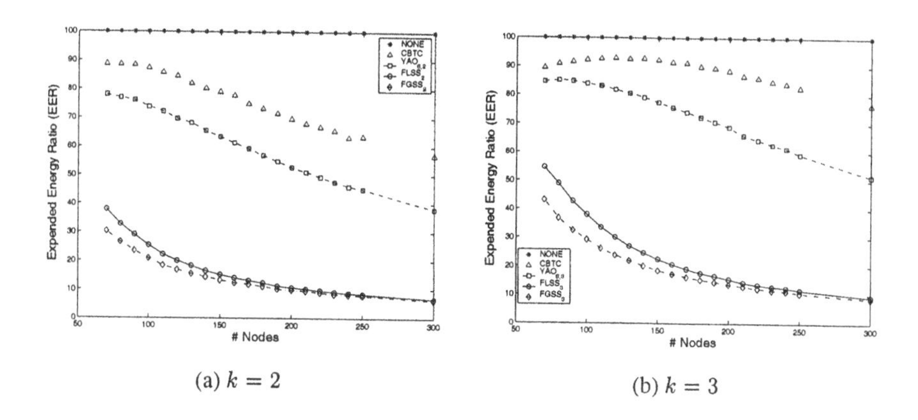

(a) $k = 2$

(b) $k = 3$

Figure 14: Comparison of CBTC($\frac{2\pi}{3k}$), $\text{YAO}_{6,k}$, FLSS_k, and FGSS_k with respect to EER.

Energy Saving. We compare the various algorithms with respect to the average expended energy ratio (EER) defined as

$$EER = \frac{E_{\text{ave}}}{E_{\text{max}}} \times 100,$$

where E_{ave} is the average transmission power over all the nodes in the network, and E_{max} is the maximal transmission power that can reach the transmission range of 261.195 m. Here we use the free-space propagation model to calculate the transmission power. Figure 14 gives the comparison results for both $k = 2$ and $k = 3$. FLSS_k clearly has the advantage, due to the fact that the average node degree of FGSS/FLSS is much smaller.

Capacity and Energy Efficiency. In the second set of simulations, we compare CBTC($\frac{2\pi}{3k}$), $\text{YAO}_{6,k}$, FLSS_k, and FGSS_k with respect to network capacity and energy efficiency using the *ns-2* simulator [30]. A total of n nodes are randomly distributed in a 1500 m × 200 m region, with half of them being sources and the other half being destinations. To observe the effect of spatial reuse, the deployment region should be large compared to the transmission/interference range. To reduce the number of nodes and to expedite simulation, we use a rectangular region, rather than a square region. In the simulation, the propagation model is the two-ray ground model, the MAC protocol is IEEE 802.11, the routing protocol is AODV, and the traffic sources are CBR and TCP traffic with bulk FTP sources. The start time of each connection is chosen randomly from [0 s, 10 s]. Each simulation run lasts for 100 seconds.

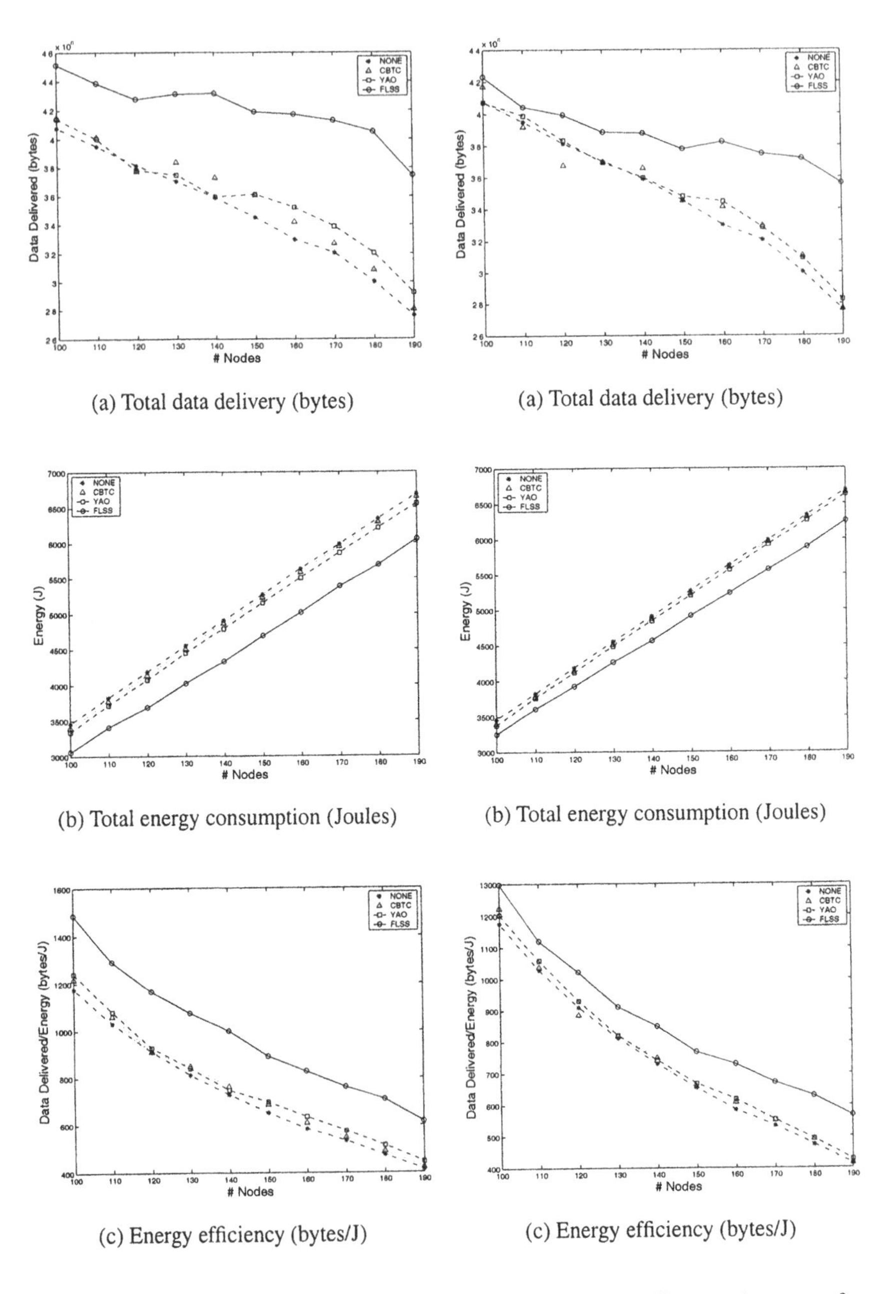

(a) Total data delivery (bytes)

(b) Total energy consumption (Joules)

(c) Energy efficiency (bytes/J)

Figure 15: Comparison of CBTC($\frac{2\pi}{3k}$), $YAO_{6,k}$, and $FLSS_k$ with respect to the network capacity and the energy efficiency under CBR traffic ($k = 2$).

(a) Total data delivery (bytes)

(b) Total energy consumption (Joules)

(c) Energy efficiency (bytes/J)

Figure 16: Comparison of CBTC($\frac{2\pi}{3k}$), $YAO_{6,k}$, and $FLSS_k$ with respect to the network capacity and the energy efficiency under CBR traffic ($k = 3$).

Figure 15 and Figure 16 compare the total amount of data delivered, the total energy consumption, and the energy efficiency, for $k = 2$ and $k = 3$, for CBR traffic. (The results for TCP/FTP traffic exhibit similar trends, and hence are not shown.) There is not much difference between the results by CBTC/YAO and the results by NONE. On the other hand, $FLSS_k$ not only improves the network capacity significantly, but also is the most power efficient.

7 Conclusion

Topology control is an important approach to improving energy efficiency and network capacity. In this chapter, we present geometric structures that are based on the MST and the minimum spanning graph for topology control. We introduce the *Local Minimum Spanning Tree* (LMST) [9], a localized topology control algorithm for homogeneous wireless networks, and present several of its nice properties. This is followed by a discussion of the node heterogeneity issue, in which we give several examples to show topology control algorithms devised under the homogeneity assumption (e.g., CBTC, RNG, and LMST) may render disconnectivity in heterogeneous networks. Then we discuss how to extend both LMST and RNG, to accommodate the case of node heterogeneity and devise their counterparts, DLSS and DRNG, that operate correctly in heterogeneous networks. We also consider the fault tolerance issue, elaborate on how to figure in route redundancy by considering k-connectivity in the process of topology construction, and devise a centralized algorithm, $FGSS_k$, and a localized algorithm, $FLSS_k$.

There are several important research avenues that should be pursued for technical advances in topology control. First, most of the topology control work determines the transmission power only based on the geometric information, but not the traffic information. The traffic information can actually be used to further refine the topology in several aspects. For example, nodes with high volume traffic should be allowed to use low transmission power, in order to save energy as well as to improve spatial reuse. For another example, the notion of connectivity can be modulated with the traffic information—two areas need not be connected if nodes in one do not communicate with those in the other. Second, the unit disk graph (UDG) model [31] has been widely used to characterize the transmission range that a node with a certain level of power can cover. However, this model does not capture the probabilistic nature of wireless communication. Use of a probabilistic model that accounts for physical layer characteristics such as shadowing and multipath fading in the process of topology construction would be one step further toward fully utilizing topology control in real environments.

References

[1] C. E. Jones, K. M. Sivalingam, P. Agrawal, and J. C. Chen, A survey of energy efficient network protocols for wireless networks, *Wireless Networks*, vol. 7, no. 4, pp. 343–358, Aug. 2001.

[2] P. Gupta and P. R. Kumar, The capacity of wireless networks, *IEEE Trans. Inform. Theory*, vol. 46, no. 2, pp. 388–404, Mar. 2000.

[3] S. Narayanaswamy, V. Kawadia, R. S. Sreenivas, and P. R. Kumar, Power control in ad-hoc networks: Theory, architecture, algorithm and implementation of the COMPOW protocol, in *Proc. European Wireless 2002, Next Generation Wireless Networks: Technologies, Protocols, Services and Applications*, Florence, Italy, Feb. 2002, pp. 156–162.

[4] P. Santi, D. M. Blough, and F. Vainstein, A probabilistic analysis for the range assignment problem in ad hoc networks, in *Proc. 1st ACM Symposium on Mobile Ad Hoc Networking and Computing (MOBIHOC)*, Long Beach, CA, USA, Aug. 2000, pp. 212–220.

[5] R. Ramanathan and R. Rosales-Hain, Topology control of multihop wireless networks using transmit power adjustment, in *Proc. IEEE INFOCOM 2000*, vol. 2, Tel Aviv, Israel, Mar. 2000, pp. 404–413.

[6] S. A. Borbash and E. H. Jennings, Distributed topology control algorithm for multihop wireless networks, in *Proc. International Joint Conference on Neural Networks, (IJCNN '02)*, Honolulu, HI, USA, May 2002, pp. 355–360.

[7] X.-Y. Li, G. Calinescu, and P.-J. Wan, Distributed construction of planar spanner and routing for ad hoc networks, in *Proc. IEEE INFOCOM 2002*, vol. 3, New York, NY, USA, June 2002, pp. 1268–1277.

[8] X.-Y. Li, P.-J. Wan, Y. Wang, and C.-W. Yi, Fault tolerant deployment and topology control in wireless networks, in *Proc. ACM Symposium on Mobile Ad Hoc Networking and Computing (MOBIHOC)*, Annapolis, MD, USA, June 2003, pp. 117–128.

[9] N. Li, J. C. Hou, and L. Sha, Design and analysis of an MST-based topology control algorithm, in *Proc. IEEE INFOCOM 2003*, vol. 3, San Francisco, CA, USA, Apr. 2003, pp. 1702–1712.

[10] N. Li and J. C. Hou, Topology control in heterogeneous wireless networks: Problems and solutions, in *Proc. of IEEE INFOCOM*, Hong Kong, China, Mar. 2004. Also available as a technical report in the Department of Computer Science, University of Illinois at Urbana Champaign, no. UIUCDCS-R-2004-2412, March 2004.

[11] N. Li and J. C. Hou, FLSS: A fault-tolerant topology control algorithm for wireless networks, in *Proc. ACM International Conference on Mobile Computing and Networking (MOBICOM)*, Philadelphia, PA, USA, Sept. 2004.

[12] C. Savarese, J. M. Rabaey, and J. Beutel, Location in distributed ad-hoc wireless sensor networks, in *Proc. IEEE International Conference on Acoustics, Speech, and Signal Processing*, vol. 4, Salt Lake City, UT, USA, May 2001, pp. 2037–2040.

[13] T. He, C. Huang, B. M. Blum, J. A. Stankovic, and T. Abdelzaher, Range-free localization schemes for large scale sensor networks, in *Proc. ACM International Conference on Mobile Computing and Networking (MOBICOM)*, San Diego, CA, USA, Sept. 2003, pp. 81–95.

[14] H. Lim and J. C. Hou, Localization for anisotropic sensor networks, in *Proc. IEEE INFOCOM*, Miami, FL, USA, Mar. 2005.

[15] V. Rodoplu and T. H. Meng, Minimum energy mobile wireless networks, *IEEE J. Select. Areas Commun.*, vol. 17, no. 8, pp. 1333–1344, Aug. 1999.

[16] R. Prim, Shortest connection networks and some generalizations, *The Bell System Technical Journal*, vol. 36, pp. 1389–1957, 1957.

[17] S. Pettie and V. Ramachandran, An optimal minimum spanning tree algorithm, *Journal of the ACM*, vol. 49, no. 1, pp. 16–34, Jan. 2002.

[18] C. Monma and S. Suri, Transitions in geometric minimum spanning trees, in *Proc. ACM Symposium on Computational Geometry*, North Conway, NH, USA, 1991, pp. 239–249.

[19] G. Toussaint, The relative neighborhood graph of a finite planar set, *Pattern Recognition*, vol. 12, pp. 261–268, 1980.

[20] K. J. Supowit, The relative neighborhood graph, with an application to minimum spanning trees, *Journal of the ACM*, vol. 30, no. 3, pp. 428–448, July 1983.

[21] L. Li, J. Y. Halpern, P. Bahl, Y.-M. Wang, and R. Wattenhofer, Analysis of a cone-based distributed topology control algorithm for wireless multi-hop networks, in *Proc. ACM Symposium on Principles of Distributed Computing (PODC)*, Newport, RI, USA, Aug. 2001, pp. 264–273.

[22] Y. J. Chu and T. H. Liu, On the shortest arborescence of a directed graph, *Science Sinica*, vol. 14, pp. 1396–1400, 1965.

[23] J. Edmonds, Optimum branchings, *Journal of Research of the National Bureau of Standards*, vol. 71B, pp. 233–240, 1967.

[24] F. Bock, An algorithm to construct a minimum spanning tree in a directed network, in *Developments in Operations Research*. New York, NY: Gordon and Breach Science Publishers, 1971, pp. 29–44.

[25] J. B. Kruskal, On the shortest spanning subtree of a graph and the traveling salesman problem, *Proceedings of the American Mathematical Society*, vol. 7, pp. 48–50, 1956.

[26] S. Even and R. E. Tarjan, Network flow and testing graph connectivity, *SIAM Journal on Computing*, vol. 4, pp. 507–518, 1975.

[27] G. D. Battista, R. Tamassia, and L. Vismara, Output-sensitive reporting of disjoint paths, *Algorithmica*, vol. 23, no. 4, pp. 302–340, 1999.

[28] M. Bahramgiri, M. Hajiaghayi, and V. S. Mirrokni, Fault-tolerant and 3-dimensional distributed topology control algorithms in wireless multi-hop networks, in *Proc. Eleventh International Conference on Computer Communications and Networks (ICCCN)*, Oct. 2002, pp. 392–397.

[29] H.-Y. Tyan, Design, realization and evaluation of a component-based software architecture for network simulation, Ph.D. dissertation, The Ohio State University, Columbus, OH, USA, 2001, `http://www.j-sim.org`.

[30] S. McCanne and S. Floyd, ns Network Simulator, `http://www.isi.edu/nsnam/ns/`.

[31] B. N. Clark, C. J. Colbourn, and D. S. Johnson, Unit disk graphs, *Discrete Mathematics,* 86:165–177, 1990.

Chapter 2

Combinatorial Evolutionary Methods in Wireless Mobile Computing

Geetali Vidyarthi
School of Computer Science
University of Windsor, Windsor, Canada, N9B 3P4
E-mail: `geetali_dutta@yahoo.com`

Alioune Ngom
School of Computer Science
University of Windsor, Windsor, Canada, N9B 3P4
E-mail: `angom@cs.uwindsor.ca`

Ivan Stojmenović
Computer Science Department
University of Ottawa, Ottawa, Canada, K1N 6N5
E-mail: `ivan@site.uottawa.ca`

1 Introduction

Like other technological developments, the development in wireless mobile communication has passed through several stages. The pioneering experiments in land mobile communication date back to the 1920s in Detroit, Michigan, USA. In Michigan 1946, the first interconnection of mobile users to the public telephone network was done to allow calls from fixed stations to mobile users. The system used a central high-power transmitter to cover a metropolitan area up to 50 miles or more from the transmitter. With this concept it was difficult to reuse the same frequency and hence resulted in limited system capacity.

A solution to this problem emerged in the 1970s when researchers at Bell Laboratories in the USA developed the concept of a cellular telephone system, which appeared in a Bell system proposal during the late 1940s. The cellular concept replaced the use of a large geographical area (where a high-power transmitter is placed at a high elevation at the center of the area) with a number of non overlapping smaller geographical areas, called cells, equipped with low-power transmitters. A cellular organization allows frequency reuse among geographically distant cells, thus greatly expanding the system capacity [50, 68]. It also allows cells to be sized according to subscriber density and traffic demand of a given area.

The developments in cellular systems can be divided into three stages: first generation cellular systems, second generation cellular systems, and third generation cellular systems.

1.1 First Generation Cellular Systems

The first generation cellular systems include the introduction of Nordic Mobile Telephone (NMT) in 1981, Advanced Mobile Phone Service (AMPS) in 1983, and Total Access Communications System (TACS) in 1985. Several other technologies were developed but AMPS, NMT, and TACS were the most successful [79]. All these "first-generation" cellular systems were analog systems and provided only basic speech services.

1.2 Second Generation Cellular Systems

One of the challenges faced by analog systems was the inability to handle the growing capacity needs in a cost-efficient manner. Moreover, each system followed different standards, which made it impossible for a person to use the same cellular phone in different countries. As a result, standardization committees for "second-generation" cellular systems worldwide adopted the digital technology, which conformed to at least three standards: one for Europe and international applications known as Global Mobile Systems (GSM); one for North America, IS-54 (North American Digital Cellular); and one for Japan, Japanese Digital Cellular (JDC) [23]. The advantages of digital systems over analog systems include ease of signaling, lower levels of interference, integration of transmission and switching, higher capacity potentials, and inclusion of new services (data services, encryption of speech and data, and Integrated Services Digital Network) [23]. Second generation cellular systems are, however, still optimized for voice service and they are not well suited for data communications.

1.3 Third Generation Cellular Systems

Data communication is an important requirement in the current environment of the internet, electronic commerce, and multimedia communications. The third generation systems referred to as Personal Communication Systems (PCS), aim at providing integrated services such as data, voice, image, and video to stationary and non stationary subscribers without temporal and spatial restrictions. Examples of PCS include Personal Handphone System, and Digital Enhanced Cordless Telecommunications.

The number of subscribers to mobile services is expected to increase in the near future; hence a lot of research has been focused on the efficient use of available resources to maximize the system capacity. Many resource management problems in cellular networks can be modeled as graphs and hence efficient solutions can be found using graph theory techniques. Efficient resource management also involves the use of optimization tools. Traditional methods of search and optimization progress through every point in the entire search space thereby rendering them not only time-consuming but also wasteful of computational resources when implemented in a very complex search space. Therefore, although optimization techniques are in abundance, Evolutionary Algorithms (EA) outperform most of them (in terms of quality of solution) in such problems because of their ability to progress through a population of points without actually going through every point in the entire search space. This chapter provides a comprehensive survey on the application of EA to some of the well-known optimization problems in wireless mobile computing. It also shows how combinatorial approaches such as graph theory methods are used to address the same.

The remainder of this chapter is organized as follows. Section 2 defines some of the fundamental concepts involved in the field of wireless mobile communication; Section 3 gives an overview of two most commonly used EAs. In Sections 4–12, we review some of the papers describing the application of EA and the use of graph theory methods to some of the difficult problems in wireless mobile communication. Conclusions and directions for future work are given in Section 13.

2 Cellular Radio Systems

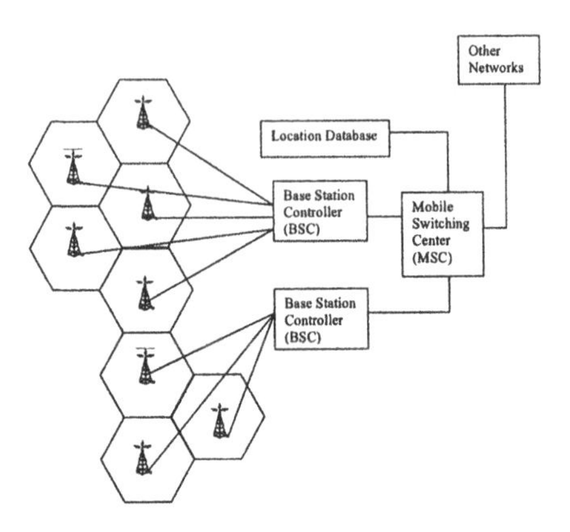

Figure 1: Network architecture.

The advent of the cellular concept was a major breakthrough in the development of wireless mobile communication. The cellular principle divides the covered geographical area into a set of smaller service areas called cells. During the early part of the evolution of the cellular concept, the system designers recognized the concept of all cells having the same shape to be helpful in systematizing the design and layout of the cellular system [49]. The 1947 Bell Laboratories paper [49] discussed four possible geometric shapes: the circle, the square, the equilateral triangle, and the regular hexagon. The regular hexagon was found to be the best over the other shapes [49]. In practice, the cell sizes are irregular and depend on the terrain and propagation conditions. Figure 1 (Modified from [1], figure 1, pp. 139) shows a typical mobile communication network.

Each cell has a base station and a number of mobile terminals (e.g., mobile phone, palms, laptops, or other mobile devices). The base station is equipped with radio transmission and reception equipment. The mobile terminals within a cell communicate through wireless links with the base station associated with the cell. A number of base stations are connected to the Base Station Controller (BSC) via microwave links or dedicated leased lines. The BSC contains logic for radio resource management of the base stations under its control. It is also responsible for transferring an ongoing call from one base station to another as a mobile user moves from cell to cell. A number of BSCs are connected to the Mobile Switching Center (MSC) also known as the Mobile Telephone Switching Office (MTSO). The MSC/MTSO is responsible for setting up and tearing down of calls to and from mobile subscribers. The MSC is connected to the backbone wire-line network such as the public switched telephone network (PSTN), Integrated Service Digital Network (ISDN), or any LAN/WAN-based network. MSC is also connected to a location database, which keeps information about the location of each mobile terminal. The base station is responsible for the communication between the mobile terminal and the rest of the information network. A base station can communicate with mobiles as long as they are within its operating range. The operating range depends upon the transmission power of the base station.

2.1 Channel Allocation

In order to establish communication with a base station, a mobile terminal must first obtain a channel from the base station. A channel consists of a pair of frequencies: one frequency (the forward link/ downlink) for transmission from the base station to the mobile terminal, and another frequency (the reverse link/ uplink) for the transmission in the reverse direction. An allocated channel is released under two scenarios: the user completes the call or the mobile user moves to another cell before the call is completed. The capacity of a cellular system can be described in terms of the number of available channels, or the number of users the system can support. The total number of channels made available to a system depends on the allocated spectrum and the bandwidth of each channel. The available frequency spectrum is limited and the number of mobile users is increasing day by day, hence the channels must be reused as much as possible to increase the system capacity. The assignment of channels to cells or mobiles is one of the fundamental resource management issues in a mobile communication system. The role of a channel assignment scheme is to allocate channels to cells or mobiles in such a way as to minimize the probability that the incoming calls are blocked, the probability that ongoing calls are dropped, and also to minimize the probability that the carrier-to-interference ratio of any call falls below a prespecified value. The channel assignment problem first appeared in [56].

2.1.1 Channel Assignment Schemes

In literature, many channel assignment schemes have been widely investigated with a goal to maximize the frequency reuse. The channel assignment schemes in general can be classified into three strategies: Fixed Channel Assignment (FCA) [49, 21, 93, 90, 44], Dynamic Channel Assignment (DCA) [12, 93, 18, 78, 11], and Hybrid Channel Assignment (HCA) [41, 93]. In FCA, a set of channels is permanently allocated to each cell based on a pre-estimated traffic intensity. In DCA, there is no permanent allocation of channels to cells. Rather, the entire set of available channels is accessible to all the cells, and the channels are assigned on a call-by-call basis in a dynamic manner. Cox and Reudink [12] proposed the DCA scheme. One of the objectives in DCA is to develop a channel assignment strategy, which minimizes the total number of blocked calls [78]. The FCA scheme is simple but does not adapt to changing traffic conditions and user distribution. Moreover, frequency planning becomes more difficult in a microcellular environment as it is based on accurate knowledge of traffic and interference conditions. These deficiencies are overcome by DCA but FCA outperforms most known DCA schemes under heavy load conditions [44]. To overcome the drawbacks of FCA and DCA, HCA was proposed by Kahwa and Georgans [41], which combines the features of both FCA and DCA techniques. In HCA one set of channels is allocated as per the FCA scheme, and the other set is allocated as per the DCA scheme. A comprehensive survey of various channel assignment schemes can be found in [42].

DCA schemes can be implemented as centralized or distributed. In the centralized approach [46, 20, 93, 92, 54, 9], all requests for channel allocation are forwarded to a central controller that has access to system-wide channel usage information. The central controller then assigns the channel by maintaining the required signal quality. In distributed DCA [10, 11, 51, 67], the decision regarding the channel acquisition and release is taken by the concerned base station based on the information from the surrounding cells. As the decision is not based on the global status of the network, it can achieve suboptimal allocation as compared to the centralized DCA and may cause forced termination of ongoing calls.

2.1.2 Channel Assignment Constraints

Radio transmission is such that the transmission in one channel causes interference with other channels. Such interference may degrade the signal quality and the quality of service. The potential sources of radio interference to a call are:

1. Co-channel interference: This radio interference is due to the allocation of the same channel to a certain pair of cells close enough to cause interference, (i.e., a pair of cells within the reuse distance).

2. Adjacent channel interference: This radio interference is due to the allocation of adjacent channels (e.g., f_i and f_{i+1}) to a certain pair of cells simultaneously.

3. Co-site interferences: This radio interference is due to the allocation of channels in the same cell that are not separated by some minimum spectral distance.

These constraints are known as Electromagnetic Compatibility Constraints (EMC) [62]. EMC can be represented by a minimum channel separation between any pair of channels assigned to a pair of cells or the cell itself [86]. If there are F channels to serve C cells in the system, the minimum channel separation required for an acceptable level of interference is described by a symmetric compatibility matrix $X[C, C]$. Each element $X_{i,j}(i, j = 1 \ldots C)$ represents the minimum separation required between channels assigned to cells i and j for an acceptable level of interference.

The reuse of channels in a cellular system is inevitable and at the same time it is directly related to co-channel interference. Co-channel interference is measured by a required Signal to Interference Ratio (SIR). As a result of co-channel interference, all channels may not be reused in every cell. In an AMPS System, when SIR is equal to 18 dB, most of the users call the system good or excellent [49]. However, the concept of a cellular system enables the discrete channels assigned to a specific cell to be reused in different cells separated by a distance sufficient to bring the value of co-channel interference to a tolerable level thereby reusing each channel many times. The minimum distance required between the centers of two cells using the same channel to maintain the desired signal quality is known as the reuse distance (D_s). The cells with center-to-center distance less than D_s belong to the same cluster. No channels are reused within a cluster.

Another basic requirement of channel assignment is the demand of channels in each cell. This channel demand or traffic demand can be modeled by a vector $\mathbf{T}$ of length C where an element T_i denotes the number of channels used in cell i. This vector can be obtained by analyzing the traffic at each cell. In reality, the value of $\mathbf{T}$ should be a function of time due to arrival of new calls, termination of ongoing calls, and handovers.

The process of channel assignment must satisfy the EMC constraints and the demand of channels in a cell. These constraints are also known as hard constraints. This requires a proper channel assignment scheme. The channel assignment problem has been shown to be NP-hard [32].

Besides the hard constraints and traffic demand constraints, other conditions that may be violated to improve the performance of the dynamic channel allocation technique include the packing condition, resonance condition, and limitation of reassignment operations [72]. These conditions are called soft constraints and were introduced in [18]. These constraints allow further lowering of the call blocking.

1. Packing condition: The packing condition tries to use the minimum number of channels every time a call arrives [72]. It allows the selection of those channels that are already in use in other cells as long as the co-channel interference constraint is satisfied.

2. Resonance: With the resonance condition, the same channels are assigned to cells that belong to the same reuse scheme [72]. This reduces the call blocking probability. The objective function should help select a combination of channels that makes maximum use of channels already in use in the reuse scheme to which the cell involved in call arrival belongs.

3. Limiting rearrangement: Channel reassignment improves the quality of service in terms of lowering call blocking probability. Hence it is an important process in dynamic channel allocation. It is the process of transferring an ongoing call to a new channel without call interruption [8]. The reassignment process greatly affects the call blocking probability. Reassignment in the entire cellular network upon a new call arrival will obviously result in lower call blocking, but it is complex both in terms of time and computation [72]. Therefore, the reassignment process is limited to the cell involved in the new call arrival. However, excessive reassignment in a cell may lead to an increase in blocking probability [72]. So a process called limiting rearrangement is considered that tries to assign, where possible, the same channels assigned before, thus limiting the reassignment of channels. Reassignment is done together

with the new call.

2.2 Multiple Access Techniques

The radio spectrum is one of the finite resources in the wireless environment. Thus one of the goals in cellular systems design is to be able to handle as many calls as possible in a given bandwidth with some reliability. The Multiple Access Technique (MAT) allows many users to share the common transmission resource, the radio spectrum, without causing interference between users. A good MAT can improve the system capacity. Different types of cellular systems employ various methods of multiple accesses. The traditional analog cellular systems, such as AMPS and TACS standards, use Frequency Division Multiple Access (FDMA). With FDMA, the available spectrum is divided into a number of radio channels of a specified bandwidth, and a selection of these channels is used within a given cell. For example, in AMPS the available spectrum is divided into blocks of 30 kHz [79], and in TACS channels are 25 kHz wide. In this technique only one subscriber at a time is assigned to a channel. The channel cannot be assigned to another user in the same cell during the call.

The second generation digital cellular systems such as IS-54 and GSM use Time Division Multiple Access (TDMA). With TDMA the available radio spectrum is divided into time slots, and only one user is allowed to either transmit or receive at each time slot. Thus, TDMA increases the throughput by sending multiple calls over a single frequency distributed over time. The channel can be assigned to another user in the same cell except that they must transmit at different times.

Due to the increasing demand for mobile communications services, the use of highly efficient multiple access techniques have become imperative. The Code Division Multiple Access (CDMA) is currently considered efficient in achieving high capacity in such systems [65]. So the third generation cellular system uses CDMA. With CDMA, several users can simultaneously access a given frequency allocation thereby increasing the spectral efficiency several times as compared to that achievable by TDMA and FDMA [28]. Beside this, it also supports users with different communication demands, e.g., different data rate requirements. Unique digital codes called pseudo-random code sequences are used to differentiate the subscribers.

2.3 Location Management

Wireless networks enable mobile users to communicate regardless of their locations. In contrast to the telephone number in traditional telecommunication systems that specifies the location of the end users, the mobile subscriber (one who subscribes to mobile services) number does not provide the location of the mobile user. Therefore, to handle the mobility of terminals in a wireless network there is a need to address two basic issues: tracking the position of a mobile user (with registrations), and to set up connections or deliver data to mobile users. A location management mechanism is composed of three components: a system database for mapping subscriber numbers to locations, an update operation for informing the system database about the changes in the mobile user's locations known as Location Update (LU), and a search operation for locating the mobile with the help of the system database known as terminal paging. In the current cellular system the coverage area of the system is divided into Location Area (LA) or Registration Area (RA), each consisting of a group of cells that forms a contiguous geographic area. Each LA has its unique identifier (ID). Each LA has a database that keeps a record of the location information of a mobile terminal. For each mobile terminal, the database entry is updated only when the mobile terminal performs a location update. The mobile terminal initiates the location update procedure by sending an update message over the uplink control channel (see Section 2.0). This is followed by some signaling procedures, which update the database. A good survey of different update schemes can be found in [89].

Upon call arrival, the system has to determine in which cell the mobile terminal is currently. Because the system knows the mobile terminal's LA, the search can be confined to the cells within the LA. Thus, a paging operation is triggered to find a target mobile. In paging, polling signals are sent over the downlink control channel to a group of cells where the mobile terminal is likely to be present. All the mobile terminals within the group of cells receive the page message but a response message is sent back over the uplink control channel by only the target mobile terminal. If the response is sent before a predetermined timeout period the paging process is terminated. Otherwise, another group of cells is chosen in the next search iteration. In order to avoid call dropping, the mobile terminal must be located within an allowable time constraint. A comprehensive survey of the various paging schemes proposed in the literature can be found in [89].

3 Introduction to Evolutionary Algorithm

During the 1950s and 1960s many scientists were inspired to use the principles of natural evolution based on Darwin's ideas [13] of natural selection as an optimization tool for engineering problems. In 1990, the term Evolutionary Algorithm (EA) was proposed to describe all algorithms based on natural evolution for problem solving.

All EAs work within a similar framework. A representation scheme is chosen to define the set of solutions that represent unique points within the search space. A candidate solution is called a *chromosome*. The pool of candidate solutions is called a *population*. The total number of individuals in a population is called the *population size*. Each individual solution is associated with an *objective value*. The objective value is problem specific and is representative of the individual solution's performance in relation to the parameter being optimized. It also reflects an individual solution's fitness in relation to other potential solutions in the search space.

EA is an iterative process and its goal is to continually improve the fitness of the best solution. It tries to do so by emulating the natural process of biological evolution through the use of natural processes, such as *reproduction*, *recombination*, and *mutation*. Reproduction selects members of a current population to use as parents for the generation of a new population [4]. The mechanisms to determine which and how many parents to select, how many new solutions called offspring to generate, and which individuals will survive into the next generation together represent the selection method. It mimics nature's survival of the fittest mechanism [82]. The selection of parents is based on fitness of each chromosome measured by a fitness evaluating function. The fitness evaluating function is a measure of how good is each member relative to other members of the population [63, 82]. Selection alone cannot create new solutions in the population. Genetically inspired operators, recombination and mutation, generate new solutions in the population. Recombination attempts to create offspring by combining the features of existing parents. The purpose of the mutation operator is to prevent the loss of useful genetic information and hence to maintain diversity within the population [82]. The process of selection and application of reproduction operators is repeated until some terminating criterion is reached. The solution to the problem is represented by the best individual so far in all generations.

The Genetic Algorithm (GA) and Evolutionary Strategy (ES) are the two most common variants of EAs. An introductory survey on GA and ES can be found in [24].

3.1 Genetic Algorithm

Genetic algorithms are stochastic parallel search algorithms based on the principle of evolution and natural selection [34]. They have proven to be robust search algorithms [29]. The implementation of GA in a specific problem starts with six fundamental issues: encoding of solution, creation of initial population, fitness evaluation, selection of progenitors, generation of progeny by genetic operators, and termination criteria. They were pioneered by John Holland [35, 34]. Holland's genetic algorithm is commonly called the Simple Genetic Algorithm (SGA). The general structure of the SGA is as follows [63].

```
Procedure SGA:
  Begin
    Set control parameters;
    {crossover and mutation probabilities, population size, and chromosome length}
    Initialize population;                                      {at random}
    Evaluate initial population;               {fitness of initial individuals}
    Repeat
      Select individuals for next generation;                 {reproduction}
      Perform crossover and mutation;                {obtain new population}
      Evaluate population;                       {fitness of new individuals}
    Until termination condition is reached;
  End;
```

Genetic parameters (population size, crossover probability, and mutation probability) should be carefully selected for optimal performance [82]. The choice of these parameters is problem specific and no exact rule exists to determine a suitable combination of these parameters [4]. A joint effect of population size, crossover probability, mutation probability, and number of crossover points in each recombination influences the performance. Goldberg [30] has suggested a population size equal to $1.65 * 2(0.21 * l)$ where l is the length of the chromosome for optimal performance. Schaffer et al. [74] have concluded that a small population size 20 to 30, a crossover probability in the range 0.75 to 0.95, and a mutation probability in the range 0.005 to 0.01 perform well.

In recent years, many variations of genetic algorithms have been proposed that bear little resemblance to Holland's original formulation.

Many methods of encoding, crossover, mutation, reproduction, and selection (roulette wheel selection, scaling technique, tournament selection, elitist model, ranking) have been used and have become a subject area of research. Interested readers may refer to [17, 22], and Krishnakumar [43]. The books by Holland [34], Goldberg [29], Davis [15], Michalewicz [57], De Jong [16], and Mitchel [58] provide detailed discussion of genetic algorithms.

3.2 Evolution Strategy

Rechenberg [69] pioneered ES. ES was proposed as an optimization method for real-valued vectors. It works on an encoded representation of the solution. ES is a random guided hill-climbing technique in which a candidate solution is produced by applying mutations to a given parent solution. The best solution generated in one generation becomes the parent for the next generation. ES is an iterative method so the process of selection and application of mutation is repeated until some terminating criterion is reached. When the termination criterion is reached, the solution to the problem is represented by the best individual so far in all generations. The basic steps of an ES algorithm can be summarized as follows,

1. Generate an initial population of λ individuals.
2. Evaluate each individual according to a fitness function.
3. Select μ best individuals called the parent population and discard the rest.
4. Apply the reproduction operator, i.e., mutation, to create λ offspring from μ parents.
5. Go to step 2 unless a desired solution has been found or a predetermined number of generations have been produced and evaluated.

In GA, the fitness of an individual determines its chances of survival into the next generation, whereas in ES only the best individuals survive. The two common variations of ES introduced by Schwefel [75] are the $(\mu + \lambda)$-ES and (μ, λ)-ES. In both approaches μ parents produce λ offspring. These two approaches differ in the selection of individuals for the next generation. In $(\mu+\lambda)$-ES, μ best individuals from all the $(\mu+\lambda)$ individuals are selected to form the next generation, but in (μ, λ)-ES, μ

best individuals from the set of λ offspring are selected to form the next generation.

Sections 4–12 discuss the use of the evolutionary algorithm and graph theory methods in the area of wireless mobile communication.

4 Base Station Placement

The infrastructure cost and planning complexity of a cellular network is closely related to the number of base stations required to achieve the desired level of coverage (locations covered by the selected number of base stations) and capacity [45]. Therefore one of the most challenging design problems in cellular networks is to decide on the location and the minimum number of base stations required to serve a given area while providing an acceptable quality of service to the mobile users. In the literature many practical approaches have been proposed to solve this problem. These include the use of GA [5, 33], simulated annealing [2], [36], and tabu search [47]. For finding precise base station locations, numerous factors such as traffic density, channel condition, interference scenario, the number of base stations, and other network planning parameters [33] must be taken into account. In the base station placement problem, the goal is to the select minimum number of base station locations from a list of potential sites while maximizing coverage in the area taking into account the radio propagation characteristics of the area. The radio propagation characteristics can be determined using ray-tracing software or by using empirical propagation models for path loss. There exists a tradeoff between coverage and the number of base stations. The higher is the number of base stations, the greater is the coverage, but there is also correspondingly greater radio interference and network cost. Determining the location of base stations is known to be NP-hard [5].

4.1 Graph-Theoretical Methods in Base Station Placement

Mathar and Niessen [55] have shown that the base station placement problem in its simplest form is equivalent to minimum dominating set problem in graph theory. Given an undirected graph $G(V, E)$, where V is the set of vertices and E is the set of edges, a set S is said to be dominating if for every $u \in V - S$ there exists a $v \in S$ such that

$(u, v) \in E$. The objective is to find a dominating set S with the minimum cardinality.

4.2 EA in Base Station Placement

Some of the papers that describe the application of EA in the base station placement problem are briefly described below.

1. Calégari et al. [5] presented a GA-based approach to address this problem. The paper assumes that a list of N possible locations that guarantee 100% radio coverage is known beforehand. The candidate solutions are represented using an N bit binary string. A value of 1 indicates the presence of a base station at the location corresponding to that bit, and zero otherwise. The chromosomes are evaluated by the fitness function shown in equation (1):

$$f = \frac{CoverRate^{\alpha}}{NB} \tag{1}$$

where $CoverRate^{\alpha}$ is the radio coverage (the percentage of locations covered by the selected base stations, and α is a parameter that is tuned to favor coverage with respect to the number of transmitters and is assigned a value of 2 in this paper) and NB is the number of selected base stations. This fitness function maximizes the coverage and minimizes the number of transmitters. Selection based on fitness value, one-point crossover, and mutation operators (flipping of the value of a randomly chosen bit of the string with a probability of 0.9) is employed.

Experimental results were carried out on a 70 km $\times$ 70 km digital terrain discretized on a 300 $\times$ 300 points grid with $N = 150$. The radio coverage of each base station was computed using a radio wave propagation simulation tool. A solution with 52 base stations covering 80.04% was found after 160 generations with a population size of 80 chromosomes. The algorithm was found to be very slow. So in order to speed up the execution, the parallelization of the algorithm with the island concept was investigated. In this concept, the population is partitioned into subpopulations called islands that evolve independently towards the solution. In order to benefit from the information found in an another island, it also supports the migration of individuals from one island to another. The computation time was reduced by running each island of 20 individuals on 4 different processors. They were also able to find a better quality of solution with 41 base stations covering 79.13% of the initial covered surface.

2. Han et al. [33] described the base station placement problem using GA with real number representation. Binary string representation (considered in [5]), suffers from representation limit (can represent discrete locations) and hence cannot guarantee anoptimal solution. The real number representation describes both the base station location and its number. The chromosome represents the x and y coordinates of a base station. A genome g is a vector of the form $g = (c_1, \cdots, c_k)$, where $c_k = (x_k, y_k)$ is the chromosome for the k-th base station position, and x_k and y_k represent the x and y coordinates of the k-th base station position. The value of k is in the range $1 \leq k \leq K$, where K is the maximum number of base stations. Each genome is evaluated using equation (2) as follows.

$$f(g) = W_c \left(\frac{ct}{tot} \right) + W_e \left(\frac{K - n(g)}{K} \right) \tag{2}$$

where *ct* is the covered traffic and *tot* is the total offered traffic. In equation (2), the first term is the objective function for coverage and it increases as the covered traffic area increases corresponding to g. The second term is the objective function for economy (cost of the network) and increases as the number of base stations decreases. W_c and W_e are weight ($W_c + W_e = 1$), and the values depend upon whether coverage or fewer base stations are preferred. $n(g)$ is the number of base stations in the genome g.

The paper defined an appropriate crossover and mutation operator. In the crossover operator, a single chromosome is generated from two chromosomes (parents). If the chromosome for the k-th base station position is defined in both the parents, the corresponding chromosome in the child is the mean of the corresponding coordinates of the parent. Otherwise the child inherits the chromosome from the parent with the defined positions. The selection of individuals for reproduction is done using tournament selection.

Simulation results were carried out on a hexagonal cellular environment with cell radius of 2.5 km and uniform and nonuniform traffic distribution. Under uniform traffic distribution with $K = 7$ and $W_c = 1$, the algorithm after the 700-th generation was able to find base station locations that provided 97.8% coverage. The algorithm was also tested for economy by varying W_e. With $K = 8$ after 700 iterations, $W_e = 0.3$ provided 81.8% coverage with 6 base stations whereas $W_e = 0.27$ provided 88.1% coverage with 7 base stations. By varying the weights the user can get the proper number of base stations and coverage. With nonuniform traffic distribution and $W_c = 0.8$ and $K = 12$, a coverage of 99% is achieved after the 1000-th generation.

5 Combinatorial Heuristics for Fixed Channel Assignment

The fixed channel assignment problem can be modeled as an undirected graph $G(V, E)$, where the set of vertices V represents the set of radio transmitters or base stations, and the set of edges represents the interference constraints for frequencies assigned to neighboring base stations. Each vertex v_i is associated with a weight w_i that represents the demand of channels in the corresponding base stations. Metzger [56] first pointed out that the channel assignment problem is equivalent to the graph coloring problem. A feasible graph coloring solution assigns w_i different colors to the vertices v_i in such a way that no two adjacent vertices (i.e., vertices connected by an edge) have the same color. The objective is to minimize the number of colors used. In the channel assignment problem each color corresponds to a channel. Colors are represented by a set of positive integers. The performance of a coloring algorithm is measured in terms of the ratio of the number of colors used to the minimum number of colors required to color G. This ratio is known as the competitive ratio. Narayanan and Shende [60] proposed an approximation algorithm which is 4/3 competitive. The fixed preference allocation algorithm proposed by Janssen et al. [38] is 3/2 competitive.

Hale [32] formulated the channel assignment problem as the T-coloring problem. The T-coloring problem is a generalization of the graph coloring problem. T-coloring assigns a nonnegative integer (channel) to each vertex of G so that the adjacent vertices are assigned channels whose separation is not in a set T of unallowable separations.

Griggs and Yeh [31] proposed an $L(2,1)$-coloring of a graph for channel assignment. The $L(2,1)$-coloring problem is a generalization of the T-coloring problem. It assigns a nonnegative integer to each vertex of G so that the channels assigned to two vertices joined by an edge differ by at least two and channels differ if the two vertices have a common neighbor.

Roxborough et al. [70] formulated the fixed channel assignment problem with both co-channel and adjacent channel constraints as the k-band chromatic bandwidth problem and with only the co-channel constraint as the distance k-chromatic number problem of cellular graphs. The distance k-chromatic number problem of a graph $G(V, E)$ with an integer k is the minimum number of colors required to color the vertices so that no two vertices separated by a distance not exceeding k have the same color. This is known as proper coloring. In the k-band chromatic bandwidth problem, a weight $w(i, j)$ is associated with an edge for all $i, j \in V$. The corresponding graph is known as an edge-weighted graph. This weight represents the frequency separation required between the channels used in the corresponding base stations. In the distance k-chromatic number problem, it has been assumed that the interference does not extend beyond k number of cells away from the call -originating cell. They have shown that by exploring the regular structure of a cellular network many instances of channel assignment problems can be solved in polynomial time. However, the algorithms presented were not optimal. Sen et al. [76] proposed an optimal algorithm for the distance-k chromatic bandwidth number problem presented in [70], and a near-optimal algorithm for the chromatic bandwidth problem with a performance bound of 4/3.

6 EA in Fixed Channel Assignment

1. Smith [80] proposed a GA-based approach to solve the FCA problem. The objective function treats the interference constraints (co-channel, adjacent channel, and co-site) as soft constraints and traffic demand satisfaction as a hard constraint. With this approach, a solution that minimizes the severity of any interference is always found. This helps to find a solution in situations where demand and interference constraints are such that no interference-free solutions are available for the network. Thus the formulation attempts to minimize the severity of any interference. The genetic representation of the solution is a binary channel assignment matrix $A[C, F]$ where $A_{i,k} \in \{0, 1\}$ for $i = 1 \ldots C$, and $k = 1 \ldots F$. The fitness of

the chromosome is measured by equation (3):

$$\text{Min} \quad f(A) = \sum_{j=1}^{C}\sum_{k=1}^{F} A_{j,k} \sum_{i=1}^{C}\sum_{l=1}^{F} P_{j,i,(|k-l|+1)A_{i,l}} \tag{3}$$

$$\text{s.t.} \quad \sum_{k=1}^{F} A_{j,k} = d_j \qquad \forall j$$

where d_j is the traffic demand of cell j and P is a factor that assigns a penalty to each assignment according to the following recursive relation

$$\begin{aligned} P_{j,i,m+1} &= \max(0, P_{j,i,m-1}) \\ P_{j,i,1} &= X_{j,i} \\ P_{j,i,1} &= 0 \end{aligned}$$

for $m = 1 \ldots F-1$ and for all $j, i \neq j$. Here X is a compatibility matrix of dimension $C \times C$. The paper has designed a crossover and mutation operator in such a way that the feasibility of the solution is guaranteed. The paper also provides an insight into the roles of the crossover and mutation operators: the crossover operator improves co-channel and adjacent channel interference and the mutation operator eliminates co-site interference. The algorithm was tested on a problem set given in [81]. It was found that mutation probability of $p_m = 0.254$ and 500 generations were required to achieve an average interference level of zero with problems of similar complexity but of size $C = 5$ and $F = 11$.

2. Ngo and Li [62] proposed a GA-based approach called the modified genetic-fix algorithm to find an optimal channel assignment matrix in FCA problems. They have considered three interference constraints (co-channel, adjacent channel, and co-site constraints), and the traffic demand constraint with nonuniform traffic distribution among the cells. The proposed algorithm creates and manipulates chromosomes with fixed size (i.e., in binary representation, the number of ones is fixed) and utilizes an encoding scheme called the minimum-separation encoding. A chromosome is a binary string that represents the channel assignment matrix $A[C, F]$ through concatenation of rows, where C represents the number of cells and F represents the number of channels. The chromosome structure incorporates both the traffic demand and co-site

constraint. If $d_{\min}$ is the minimum number of frequency bands by which channels assigned to cell i must differ to prevent co-site constraint then the minimum-separation encoding scheme works by eliminating $(d_{\min} - 1)$ zeros following each 1 in each row of the channel assignment matrix. This compression reduces the search space. A chromosome is evaluated by an objective function that includes only the co-channel and adjacent channel constraint. For the co-channel and adjacent channel interference constraint to be satisfied, if channel q is in cell j then channel p cannot be assigned to cell i if $|p-q| < X_{i,j}$ $(p \neq q,\ i \neq j)$ where X is the compatibility matrix of dimension $C \times C$. Mathematically, it can be formulated as equation (4):

$$f = \sum_{i=1}^{C} \sum_{p=1}^{F} \left(\sum_{j=1, j \neq i, X_{i,j} > 0}^{C} \quad \sum_{q=p-(X_{i,j}-1), 1 \leq q \leq F}^{F} A_{j,q} \right) A_{i,p} \qquad (4)$$

The chromosome is evaluated by equation (4). The genetic-fix algorithm defines its own mutation and crossover operator in such a way that the fixed number of ones is always preserved. The algorithm was tested on 5 sets of problem with cells ranging from 4 to 25, number of channels ranging from 11 to 309, with 5 different compatibility matrices and 3 different demand vectors. The simulation parameters for GA were set to $p_c = 0.95$, $p_m = 0.0005$, population size $= 10$. The results were compared with the neural network-based approach of [25]. The frequency of convergence of the algorithm for all the problem sets was found to be better than the neural network approach. In the fifth problem, they were able to find better solutions with a shorter channel span than previously reported in the literature.

3. Chakraborty and Chakraborty [6] used GA to find the minimum required bandwidth that satisfies a given channel demand without violating interference constraints (co-channel, adjacent channel, and co-site interference constraints). The chromosome is a frequency assignment matrix $A[F, C]$ where $A_{ij}(i = 1 \ldots F$ and $j = 1 \ldots C)$ is either -1, 0, 1, or 9.

   (a) $A_{ij} = -1$ indicates that the i-th channel is not used in the j-th cell and the i-th channel cannot be used in the j-th cell.

(b) $A_{ij} = 0$ indicates that the i-th channel is not used in the j-th cell and the use of the i-th channel in the j-th cell will not result in any interference.

(c) $A_{ij} = 1$ indicates that the i-th channel is used in the j-th cell.

(d) $A_{ij} = 9$ indicates that the i-th channel is not used in the j-th cell.

The paper considered the value of F to be sufficiently large, so that some channels are left unused even after adequate channels have been allocated to all cells. The fitness of the chromosome is measured by the frequency bandwidth a chromosome uses, i.e., by its F value. In the case of chromosomes with the same value of F, the chromosome with the highest number of 0s (a chromosome that allows more channels to be added without violating interference) is considered the fittest. The paper presents an algorithm to generate the initial population, and also defines a genetic mutation operator on those valid chromosomes such that the resulting chromosome is also a valid solution.

Experiments were carried out for three sets of problems with the number of cells ranging from 21–25, two different sets of compatibility matrices, and three sets of traffic demand vectors. For the network with 25 cells, they were able to obtain the best result reported so far in the literature.

4. Jin et al. [40] considered a cellular network with a fixed number of available frequencies. They formulated a cost model for the FCA problem with available bandwidth as one of the hard constraints. The proposed cost function analyzes each assignment in terms of damage caused to the quality of service by blocked calls and interference between cells. They provided a certain degree of relaxation to the demand constraint and the co-site constraint of EMC. A GA-based approach was proposed to minimize the cost function. The main objective is to find an assignment of channels that minimizes the total amount of blocked calls in the network and the amount of interference experienced by calls. A chromosome represents an allocation matrix $A[C, F]$, where C is the number of cells and F is the number of available channels. The damage caused by

blocked calls in the network is given by

$$\sum_{i=1}^{C} \sum_{j=F_i+1}^{n_i} P(Q_i = j)(j - F_i)$$

The term $\sum_{j=F_i+1}^{n_i} P(Q_i = j)(j - F_i)$ represents the expected amount of blocked calls in cell i, where n_i is the amount of mobiles in cell i, $F_i = \sum_{p=1}^{F} A_{i,p} \geq d_i$ is the number of channels allocated to cell i (d_i is the demand of channels in cell i), and Q_i represents the random variable of required channels in cell i. The damage due to interference between channel p assigned to cell i and channel q assigned to cell j is in negative proportion to the distance in channels between p and q. Hence the damage due to interference is defined by $f(i,j,p,q)$ where

$$f(i,j,p,q) = \begin{cases} 0 & \text{if} \quad |p-q| \geq X_{i,j} \\ f_{i,p} f_{j,q} \Psi_C(X_{i,j} - |p-q|) & \text{if} \quad |p-q| < X_{i,j} \quad \text{and} \quad i = j \\ f_{i,p} f_{j,q} \Psi_A(X_{i,j} - |p-q|) & \text{otherwise} \end{cases}$$

where $\Psi_C(x)$ and $\Psi_A(x)$ are strictly increasing functions in x. X is the compatibility matrix and $X_{i,j}$ represents the minimum channel separation required between channels assigned to cells i and j to avoid interference. The total damage due to blocked calls and interference is given by equation (5).

$$\sum_{i=1}^{C} \sum_{j=1}^{C} \sum_{p=1}^{F} \sum_{q=1}^{F} f(i,j,p,q) + w \sum_{i=1}^{C} \sum_{j=F_i+1}^{n_i} P(Q_i = j)(j - F_i) \qquad (5)$$

where w represents the relative weight of damages due to blocked calls to the damages caused by interference.

A heuristic was proposed to generate an initial population with low interference between cells. The paper also defines appropriate an crossover operator. Eight benchmark problems with 21 cells, frequencies ranging from 40 to 64, with three different compatibility matrices, and three sets of communication loads (a: ($\mu = 5$ to 45, $\sigma = 1.1$ to 9.11), b: ($\mu = 8$ to 57, $\sigma = 1.25$ to 11.62), and c: ($\mu = 33.9$ to 191.81, $\sigma = 3.52$ to 21.62)) have been examined. The terminating criterion was either the 1000-th generation or 10 contiguous runs with no improvement. They were able to obtain better solutions with the average number of generations ranging from 52 to 80.

7 Combinatorial Methods in Dynamic Channel Assignment

In dynamic channel assignment, the graph to be colored changes over time. Thus, the problem of dynamic channel assignment can be modeled with an ordered sequence of graphs $G(V, E, w_t), t \geq$, where w_t is the set of calls to be served by the network at time t [37]. The algorithm should color graph G_t before G_{t+1}. Janssen et al. [37] have presented an algorithm which is 4/3 competitive.

8 EA in Dynamic Channel Assignment

In the literature, a number of DCA algorithms have been proposed [12, 93, 78, 85, 92, 9, 19, 8, 18, 77, 73, 72, 64, 71, 39]. These algorithms can be classified into two classes of DCA schemes based on the type of information used in allocating a channel [64]: (1) interference adaptive scheme, and (2) traffic adaptive scheme. In the interference adaptive scheme, the decision regarding the allocation of a channel is based on the measurement of the carrier-to-interference ratio. In the traffic adaptive scheme, the channel allocation decision is based on the traffic conditions in neighboring cells of a cell involved in call arrival. The interference adaptive scheme has been described in [26, 61]. Here the propagation measurements from each base station to mobile and vice versa are made. A channel l is allocated to a new call if it does not cause any interference to the calls already in progress on l and at the same time does not receive any interference from the existing calls in the system.

8.1 Traffic Adaptive Scheme

The traffic adaptive scheme in which an available channel is associated with a cost is called exhaustive searching DCA [12, 93, 78, 92, 19, 18, 73, 72, 71]. The cost of a channel reflects the impact of allocating the channel on the ongoing calls in the system. When a call arrives, the system tries to allocate the channel with the minimum cost. Based on this concept, a GA approach was proposed in Sandalidis et al. [73, 71], and the application of ES was studied in Sandalidis et al. [72]. Sandalidis et al. [71, 72], modeled the channel assignment constraints (soft constraints, hard constraints, traffic demand) discussed in Section 2.1.2 as an energy function whose minimization gives the optimal allocation. The energy function was formulated for the cell involved in call arrival.They have proposed a binary chromosome to represent a cell involved in call arrival. A gene represents a channel (0: the channel is free; 1: the channel is occupied), and the length of the chromosome is always equal to the total number of channels available to the system. The fitness of the chromosome is measured by an energy function.

1. Sandalidis et al. [71] modeled an energy function that takes care of EMC constraints (co-channel, adjacent channel, and co-site interference), traffic demand, and soft constraints (packing, resonance, and limiting rearrangement discussed in Section 2.1.2) as shown in equation (6).

$$\begin{aligned}
E \;=\; & \frac{W_1}{2}\sum_{j=1}^{F}\sum_{i=1,i\neq k}^{C} V_{k,j}\cdot A_{i,j}\cdot interf(i,k) + \frac{W_2}{2}\sum_{j=1}^{F} V_{k,j}\cdot CSC_{k,j} \\
& + \frac{W_3}{2}\sum_{j=1}^{F} V_{k,j}\cdot ADJ_{k,j} - \frac{W_4}{2}\sum_{j=1}^{F}\sum_{i=1,i\neq k}^{C} V_{k,j}\cdot A_{i,j}\cdot \frac{(1-interf(i,k))}{dist(i,k)} \\
& + \frac{W_5}{2}\sum_{j=1}^{F}\sum_{i=1,i\neq k}^{C} V_{k,j}\cdot A_{i,j}\cdot (1-res(i,k)) \\
& + \frac{W_6}{2}\left\{\sum_{j=1}^{F} V_{k,j} - traf(k)\right\} - \frac{W_7}{2}\sum_{j=1}^{F} V_{k,j}\cdot A_{i,j} \qquad (6)
\end{aligned}$$

k	:	Cell where a call arrives
F	:	Number of channels available in the network
C	:	Number of cells in the network
$A_{i,j}$	:	Allocation matrix, $A_{i,j} = 1$ if channel j is assigned to cell i, and 0 otherwise
V_k	:	Output vector (the solution) for cell k with dimension F
$V_{k,j}$	:	j-th element of vector V_k, $V_{k,j} = 1$ if channel j is assigned to cell k, and 0 otherwise
$interf(i,k)$	:	Function whose value is 1 if there is co-channel interference between cells i and k, otherwise 0
$CSC_{i,j}$	:	The ij-th element of matrix $CSC, CSC_{i,j} = 1$ if co-channel interference exists between cells i and j, otherwise 0
$ADJ_{i,j}$	:	The ij-th element of matrix $ADJ, ADJ_{i,j} = 1$ if adjacent channel interference exists between cells i and j, otherwise 0
$dist(i,k)$	:	Distance (normalized) between cells i and k
$res(i,k)$	:	Function that returns a value of one if cells i and k belong to the same reuse scheme, otherwise zero
$traf(k)$	:	The number of channels that cell k must serve, i.e., the traffic demand of cell k

In equation (6), the first term takes care of co-channel interference, the second term takes care of co-site interference, the third term takes care of adjacent channel interference, the fourth term takes care of packing condition, the fifth term takes care of the resonance condition, the sixth term takes care of traffic demand, and the seventh term takes care of the limiting rearrangement condition. $W_1, W_2, W_3, W_4, W_5, W_6$, and W_7 are positive constants that determine the significance of the conditions. In this equation, the rearranging operation is considered only in the cell involved in the new call arrival.

The chromosome with the minimum energy gives the desired solution. The call is blocked if the desired solution causes co-channel interference and does not satisfy the traffic requirement of the cell at that time. Otherwise, the call is successful and the channel usage information of the cell is updated according to the desired solution. The performance of the algorithm was measured in terms of probability of blocking of new calls. Experimental results were carried out in a parallelogram topological network of 49 hexagonal cells and 70 channels given in [93] under uniform and two nonuniform traffic distribution patterns given in [93]. Roulette wheel selection, two-point crossover, standard mutation operator, population size of 50, crossover probability of 0.75, and mutation probability of 0.05 were used. The GA was terminated after 100 generations. The results were compared with FCA, Borrowing with Channel Ordering (BCO), Borrowing with Directional Channel Locking strategy (BDCL), and Locally Optimized Dynamic Assignment strategy (LODA) schemes proposed in [93] and the Hopfield neural network approach proposed in [18]. The proposed algorithm was compared under three different cases: 1) with co-channel interference constraint only, 2) with co-channel and co-site interference constraints only, and 3) with all the three interference constraints. With uniform traffic distribution the algorithm outperformed all of them in case 1 but showed a poor performance for the other two cases. With nonuniform traffic distribution only case 1 was considered. The performance of the algorithm was better than FCA, BCO, and LODA but poorer than BDCL and the Hopfield neural network approach.

2. Sandalidis et al. [72] proposed an ES-based approach called Combinatorial Evolutionary Strategy DCA (CES DCA) to solve the DCA problem. CES DCA is a $(1, \lambda)$-ES. They modeled an energy function as shown in equation (7) that takes care of the packing condition, resonance condition, limiting rearrangement, and co-channel interference.

$$\begin{aligned} E = {} & \frac{W_1}{2}\sum_{j=1}^{F}\sum_{i=1, i\neq k}^{C} V_{k,j}\cdot A_{i,j}\cdot interf(i,k) - \\ & \frac{W_2}{2}\sum_{j=1}^{F}\sum_{i=1, i\neq k}^{C} V_{k,j}\cdot A_{i,j}\cdot \frac{(1-interf(i,k))}{dist(i,k)} + \\ & \frac{W_3}{2}\sum_{j=1}^{F}\sum_{i=1, i\neq k}^{C} V_{k,j}\cdot A_{i,j}\cdot (1-res(i,k)) - \\ & \frac{W_4}{2}\sum_{j=1}^{F} V_{k,j}\cdot A_{i,j} \end{aligned} \tag{7}$$

In equation (7), the first term takes care of co-channel interference, the second term takes care of the packing condition, the third term takes care of the resonance condition, and the last term takes care of the limiting rearrangement condition. The weights W_1, W_2, W_3, and W_4 are positive constants that determine the significance of the conditions, and the other variables carry the same meaning as in Section 8.1.1. The traffic requirement is incorporated into the problem representation. Thereby the fitness function is simplified as compared to [71]. The number of ones in the chromosome is equal to the traffic requirement of the cell at that instant. The energy function determines the fitness of the chromosome. The chromosome with the minimum energy gives the desired solution. If the desired solution causes co-channel interference the call is blocked. Otherwise, the call is successful, and the channel usage information of the cell is updated according to the fittest chromosome. The performance of the algorithm was measured in terms of probability of blocking of new calls. Experimental results were carried out in a parallelogram topological network of 49 hexagonal cells and 70 channels given in [93] under uniform and two nonuniform traffic distribution patterns given in [93]. The values of W_1, W_2, W_3, andW_4 were set to 10, 3, 1, and 2, respectively. The algorithm was terminated when the destabilization process occurred for the second consecutive time. The population size was set to 50. The performance of the algorithm was compared with FCA, BCO, BDCL, and LODA schemes proposed in [93], the genetic algorithm-based DCA approach (GA DCA) proposed in [73], and the Hopfield neural network approach proposed in [18]. Under uniform traffic distribution, CES DCA outperformed all the algorithms. Under nonuniform traffic distribution CES DCA was compared with FCA, BCO, BDCL, and LODA and it outperformed all of them.

3. Lima et al. [48] proposed two GA-based strategies called GAL and GAS to solve the DCA problem. GAL looks for an idle channel to serve an incoming call, and the assigned channel serves the call until it is terminated, whereas in GAS, an ongoing call can be switched to a different channel. The proposed algorithms take care of co-channel interference based on a fixed reuse distance (three cells) concept. The main objective is to find a channel assignment matrix that minimizes the total amount of blocked calls in the network. A chromosome is represented by a vector $A_i = 1 \times F$, where F

is the number of available channels and the j-th element of A_i is 1 if channel j is assigned to cell i, and 0 otherwise. Assignment of channels to a cell is associated with a cost function defined by equation (8).

$$f = \sum_{i=1}^{C} \sum_{j=1}^{F} fit(i,j) \tag{8}$$

where C represents the number of cells in the system and $fit(i,j) = n_1(j).r_1 + n_2(j).r_2 + n_3(j).r_3 + n_4(j).r_4$ where $n_1(j)$ represents the number of compact cells of cell i using channel j, $n_2(j)$ represents the number of co-channel cells of i located in the third tier of cells that do not use j, $n_3(j)$ represents the number of other co-channel cells of i using j, and $n_4(j)$ represents the number of channels to be blocked due to assigning j. Compact cells are the cells located at a minimum average distance between co-channel cells. r_1, r_2, r_3, r_4 are weights that determine the significance of each term with a value of 5, 1, -1, and -15, respectively. In GAL only the idle channels undergo mutation. The algorithms are evaluated with a parallelogram topological network of 49 hexagonal cells and 70 channels under uniform and nonuniform traffic distribution [93] and time-varying traffic pattern. GAL and GAS were found to provide lower average call blocking probability as compared to the schemes reported in [93] and [64].

9 EA in Hybrid Channel Assignment

1. Sandalidis et al. [72] applied their CES heuristic on the HCA problem. Their method, we call CES HCA, is a $(1,\lambda)$-ES. The problem representation and energy function are the same as discussed in section 8.1.2. Experiments were also carried out as discussed in Section 8.1.2. Here three representative ratios were considered: 21:49 (FCA is 21 channels and DCA is 49 channels), 35:35, and 49:21. In all these cases, the algorithm tries to assign a channel from the DCA set only when all the channels in the FCA set are busy. Under uniform traffic distribution, 21:49 CES HCA outperformed FCA and provided better performance than LODA up to 9.5% increase in traffic load, 35:35 CES HCA outperformed FCA and LODA, and 49:21 CES HCA outperformed FCA and LODA and provided almost the same performance as BCO. Under one

nonuniform traffic distribution, 21:49 CES HCA and 35:35 CES HCA outperformed FCA only, and 49:21 CES HCA outperformed FCA and provided almost the same performance as BCO; with another non-uniform traffic distribution, 21:49 CES HCA outperformed FCA and provided almost the same performance as LODA, 35:35 CES HCA outperformed FCA and LODA, and 49:21 CES HCA outperformed FCA and LODA and provided almost the same performance as BCO.

2. Vidyarthi et al. [87] proposed a new HCA strategy, called the D-ring strategy, using the distributed dynamic channel assignment strategy based on the fixed reuse distance concept. Here, D is the fixed reuse distance. D rings of cells around a given cell form the interference region. The channels are allocated to the host cell from a set of channels that excludes all those channels which are in use in the interference region. As such the selected channels always satisfy the co-channel interference constraint. They proposed an ES-based approach for the solution of the HCA using integer vector representation (as chromosome), where each integer at position i in the vector represents a channel number assigned to a call in the i-th cell. The disadvantage of the binary string representation considered in [73], [72], and [71] is that although we are interested only in d channels, extra memory is consumed in storing the information about the other unused channels. Here d represents the current traffic demand in the cell involved in call arrival. Binary string representation also yields slower evaluation and manipulation of candidate solutions, due to the size of the binary representation. The other advantage of the integer representation is that the size of the solution vector is short and thus it is easier and faster to process.

   [87] also proposed a novel way of generating the initial parent and the initial population. Instead of starting from a totally random combination of channel numbers, they start with solution vectors with $(d-1)$ channels allocated to the cell by the algorithm in its last call arrival. This way of generating initial parent and initial population will reduce the number of channel reassignments and therefore yields a faster running time. The initial parent is also a potentially good solution because channels for ongoing calls are already optimized.

Compared to Sandalidis et al. [72], the fitness function is simpler. This because one major hard constraint, the co-channel interference, is taken care of by the D-ring-based strategy. This also leads to a simpler and faster fitness calculation than that of Sandalidis et al. [72]. The problem representation also takes care of the traffic demand constraint as the number of channels in a solution vector equals the demand of channels in the cell. The soft constraints are modeled as an energy function as in [72]. It is shown in equation (9). The minimization of this function gives a near-optimal channel allocation.

$$
\begin{aligned}
E \;=\; & -W_1 \sum_{j=1}^{d_k} \sum_{i=1, i\neq k}^{C} A_{i,V_{k,j}} \cdot \frac{1}{dist(i,k)} + W_2 \sum_{j=1}^{d_k} \sum_{i=1, i\neq k}^{C} A_{i,V_{k,j}} \cdot (1 - res(i,k)) \\
& - W_3 \sum_{j=1}^{d_k} A_{k,V_{k,j}} \qquad (9)
\end{aligned}
$$

k	:	Cell where a call arrives
d_k	:	Number of channels allocated to cell k (traffic demand in cell k)
C	:	Number of cells in the network
V_k	:	Output vector (the solution) for cell k with dimension d_k
$V_{k,j}$	:	The j-th element of vector V_k
$A_{i,V_{k,j}}$	:	The element located at the i-th row and $V_{k,j}$-th column of the allocation matrix A
$dist(i,k)$	:	Distance (normalized) between cells i and k
$res(i,k)$	:	Function that returns a value of one if the cells i and k belong to the same reuse scheme, otherwise zero

W_1, W_2, and W_3 are positive constants. The first term expresses the packing condition. The second term expresses the resonance condition. The last term expresses the limiting reassignment. This term results in a decrease in the energy if the new assignment for the ongoing calls in cell k is the same as the previous allocation. The value of the positive constants determines the significance of the different terms. The energy function determines the fitness of the chromosome. The fittest chromosome in all the generations is the desired solution. Simulations were carried out as in [72] for nonuniform traffic distribution only. The performance of the algorithm was compared with those obtained in [72]. Under one traffic distribution, all the representative ratios outperformed their counterpart in [72]. Under another traffic pattern, all the representative ratios outperformed their counterpart in [72] up to 60% increase in traffic load. Beyond 60% their performance was more or less the same.

10 Graph Theoretical Methods in Mobility Management

Naor et al. [59] proposed a Cell Identification Code (CIC) for tracking mobile users based on the movement-based location update scheme suggested in [3]. The proposed scheme provides mobile users with location information necessary to reduce the cost of tracking a mobile user. The idea is to identify each cell with a code different from its nearest neighbors in order to facilitate the detection of cell boundary crossing. Thus the movement-based update scheme using CIC is a special case of the graph coloring problem. The objective is to find the minimum number of codes needed to code the vertices such that no two adjacent vertices have the same code. For hexagon-shaped cells only three colors are needed where a color corresponds to a code.

11 EA in Mobility Management

In a cellular network, in order to route a call, the mobile terminal needs to be correctly located within a fixed time delay. The location management involves two types of activities: paging and location update (LU). Both paging and LU increase network traffic overhead and consume the scarce radio resource. Therefore, during a certain period of time, the total cost of location management involves the sum of two orthogonal cost components: paging cost and LU cost. The two costs are orthogonal because the higher the frequency of LU is, the lesser is the frequency of paging attempts required to locate the mobile terminal [14]. Thus there exists a tradeoff between paging cost and LU cost that varies with the size of the location area (LA). If the LA is large, there are fewer inter-LA crossings resulting in a lower LU cost but the number of base stations needed to be paged increases correspondingly. Therefore one way of reducing LU cost is by effective planning of LA.

LA planning was considered in [27] using a graph theoretic approach, in [53, 52] using two heuristic algorithms, in [66] using a greedy algorithm, and in [88] using genetic algorithm. LA planning decomposes a group of cells into LAs in which LU traffic is minimized without violating the paging bound (bandwidth available for paging). In general LU traffic is proportional to the number of mobile terminals crossing the LA border and paging cost is proportional to the number of calls to all mobile terminals in the LA. Some of the papers that describe the use of evolutionary algorithms to mobility management are described below.

1. Das and Sen [14] proposed a GA approach for the optimization of the total location management cost. The proposed scheme provides an improvement over the existing zone-based schemes by minimizing the total location management cost for individual users by devising an update strategy for each mobile that considers per-user mobility and call arrival pattern on top of the conventional zone based approach. In the conventional zone-based approach, all mobile users are made to update whenever they cross a LA boundary. Because these LAs are formed based on a common mobility pattern for all the users in the system, this approach leads to a significant number of redundant updates. In the proposed scheme, each user updates only in preselected LAs called his reporting areas. GA has been used to find the optimal update strategy. First, the LAs in the service area are numbered sequentially. Then the update strategy is represented by a binary chromosome $\{d_n \cdots d_2 d_1\}$ where i is the sequence number of the LA, n is the total number of LAs in the service area, and d_i is the decision variable for the user in LA i such that

$$d_i = \begin{cases} 1: & \text{if update occurs in LA } i \\ 0: & \text{otherwise} \end{cases}$$

The total location management cost LMC has been defined as in equation (10).

$$LMC = \sum_{i=1}^{N} \pi_i LM_i^{d_i} \tag{10}$$

where π_i is the location probability of the user in the LA i, and $LM_i^{d_i}$ is the average location management cost of the user in the LA i, and is calculated assuming call arrival as a poisson distribution. The chromosome is evaluated using fitness function $\frac{1}{LMC}$. Roulette wheel selection mechanism, crossover probability of 0.8, mutation probability of 0.001, and population size of 20 are used.

The algorithm was tested for various values of call arrival rate and ratio of update and paging cost for 10 location areas (with location probabilities $\pi_0 = \pi_9 = 0.08333$, and $\pi_i = 0.1041$ for $1 \leq i \leq 8$). It was observed that for low user residing probability in LAs, low call arrival rate, and high update cost for a user, skipping of updating in several LAs leads to the minimization of the overall location management cost.

2. One way to reduce the paging cost is to partition the LA into paging zones for each user based on the predetermined probability of locating the mobile user at different locations within the LA. Sun and Lee [84] proposed a GA approach for the optimal planning of such paging zones. For each mobile user a multilayered model is developed based on the mobile-phone usage for different times of activity during a day—home, work, social. LA is decomposed into a set of multiple location layers $\{L_1, L_2, \ldots, L_n\}$ where $1 \leq i \leq n$ and n is the number of layers in the multilayered model, based on the mobility patterns that describe the likelihood of locating the user in a particular cell during a particular time of the day. For example, L_1 and L_2, may refer to the home and working area, respectively, of a mobile user. Then for each user k the cells in the LA are partitioned into paging zones for each activity layer j such that each zone consists of cells with similar probability of locating K. When a call is received for a particular user, a paging message is first sent to the paging zone with the highest probability of locating that user at that particular time of the day. If there is no response for the first paging message, then the paging zone with next highest probability is paged, and so on. Thus paging cost is incurred if and only if the paging zone is paged. The paging cost to locate a mobile user has been formulated as defined in equation (11):

$$f = \alpha \cdot \beta \cdot \left[N(P_{k,j,1}) + \sum_{l=2}^{T} \left\{ \left(1 - \sum_{i=1}^{l-1} prob(P_{k,j,i}) \right) N(P_{k,j,l}) \right\} \right] \tag{11}$$

where $prob(P_{k,j,l})$ is the probability of locating the k-th user in the j-th activity layer of l-th zone, T is the total number of paging zones, $N(P_{k,j,l})$ is the number of cells in the l-th zone, α denotes the consumption cost in the forward control channel per paging message, and β denotes the consumption cost in the fixed link channel consumption in the mobile switching center per paging message. The objective is to find a paging zone with the lowest cost. In this scheme each user has unique paging zones. Hence, optimization must be carried out separately for each individual mobile user.

The candidate solutions were encoded with integer representation. The cells and paging zones in the LA are numbered sequentially. The length of the chromosome equals the number of cells in the LA, gene position corresponds to the cell number, and the value of a gene at a particular position corresponds to the paging zone to which the cell number belongs. For example, if there are 5 cells (1, 2, 3, 4, 5) and 3 paging zones (1, 2, 3) in an LA, and paging zone 1 contains cell numbers 1 and 2, paging zone 2 contains cell numbers 4 and 5, and paging zone 3 contains cell number 3, then the corresponding chromosome representation is "1322". The tournament selection mechanism and standard crossover and mutation operators were used. Experiments were carried out with different paging zones: system-wide paging zone at different location layers, two-paging zone, three-paging zone, and maximum paging zone (number of paging zone is equal to the number of cells). Paging cost was observed to decrease with the increase in the number of paging zones.

3. Wang et al. [88] proposed a GA approach for the optimal planning of LA to reduce the LU cost. It was assumed that cell planning, LU, and paging traffic for each cell are known beforehand, each LA contains a disjoint set of cells, and the paging bound is fixed for each LA. The LA planning problem was encoded using a binary chromosome with a border-oriented representation. In border-oriented representation, all borders are numbered sequentially and the corresponding bit in the chromosome is 1 if that particular cell border is to be a border between two adjoining LAs, and 0 otherwise.

   Figure 2 (adapted from [88], Figure 2, pp. 989) shows the numbering of borders (b_i) and numbering of cells (c_i) in a system with 7 cells. Figure 3 (adapted from [88], Figure 3, pp. 989) shows an LA planning result of the 7 cell system shown in Figure 2. Figure 4 (adapted from [88], Figure 10 (b), pp. 991) shows the border-oriented chromosome structure V of the LA planning shown in

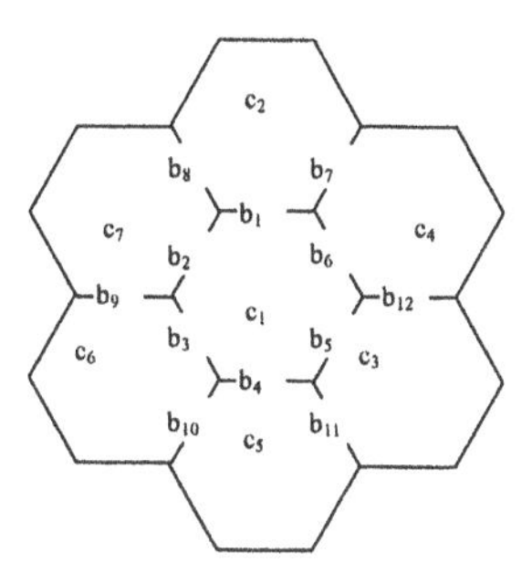

Figure 2: Numbering of cells and border.

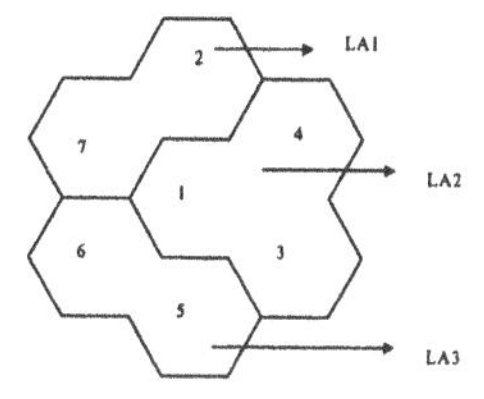

Figure 3: An LA planning result.

1	1	1	1	0	0	1	0	1	0	1	0

Figure 4: Border-oriented chromosome structure.

Figure 3. In Figure 4, $V_i = 1$ if the border b_i exists in the LA planning shown in Figure 3, otherwise $V_i = 0$. For example, the border b_1 between the cells c_1 and c_2 exists in the LA planning shown in Figure 3, hence $V_i = 1$.

The chromosomes are evaluated using the fitness function shown in equation (12):

$$f = \frac{\alpha_1}{2} \sum_{j=1}^{n} v_j w_j + \alpha_2 \sum_{i=1}^{m} \max\{0, (P_i - B)\} \qquad (12)$$

The first term denotes the LU cost and the second term denotes the cost due to the violation of the paging bound. In this equation, n is the total number of borders, w_j is the crossing intensity of border of j, v_j is the j-th bit of the chromosome being evaluated, m is the total number of LAs, P_i is the paging traffic in LA i, B is the paging bound that has been considered fixed for each LA, and α_1 and α_2 are constants used to weigh the relative importance of LU cost and paging bound violation. The objective is to find an LA with minimum cost.

Simulations were carried out with five hexagonal systems with cells ranging from 19–91. The cells were configured as an H-mesh. The cell paging cost and border crossing intensity were generated with normal distribution ($\mu = 100$, $\sigma^2 = 20$). The results were compared with a hill-climbing approach after 100 simulation runs. Selection based on fitness value, standard crossover, and mutation operators were used. Population size, crossover probability, mutation probability, and maximum number of generations were set to 20, 1, 0.02, and 1000, respectively. In all the cases, GA outperformed the hill-climbing approach with a percentage improvement of 10%–29%. Hence the GA approach was concluded to be a robust technique for LA planning.

12 EA in Code Division Multiple Access

1. Chan et al. [7] proposed a GA-based approach to solve the resource management problem in direct sequence CDMA systems of a mobile communication network. Transmission power and transmission rate of all connecting users are two important resource management issues. The objective is to minimize the total transmission power and maximize the total transmission rate of all users simultaneously while satisfying the required transmission power, transmission rate, and bit energy-to-noise density ratio of each user. The total cost to be minimized is formulated as shown in equation (13).

$$
\begin{aligned}
\text{Min} \quad f = \; & \alpha \left[\lambda_p \sum_{i=1}^{N} p_i + \lambda_r \left(T - \sum_{i=1}^{N} r_i \right) \right] \\
& + \beta \left(\lambda_x \sum_{i=1}^{N} x_i \right) \qquad (13) \\
\text{s.t.} \quad & P_i^{\min} \le p_i \le P_i^{\max} \qquad \forall i \\
& R_i^{\min} \le r_i \le R_i^{\max} \qquad \forall i \\
& \left(\frac{E_b}{N_0} \right)_i \ge \gamma_i \qquad \forall i \\
\text{where} \quad & \left(\frac{E_b}{N_0} \right)_i = \frac{\frac{g_{bi} \cdot p_i}{r_i}}{\frac{1}{W} \sum_{j=1, j \ne i}^{N} g_{bj} \cdot p_j} + \eta \qquad \forall i
\end{aligned}
$$

N	:	Total number of connecting users in the system
p_i	:	Transmission power of user i
$P_i^{\min}$	:	Minimum allowed transmission power of user i
$P_i^{\max}$	:	Maximum allowed transmission power of user i
r_i	:	Transmission rate of user i
$R_i^{\min}$	:	Minimum allowed transmission rate of user i
$R_i^{\max}$	:	Maximum allowed transmission rate of user i
$\frac{E_b}{N_0}$	:	Received bit energy-to-noise ratio of user i
γ_i	:	Threshold value of bit energy-to-noise ratio of user i
x_i	:	$x_i = 1$ if user i violates the QoS requirement and 0 otherwise
T	:	A constant larger than the total transmission rate of N users
W	:	Total spread spectrum bandwidth
η	:	Background noise power
g_{bi}	:	Link gain between mobile user i and base station b

λ_p and λ_r represent the fixed cost per unit power in watts and per unit rate in kilo bits per second, respectively. λ_x represents the fixed cost per user who violates her QoS requirement. QoS is a measure of signal quality measured in terms of bit energy-to-noise ratio. α and β are weights. $\alpha = 1, \lambda_p = 1, \lambda_r = 1, \lambda_x = 1, \alpha = 1, \beta = 5000, N = 50$, and $T = 15000$. The first term represents the minimization of total transmission power, the second term represents the minimization of total transmission rate, and the third term takes care of the number of users violating their QoS requirement. The proposed method looks for a solution with the minimum power, maximum rate, and with minimum number of users violating the QoS requirement. Each chromosome is represented by real floating point values. The length of the chromosome is $2N$, with two genes per user. One gene represents the transmission power and the other gene represents the transmission rate. The GA is characterized by two-point crossover, roulette wheel selection, population size of 30, $P_c = 0.8$, and $P_m = .01$. Two approaches have been considered: single objective and multiobjective. The multiobjective approach considers each term of equation (13) as a single function. The GA is terminated after 150 generations in the single objective case, and after 110 in the multiobjective case. The results were compared with simulated annealing, tabu search, and generalized reduced gradient method. GA provided the best results both for single and multiobjective approaches.

13 Conclusions and Future Directions

In this chapter, we presented several streams of research on optimization problems in wireless mobile communication that have been conducted by researchers in the past. We also discussed some details on the implementation of the genetic algorithm and evolutionary strategy to these problems. Wireless mobile communications are evolving fast. It is already certain that over the coming years these EAs will play an increasingly important role in solving the optimization problems that come up in various areas of wireless mobile communication. The papers considered in this chapter, have focused on only one problem at a time. But some of the problems in wireless mobile communication are highly correlated. One such example is the problem of minimal assignment of channels while minimizing the allocation of transmitter and/or receiver powers to the mobile terminals. So one can explore the possibility of solving such integrated problems using EAs.

EAs provide near-optimal solutions but solution quality cannot be measured in terms of optimality only. Hence in order to better explore the search space, one of the future research directions could be to use EAs in conjunction with large-scale linear optimization techniques such as Lagrangian relaxation, Benders decomposition, or Dantzig–Wolf decomposition. Although the linear programming relaxation of the problem would provide a lower bound, efficient heuristic solution procedures based on EAs can be developed to generate feasible solutions to the problem. This would provide us with an indication of how far the obtained solution is from the optimal solution.

References

[1] I.F. Akyildiz and S.M. Ho, "On Location Management for Personal Communications Networks", *IEEE Communications Magazine*, vol. 34, no. 9, pp. 138–145, 1996.

[2] H.R. Anderson and J.P. Mcgeehan, "Optimizing Microcell Base Station Locations using Simulated Annealing Techniques", *in Proceedings of the IEEE 44th Vehicular Technology Conference*, vol. 2, pp. 858–862, 1994.

[3] A. Bar-Noy, I. Kessler, and M. Sidi,"Mobile Users: To Update or Not to Update?", *ACM/Baltzer J. Wireless Networks*, vol. 1, no. 2, pp. 175–95, 1995.

[4] D. Beckmann and U. Killat, "A New Strategy for The Application of Genetic Algorithms to The Channel Assignment Problem", *IEEE Transactions on Vehicular Technology*, vol. 48, pp. 1261–1269, 1999.

[5] P. Calégari, H. Guidec, H.P. Kuonen, and D. Wagner, "Genetic Approach to Radio Network Optimization for Mobile Systems", *in Proceedings of IEEE Vehicular Technology Conference* , vol. 2, pp. 755–759, 1997.

[6] G. Chakraborty and B. Chakraborty, "A Genetic Algorithm Approach to Solve Channel Assignment Problem in Cellular Radio Networks", *in Proceedings of the IEEE Workshop on Soft Computing Methods in Industrial Applications*, Finland, pp. 34–39, 1999.

[7] T.M. Chan, S. Kwong, and K.F. Man, "Multiobjective Resource Optimization in Mobile Communication Network", *in Proceedings of the 29th Annual Conference of the IEEE Industrial Electronics Society*, vol. 3, pp. 2029–2034, 2003.

[8] P.T.H. Chen, M. Palaniswami, and D. Everitt, "Neural Network-Based Dynamic Channel Assignment for Cellular Mobile Communication Systems", *IEEE Transaction on Vehicular Technology*, vol. 43, no. 2, pp. 279–288, 1994.

[9] J.C.-I. Chuang, "Performance Issues and Algorithms for Dynamic Channel Assignment", *IEEE Journal on Selected Areas in Communications*, vol. 11, no. 6, pp. 955–963, 1993.

[10] L.J. Cimini, Jr., G.J. Foschini, "Distributed Dynamic Channel Allocation Algorithms for Micro-cellular Systems", in J.M. Holtzman and D.J. Goodman (eds.), *Wireless Communications—Future Directions*, Hingham, MA: Kluwer Academic Publishers, pp. 219–241, 1993.

[11] L.J. Cimini and G.J. Foschini, "Distributed Algorithms for Dynamic Channel Allocation in Microcellular Systems", *in Proceedings of the IEEE Vehicular Technology Conference*, pp. 641–644, 1992.

[12] D.C. Cox and D.O. Reudink, "Increasing Channel Occupancy in Large Scale Mobile Radio Systems: Dynamic Channel Reassignment", *IEEE Transanctions on Vehicular Technology*, vol. VT-22, pp. 218–222, 1973.

[13] C. Darwin, *On the origin of Species by Means of Natural Selection*, London: John Murray, 1959.

[14] S.K. Das and S.K. Sen, "A New Location Update Strategy for Cellular Networks and its Implementation using a Genetic Algorithm", *in Proceedings of the ACM/IEEE Third Annual International Conference On Mobile Computing and Networking (MOBICOM)*, Budapest, Hungary, pp. 185–194, 1997.

[15] L. Davis, *Handbook of Genetic Algorithms.* New York: Van Nostrand Reinhold, 1991.

[16] K.A. De Jong, "Genetic-Algorithm-Based Learning", *in Machine Learning: An Artificial Intelligence Approach*, vol. 3, Y. Kodratoff and R. Michalski (eds.), San Francisco: Morgan Kaufman, pp. 611–638, 1990.

[17] K. Deb and A. Kumar, "Real-Coded Genetic Algorithms with Simulated Binary Crossover: Studies on Multi-Modal and Multi-Objective Problems", *Complex Systems*, vol. 9, no. 6, pp. 431–454, 1995.

[18] E. Del Re, R. Fantacci, and G. Giambene, "A Dynamic Channel Allocation Technique based on Hopfield Neural Networks", *IEEE Transanctions on Vehicular Technology*, vol. VT-45, no. 1, pp. 26–32, 1996.

[19] D. Dimitrijevic and J. Vucetic, "Design and Performance Analysis of The Algorithms for Channel allocation in Cellular Networks", *IEEE Transactions on Vehicular Technology*, vol. 42, pp. 526–534, 1993.

[20] S.A. El-Dolil, W.C. Wong, and R. Steel, "Teletraffic Performance of Highway Microcells with Overlay Macrocell", *IEEE Journal on Selected Areas in Communications*, vol. 7, no. 1, pp. 71–78, 1989.

[21] S. M. Elnoubi, R. Singh, and S.C. Gupta, "A New Frequency Channel Assignment Algorithm in High Capacity Mobile Communication Systems", *IEEE Transactions on Vehicular Technology*, vol. VT-21, no. 3, pp. 125–131, 1982.

[22] L. Eshelman and J.D. Schaffer, "Real-coded Genetic Algorithms and Interval-Schemata", *Foundations of Genetic Algorithm II*, pp. 187–202, 1993.

[23] K. Feher, *Wireless Digital Communications: Modulation and Spread Spectrum Applications*, New York: Prentice Hall, 1995.

[24] L.J. Fogel, "An Introduction to Simulated Evolutionary Optimization",*IEEE Transactions on Neural Networks*, vol. 5, no. 1, pp. 3–14, 1994.

[25] N. Funabiki and Y. Takefuji, "A Neural Network Parallel Algorithm for Channel Assignment Problems in Cellular Radio Networks", *IEEE Transactions on Vehicular Technology*, vol. 41, no. 4, pp. 430–437, 1992.

[26] Y. Furuya and Y. Akaiva, "Channel Segregation: A Distributed Adaptive Channel Allocation Scheme for Mobile Communication systems", *DMR II*, Stockholm, pp. 311–315, 1987.

[27] A. Gamst, "Application of Graph Theoretical Models to GSM Radio Network Planning", *in Proceedings of the IEEE Symposium on Circuits and Systems*, pp. 942–945, 1991.

[28] K.S. Gilhousen, I.M. Jacobs, R. Padovani, A.J. Viterbi, L.A.J. Weaver, and C.E. Wheatley , "On the Capacity of a Cellular CDMA System", *IEEE Transactions on Vevicular Technology* , vol. 40, no. 2, pp. 303–312, 1991.

[29] D.E. Goldberg, *Genetic Algorithms in Search, Optimization, and Machine Learning*, New York: Addison-Wesley, 1989.

[30] D.E. Goldberg, "Optimal Initial Population Size for Binary-coded Genetic Algorithms", *TCGA Report No. 850001. The clearinghouse for Genetic Algorithms. Tuscaloosa: University of Alabama*

[31] J.R. Griggs, R.K. Yeh, "Labelling Graphs with a Condition at Distance 2," *SIAM Journal on Discrete Mathematics*, vol. 5, pp. 586–595, 1992.

[32] W.K. Hale, "Frequency Assignment: Theory and Applications," *in Proceedings IEE*, vol. 68, no. 12, pp. 1497–1514, 1980.

[33] J.K. Han, B.S. Park, Y.S. Choi, and H.K. Park, "Genetic Approach with a new Representation for Base Station Placement in Mobile Communications", *in Proceedings of the IEEE Vehicular Technology Conference*, VTC 2001, vol. 4, pp. 2703–2707, 2001.

[34] J.H. Holland, *Adaptation in Natural and Artificial systems*, Ann Arbor: University of Michigan Press, 1975.

[35] J.H. Holland, Outline for a Logical Theory of Adaptive systems, *Journal of the Association of Computing Machinery*, vol. 3, pp. 297–314, 1962.

[36] S. Hurley, "Planning Effective Cellular Mobile Radio Networks", *IEEE Transactions on Vehicular Technology*, vol. 51, no. 2, pp. 254–253, 2002.

[37] J. Janssen, D. Krizanc, L. Narayanan, and S. Shede, "Distributed Online Frequency Assignment in Cellular Networks ", *Journal of Algorithms*, vol. 36, pp. 119–151, 2000.

[38] J. Janssen, K. Kilakos, and O. Marcotte, "Fixed Preference Channel Assignment for Cellular Telephone Systems", *IEEE Transactions on Vehicular Technology*, vol. 48, no. 2, pp. 533–541, 1999.

[39] J. Jiang, T.-H. Lai, and N. Soundarajan, "On Distributed Dynamic Channel Allocation in Mobile Cellular Networks", *IEEE Transactions on Parallel and Distributed Systems*, vol. 13 , no. 10, pp. 1024–1037, Oct. 2002.

[40] M.-H. Jin, H.-K. Wu, J.-T. Horng, and C.-H. Tsai, "An Evolutionary Approach to Fixed channel Assignment Problems with Limited Bandwidth Constraint", *in Proceedings of the IEEE International Conference on Communications* vol. 7, pp. 2100–2104, 2001.

[41] T.J. Kahwa and N.D. Georgans, "A Hybrid Channel Assignment Schemes in Large-Scale, Cellular Structured Mobile Communication Systems", *IEEE Transactions on Communications*, vol. 26, pp 432–438, 1978.

[42] I. Katzela and M. Naghshineh, "Channel Assignment Schemes for Cellular Mobile Telecommunication Systems: A Comprehensive Survey", *IEEE Personal Communications Magazine*, vol. 3, no. 3, pp. 10–31, 1996.

[43] K. Krishnakumar, "Micro-Genetic Algorithms for Stationary and Non-Stationary Function Optimization", *in SPIE Proceedings on Intelligent Control and Adaptive systems, Philadelphia, PA*, vol. 1196, pp. 289–296, 1989.

[44] W.K. Lai and G.C. Coghill, "Channel Assignment through Evolutionary Optimization," *IEEE Transactions on Vehicular Technology*, vol. 45, no. 1, pp. 91–96, 1996.

[45] W.C.Y. Lee, *Mobile Cellular Telecommunications Systems*, New York: McGraw Hill, 1995.

[46] W.Y.C. Lee, "New Cellular Schemes for Spectral Efficiency", *IEEE Transactions on Vehicular Technology*, vol. VT-36, 1987.

[47] W.C.Y. Lee and H.G. Kang, "Cell Planning with Capacity Expansion in Mobile Communications: A Tabu Search", *IEEE Transactions on Vehicular Technology*, vol. 49, no. 5, pp. 1678–1691, 2000.

[48] M.A.C. Lima, A.F.P. Araujo, and A.C. Cesar, "Dynamic Channel Assignment in Mobile Communications based on Genetic Algorithms", *in Proceedings of the 13th IEEE International Symposium on Personal, Indoor and Mobile Radio Communications*, vol. 5, pp. 2204–2208, 2002.

[49] V.H. MacDonald, "The cellular Concepts," *The Bell System Technical Journal*, vol. 58, pp. 15–42, 1979.

[50] V.H. MacDonald, "The Cellular Concepts", *The Bell System Technical Journal*, vol. 44, pp. 547–588,1965.

[51] K. Madani and H.A. Aghvami, "Performance of Distributed Control Channel Allocation (DCCA) under Non-Uniform Traffic Condition in

Microcellular Radio Communications", *in Proceedings of The IEEE International Conference on Communications, ICC'94*, pp. 206–210, 1994.

[52] J.G. Markoulidakis and J. Dost, "Heuristic Algorithms for Optimal Planning of Location Areas in Future Mobile Telecommunication Networks", National Technical University, Athens, Technical Report, 1994.

[53] J.G. Markoulidakis and E.D. Sykas, "Method for Efficient Location Area Planning in Mobile Communication", *Electronic Letter*, vol. 29, no. 25, pp. 2165–2166, 1993.

[54] R. Mathar and J. Mattfeldt, "Channel Assignment in Cellular Radio Networks", *IEEE Transactions on Vehicular Technology*, vol. 42, no. 4, pp. 647–656, 1993.

[55] R. Mathar and T. Niessen, "Optimum Positioning of Base Stations for Cellular Radio Networks", *Wireless Networks*, vol. 6, pp. 421–428, 2000.

[56] B.H. Metzger, "Spectrum Management Technique", Presentation at 38th National ORSA meeting (Detroit, MI), Fall 1970.

[57] Z. Michalewicz, *Genetic Algorithms + Data Structures = Evolution Programs*, Berlin: Springer-Verlag, 1996.

[58] M. Mitchel, *An Introduction to Genetic Algorithm*, Cambridge, MA: MIT Press, 1997.

[59] Z. Naor, H. Levy, and U. Zwick, "Cell Identification Codes for Tracking Mobile Users", *Wireless Networks*, vol. 8, pp. 73–84, 2002.

[60] L. Narayanan and S. Shende, "Static Frequency Assignment in Celluar Networks", *SIROCCO 97*, 1997.

[61] R.W. Nettleton and G.R. Schloemer, "A High Capacity Assignment Method for Cellular Mobile Telephone Systems", *in Proceedings of the IEEE 39th Vehicular Technology conference*, pp. 359–367, 1989.

[62] C.Y. Ngo and V.O.K. Li, "Fixed Channel Assignment in Cellular Radio Networks using A Modified Genetic Algorithm", *IEEE Transactions on Vehicular Technology*, vol. 47, no. 1, pp. 163–72, 1998.

[63] A. Ngom, "Genetic Algorithm for the Jump Number Scheduling Problem", *Order*, vol. 15, pp. 59–73, 1998.

[64] J. Nie and S. Haykin, "A Q-Learning-Based Dynamic Channel Assignment Technique for Mobile Communication Systems", *IEEE Transactions on Vehicular Technology*, vol. 48, no. 5, pp. 1676–1687, September, 1999.

[65] R.L. Pickholtz, L.B. Milstein, and D.L. Schilling, "Spread Spectrum for Mobile Communications", *IEEE Transactions on Vehicular Technology*, vol. VT-40, no. 2, pp. 313–322, 1991.

[66] J. Plehn, "The Design Of Location Areas in A GSM-Network", *in Proceedings of the IEEE 45th Vehicular Technology Conference*, vol. 2, pp. 871–875, 1995.

[67] R. Prakash, N. Shivarati, and M. Singhal, " Distributed Dynamic Channel Allocation for Mobile Computing", *in Proceedings of the 14th ACM Symposium on Principles of Distributed Computing*, Ottawa, Canada, pp. 47–56, 1995.

[68] T.S. Rappaport, *Wireless Communications: Principles and Practices*, Englewood Cliff, NJ: Prentice Hall, 1996.

[69] I. Rechenberg, *Evolutionsstrategie: Optimierung Technischer Systeme nach Prinzipien der Biologischen Evolution*, Stuttgart: Frommann-Holzboog Verlag, 1973.

[70] T. Roxborough, S. Medidi, and A. Sen, "On Channel Assignment Problem in Cellular Networks", *in Conference Record of the Thirty-First Asilomar Conference on Signals, Systems & Computers*, vol. 1, pp. 630–634, 1997.

[71] H.G. Sandalidis, P. Stavroulakis, and J. Rodriguez-Tellez, "Genetic Inspired Channel Assignment Schemes for Cellular Systems", *Mathematics and Computers in Simulation*, vol. 51, pp. 273–286, 2000.

[72] H.G. Sandalidis, P. Stavroulakis, and J. Rodriguez-Tellez, "An Efficient Evolutionary Algorithm for Channel Resource Management in Cellular Mobile Systems", *IEEE Transactions on Evolutionary Computation*, vol. 2, no. 4, pp. 125–137, 1998.

[73] H.G. Sandalidis, P. Stavroulakis, and J. Rodriguez-Tellez, "Implementation of Genetic Algorithms to a Channel Assignment Problem in Cellular communicationsA", *in Proceedings of the 6th Internation Conference on Advances in Communications and Control (COMCON 6)*, pp. 453460, June, 1997.

[74] J.D. Schaffer, J.D. Caruana, L.J. Eshelman, and R. Das, "A study of Control Parameters Affecting Online Performance of Genetic Algorithms for Function Optimization", *in Proceedings of the Third International conference on Genetic Algorithms, San Mateo, CA: Morgan Kaufmann*, pp. 51–60, 1989.

[75] H.P. Schwefel, *Numerical Optimization of Computer Models*, New York: Wiley, 1981.

[76] A. Sen, T. Roxborough, and B.P. Sinha, "On an Optimal Algorithm for Channel Assignment in Cellular Networks", *in Proceedings of the IEEE International Conference on Communications*, vol. 2, pp. 1147–1151, 1999.

[77] S. Singh and D. Bertsekas, "Reinforcement Learning for Dynamic Channel Allocation in Cellular Telephone Systems", *in Advances in Neural Information Processing Systems*, 1997.

[78] K.N. Sivarajan, R.J. McEliece, and J.W. Ketchum,"Dynamic Channel Assignment in Cellular Radio", *IEEE 40th Vehicular Technology Conference*, pp. 631–637, 1990.

[79] C. Smith, *3G Wireless Networks*, New York: McGraw-Hill Professional, 2001.

[80] K.A. Smith, " Genetic Algorithm for The Channel Assignment Problem", *in Global Telecommunications Conference*, GLOBECOM 1998, vol. 4. pp. 2013–2018, 1998.

[81] K.A. Smith and M. Palaniswami, " Static and Dynamic Channel Assignment using Neural Networks", *IEEE Journal on Selected Areas in Communications*, vol. 15, no. 2, pp. 238–249, 1997.

[82] M. Srinivas and L.M. Patnaik, "Genetic Algorithms: A Survey", *IEEE Computer*, vol. 27, no. 6, pp. 17–26, 1994.

[83] I. Stojmenović, *Handbook of Wireless Networks and Mobile Computing*, New York: John Willey & Sons, 2003.

[84] J. Sun and H.C. Lee, "Optimal Mobile Location Tracking By Multilayered Model Strategy", *in Proceedings of the Third IEEE International Conference on Engineering of Complex Computer Systems*, pp. 86–95, 1997.

[85] S. Tekinay and B. Jabbari, "Handover and Channel Assignment in Mobile Cellular Networks", *IEEE Communications Magazine*, vol. 29, no. 11, pp. 42–46, 1991.

[86] A. Thavarajah and W.H. Lam, "Heuristic Approach for Optimal Channel Assignment in Cellular Mobile Systems", *in IEE Proceedings on Communications*, vol. 146, no. 3, pp. 196–200, 1999.

[87] G.D. Vidyarthi, A. Ngom, and I. Stojmenovic, "A Hybrid Channel Assignment Approach using an Efficient Evolutionary Strategy in Wireless Mobile Networks", *IEEE Transactions on Vehicular Technology*, vol. 54, no. 5, pp. 1887–1895, 2005.

[88] T. Wang, S. Hwang, and C. Tseng, "Registration Area Planning for PCS Networks using Genetic Algorithms", *IEEE Transaction On Vehicular Technology*, vol. 47, no. 3, pp. 987–995, 1998.

[89] V.W. Wong and V.C.M. Leung, "Location Management for Next-Generation Personal Communications Networks", *IEEE Network*, vol. 14, no. 5, pp. 18–24, 2000.

[90] Z. Xu and P.B. Mirchandani, "Virtually Fixed Channel Assignment for Cellular Radio-Telephone Systems: A Model and Evaluation", *in Proceedings of the IEEE International Conference on Communications, ICC'92, Chicago*, vol. 2, pp. 1037–1041, 1982.

[91] A. Yener and A. Rose, "Genetic Algorithms Applied to Cellular Call Admission: Local Policies", *IEEE Transactions on Vehicular Technology*, vol. 46, no. 1, pp. 72–79, 1995.

[92] M. Zhang, and T.S. Yum, "The Nonuniform Compact Pattern Allocation Algorithm for Cellular Mobile Systems", *IEEE Transactions on Vehicular Technology*, vol. 40, no. 2, pp. 387–391, 1991.

[93] M. Zhang, and T.S. Yum, "Comparisons of Channel Assignment Strategies in Cellular Mobile Telephone Systems", *IEEE Transactions on Vehicular Technology*, vol. 38, no. 4, pp. 211–215, 1989.

Chapter 3

Optimal Server Allocation in Wireless Networks: The Use of Index Policies

Navid Ehsan
Electrical Engineering and Computer Science Department
University of Michigan, Ann Arbor, MI 48109
E-mail: nehsan@eecs.umich.edu

Mingyan Liu
Electrical Engineering and Computer Science Department
University of Michigan, Ann Arbor, MI 48109
E-mail: mingyan@eecs.umich.edu

1 Introduction

In wireless communication systems resources are costly and scarce. Therefore the efficient use and sharing of these resources is an essential issue in designing such systems and it directly affects the resulting performance. The time-varying nature of wireless channels as well as diverse quality of service requirements presented by a variety of applications are some of the challenges in obtaining efficient resource allocation strategies.

In this chapter we focus on the problem of sharing a single server or channel among multiple users/queues, and survey a sequence of related results. Time is slotted and in each slot the server is allocated to at most one of the users. The allocation decision is made in a centralized way at the beginning of each slot based on the state information (including backlog and channel condition) of each queue. This model can be the result of a system that is discrete in nature; i.e., decisions are made at predetermined instants of time, or can be the result of applying uniformization to continuous time models (for equivalence between discrete and continuous time models see [1, 2, 3]).

There are many scenarios of wireless systems where this model applies. Examples include a cellular system where all users transmit and receive via a base station, in which users may advertise their backlog and fading state to the base station via a control channel. Based on such information the base station decides on how to allocate the channel. A satellite communication system has a similar abstraction in that all ground terminals transmit and receive via the satellite. Ground terminals transmit the queue backlog information to the satellite, which in turn decides and broadcasts the allocation decision. In this case the backlog information could be delayed due to excessive propagation delay.

Our primary interest in this chapter is in deriving allocation strategies that allow the system to perform in the most efficient way. Specifically, we assume that backlogged packets incur a cost, and consider the aforementioned optimal server allocation problem with the objective of minimizing the expected average packet holding cost or expected total discounted packet holding cost over a finite (or infinite) time horizon. Although in general reducing holding cost has the effect of reducing packet delay, different forms of the cost function lead to different performance criteria. For example, under a linear cost function identical to all queues (i.e., each packet incurs a constant unit cost), minimizing holding cost is equivalent to maximizing system throughput. Similarly, a linear cost function with different unit costs for different queues leads to a weighted throughput criterion. Different cost functions also lead to different optimal strategies, which is further discussed in this chapter.

This problem has been extensively studied in the literature. In [4, 5] the problem of parallel queues with different holding costs and a single server was considered, and the simple $c\mu$ rule was shown to be optimal. The studies in [6, 7, 8] considered the server allocation problem to multiple queues with varying connectivity but of the same service class. Each of these studies determined policies that maximize throughput over an infinite horizon. In particular, [6] derived the sufficient condition for stability and showed that the Longest Connected Queue (LCQ) policy stabilizes the system if it is stabilizable. The same policy minimizes the delay in the special case of symmetric queues. A similar problem was also considered in [9] but with differentiated service classes where different queues have different holding costs. In most of the above work the state of the system, i.e., connectivity and the number of packets in each queue, is precisely known before server allocation is made.

In [10, 11] the server allocation problem with delayed state observation was considered. The problem of optimally routing to two queues with imperfect and noisy information was studied in [12].

In some other work the focus has been on maximizing the throughput of the system given infinite sources. A user is considered an infinite source if it always has packets to send. Under such a scenario when there is only a single user, fading naturally degrades the performance. However, in a multiuser system with independent fading for different users, as shown by [13] via an information theory argument, the performance (total throughput) of the system increases as the number of users increases and the optimal policy in this case is to transmit to/from the user in the best state (see also [14, 15]).

The single-server allocation problem described above can also be viewed as a special case of the *restless bandit problem* studied in [16, 17, 18, 19]. A general optimal solution is not known for this class of problems. The asymptotic behavior of this class of problems was studied in [16] and [17] when the numbers of queues and servers go to infinity with a fixed ratio. An index policy can be defined based on Whittle's heuristic, which is suboptimal in the finite (number of servers and queues) case and asymptotically optimal in the infinite case.

In this chapter we present some of the above cited results. In particular, we largely focus on a class of policies known as *index policies*. Under an index policy, each user is assigned an *index*, which is a function of its own backlog and/or channel state, and does not depend on other users' states. The index is also updated as the user's state changes. An index policy is then defined as the policy that serves the user with the highest index at each instance of time. If the indices are chosen properly, the index policy can be shown to be optimal in certain scenarios or under certain conditions.

Our intention in this chapter is primarily to provide a relatively in-depth survey on results relevant to the problem outlined above. It is therefore inevitably nonexhaustive. Rather we focus on a few representative problem instances and give a more extensive discussion on those. Most of the results presented here are from the existing literature where references are provided. The amount of original contribution in this chapter is quite modest, although in some parts we provide extra examples or alternative proofs to those available in the references, as indicated. We omit most of the lengthy technical proofs that can be found in the references and instead focus more on the conceptual understanding and intuitive interpretation of the results. Throughout the chapter the form of address *we* is meant conversationally to suggest us and the reader.

The rest of the chapter is organized as follows. In the next section we describe the model we use in this chapter and formulate the problem. In Section 3 we examine a heuristic policy based on Whittle's indices for this problem. In Section 4 we show the optimality of an index policy for static channels. In Section 5 we consider dynamic channels and show that the index policy is optimal under sufficient separation. Section 6 presents a server allocation problem where the backlog information of the users is delayed. In Section 7 we show some properties of the value function for a two-user system for a class of more general cost functions. Section 8 concludes the chapter.

2 Problem Formulation

In this section we describe the network model we adopt as an abstraction of the bandwidth allocation problem described in the previous section, and formally present the optimization problem along with a summary of notations.

Consider a set of queues $\mathcal{N} = \{1, 2, \ldots, N\}$ that transmit or receive packets to/from a single server and in doing so compete for shares of a common channel that consists of time slots. Packets arrive at each queue according to arbitrary random processes. The allocation of the channel is done for each time slot.

We consider time evolution in discrete time steps indexed by $t = 0, 1, \ldots, T-1$, with each increment representing a slot length. Slot t refers to the slot defined by the interval $[t, t+1)$. One slot time equals one packet transmission time if the transmission is successful. In subsequent discussions we use the terms *steps*, *stages*, *bandwidth*, and *slots* interchangeably.

At each time slot a user can be in any channel state from the set $\mathcal{S} \subseteq \mathbb{Z}_+$. Let $S = |\mathcal{S}|$. We assume that the channel remains in the same state during a time slot. We assume that a user's channel state is independent of all the other users'. If a user is in state $s \in \mathcal{S}$ at time t then at time $t+1$ it moves to state $r \in \mathcal{S}$ with probability $p_i(s, r)$. Whenever user i is in state $s \in \mathcal{S}$ and is nonempty, the transmission from that user will be successful with probability μ_s if the channel is allocated to it.

The allocation decision is made based on the backlog information of each queue (number of packets waiting/existing in the queue). This information is provided by the queues either at the beginning of a slot or via a separate control channel. We ignore the transmission time of such information, which does not affect our analysis. Based on this information an allocation decision is made by the server and broadcast to all queues over a noninterfering channel.

Below we summarize key notations used in subsequent sections. In general boldface letters are vectors and normal letters are scalars/random variables.

Let $b_{i,t}$ be the backlog of queue/user i at the beginning of frame t (more precisely this is the backlog of queue i at time instant t_-). Denote by $\mathbf{b}_t$ the vector $(b_{1,t}, b_{2,t}, \dots, b_{N,t})$. We use the same convention for other quantities as defined below.

Denote by $\mathbf{b}_t^{i+} = \mathbf{b}_t + \mathbf{e}_i$, where $\mathbf{e}_i$ is the ith N-dimensional unit vector, i.e., a vector with all elements being zero except a one in the ith position.

$\mathbf{a}_t = (a_{1,t}, \dots, a_{N,t})$: Random arrivals during $[t, t+1)$ to each queue.

$p(\mathbf{a})$: The joint probability mass function for having $\mathbf{a}$ arrivals in a time slot.

$\mathbf{s}_t$: The set of all channel states of the users at time t.

$\mathcal{S}$: The set of all possible channel states of the users.

μ_s: The probability of success for transmitting/receiving the packet when the user is in state $s \in \mathcal{S}$.

$p_i(s,r) = \mathbb{P}[s_{i,t+1} = r | s_{i,t} = s]$, $\forall s, r \in \mathcal{S}$.

$C_u = \sum_{t=u}^{T-1} \beta^t \sum_{i=1}^{N} c_i(b_{i,t})$: The cost-to-go, from time u on (note that $C_0 = C$).

$\mathcal{F}_t$: The σ-field of the information available up to time t.

For any scalar x define $x^+ = [x]^+ = x$ if $x \geq 0$ and is equal to zero otherwise. For a vector $\mathbf{x}$, we define $\mathbf{x}^+ = [\mathbf{x}]^+$ the same way component-wise. Also for two vectors $\mathbf{x}$ and $\mathbf{y}$ by $\mathbf{x} \leq \mathbf{y}$ we mean that inequality holds component by component.

The objective is to find an allocation policy π that minimizes one of the following two cost measures.

$$J_T^\pi = E^\pi[C|\mathcal{F}_0] \ , \quad \text{where} \quad C = \sum_{t=0}^{T-1} \beta^t c(\mathbf{b}_t) \ , \tag{1}$$

$$\bar{J}_T^\pi = E^\pi[\bar{C}|\mathcal{F}_0] \ , \quad \text{where} \quad \bar{C} = \frac{1}{T} \sum_{t=0}^{T-1} c(\mathbf{b}_t) \ , \tag{2}$$

where the former is a total discounted cost measure and the latter an average cost measure. For finite T we let $\beta \in (0,1]$; in this case for $\beta = 1$ the two performance measures are essentially the same. As $T \to \infty$ (infinite horizon) we require β to be strictly less than one, i.e., $\beta \in (0,1)$. In this case the two performance measures may lead to different optimal policies. For now the packet holding cost $c(\mathbf{b}) : \mathbb{Z}_+^N \to \mathbb{Z}_+$ is an arbitrary nonnegative function. Later we restrict c to certain classes of functions. We only consider synchronized transmissions, i.e., users can only transmit at the beginning of time slots that are allocated/specified by the server. For server allocation problems under asynchronous transmission times see [20, 21] and references therein.

3 Formulation as a Restless Bandit Problem

In this and the next three sections we consider the special case of linear cost functions; i.e.,

$$c(\mathbf{b}_t) = \sum_{i=1}^{N} c_i b_{i,t} , \tag{3}$$

where $c_i > 0$, $\forall i$. Consider a system consisting of N controlled Markov processes. At the beginning of each time slot one can activate one of the processes. A reward is given as a function of the current state of the activated process and the activated process changes its state according to a Markov chain. The goal is to maximize the total discounted reward over time. This problem was solved by Gittins for the case when a process does not undergo state transition unless it is activated, also known as the *multiarmed bandit problem*, and an index policy was shown to be optimal [22, 23]. For many other problems the state of a process may change regardless of whether it is activated. This class of problems, also referred to as the *restless bandit* problem, was studied in numerous studies; see for example [16, 17, 18, 19]. The problem outlined in the previous section can be viewed as a special case of the restless bandit problem. This is because due to new arrivals and channel state changes, the state of a user $(b_{i,t}, s_{i,t})$ changes even if it is not served. An optimal solution for the general restless bandit problem is not known.

Whittle introduced an index in [16] for the infinite horizon case. The corresponding index policy is in general not optimal, but it performs well in many cases (see, for example, [24]). The index introduced in [16] is for the general restless bandit problem. In what follows we derive this index for the special case of the problem outlined in Section 2.

Consider a single queue with cost c_i that can be in any state in the set $\mathcal{S}$. When in state $s \in \mathcal{S}$ it can transmit with a probability of success μ_s if the channel is allocated to it. On the other hand, the passive action is subsidized; i.e., if the channel is not allocated to the user, the user gets γ units of credit. Note that the cost measure to be minimized in this case is defined as

$$C' = \sum_{t=0}^{\infty} \beta^t (c_i b_{i,t} - \gamma \cdot 1[\text{channel is not allocated at time } t]) ,$$

where $1[\cdot]$ is the indicator function. The cost-to-go C'_t is also defined accordingly. The Whittle's index of a user is defined as follows.

Definition 3.1 *The index of a user* γ^* *is*

$$\gamma^* = \sup\{\gamma | \textit{It is optimal to allocate the server when the subsidy is } \gamma\} .$$

If the index is γ^* *then for* $\gamma = \gamma^*$ *it is optimal either to allocate the slot or to let the slot remain idle.*

The following lemma shows the properties of the index defined above.

Lemma 3.2 *Suppose the user is in state* (b, s) *at time* t.

(a) Suppose that it is optimal to allocate the slot for some value γ. *Then it is optimal to allocate the slot for all* $\hat{\gamma} \leq \gamma$.

(b) Suppose that it is optimal not to allocate the slot for some value γ. *Then it is optimal not to allocate the slot for all* $\hat{\gamma} \geq \gamma$.

This lemma implies the existence of a threshold γ^* such that for all $\gamma \leq \gamma^*$ it is optimal to allocate the slot and for all $\gamma \geq \gamma^*$ it is optimal not to allocate the slot (note that γ^* can be infinity). Instead of proving Lemma 3.2 directly, we calculate the value γ^* from which the above lemma directly follows.

It is clear that when the queue is empty then $\gamma^* = 0$. Using the following lemma we can calculate the index of the user when the queue is not empty.

Lemma 3.3 *Suppose the queue is in state* (b_t, s) *at the beginning of slot* t *where* $b_t > 0$.

(a) If $\gamma \leq \frac{\mu_s c_i \beta}{1-\beta}$ *then it is optimal to allocate the slot to the queue.*

(b) If $\gamma \geq \frac{\mu_s c_i \beta}{1-\beta}$ *then it is optimal not to allocate the slot to the queue.*

(c) For $\gamma = \frac{\mu_s c_i \beta}{1-\beta}$ *the user is indifferent between the two actions.*

Proof. Let π be the policy that allocates the slot at time t to the user and allocates optimally thereafter and let π' be the policy that keeps the slot idle and allocates optimally thereafter.

(a) Let $t' > t$ be the first (random) time that policy π' allocates a slot to the user after time t when the user is in state s. Define policy $\hat{\pi}$ as follows. Allocate the slot to the user, follow the same policy as π' up to time t', but do not allocate the slot at time t'. From time $t'+1$ on the expected number of packets is the same under both policies π' and $\hat{\pi}$. Therefore we have

$$\begin{aligned} & E^{\pi}[C_t'|(b_t,s),\mathcal{F}_t] - E^{\pi'}[C_t'|(b_t,s),\mathcal{F}_t] \\ \leq\ & E^{\hat{\pi}}[C_t'|(b_t,s),\mathcal{F}_t] - E^{\pi'}[C_t'|(b_t,s),\mathcal{F}_t] \quad \text{a.s.} \\ \leq\ & \gamma(\beta^t - \beta^{t'}) - \mu_s c_i \sum_{u=t+1}^{t'} \beta^u \quad \text{a.s.} \\ =\ & \gamma(\beta^t - \beta^{t'}) - \frac{\mu_s c_i \beta(\beta^t - \beta^{t'})}{1-\beta} \leq 0\,. \end{aligned} \tag{4}$$

Therefore it is optimal to allocate the slot.

(b) Redefine policy $\hat{\pi}$ as follows. Policy $\hat{\pi}$ does not allocate the slot. After that it allocates the same as policy π. Therefore if the transmission at time t is successful under policy π, in the *worst case* it will always have one more packet in the queue compared to policy π. Therefore we have

$$\begin{aligned} & E^{\pi}[C_t'|(b_t,s),\mathcal{F}_t] - E^{\pi'}[C_t'|(b_t,s),\mathcal{F}_t] \\ \geq\ & E^{\pi}[C_t'|(b_t,s),\mathcal{F}_t] - E^{\hat{\pi}}[C_t'|(b_t,s),\mathcal{F}_t] \quad \text{a.s.} \\ \geq\ & \gamma\beta^t - \mu_s \sum_{u=t+1}^{\infty} \beta^u c_i \quad \text{a.s.} \\ =\ & \beta^t(\gamma - \frac{\mu_s c_i \beta}{1-\beta}) \geq 0\,. \end{aligned} \tag{5}$$

Therefore it is optimal not to allocate the slot at time t.

(c) This is a direct result of parts (a) and (b). □

Theorem 3.4 *The Whittle's index of user i in state $(b_{i,t}, s_{i,t})$ is given by*

$$I_{i,t} = \begin{cases} \frac{\beta c_i \mu_{s_{i,t}}}{1-\beta}, & \text{if } b_{i,t} > 0\,; \\ 0, & \text{if } b_{i,t} = 0\,. \end{cases} \tag{6}$$

Proof. This follows directly from Lemma 3.3 and Definition 3.1. □

The *Whittle's heuristic index policy* based on the above index is defined as one that allocates the slot to the user with the highest index. We see that the index policy defined above is essentially the same as the following policy. At each time t when the state of the system is $(\mathbf{b}_t, \mathbf{s}_t)$ allocate the slot to the user i such that

$$i = \arg \max_{j: b_{j,t} \neq 0} c_j \mu_{s_{j,t}} \ . \tag{7}$$

This policy (which we call the $c\mu$ rule for the remainder of this chapter) is not always optimal (see [16, 17, 18, 25]). In the next two sections we present results that show in certain specific problem scenarios this policy is optimal; and in some other instances it can be optimal under certain sufficient conditions.

4 Optimality of the $c\mu$ Rule in Static Channels

In this section we examine the optimality of the $c\mu$ rule in a static channel.

Definition 4.1 Static vs. Dynamic Channel. *We say that the channel is static if the quality of the channel (i.e., transmission rate of a user when it is allocated the channel) does not vary over time. In contrast, in a dynamic channel the channel quality of a user can change over time.*

It was shown in a number of studies that the $c\mu$ rule is optimal in static channels ([4, 5, 26]). More precisely, the channel may be viewed as having N states and each user can only be in one state. Thus we have $\mathcal{S} = \mathcal{N}$. User i (if it is nonempty) can transmit with a fixed success probability μ_i if it is allocated the channel. Packets arrive according to an arbitrary random process.

With these assumptions, the index for each queue can be defined as follows.

$$I_{i,t} = \begin{cases} c_i \mu_i, & \text{if } b_{i,t} > 0 \ ; \\ 0, & \text{if } b_{i,t} = 0 \ . \end{cases} \tag{8}$$

A dynamic programming argument is used in [5] to show that the above index policy is optimal when $N = 2$ for infinite horizon. In [4] an interchange argument is used to further show its optimality for $N \geq 2$ in both finite and infinite horizons and for arbitrary arrival processes. Later [26] showed that the region corresponding to the admissible policies is a polymatroid. Using this argument and the results from [27] they proved the optimality of the $c\mu$ rule. These are quite different techniques. Below we reproduce the same argument as in [25, 26, 28] to show the optimality of the $c\mu$ rule using the strong conservation laws.

Let $\psi = (\psi_1, \ldots, \psi_N)$ be a permutation of the set $\mathcal{N}$ and let $\pi(\psi)$ be the following policy. Strict priority is given to queues according to the permutation ψ; i.e., ψ_1 has the highest priority and ψ_N has the lowest priority.

Definition 4.2 Strong Conservation Laws. *Let Π be a set of policies and let $\mathbf{x}$ be the vector of some performance measure. The set of performance vectors $\mathbf{x}^\pi, \pi \in \Pi$ is said to satisfy strong conservation laws if there exists a set function $f : 2^{\mathcal{N}} \rightarrow \mathcal{R}_+$ such that,*

$$\sum_{i \in \mathcal{N}} x_i^\pi = f(\mathcal{N}), \quad \forall \pi \in \Pi, \tag{9}$$

and for all $A \subset \mathcal{N}$, we have

$$\sum_{i \in A} x_i^{\pi(\psi)} = f(A), \quad \forall \psi \text{ s.t. } \{\psi_1, \ldots, \psi_{|A|}\} = A \ ,$$
$$\sum_{i \in A} x_i^\pi \geq f(A), \quad \forall \pi \in \Pi \ . \tag{10}$$

The following theorem was proven in [28, 29].

Theorem 4.3 *Given a set of performance vectors $\{\mathbf{x}^\pi | \pi \in \Pi\}$ that satisfy strong conservation laws, the objective*

$$\min_{\pi \in \Pi} \sum_{i \in \mathcal{N}} \rho_i x_i^\pi$$

is optimized by an index policy where the index of queue i is ρ_i.

A policy π is called *work conserving* if it does not idle the server when there are nonempty queues in the system. Let Π be the set of all work conserving policies. It can be shown that the following two measures satisfy the strong conservation laws outlined above for both finite and infinite T (details can be found in [26]).

$$x_i^\pi = \sum_{t=0}^{T-1} \beta^t \frac{b_{i,t}^\pi}{\mu_i}, \quad i \in \mathcal{N} \ ,$$
$$\bar{x}_i^\pi = \frac{1}{T} \sum_{t=0}^{T-1} \frac{b_{i,t}^\pi}{\mu_i}, \quad i \in \mathcal{N} \ .$$

On the other hand, the cost measures given in (1) and (2) can be expressed as:

$$J_T^\pi = \sum_{i=1}^{N} c_i \mu_i x_i^\pi ,$$
$$\bar{J}_T^\pi = \sum_{i=1}^{N} c_i \mu_i \bar{x}_i^\pi .$$

Therefore the optimal policy is to give priority to the queue with the largest $c\mu$, which is to always serve the highest indexed nonempty queue. This is essentially the index policy where the indices are defined by (8).

5 Optimality of the $c\mu$ Rule in Dynamic Channels

The same index policy (i.e., the $c\mu$ rule that was shown to be optimal in a static channel) is in general not optimal given a dynamic channel. In this section we first illustrate this via an example. We then consider two specific cases. The first involves symmetric queues and an index policy (a subclass of the $c\mu$ rule) was shown to be optimal. In the second case, which involves differentiated service costs, the $c\mu$ rule was shown to be optimal under a certain sufficient condition.

5.1 An Example

We show by the following example that serving the queue with the highest index as defined in (6) is not necessarily optimal when the channel quality varies with time. We consider the average cost criterion for a two-state channel where a queue is either connected or disconnected.

Consider two queues with Bernoulli arrivals. The probability of having an arrival in queue i during any time slot is $p_i = 0.4, i \in \{1, 2\}$. Let $\mathcal{S} = \{0, 1\}$ and suppose $\mu_0 = 0$ and $\mu_1 = 1$. That is, a transmission is successful with probability 1 when the queue is connected and 0 otherwise. Let $q_i = \mathbb{P}[s_{i,t+1} = 1 | s_{i,t} = s, s \in \mathcal{S}]$, and we have $q_1 = 1$, $q_2 = 0.5$; i.e., queue 1 is always connected whereas queue 2 has an independent equal probability of being either connected or disconnected in each slot. Also let $c_1 = 20$ and $c_2 = 10$. Now consider the problem of minimizing the average cost over an infinite horizon.

Let π be the policy that uses the $c\mu$ rule. Therefore it always serves queue 1 if it is nonempty (note that this queue is always connected). The probability that queue 2 is served if it is nonempty is equal to the probability that it is connected multiplied by the probability that queue 1 is empty, implying at least there was no arrival to queue 1 in the previous slot. This results in the following.

$\mathbb{P}^{\pi}$[queue 2 is served if it is nonempty] $\leq q_2(1-p_1) = 0.3$.

However, the probability of an arrival in queue 2 is 0.4 which is greater than its service rate. This results in a transient Markov chain with infinite mean. Therefore the average cost for policy π is infinite (for details on the analysis of Markov chains see [30]).

Next we consider policy π' defined as follows. π' serves queue 2 whenever this queue is connected and nonempty. We show that under this policy the backlog in both queues forms a recurrent Markov chain with finite mean resulting in a finite average cost. In particular, for $l \geq 1$ we have:

$$\mathbb{P}^{\pi'}(b_{2,t+1} = k | b_{2,t} = l) = \begin{cases} q_2(1-p_2), & \text{if } k = l-1; \\ p_2(1-q_2), & \text{if } k = l+1; \\ 1 - p_2 - q_2 + 2p_2 q_2, & \text{if } k = l. \end{cases} \quad (11)$$

For $l = 0$ we have:

$$\mathbb{P}^{\pi'}(b_{2,t+1} = k | b_{2,t} = l) = \begin{cases} 1 - p_2, & \text{if } k = l; \\ p_2, & \text{if } k = l+1. \end{cases} \quad (12)$$

We can calculate the steady-state probabilities for the chain defined above. Let $\gamma_{2,k}$ be the steady-state probability of having k packets in the second queue. Then we have:

$$\begin{aligned} \gamma_{2,0} &= \frac{(q_2 - p_2)}{q_2}; \\ \gamma_{2,1} &= \frac{p_2(q_2 - p_2)}{q_2^2(1-p_2)}; \\ \gamma_{2,k} &= \gamma_{2,1}\left(\frac{p_2(1-q_2)}{q_2(1-p_2)}\right)^{k-1}, \quad k \geq 2. \end{aligned}$$

The expected value of the queue size of queue 2 in steady-state is therefore

$$\lim_{t\to\infty} E[b_{2,t}] = \sum_{k=1}^{\infty} k\gamma_{2,k} = \frac{p_2(1-p_2)}{q_2 - p_2} = 2.4.$$

The probability that queue 2 is empty is $\gamma_{2,0} = 0.2$. Suppose the first queue is nonempty. It will not be served if queue 2 is connected and nonempty. Denote this probability by κ; then we have $\kappa = q_2(1 - \gamma_{2,0})$. In steady-state for $l \geq 1$ we have:

$$\mathbb{P}^{\pi'}(b_{1,t+1} = k | b_{1,t} = l) = \begin{cases} (1-\kappa)(1-p_1), & \text{if } k = l-1; \\ \kappa p_1, & \text{if } k = l+1; \\ \kappa + p_1 - 2\kappa p_1, & \text{if } k = l. \end{cases} \quad (13)$$

For $l = 0$ we have:

$$\mathbb{P}^{\pi'}(b_{1,t+1} = k | b_{1,t} = l) = \begin{cases} 1 - p_1, & \text{if } k = l; \\ p_1, & \text{if } k = l+1. \end{cases} \quad (14)$$

One can again calculate the steady-state probabilities and find the expected value of the queue size. The expected queue size in steady-state turns out to be $\lim_{t\to\infty} E^{\pi'}[b_{1,t}] = 1.2$. Therefore we have:

$$\begin{aligned} \lim_{T\to\infty} \bar{J}_T^{\pi'} &= \lim_{t\to\infty} \sum_{i=1}^{2} c_i \sum_{k=0}^{\infty} k \mathbb{P}^{\pi'}[b_{i,t} = k] \\ &= \lim_{t\to\infty} \sum_{i=1}^{2} c_i E^{\pi'}[b_{i,t}] = 48. \end{aligned} \quad (15)$$

It can be seen that policy π' leads to a finite cost and therefore it performs better than policy π which is the $c\mu$ rule.

5.2 Optimality of the Longest Connected Queue Policy – Symmetric Queues

The authors of [6, 31] considered a class of server allocation problems with dynamic channels. In particular, in [6] the problem of optimal server allocation to N users was studied where each user can be in one of only two states, connected (in which case user i transmits with success probability μ_i) or disconnected (in which case the user cannot transmit). User queues are considered *symmetric* in that they all have the same linear holding cost with unit cost $c_i = c$ for some constant c and that all queues transmit with the same success probability $\mu_i = \mu$ for some constant μ. It was proved that when the queues are symmetric the optimal policy (with respect to the cost objective given earlier) is to serve the longest (connected) queue, which is an index policy with the index defined as follows.

$$I_{i,t} = \begin{cases} b_{i,t}, & \text{if } i \text{ is connected;} \\ 0, & \text{if } i \text{ is not connected.} \end{cases} \quad (16)$$

This index policy is often referred to as the *Longest Connected Queue* (LCQ) policy. Note that by the symmetry assumption all nonempty queues have the same index under the $c\mu$ rule. Therefore the LCQ policy is an element of the class of policies defined by the $c\mu$ rule. This policy gives priority to the users based on their backlog.

The problem of jointly routing and serving N queues where the arrival and service rates may depend on the queue size was studied in [31]. It was shown that if the arrival and service rates at the stations satisfy certain symmetry and monotonicity conditions, then the policy that serves the longest queue is optimal. In both cases the optimality of LCQ policy was shown in a much stronger sense than the expectation measure we are considering.

5.3 Two State Channel – Differentiated Services

The reason behind the optimality of the LCQ policy is that it minimizes the probability that all nonempty queues are disconnected, in which case the server is forced to be idle even though there are packets to be transmitted. When all queues are equally costly and have the same connectivity probability, to minimize the above probability it is intuitively clear that we should serve the longest connected queue first.

When queues are asymmetric both in terms of the holding cost and channel state variation then there exists a tradeoff between serving queues with higher cost and/or better channel conditions, and serving queues with larger backlog or queues that are less likely to be connected in the future (see the example in Section 5.1). Due to this tradeoff, it is conceivable that an index policy defined solely based on a queue's backlog is not in general optimal. For the same reason the $c\mu$ rule is also in general not optimal. In [9] such a problem was studied with different holding costs and transmission success probabilities. It was shown that the $c\mu$ rule is not optimal as expected, but would be optimal if the indices were sufficiently separated. We describe this result in more detail below.

The channel is modeled as follows. Each queue is either connected or disconnected. Different queues may have different success probabilities when they are connected. Therefore the state space of the channel may be viewed as such that $\mathcal{S} = \mathcal{N} \cup \{0\}$. Queue i can be either in state $s = i$ (connected) with transmission success probability μ_i or it can be in state $s = 0$ (disconnected) in which case it cannot transmit, i.e., $\mu_0 = 0$. For now, we make the following assumption on the channel (this assumption is relaxed later in this section). The channel state of a queue within each time slot is independent of its previous state and the state of all the other queues. Queue i is connected to the server with probability q_i and disconnected with probability $1 - q_i$.

We then have the following sufficient condition for the optimality of the $c\mu$ rule.[1]

Theorem 5.1 *Suppose the time horizon is $T > 1$ and that at $t = 0$ queue i is connected and nonempty and we have*

$$\begin{aligned} &\frac{c_i\mu_i(1-((1-q_i\mu_i)\beta)^{T-1})}{1-(1-q_i\mu_i)\beta} \geq \frac{\mu_j c_j(1-\beta^{T-1})}{1-\beta}, \\ &\forall j \neq i \ \text{ s.t. } \ s_{j,t} \neq 0, \ b_{j,t} > 0 \ . \end{aligned} \tag{17}$$

Then it is optimal to allocate the next slot to queue i.

This is essentially the same result as in [9] when $T \to \infty$. We use a different method here to prove the theorem, starting with a few lemmas.

Lemma 5.2 *Let T be the time horizon. Let π be the optimal policy when the state at time t is $\mathbf{b}_t$ and let π' be the optimal policy when the state at time t is $\mathbf{b}_t^{i+}$. Then we have*

$$E^{\pi'}[C_t|\mathbf{b}_t^{i+}, \mathcal{F}_t] - E^{\pi}[C_t|\mathbf{b}_t, \mathcal{F}_t] \leq \frac{\beta^t c_i(1-\beta^{T-t})}{1-\beta} \quad \text{a.s.} \tag{18}$$

Proof. Suppose from time t on when the state is $\mathbf{b}_t^{i+}$ we use policy $\hat{\pi}$ instead of π', and define $\hat{\pi}$ as follows. $\hat{\pi}$ allocates to the same queue as policy π does starting from $\mathbf{b}_t$. Note that the difference on the left-hand side of (18) is maximized when queue i is never served under policy π (and thus $\hat{\pi}$). Therefore we have:

$$\begin{aligned} & E^{\pi'}[C_t|\mathbf{b}_t^{i+}, \mathcal{F}_t] - E^{\pi}[C|\mathbf{b}_t, \mathcal{F}_t] \\ \leq\ & E^{\hat{\pi}}[C_t|\mathbf{b}_t^{i+}, \mathcal{F}_t] - E^{\pi}[C|\mathbf{b}_t, \mathcal{F}_t] \\ \leq\ & \sum_{u=t}^{T-1} \beta^u c_i \leq \frac{\beta^t c_i(1-\beta^{T-t})}{1-\beta} \quad \text{a.s.,} \end{aligned} \tag{19}$$

[1] This sufficient condition is slightly tighter than the one given in [9] for the finite horizon case, and becomes identical under an infinite horizon. For this reason the proof of the result is also different, which we present here.

where the first inequality is due to the fact that $\hat{\pi}$ is not necessarily optimal for the initial state $\mathbf{b}_t^{i+}$. This completes the proof. □

Lemma 5.3 *Let T be the time horizon. Let π be the optimal policy when the state at time t is $\mathbf{b}_t$ and let π' be the optimal policy when the state at time t is $\mathbf{b}_t^{i+}$. Then we have*

$$E^{\pi'}[C_t|\mathbf{b}_t^{i+}, \mathcal{F}_t] - E^{\pi}[C_t|\mathbf{b}_t, \mathcal{F}_t] \geq \frac{\beta^t c_i(1-((1-q_i\mu_i)\beta)^{T-t})}{1-(1-q_i\mu_i)\beta} \quad \text{a.s.} \tag{20}$$

Proof. Suppose for $\mathbf{b}_t$ we use policy $\hat{\pi}$ instead of π', and define $\hat{\pi}$ as follows. $\hat{\pi}$ allocates to the same queue as policy π does starting from $\mathbf{b}_t^{i+}$. Note that the left-hand side of (20) is minimized if the slot is always allocated to queue i under policy π' (and thus $\hat{\pi}$). Therefore we have:

$$\begin{aligned} & E^{\pi'}[C_t|\mathbf{b}_t^{i+}, \mathcal{F}_t] - E^{\pi}[C_t|\mathbf{b}_t, \mathcal{F}_t] \\ \geq\ & E^{\pi'}[C_t|\mathbf{b}_t^{i+}, \mathcal{F}_t] - E^{\hat{\pi}}[C_t|\mathbf{b}_t, \mathcal{F}_t] \\ \geq\ & \sum_{u=t}^{T-1} \beta^t (1-q_i\mu_i)^{t-u} c_i \geq \frac{\beta^t c_i(1-((1-q_i\mu_i)\beta)^{T-t})}{1-(1-q_i\mu_i)\beta} \quad \text{a.s.}, \end{aligned} \tag{21}$$

where again the first inequality is due to the fact that $\hat{\pi}$ is not necessarily optimal for the initial state $\mathbf{b}_t$. This completes the proof. □

Proof of Theorem 5.1. Suppose at time $t = 0$ queues i and j are both connected and nonempty. Let π be the policy that assigns the slot at $t = 0$ to queue i and allocates optimally thereafter. Let π' be the policy that assigns the slot at time $t = 0$ to queue j and then follows the optimal policy. Let $\mathbf{a}_0$ be the number of arrivals during the interval $[0, 1)$, thus we have:

$$\begin{aligned} & E^{\pi}[C|\mathbf{b}_0, \mathcal{F}_0] - E^{\pi'}[C|\mathbf{b}_0, \mathcal{F}_0] \\ =\ & (1-\mu_i)(1-\mu_j)(E^{\pi}[C_1|\mathbf{b}_0 + \mathbf{a}_0, \mathcal{F}_1] - E^{\pi'}[C_1|\mathbf{b}_0 + \mathbf{a}_0, \mathcal{F}_1]) \\ & +\mu_i(1-\mu_j)(E^{\pi}[C_1|\mathbf{b}_0 + \mathbf{a}_0 - \mathbf{e}_i, \mathcal{F}_1] - E^{\pi'}[C_1|\mathbf{b}_0 + \mathbf{a}_0, \mathcal{F}_1]) \\ & +(1-\mu_i)\mu_j(E^{\pi}[C_1|\mathbf{b}_0 + \mathbf{a}_0, \mathcal{F}_1] - E^{\pi'}[C_1|\mathbf{b}_0 + \mathbf{a}_0 - \mathbf{e}_j, \mathcal{F}_1]) \\ & +\mu_i\mu_j(E^{\pi}[C_1|\mathbf{b}_0 + \mathbf{a}_0 - \mathbf{e}_i, \mathcal{F}_1] - E^{\pi'}[C_1|\mathbf{b}_0 + \mathbf{a}_0 - \mathbf{e}_j, \mathcal{F}_1]) \\ =\ & \mu_i(E^{\pi}[C_1|\mathbf{b}_0 + \mathbf{a}_0 - \mathbf{e}_i, \mathcal{F}_1] - E^{\pi'}[C_1|\mathbf{b}_0 + \mathbf{a}_0, \mathcal{F}_1]) \\ & +\mu_j(E^{\pi}[C_1|\mathbf{b}_0 + \mathbf{a}_0, \mathcal{F}_1] - E^{\pi'}[C_1|\mathbf{b}_0 + \mathbf{a}_0 - \mathbf{e}_j, \mathcal{F}_1])\ . \end{aligned} \tag{22}$$

Therefore by Lemmas 5.2 and 5.3 we can see that

$$\begin{aligned} & E^{\pi}[C|\mathbf{b}_0, \mathcal{F}_0] - E^{\pi'}[C|\mathbf{b}_0, \mathcal{F}_0] \\ \leq\ & \beta^t\left(\frac{\mu_j c_j(1-\beta^{T-1})}{1-\beta} - \frac{c_i\mu_i(1-((1-q_i\mu_i)\beta)^{T-1})}{1-(1-q_i\mu_i)\beta}\right)\ . \end{aligned} \tag{23}$$

Thus if (17) is satisfied, then $E^{\pi}[C|\mathbf{b}_0, \mathcal{F}_0] - E^{\pi'}[C|\mathbf{b}_0, \mathcal{F}_0] \leq 0$ and therefore it is optimal to allocate to queue i. □

Remark 5.4 *Note that in Theorem 5.1, the time horizon T is essentially the "time to go" from when the decision is made. If queues i and j are connected at time t, then the sufficient condition for allocating to i is the same as (17) with T replaced by $T - t$. Also note that if the sufficient condition is satisfied for horizon T, then it is satisfied for all $T' < T$.*

The above theorem holds for all $T \geq 0$, therefore we can let $T \to \infty$ to get the following sufficient condition for the infinite horizon.

Theorem 5.5 *Consider an infinite time horizon and suppose at time t queue i is connected and nonempty, and we have*

$$\frac{\mu_i c_i}{1 - (1 - q_i \mu_i)\beta} \geq \frac{\mu_j c_j}{1 - \beta} \quad \forall j \neq i \ \text{ s.t. } \ s_{j,t} \neq 0, \ b_{j,t} > 0. \tag{24}$$

Then it is optimal to allocate the next slot to queue i.

The intuition of the above results, as pointed out in [9] is that due to different holding costs, transmission success probabilities, and different connectivity probabilities, allocation to more costly queues or queues in better conditions (which then runs the risk of emptying the queue) is only justified (or compensated) if it is sufficiently more expensive than a less costly queue or a queue in a worse condition. One can also show via examples that if the sufficient condition is not satisfied, the index policy may or may not be optimal.

So far we have assumed that the channel state at each time slot is independent of all other time slots and of the state of all other users. To relax this assumption, we can follow the same procedure as above to show the following result without any independence assumption on the channel.

Theorem 5.6 *Suppose the time horizon is T and that at $t = 0$ queue i is connected and we have:*

$$\frac{\mu_i c_i (1 - ((1 - \mu_i)\beta)^{T-1})}{1 - (1 - \mu_i)\beta} \geq \frac{\mu_j c_j (1 - \beta^{T-1})}{1 - \beta}, \tag{25}$$
$$\forall j \neq i \ \text{ s.t. } \ s_{j,t} \neq 0, \ b_{j,t} > 0 \ .$$

Then it is optimal to allocate the next slot to queue i.

The intuition of this result remains the same. Note that if we do not use any information on how the state changes, then the required separation between the indices becomes larger.

6 Server Allocation with Imperfect Information

In this section we consider scenarios where the state information is not perfectly observed. In [32] the problem of allocating the channel to one of two users was considered where the channel states change according to a two-state Markov chain [33]. It was shown that under certain conditions the optimal policy is to allocate the channel to the user that is more likely to be in the good state. In [11] a server allocation problem in a static channel with delayed state information was considered. Below we discuss in more detail this problem scenario and related results.

Specifically, each queue advertises to the server its buffer size at the beginning of the tth slot, $\mathbf{b}_t$, and the server uses this information to allocate the next slot. Thus, unlike the previous problems, in this case the server does not have the precise backlog information $\mathbf{d}_t$ for the allocation of slot t. Rather it knows the backlog of the previous slot $\mathbf{b}_{t-1}$ and its allocation for the previous slot, denoted by the vector $\mathbf{w}_{t-1}$, assuming that the server recalls its last allocation decision. Consequently the server knows the following quantity at time t: $\mathbf{d}_t = [\mathbf{b}_{t-1} - \mathbf{w}_{t-1}]^+$, which is completely determined from the buffer occupancy and allocation information of the $(t-1)$th slot. This is referred to as the *existing backlog* because this is the amount carried over from the previous slot due to under-allocation (as opposed to new arrivals that occurred during the previous slot). Alternatively it is also called the amount of *deterministic packets* to be distinguished from the random arrivals that occurred during that frame. We have $\mathbf{b}_t = \mathbf{d}_t + \mathbf{a}_{t-1}$.

The objective is to find an allocation policy π that minimizes the following cost function.

$$J_T^\pi = E^\pi[C|\mathcal{F}_0] , \quad \text{where} \quad C = \sum_{t=1}^{T} \beta^{t-1} \sum_{i=1}^{N} c_i b_{i,t} , \tag{26}$$

which is essentially the same as the one presented earlier, with the minor difference of starting at time $t = 1$ due to the one-step delay.

By restricting to Bernoulli arrival processes with arrival probability p_i for queue i, [11] introduced the following index and examined the optimality of the corresponding index policy. The index of queue i at time t is defined as

$$I_{i,t} = \begin{cases} c_i & \text{if} \quad d_{i,t} > 0 , \\ c_i \cdot p_i & \text{if} \quad d_{i,t} = 0 . \end{cases} \tag{27}$$

This definition essentially suggests that when a queue is for sure not empty (if $d_{i,t} > 0$ then $b_{i,t}$ must be positive), its priority is determined by its unit holding cost. When a queue may be empty (if $d_{i,t} = 0$ then $b_{i,t} = 0$ with probability $1 - p_i$) on the other hand, its priority is determined by the *expected* holding cost, the original holding cost discounted by the arrival probability. Note that the indices are defined based on the deterministic part of the queue, $d_{i,t}$, because we do not know the actual queue size $b_{i,t}$.

The main result of [11] is the following theorem.

Theorem 6.1 *Let the time horizon be T. Suppose that at time t ($1 \leq t \leq T-1$) for some queue i, $I_{i,t} \geq I_{j,t}$, $\forall j \neq i$.*

1. *If $T = 2$ then it is optimal to allocate the slot at time t to queue i;*
2. *For arbitrary T, if $d_{i,t} > 0$, then it is optimal to allocate the slot at time t to queue i;*
3. *For arbitrary T, if $d_{i,t} = 0$, then it is optimal to allocate the slot at time t to queue i if for all $j \neq i$ we have:*

$$I_{i,t}(\frac{1-(p_i\beta)^{T-t}}{1-p_i\beta}) \geq I_{j,t}(\frac{1-\beta^{T-t}}{1-\beta}) \ . \tag{28}$$

This theorem says that it is always optimal to serve the highest indexed queue, if this queue is also not empty (deterministically). We consider a system in this state to be *away* from the boundary. However, when the state of the system is on the boundary, i.e., the queue with the highest index has zero deterministic packets, then the index policy is optimal, if the highest index is sufficiently separated from (larger than) all other indices. This separation is given by (28), and is reminiscent of the separation condition derived in [9] presented in the previous section. The intuition behind this sufficient condition is that due to the randomness in packet arrival, which is unobservable at the time of the decision, assigning the slot to an empty queue, rather than another nonempty or empty queue, can be optimal if this queue is sufficiently "costly", so that the gain sufficiently compensates the loss due to potential overallocating (i.e., a wasted slot if there is no packet arrival and the deterministic part is also zero).

This result can be shown using an argument very similar to the one we presented in the previous section. The techniques used in [11] are slightly different, which we summarize below in a sequence of lemmas.

For simplicity of discussion, we refer to queues as *empty* if their deterministic part is zero.

Lemma 6.2 *Suppose* $T = 2$ *and* $I_{i,1} \geq I_{j,1}$. *Let* π *be the policy that assigns the slot at time* $t = 1$ *to queue* j *and let* π' *be the policy that assigns the slot to queue* i. *Then* $J_T^{\pi} \geq J_T^{\pi'}$ *a.s.*

Lemma 6.3 *Let the time horizon be* T *and suppose* $I_{i,1} \geq I_{j,1}$ *and* $d_{i,1} > 0$. *Let* π *be a policy that serves queue* j *at time* $t = 1$ *and then follows an optimal policy given* $\mathbf{d}_2^{\pi}$ *thereafter. Let* π' *be a policy that serves queue* i *at time* $t = 1$ *and then follows an optimal policy given* $\mathbf{d}_2^{\pi'}$ *thereafter. Then we have* $J_T^{\pi} \geq J_T^{\pi'}$ *a.s.*

This lemma shows that if a nonempty queue has a higher index than other queues (empty or not) at a particular time step, then it is optimal to serve this queue, given that the allocation made for the remaining steps is optimal.

The next two lemmas show what happens if the highest indexed queue happens to be empty.

Lemma 6.4 *Let* π *be an optimal policy given the initial state* $\mathbf{d}_1$ *and let* π' *be an optimal policy given the initial state* $\mathbf{d}_1^{i+}$. *Then*

$$E^{\pi'}[C|\mathcal{F}_0, \mathbf{d}_1^{i+}] - E^{\pi}[C|\mathcal{F}_0, \mathbf{d}_1] \leq \frac{c_i(1-\beta^T)}{1-\beta} \quad \text{a.s.} \tag{29}$$

Lemma 6.5 *Let* π *be an optimal policy given initial state* $\mathbf{d}_1$ *and we have* $d_{i,1} = 0$. *Let* π' *be an optimal policy given initial state* $\mathbf{d}_1^{i+}$. *Then*

$$E^{\pi'}[C|\mathcal{F}_0, \mathbf{d}_1^{i+}] - E^{\pi}[C|\mathcal{F}_0, \mathbf{d}_1] \geq \frac{c_i(1-(p_i\beta)^T)}{1-p_i\beta} \quad \text{a.s.} \tag{30}$$

The next two lemmas give sufficient conditions under which a higher indexed but empty queue should be served, given all subsequent allocations are done optimally.

Lemma 6.6 *Let the time horizon be* T *and suppose we have* $I_{i,1} \geq I_{j,1}$, $d_{i,1} = 0$, *and* $d_{j,1} > 0$. *Let* π *be a policy that serves queue* j *at time* $t = 1$ *and then follows an optimal policy given* $\mathbf{d}_2^{\pi}$ *thereafter. Let* π' *be a policy that serves queue* i *at time* $t = 1$ *and then follows an optimal policy given* $\mathbf{d}_2^{\pi'}$ *thereafter. Then we have;*

$$E^{\pi}[C|\mathcal{F}_0, d_{i,1} = 0, d_{j,1} \neq 0] \geq E^{\pi'}[C|\mathcal{F}_0, d_{i,1} = 0, d_{j,1} \neq 0] \quad \text{a.s.,}$$

if

$$p_i c_i \frac{(1-(\beta p_i)^{T-1})}{(1-\beta p_i)} \geq c_j \frac{1-\beta^{T-1}}{1-\beta} . \tag{31}$$

Lemma 6.7 *Let the time horizon be T and suppose we have $I_{i,1} \geq I_{j,1}$ and $d_{i,1} = d_{j,1} = 0$. Let π be a policy that serves queue j at time $t = 1$ and then follows an optimal policy given $\mathbf{d}_2^{\pi}$ thereafter. Let π' be a policy that serves queue i at time $t = 1$ and then follows an optimal policy given $\mathbf{d}_2^{\pi'}$ thereafter. Then we have*

$$E^{\pi}[C|\mathcal{F}_0, d_{i,1} = d_{j,1} = 0] \geq E^{\pi'}[C|\mathcal{F}_0, d_{i,1} = d_{j,1} = 0] \text{ a.s.},$$

if

$$p_i c_i \frac{(1-(\beta p_i)^{T-1})}{(1-\beta p_i)} \geq p_j c_j \frac{1-\beta^{T-1}}{1-\beta} . \tag{32}$$

A few facts from the above results:

1. If (31) is satisfied, then (32) is also satisfied, i.e., (31) is a stronger condition than (32).

2. If (31) is satisfied for some horizon T then it is also satisfied for any horizon $T' \leq T$; i.e., the condition becomes weaker and weaker as T decreases. The same applies to (32).

3. All the above lemmas remain valid for the allocation of an arbitrary time slot t within the horizon ($1 \leq t \leq T-1$ as opposed to $t = 1$), given $\mathbf{d}_t$, when T is replaced by $T - t + 1$ in equations (29), (30), (31), and (32). This is because in all these lemmas what matters is the time to go, which is $T - t + 1$ for an allocation made at t and a horizon of T.

Combining Lemmas 6.2, 6.3, 6.6, and 6.7 we are now able to prove Theorem 6.1 as follows.

Proof of Theorem 6.1. The case of $T = 2$ is directly given by Lemma 6.2.

If $d_{i,t} > 0$ then it follows directly from Lemma 6.3 that it is optimal to serve queue i.

Suppose $d_{i,t} = 0$ and its index satisfies (28) with respect to all other queues. Consider now some other queue j such that $d_{j,t} \neq 0$, and queue k such that $d_{k,t} = 0$. Then by Lemma 6.6 the policy that serves queue

i at time t and then follows an optimal policy is at least as good as any policy that serves queue j at time t. Similarly by Lemma 6.7 the policy that serves queue i at time t and then follows an optimal policy is at least as good as any policy that serves queue k at time t. Thus a policy that serves queue i at time t and then follows an optimal policy is at least as good as any other policy. Therefore serving queue i under (28) when $d_{i,t} = 0$ is optimal. □

Remark 6.8 *Theorem 6.1 gives a sufficient condition for the optimality of the index rule for one step when the state of the system is on the boundary (the queue with the highest index is empty). A straightforward induction argument shows that if the conditions of Theorem 6.1 hold in every time step (this requires the indices to be separated as defined by condition (28)), then the index policy is optimal for the entire horizon.*

7 Properties of Optimal Allocation with a General Cost Function

All previous sections used linear cost functions as defined in (3). In [34] more general forms of the cost function were considered, and certain properties of the optimal policy were shown for the simple case of two users and a static single-state channel where all users have the same channel quality of $\mu_i = 1$. In [34] the state observation is one-step delayed as in the previous section, but very similar arguments and results hold for the case with no delay. Below we summarize these properties in the context of a system without observation delay.[2] It is shown that the optimal policy is of the threshold type. Therefore it may or may not be realized as an index policy. However, this part is included for completeness as it does reveal interesting features of the problem when costs are not restricted to linear functions.

Define $V_t(\mathbf{b})$ as follows.

$$V_t(\mathbf{b}) = \min_{\pi} E^{\pi}[\sum_{u=0}^{t-1} \beta^u c(\mathbf{b}_u) | \mathbf{b}_0 = \mathbf{b}] \ , \tag{33}$$

where V_t is the *value function* or the cost-to-go for time horizon t. One

[2] Although we only consider the discounted cost in this section, the results can be generalized to the average cost as well (see [35, 36, 37]).

can then establish the dynamic programming equations [38, 39]:

$$\begin{aligned} V_0(\mathbf{b}) &= 0, \\ V_t(\mathbf{b}) &= c(\mathbf{b}) + \beta \min_i\{E_{\mathbf{a}}[V_{t-1}([\mathbf{b}-\mathbf{e}_i]^+ + \mathbf{a})]\}, \end{aligned} \tag{34}$$

where $\mathbf{a}$ is the vector of random arrivals during the first time slot.

Let $W_t(\mathbf{b}) = E_{\mathbf{a}}[V_t(\mathbf{b}+\mathbf{a})]$. Define a transform T_1 as follows.

$$T_1 f(\mathbf{b}) = \min_i\{f([\mathbf{b}-\mathbf{e}_i]^+)\} ; \tag{35}$$

thus we have for $\mathbf{b} \in \mathbb{Z}_+^2$,

$$\begin{aligned} V_0(\mathbf{b}) &= 0, \\ V_t(\mathbf{b}) &= c(\mathbf{b}) + \beta T_1 W_{t-1}(\mathbf{b}). \end{aligned} \tag{36}$$

For the rest of this section we use $\mathbf{x}$ to denote an arbitrary argument.

Definition 7.1 *A function $f : \mathbb{Z}_+^2 \to \mathbb{R}$ belongs to the set $\mathcal{V}$ if $f(\mathbf{x})$ satisfies the following conditions.*

C.1 $f(\mathbf{x}) \le f(\mathbf{x}+\mathbf{e}_i), \quad i \in \{1,2\}$;
C.2 $f(\mathbf{x}+\mathbf{e}_1) + f(\mathbf{x}+\mathbf{e}_2) \le f(\mathbf{x}) + f(\mathbf{x}+\mathbf{e}_1+\mathbf{e}_2)$;
C.3.a $f(\mathbf{x}+\mathbf{e}_1) + f(\mathbf{x}+\mathbf{e}_1+\mathbf{e}_2) \le f(\mathbf{x}+\mathbf{e}_2) + f(\mathbf{x}+2\mathbf{e}_1)$;
C.3.b $f(\mathbf{x}+\mathbf{e}_2) + f(\mathbf{x}+\mathbf{e}_1+\mathbf{e}_2) \le f(\mathbf{x}+\mathbf{e}_1) + f(\mathbf{x}+2\mathbf{e}_2)$.

C.1 is the *monotonicity* condition and requires the function $f(\mathbf{x})$ to be nondecreasing in both its elements, **C.2** is the *supermodularity* condition, and **C.3** is the *superconvexity* condition following the terminology used in [40]. Note that these are rather benign conditions, and they specify a very large class of cost functions of practical interest.

Remark 7.2 *Note that conditions* **C.2** *and* **C.3.a** *result in the convexity of f in x_1. Similarly,* **C.2** *and* **C.3.b** *imply the convexity of f in x_2.*

The following sequence of results shows that if the cost function $c(\mathbf{b}) \in \mathcal{V}$, then the value function $V_t(\mathbf{b})$ will also have these properties for all values of t.

Lemma 7.3 *If $f \in \mathcal{V}$, then the function $g : \mathbb{Z}^2 \to \mathbb{R}$ defined as $g(\mathbf{x}) = f(\mathbf{x}+\mathbf{a})$ is in $\mathcal{V}$ for all $\mathbf{a} \in \mathbb{Z}_+^2$.*

The proof is straightforward by replacing $\mathbf{x}+\mathbf{a}$ instead of $\mathbf{x}$ in conditions **C.1** to **C.3**.

Lemma 7.4 *If $f_1, f_2, \ldots$ are a sequence of functions that belong to $\mathcal{V}$, then $g(\mathbf{x}) = \sum_l p_l f_l(\mathbf{x})$ also belongs to $\mathcal{V}$, where p_l are constants.*

Lemma 7.5 *If $f \in \mathcal{V}$, then $T_1 f \in \mathcal{V}$.*

The proofs for the above two lemmas are left to the reader; they can also be found in [34, 41]. We prove the main theorem below.

Theorem 7.6 *Suppose there are two users and one slot to be allocated. If the cost function $c(\mathbf{b}) \in \mathcal{V}$, then*
(a) For all time t we have $V_t(\mathbf{b}) \in \mathcal{V}$; and
(b) The optimal policy in assigning one slot is of the threshold type.

Proof.
(a) We prove the result by induction. For the induction basis we have $V_0(\mathbf{b}) = 0$, therefore $V_0(\mathbf{b}) \in \mathcal{V}$.

Next we show that if $V_{t-1}(\mathbf{b}) \in \mathcal{V}$, then $V_t(\mathbf{b}) \in \mathcal{V}$. We have $V_t(\mathbf{b}) = c(\mathbf{b}) + \beta T_1 W_{t-1}(\mathbf{b})$. By Lemmas 7.3 and 7.4 we have that $W_{t-1}(\mathbf{b}) \in \mathcal{V}$. Therefore by Lemma 7.5, we have $T_1 W_{t-1}(\mathbf{b}) \in \mathcal{V}$. Thus by using Lemma 7.4 we conclude that $V_t(\mathbf{b}) \in \mathcal{V}$.

(b) Note that when $x_i = 0$, $i \in \{1, 2\}$, it is optimal not to assign the server to queue i. Therefore assume $x_i > 0$, $i \in \{1, 2\}$. By part (a) of this theorem, $V_t \in \mathcal{V}$ for all t. Therefore $W_t \in \mathcal{V}$. Thus by property **C.3.a** we have

$$W_t(\mathbf{b} + \mathbf{e}_1) + W_t(\mathbf{b} + \mathbf{e}_1 + \mathbf{e}_2) \leq W_t(\mathbf{b} + 2\mathbf{e}_1) + W_t(\mathbf{b} + \mathbf{e}_2) .$$

By replacing $\mathbf{b}$ with $\mathbf{b} - \mathbf{e}_1 - \mathbf{e}_2$ we have

$$W_t(\mathbf{b} - \mathbf{e}_2) + W_t(\mathbf{b}) \leq W_t(\mathbf{b} + \mathbf{e}_1 - \mathbf{e}_2) + W_t(\mathbf{b} - \mathbf{e}_1) .$$

Rearranging, we get

$$W_t(\mathbf{b} - \mathbf{e}_2) - W_t(\mathbf{b} - \mathbf{e}_1) \leq W_t(\mathbf{b} + \mathbf{e}_1 - \mathbf{e}_2) - W_t(\mathbf{b}) .$$

The last inequality suggests, that if the left-hand side is nonnegative, then the right-hand side is also nonnegative. Therefore if the optimal decision is to allocate to the first queue when the state is $\mathbf{b}$ for some $\mathbf{b}$, then it is optimal to allocate the slot to the first queue when the state is $\mathbf{b} + \mathbf{e}_1$. Similarly using **C.3.b** it can be shown that if the optimal decision is to allocate to the second queue when the state is $\mathbf{b}$, then it is optimal to allocate the slot to the second queue when the state is $\mathbf{b}+\mathbf{e}_2$. We can define the threshold as follows.

$$h_{t+1}(b_1) = \min\{b_2 \geq 0 | W_t([\mathbf{b}-\mathbf{e}_2]^+) \leq W_t([\mathbf{b}-\mathbf{e}_1]^+)\}.$$

$h_{t+1}(b_1) = \infty$ when the above set is empty. Suppose the time horizon is $t+1$. If we have $b_{2,0} \geq h_{t+1}(b_{1,0})$ then the optimal policy is to assign the slot at time 0 to queue 2, otherwise the optimal decision rule is to assign the slot to queue 1 (if the set is empty then the threshold is infinity), proving the optimality of a threshold policy. □

Remark 7.7 *Although we proved the result for $\mu = 1$, it holds true for any $0 \leq \mu \leq 1$. To prove the result for this more general case, one has to redefine T_1 as follows.*

$$T_1 f(\mathbf{b}) = \min_i\{\mu f([\mathbf{b}-\mathbf{e}_i]^+) + (1-\mu) f(\mathbf{b})\} \ .$$

It can be shown that Lemma 7.5 still holds for this more general case. The rest of the proof remains the same.

8 Conclusion

In this paper we considered the problem of optimally sharing a single slot between multiple queues. The objective is to minimize the discounted cost (equation (1)) or the average cost (equation (2)) over a finite or an infinite horizon. We surveyed in detail a sequence of scenarios and corresponding results related to this problem.

First we considered linear cost functions with different coefficients to differentiate between different services. We formulated this problem as a restless bandit problem and calculated the equivalence of Whittle's indices in this problem. Although these indices are in general not optimal we showed their optimality in the static channel. For dynamic channels we considered two cases. In the symmetric case serving the longest queue is optimal. For the case of differentiated services, we studied a two-state channel, where a user can be either connected or disconnected. We showed that in this case the $c\mu$ rule is optimal if the indices are sufficiently separated. We then studied the optimality of an index policy in the case of imperfect and delayed information. We defined the indices based on the one-step reward from serving the queue and provided sufficient conditions for the optimality of the index policy.

Finally for a general cost function with a single channel state in a two-user system, we proved the propagation of monotonicity, supermodularity, and superconvexity for the value functions. Using these properties we proved that the optimal policy is of the threshold type.

References

[1] S. Lippman, "Applying a new device in the optimization of exponential queueing systems," *Oper. Res.*, vol. 25, pp. 687–710, 1975.

[2] R. Serfozo, "An equivalence between discrete and continuous time markov decision processes," *Oper. Res.*, vol. 27, pp. 616–620, 1979.

[3] C. G. Cassandras and S. Lafortune, *Introduction To Discrete Event Systems*, Kluwer Academic, Hingham, MA, 1999.

[4] C. Buyukkoc, P. Varaiya, and J. Warland, "The $c\mu$-rule revisited," *Adv. Appl. Prob.*, vol. 17, pp. 237–238, 1985.

[5] J. S. Baras, A. J. Dorsey, and A. M. Makowski, "Two competing queues with linear costs and geometric service requirements: The μc rule is often optimal," *Adv. Appl. Prob.*, vol. 17, pp. 186–209, 1985.

[6] L. Tassiulas and A. Ephremides, "Dynamic server allocation to parallel queues with randomly varying connectivity," *IEEE Trans. Information Theory*, vol. 39, no. 2, pp. 466–478, March 1993.

[7] L. Tassiulas, "Scheduling and performance limits of networks with constantly changing topology," *IEEE Trans. Information Theory*, vol. 43, no. 3, pp. 1067–73, May 1997.

[8] N. Bambos and G. Michailidis, "On the stationary dynamics of parallel queues with random server connectivities," in *Proc. IEEE Conference on Decision and Control (CDC)*, 1995.

[9] C. Lott and D. Teneketzis, "On the optimality of an index rule in multi-channel allocation for single-hop mobile networks with multiple service classes," *Probability in the Engineering and Informational Sciences*, vol. 14, no. 3, pp. 259–297, July 2000.

[10] N. Ehsan and M. Liu, "Optimal bandwidth allocation with delayed state observation and batch assignment," Tech. Rep. CGR 03-11, EECS Department, University of Michigan, Ann Arbor, 2003.

[11] N. Ehsan and M. Liu, "On the optimality of an index policy for bandwidth allocation with delayed state observation and differentiated services," in *Proc. IEEE INFOCOM*, 2004.

[12] F. J. Beutler and D. Teneketzis, "Routing in queueing networks under imperfect information: Stochastic domain and thresholds," *Stochastics and Stochastic Reports*, vol. 26, pp. 81–100, 1989.

[13] R. Knopp and P. Humblet, "Information capacity and power control in single cell multiuser communications," in *Proc. IEEE International Conference on Communications (ICC)*, 1995.

[14] P. Viswanath, D. N. C. Tse, and R. Laroia, "Opportunistic beamforming using dumb antennas," *IEEE Trans. Information Theory*, vol. 48, no. 6, June 2002.

[15] S. Borst, "User-level performance of channel-aware scheduling algorithms in wireless data networks," in *Proc. IEEE INFOCOM*, 2003.

[16] P. Whittle, "Restless bandits: Activity allocation in a changing world," *A Celebration of Applied Probability, J. Appl. Prob.*, vol. 25A, pp. 287–298, 1988.

[17] R. Weber and G. Weiss, "On an index policy for restless bandits," *J. Appl. Prob.*, vol. 27, pp. 637–648, 1990.

[18] J. Nino-Mora, "Restless bandits, partial conservation laws, and indexability," *Adv. Appl. Prob.*, vol. 33, no. 1, pp. 76–98, 2001.

[19] C. H. Papadimitriou and J. N. Tsitsiklis, "The complexity of optimal queueing network control," *Mathematics of Operations Research*, vol. 24, no. 2, pp. 293–305, May 1999.

[20] L. Tassiulas and S. Papavassiliou, "Optimal anticipative scheduling with asynchronous transmission opportunities," *IEEE Trans. Automatic Control*, vol. 40, no. 12, pp. 2052–2062, December 1995.

[21] M. Carr and B. Hajek, "Scheduling with asynchronous service opportunities with applications to multiple satellite systems," *IEEE*

Trans. Automatic Control, vol. 38, no. 12, pp. 1820–33, December 1993.

[22] J. C. Gittins, "Bandit processes and dynamic allocation indices," *J. Royal Statistical Society Series*, vol. B14, pp. 148–167, 1972.

[23] P. Whittle, "Multi-armed bandits and the gittins index," *J. Royal Statistical Society*, vol. 42, no. 2, pp. 143–149, 1980.

[24] M. Raissi-Dehkordi and J. S. Baras, "Broadcast scheduling in information delivery systems," in *Proc. IEEE GLOBECOM*, 2002.

[25] D. Bertsimas and J. Nino-Mora, "Conversion laws, extended polymatroids and multi-armed bandit problems," *Mathematics of Operations Research*, vol. 21, pp. 257–306, 1996.

[26] J. G. Shanthikumar and D. D. Yao, "Multi class queueing systems: Polymatroid structure and optimal scheduling control," *Oper. Res.*, vol. 40, pp. 293–299, 1992.

[27] J. Edmonds, "Submodular functions, matroids and certain polyhedra," in *Proc. of the Calgary International Conference on Combinatorial Structures and Their Applications*, 1970, pp. 69–87.

[28] T. C. Green and S. Stidham Jr., "Sample path conservation laws, with applications to scheduling queues and fluid systems," *Queueing Systems*, vol. 36, pp. 175–199, 2000.

[29] D. J. Welsh, *Matroid Theory*, Academic Press, London, 1976.

[30] G. Grimmett and D. Stirzaker, *Probability and Random Processes*, Oxford University Press, New York, 2001.

[31] R. Menich and R. Serfozo, "Optimality of routing and servicing in dependent parallel processing systems," *Queueing Systems*, vol. 9, pp. 403–418, 1991.

[32] G. Koole, Z. Liu, and R. Righter, "Optimal transmission policies for noisy channels," *Oper. Res.*, vol. 49, pp. 892–899, 2001.

[33] E. N. Gilbert, "Capacity of a burst-noise channel," *Bell Systems Technical Journal*, vol. 39, pp. 1253–1266, 1960.

[34] N. Ehsan and M. Liu, "Properties of optimal resource sharing in a satellite channel," in *Proc. IEEE Conference on Decision and Control (CDC)*, December 2004.

[35] R. R. Weber and S. Stidham Jr, "Optimal control of service rates in networks of queues," *Adv. Appl. Prob.*, vol. 19, pp. 202–218, 1987.

[36] B. Hajek, "Optimal control of two interacting service stations," *IEEE Trans. Automatic Control*, vol. AC-29, pp. 491–499, 1984.

[37] R. Serfozo, "Optimal control of random walks, birth and death processes, and queues," *Adv. Appl. Prob.*, vol. 13, pp. 61–83, 1981.

[38] P. R. Kumar and P. Varaiya, *Stochastic Systems, Estimation, Identification and Adaptive Control*, Prentice-Hall, Englewood Cliffs, NJ, 1986.

[39] D. P. Bertsekas, *Dynamic Programming, Deterministic And Stochastic Models*, Prentice-Hall, Englewood Cliffs, NJ, 1987.

[40] G. M. Koole, "Structural results for the control of queueing systems using event-based dynamic programming," *Queueing Systems*, vol. 30, pp. 323–339, 1998.

[41] N. Ehsan and M. Liu, "Optimal server allocation in batches," Tech. Rep. CGR 04-10, EECS Department, University of Michigan, Ann Arbor, August 2004.

Chapter 4

Performance Optimization Using Multipath Routing in Mobile Ad Hoc and Wireless Sensor Networks

Wenjing Lou
Department of Electrical and Computer Engineering
Worcester Polytechnic Institute, Worcester, MA 01609
E-mail: wjlou@ece.wpi.edu

Wei Liu
Department of Electrical and Computer Engineering
University of Florida, Gainesville, FL 32611
E-mail: liuw@ufl.edu

Yanchao Zhang
Department of Electrical and Computer Engineering
University of Florida, Gainesville, FL 32611
E-mail: yczhang@ufl.edu

1 Introduction

Multipath routing, sometimes called *traffic dispersion* [14], has been one of the most important current directions in the area of routing. The current routing is based on the single path routing: between a source and a destination, the single minimum-cost path tends to be selected although different cost metrics may yield different paths. However, in a reasonably well-connected network, there may exist several paths between a source–destination pair. The concept of multipath routing is to give the source node a choice at any given time of multiple paths to a particular destination by taking advantage of the connectivity redundancy of the underlying network. The multiple paths may be used alternately, namely, traffic taking one path at a time, or they may be used concurrently, namely, traffic flowing through multiple paths simultaneously.

Multipath routing (or *dispersity routing* as termed by the author) was first proposed by Maxemchuk to spread the traffic from a source in space rather than in time as a means for load balancing and fault handling in packet switching networks [29–31]. The method was shown to equalize load and increase overall network utilization; with redundancy, it improves the delay and packet loss properties at the expense of sending more data through the network. Since then, the multipath routing technique has been applied to various types of networks, such as the communication networks, B-ISDN, ATM networks, etc., and for various network control and management purposes, such as aggregating the bandwidth, minimizing the delay, supporting the Quality of Service (QoS) routing, smoothing the burstiness of the traffic, alleviating the network congestion, and improving the fault tolerance, etc. [3, 6, 9, 39, 40]. Interested readers are referred to [14] for a comprehensive survey of the earlier works on traffic dispersion in wired networks.

Mobile Ad hoc NETworks (MANETs) and Wireless Sensor Networks (WSNs) have received tremendous attention in the past few years. A MANET is a collection of nodes that can move freely and communicate with each other using the wireless devices. For the nodes that are not within the direct communication range, other nodes in the network work collectively to relay packets for them. A MANET is characterized by its dynamic topological changes, limited communication bandwidth, and limited battery power of nodes. The network topology of a MANET can change frequently and dramatically. One reason is that nodes in a MANET are capable of moving collectively or randomly. When one node moves out of/into the transmission range of another node, the link between the two becomes down/up. Another reason that causes the topological changes is the unstable wireless links, which might become up and down due to the signal fading (obstacles between the two end nodes), interference from other signals, or the changing of transmission power levels. Most of mobile nodes are battery powered. When nodes run out of battery power, node failure will also cause topological changes. Although a close relative to MANETs, a WSN differs from an ad hoc network in many aspects [2]. The number of nodes in a WSN is usually much larger than that in an ad hoc network. Sensor nodes are more resource constrained in terms of power, computational capabilities, and memory. Sensor nodes are typically randomly and densely deployed (e.g., by aerial scattering) within the target sensing area. The post deployment topology is not predetermined. Although in many cases the nodes are static, the topology might change frequently because the sensor nodes and the wireless channels are prone to failure.

Multipath routing has drawn extensive attention in MANETs and WSNs recently. The dense deployment of nodes in MANETs/WSNs makes the multipath routing a natural and promising technique to cope with frequent topological changes and consequently unreliable communication services. Research efforts have also been made using multipath routing to improve the robustness of data delivery [41,46], to balance the traffic load and balance the power consumption among nodes [13,45], to reduce the end-to-end delay and the frequency of route discoveries [11,33], and to improve network security [24,48], etc. Two primary technical focuses in this area are (a) the multipath routing protocols that are able to find multiple paths with the desired properties, and (b) the policies on the usage of the multiple paths and the traffic distribution among the multiple paths, which very often involve coding schemes that help to split the traffic.

Communication security and reliability are two important issues in any network and they are seemingly contradictive goals from the perspective of adding redundancies. Traditionally reliability can be achieved by sending redundant data over multiple paths. On the other hand, the redundant data give adversaries better chances to intercept the information. To address this issue, we proposed a novel *Security Protocol for REliable dAta Delivery* (SPREAD) to enhance both security and reliability and we investigated the SPREAD scheme in both MANETs and WSNs. The goal of the proposed SPREAD scheme is to provide further protection to the data delivery service, specifically, to reduce the probability that secret messages might be compromised/lost when they are delivered across the insecure/unreliable network. The basic idea is to transform a secret message into multiple shares by secret sharing schemes and then deliver the shares via multiple independent paths to the destination so that even if a small number of nodes that are used to relay the message shares are compromised/faulty, the secret message as a whole is not compromised/lost. In the SPREAD scheme, multipath routing is used to distribute the trust to multiple paths/nodes for security purposes and to diminish the effect of unreliable nodes/links for the purpose of reliability. An end-to-end multipath routing technique to find multiple node-disjoint paths between a source–destination pair for end-to-end data delivery was investigated in a MANET environment [24]. While in a WSN, noticing that a typical communication task is for every sensor node to sense its local environment and, upon request, send the data of interest back to a base station, we proposed an efficient *N-to-1* multipath discovery protocol that is able to find multiple node-disjoint paths from every sensor node to the base station simultaneously in one route discovery [26]. The SPREAD scheme combines both concurrent multipath routing and alternate multipath savaging techniques and provides more secure and more reliable data collection service. In other words, it makes the data delivery service more resistent to node/link failure and node compromise problems.

This chapter aims to introduce readers to the concept and techniques of multipath routing in a MANET/WSN, with a focus on the performance gains from multipath routing and the techniques of the construction and usage of the paths. The chapter is organized as follows. We identify some applications and examine their performance benefits from multipath routing in Section 2. We review some multipath routing protocols in the literature in Section 3. Then in Section 4 we present the SPREAD scheme with moderate details. Section 5 concludes the chapter.

2 Performance Benefits from Multipath Routing

Multipath routing has been studied for various network control and management purposes in various types of networks. In this section, we outline some of the applications of multipath routing that improve the performance of an ad hoc network and a sensor network.

2.1 Reliability

By "reliability" we mean the probability that a message generated at one place in the network can actually be routed to the intended destination. Reliability is a big challenge in MANETs/WSNs because packets transmitted are subject to loss due to frequent topological changes, severe media access conflicts, and various kinds of interference that affect the wireless transceivers in correctly decoding the wireless signals.

Multipath routing in a MANET was originally developed as a means to provide route failure protection. For example, the Dynamic Source Routing (DSR) protocol [17] is capable of caching multiple routes to a certain destination. When the primary path fails, an alternate one will be used to salvage the packet. The Temporally Ordered Routing Algorithm (TORA) [34] also provides multiple paths by maintaining a destination-oriented Directed Acyclic Graph (DAG) from the source node. Multipath extensions of some protocols that originally depend on single-path routing have also been proposed, such as the AODV-BR [27], Alternative Path Routing (APR) [36], and Split Multipath Routing (SMR) [28], etc., which improve the single-path routing protocols by providing multiple alternate routes. In these cases, the multiple paths are not

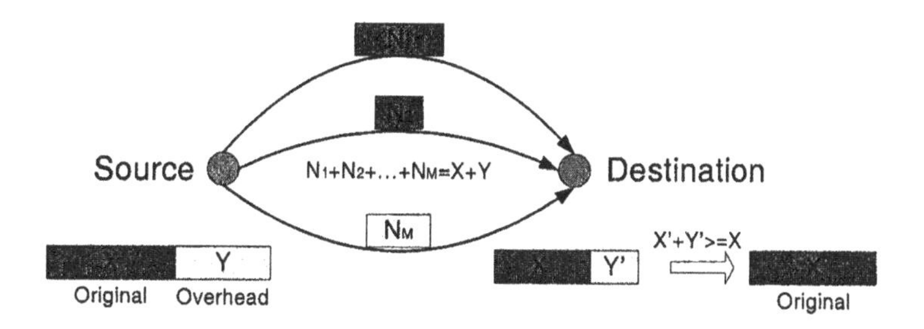

Figure 1: Concurrent multipath routing for reliability.

used simultaneously. The traffic takes one of the multiple paths at a time. Other paths are kept as backup in case the used one is broken. When all known paths are broken, a new multipath discovery procedure is initiated. Alternate path routing has also been adopted at the link layer: when multiple next hops are available, the packet is routed through the one that exhibits the best channel condition [19].

Another way of using the multiple paths is to have the traffic flow through multiple paths simultaneously. Concurrent multipath routing in a MANET has been developed to improve the throughput and reliability and achieve load balancing. Some type of source coding scheme is usually incorporated, particularly for reliability, such as the *Forward Error Correction* (FEC) codes, *Reed-Solomon* (RS) codes, etc. As shown in Figure 1, a certain amount of redundancy is coded into the data traffic such that the decoding could tolerate a certain amount of data loss or even the failure of a complete path (paths). One application of concurrent multipath routing was proposed in [41] in which overhead information is added to the original data load by *diversity coding* [1], then segments consisting of both types of data are distributed over multiple independent paths simultaneously. The diversity coding has the property that upon receiving a certain amount of data, either original or overhead, the original information can be fully recovered. With an analytical framework, the scheme was shown to be effective in increasing the overall information delivery ratio. Another interesting application [15] was proposed that combined the Multiple Path Transport (MPT) and *Multiple Description Coding* (MDC) in order to send video and image information in a MANET. The MDC has the property that the picture quality is acceptable from any one description and any additional description enhances the picture quality accumulatively. The transmission of the multiple descriptions via multiple paths helps to provide the higher bandwidth required by the video/image transmission, and on the other hand, helps to provide more robust end-to-end connection.

In [24] and [26], we combined the *threshold secret sharing* scheme and multipath routing and proposed the SPREAD idea to provide more reliable and more secure data delivery/collection services in a MANET/WSN. The secret sharing scheme has the similar property as the diversity coding from the reliability perspective: a (T, N) threshold secret sharing algorithm divides a piece of information into N segments. From any T out of N segments, one can reconstruct the original information. In addition, the secret sharing scheme has a desirable security property. That is, with less than the threshold, namely T, segments, one could learn nothing about the information and has no better chance to recover the original information than an outsider who knows nothing at all about it. The SPREAD idea was investigated in [24] for end-to-end data delivery in a MANET. It is noticed that the improved end-to-end data delivery ratio relies on the excessive information redundancy allocated to the multiple paths. The reliability of the overall end-to-end delivery is improved because it can tolerate a certain amount of packet loss or path loss, while the reliability of each path remains unimproved. This situation makes reliability and security contradict goals: high redundancy improves reliability but deteriorates the foundations of the security enhancement. The SPREAD scheme was extended in [26] by proposing an *N-to-1* multipath discovery protocol. The distinct feature of the N-to-1 multipath discovery protocol is that it is able to find from every node to a particular destination (the sink node in a WSN) multiple node-disjoint paths in one route discovery efficiently. With the multipath available at every node, the SPREAD scheme achieves both reliability and security goals with little or no information redundancies. The end-to-end concurrent multipath routing is applied at the source to spread the traffic onto multiple disjoint paths between the source and the destination for security purposes. The alternate path routing is applied while each packet is travelling on its designated path: once node/link failure is encountered, the packet is locally salvaged by using an alternate path. By this means, the reliability of each path is greatly improved thus the required end-to-end redundancy is greatly decreased. The dilemma between security and reliability is neatly resolved by the combination of concurrent multipath routing and alternate multipath routing. More details of the SPREAD scheme are presented in Section 4.

2.2 Load/Energy Consumption Balancing

Nodes in a MANET or WSN are typically powered by batteries that have a limited energy reservoir. In some application scenarios, replenishment of power supplies might not be possible. The lifetime of the nodes shows strong dependence on the lifetime of the batteries. In the multihop MANET/WSN, nodes depend on each other to relay packets. The loss of some nodes may cause significant topological changes, undermine the network operation, and affect the lifetime of the network.

Energy-efficient routing has been the subject of intensive study in recent years. One goal of the Energy Aware Routing (EAR) protocols is to select the best path such that the total energy consumed by the network is minimized [44]. A serious drawback of minimum energy routing is that nodes will have a wide difference in energy consumption. Nodes on the minimum energy paths will quickly drain out while the other nodes remain intact. This will result in the early death of some nodes. Another objective of the EAR is to maximize the system lifetime, which is defined as the duration when the system starts to work till any node runs out of energy, or till a certain number of nodes run out of energy, or till the network is partitioned, etc. For this purpose, multipath routing has been shown effective because it distributes the traffic load among more nodes and in proportion to their residual energies. When the energy consumption among nodes is more balanced, the mean time to node failure is prolonged, and the system lifetime is prolonged too [10, 13, 45].

2.3 Routing Overhead

Another benefit of multipath routing is the reduction of the routing overhead. Existing ad hoc routing protocols can be generally categorized into three classes: *table-driven* (or *proactive*, such as DSDV and WRP), *on-demand* (or *reactive*, such as DSR and AODV), and *hybrid* (the combination of the two, such as ZRP) [38]. Most of the performance studies indicate that on-demand routing protocols outperform table-driven protocols [5, 18]. The major advantage of the on-demand routing comes from the reduction of the routing overhead, as high routing overhead usually has a significant performance impact in low-bandwidth wireless networks. An on-demand routing protocol attempts to discover a route to a destination "on-demand" when it is presented a packet for forwarding to that destination but it does not already know a path. It utilizes a route discovery process to find the path(s). Discovered routes are maintained by a route maintenance procedure until either the destination becomes inaccessible along every path from the source or until the route is no longer desired. The route discovery is a costly operation and it usually involves a network wide flooding of route request packets because the node has no idea where the destination is. Typically three types of routing messages are used: the *Route Request* (RREQ) and *Route Reply* (RREP) messages are used in the route discovery process to search for a route; the *Route Error* (RERR) message is used to report the breakage of an intermediate link on a route back to the source.

On-demand multipath protocols find multiple paths between a source and a destination in a single route discovery. A new route discovery is needed only when all the found paths fail. In [33], the authors proved that the use of multiple paths in DSR can keep correct end-to-end connection for a longer time than a single path. Therefore, by keeping multiple paths to a destination, the frequency of the costly route discovery is much lower. Moreover, in a single path routing case, when a node fails to transmit a packet to its next hop, a route error message will be sent back to the source indicating the breakage of the path. With multiple alternate paths available, nodes can actively salvage the packet by sending it to an alternate path; a route error will occur only when all the available paths fail. The occurrence of route error is therefore reduced too. Although the search for multiple paths may need more route request messages and route reply messages in a single route discovery process, the number of overall routing messages is actually reduced. Similar results have been reported in [11].

2.4 Quality of Service (QoS)

An important objective of multipath routing is to provide quality of service, more specifically, to reduce the end-to-end delay, to avoid or alleviate the congestion, and to improve the end-to-end throughput, etc.

It has been shown that multipath routing helps significantly in providing QoS by reducing the end-to-end delay for packet delivery [11]. The reduction in the end-to-end delay is not that intuitive and is attributed to multiple factors. Notice that the end-to-end delay is the latency between a packet sent at the source and received at the destination. Besides the ordinary transmission delay, propagation delay, and queueing delay, which widely exist in all IP networks, there are two types of latency caused particularly by ad hoc on-demand routing protocols. One is the latency the protocol takes to discover a route to a destination when there is no known route to that destination. This type of latency is due to the on-demand behavior of the routing protocol and exists in all such protocols. As we mentioned in Section 2.3, multipath routing effectively reduces the frequency of route discovery therefore the latency caused by this reason is reduced. The other one is the latency for a sender to "recover" when a route being used breaks. The latency resulting from broken routes could be very large because the amount of latency is the addition of the following three parts: the time for a packet to travel along the route to the node immediately before the broken link, the time for that node to detect the broken link, and the time for a route error message to travel from that node back to the source node. Among them, the time to detect a broken link could be very large because the failure of the link can only be determined after having made a certain number of attempts to transmit the packet over the broken link but having failed to receive a passive or explicit acknowledgment of success. This latency caused by route errors is a significant component in the overall packet latency. Again, as we explained in Section 2.3, multipath routing avoids or reduces the occurrence of route errors therefore the packet latency is further reduced. Some other factors contribute to the reduction in the end-to-end delay as well, such as the routing around the congested area, etc.

Multipath routing has been shown effective in wired networks for providing high bandwidth by allocating traffic onto multiple independent paths simultaneously so that the bandwidth of the multiple paths can be aggregated for a request for which the bandwidth of any single path would not suffice. Its effectiveness is intuitive because the using of the multiple paths is independent of each other in wired networks. However, in a MANET or a WSN, shared wireless channels make the situation different. Wireless links are a relatively "soft" concept. When nodes are sharing a single wireless channel and using some Medium Access Control (MAC) protocol such as the IEEE 802.11 to coordinate the access to the shared channel, the communication activities among the links are no longer independent. For example, as shown in Figure 2, with the IEEE 802.11, when one node, say node 5, is transmitting to another node, say node 6, all the neighbors of both the transmitter and the receiver (i.e., nodes 1,2,3,4,7,8,12 in the example) have to keep quiet to avoid possible collision with the ongoing transmission. Therefore, node disjointness in shared channel networks does not imply the independence of the paths. Instead, the communication activities of the multiple paths affect each other very much. This problem has been referred to in [36] as the *route coupling* problem and the results showed that the effect is so severe in single-channel networks that it provides only negligible improvements in quality of service. The selection of multiple paths that cause less coupling is therefore an important challenge in the current multipath routing protocol design. In [45] the authors considered the route coupling problem by selecting less correlated multiple paths. The correlation factor of two node-disjoint paths is defined as the number of links connecting the two paths and it is an indicator of chances that the transmission along one path could interfere with the other. The results show that multipath routing is still an effective means of improving the throughput though the gain is not as significant as that from independent paths.

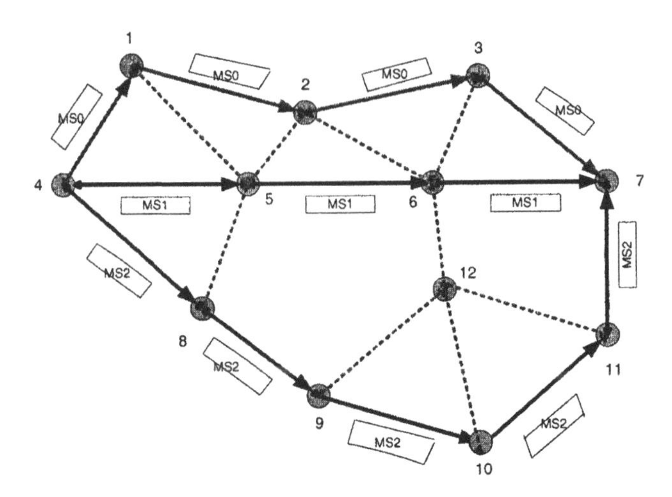

Figure 2: Route coupling problem in shared channel multipath routing.

2.5 Security

A few efforts have been made to improve network security by using multipath routing. When used for security purposes, multipath routing is often combined with secret sharing cryptography. As we mentioned in Section 2.1, a (T, N) threshold secret sharing scheme has the nice property that it divides a secret into N pieces, called *shares* or *shadows*; One can derive nothing from any less than T shares, whereas with an efficient algorithm, the original secret can be reconstructed from any T shares [43]. Therefore, schemes combining multipath routing and secret sharing techniques typically involve the splitting of a secret by secret sharing schemes and the delivery of the shares by multipath routing. By this means, the trust is distributed to multiple nodes/paths in the network and the system is made more resilient to a collusive attack by up to a certain number of compromised nodes.

Secret sharing was originally proposed for key management in information security systems [42]. Key management is also possibly the most critical and complex issue when talking about security in a MANET or a WSN. The applicability of many other security services, such as confidentiality and authentication, relies on effective and efficient key management. In [48], the authors used replication and threshold cryptography and built a more secure and more available public key management service to deal with the denial of service attacks in a MANET. The idea is to distribute the functionality of the Certificate Authority (CA) of a Public Key Infrastructure (PKI) into multiple servers (or trusted nodes). In this way, both the availability and the security of the CA can be improved. In the proposed model, the threshold cryptography is used to split the system secret into multiple shares and each server holds one share; multiple servers collectively perform functions such as signing a certificate and refreshing key shares. Multipath routing means the routing of multiple shares from multiple servers to a single combiner. This approach was further investigated in [21] where CAs are further localized by distributing the servers more evenly in the network such that collective cryptographic operations can be done locally by neighbors of the requesting node. Another key management approach based on multipath routing is a probabilistic approach for the establishing of pairwise secret keys [8, 49]. Due to the intensive computational complexity, operations based on public key algorithms are too expensive in resource constrained MANETs/WSNs. The design of the approach is based on probabilistic key sharing and threshold secret sharing techniques. By probabilistic key sharing, every node in the network will be preloaded with a certain number of initial keys. With overwhelming probability, any pair of nodes would share one or more common keys. Then using the common keys as the seeds, the source node generates a new secret key and divides it into multiple pieces using the secret sharing scheme. The multiple pieces of the secret key are then delivered to the destination via multipath routing. The multipath in their scheme is logical: multiple pieces may flow through the same physical paths while encrypted by different common keys. The SPREAD scheme we proposed is similar to the above approaches in that the scheme is based on the combination of secret sharing and multipath routing too. However, the SPREAD scheme aims to protect the data traffic delivered across the insecure network assuming that the end-to-end encryption is neither secure nor reliable. It is an enhancement to the end-to-end encryption. The multipath in our SPREAD is physically node-disjoint paths and desired path finding algorithms were proposed. The SPREAD can certainly be used to deliver the keys instead of data traffic. However, by delivering the data traffic, SPREAD improves not only security, but also the reliability which is a big challenge in MANETs/WSNs.

3 Multipath Routing Protocols

Routing in ad hoc networks presents great challenges. The challenge comes mainly from two aspects: constant node mobility causes frequent topological changes and limited network bandwidth restricts the timely topological updates at each router. On-demand routing has been widely developed in mobile ad hoc networks in response to the bandwidth constraints because of its effectiveness and efficiency. The multipath routing technique is another promising technique to combat problems of frequent topological changes and link instability because the use of multiple paths could diminish the effect of possible node/link failures. Moreover, as we have discussed in Section 2, multipath routing has been shown effective in improving the reliability, fault-tolerance, end-to-end delay, and security, as well as in achieving load balancing, etc. However, how to actually achieve those performance benefits depends on the availability of the desired multiple paths and it further depends on the capability of the multipath finding/routing techniques. In this section, we review some multipath routing protocols available in the literature. Multipath routing is a more difficult issue than single-path routing. How to find the right number of paths with the desired property effectively and efficiently remains a big challenge in multipath routing research.

3.1 Partially Disjoint Paths

As we discussed in Section 2, Alternate Path Routing (APR) is an effective and efficient approach to improve the reliability of data delivery. It also helps to reduce the routing overhead and the end-to-end delay. In APR, nodes maintain multiple paths (either complete paths as in DSR or only next hop nodes as in AODV) to the destination. When the primary route fails, packets are shifted to an alternate path. A route error occurs only when all the available paths fail. For this category of multipath applications, the paths are not necessarily completely disjoint. Partially disjoint paths can fulfill the task.

Several multipath routing protocols have been proposed in order to provide the desired alternate paths. AODV-BR (Backup Routing) [27] is one example of such routing protocols based on AODV. AODV is an on-demand single path routing protocol based on a distance vector. In AODV, a source node starts the route discovery procedure by broadcasting a route request packet. Each RREQ packet contains an unique broadcast ID of the source node, which, along with the source node's IP, uniquely identifies a RREQ packet so that nodes can detect and drop duplicate RREQ packets. The RREQ packet also contains a destination-sequence number of the destination node that indicates the freshness of the packet so that nodes can detect and drop stale routing packets. An intermediate node, upon receiving a fresh RREQ packet for the first time, needs to set up a reverse path by recording in its route table the address of the node from which the RREQ packet is received. It then rebroadcasts the RREQ packet. Duplicated RREQ packets that arrive later are simply dropped. Once the RREQ packet reaches the destination or an intermediate node that has a fresh enough route to the destination, the destination or the intermediate

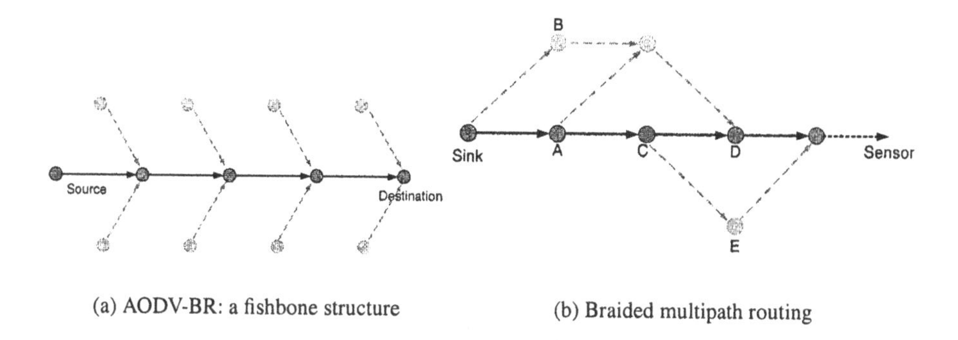

(a) AODV-BR: a fishbone structure

(b) Braided multipath routing

Figure 3: Alternate path routing.

node unicasts a route reply packet back following the reverse path established before. While the RREP packet travels along the reverse path, a forwarding route entry is set up at each node along the path. The proposed AODV-BR follows the propagation of route requests in AODV exactly, whereas the route reply phase is slightly modified to construct the alternate route. Basically, the algorithm takes advantage of the broadcast nature of wireless communication: nodes promiscuously overhear packets transmitted by their neighbors. Once a node overhears (i.e., the node is not the intended receiver of the packet) a RREP packet transmitted by its neighbor (which must be on the primary route), it records that neighbor as the next hop to the destination in its alternate route table. With this simple modification, AODV-BR is able to establish a primary route and alternate routes that look like a fishbone (see Figure 3(a)). Data packets are delivered along the primary path. Once a node detects a link breakage, it locally broadcasts the data packet to its neighbors. Neighbor nodes that have an entry for the destination help to salvage the packet by unicasting the packet to their next hop node.

Another protocol that aims to find partially disjoint paths is the braided multipath routing proposed in [13] in order to increase resilience to node failure in WSNs. The general data dissemination follows the *directed diffusion* [16] paradigm. As with the basic directed diffusion scheme, each node computes the *gradient* and locally determines its most preferred neighbor in the direction of the intended sensor node, based on some empirical information that has initially been flooded throughout the network. Braided multipath relaxes the requirement of the node-disjointness of the complete paths. Instead, it aims to find a small number of alternate paths between the sink and the intended sensor node that are partially node-disjoint with the primary path. For example, a path that differs from the primary path by one node could be an alternate path. The path-finding algorithm makes use of two *path enforcement* messages and depends on some localized techniques to construct braids at each node along the primary path. The procedure can be briefly described as follows. The sink initiates by sending out a *primary path reinforcement* message to its most preferred neighbor, say A (as illustrated in Figure 3(b)). In addition, the sink sends an *alternate path reinforcement* to its next preferred neighbor, say B. An intermediate node, say C, once receiving a primary path reinforcement, forwards the primary path reinforcement to its most preferred neighbor, say D. Therefore, the path travelled by the primary path reinforcement forms the primary path to the intended sensor node. Besides forwarding the primary path reinforcement, each node on the primary path, say C, also initiates an alternate path reinforcement to its next most preferred neighbor (in this example, E). Once a node that is not on the primary path receives an alternate path reinforcement, it forwards it to its most preferred neighbor. If the node that receives the alternate path reinforcement is on the primary path, it simply stops the propagation of the alternate path reinforcement. By this means, an alternate path reinforcement initiated at a node on the primary path provides an alternate path that routes around the next node on the primary path but tends to rejoin the primary path later. The multipath structure formed by this technique looks like a braid (see Figure 3(b)).

3.2 Disjoint Paths

In many multipath routing applications, disjoint paths are more attractive due to the independence of the paths. A number of multipath routing algorithms or protocols have been proposed in order to find disjoint paths in MANETs/WSNs.

There are two types of disjoint paths: *edge-disjoint* and *node-disjoint* (or *vertex-disjoint*). Despite the source and destination, node-disjoint paths have no node in common, and edge-disjoint paths do not share common edges. Clearly, node-disjoint paths are also edge-disjoint paths. From the reliability perspective, both the nodes and the wireless links are error-prone. The node failure could be caused by the physical node failure (e.g., physical damage or depletion of the battery) or the heavy congestion at the node which causes packet drop due to buffer overflow. The link failure could be caused by the breakage of the link due to the node's moving out of the transmission range, media access contention, multiuser interference, or any interference that causes the radio signal to not be correctly decoded at the intended receiver. Both node-disjoint paths and edge-disjoint paths help in terms of reliability. For some other applications, such as the security consideration which aims to deal with the compromised node problem, node-disjoint paths are desired.

3.2.1 Edge-Disjoint Paths

Diversity injection is one technique that has been proposed to find multiple disjoint paths between a single source–destination pair for on-demand routing protocols [35]. Typically, in a route discovery procedure of an on-demand routing protocol, route query (e.g., RREQ) messages are broadcast by every node in the network. An intermediate node only responds to the first received RREQ and simply discards the duplicated ones that come later. The diversity injection technique tries to reclaim the dropped information contained in the duplicate RREQs by recording in a temporary query cache the accumulated route information contained in all the received RREQs. Because RREQ is only forwarded once at each node, each RREQ received at a node travels a different path. By claiming the path information contained at each RREQ, a node learns a diversified route information back to the source. Then during the route reply phase, when the node receives a route reply packet, it checks its temporary query cache and selects a different reverse path for the reply packet. The selection of the reverse path favors the less "heard" path so that the route reply packets bring back more diversified path information to the source.

Split Multipath Routing (SMR) is another on-demand routing protocol aiming to build maximally disjoint multiple routes [28]. It is source-routing-based. Notice that the propagation of the RREQs essentially builds a tree structure rooted at the source. The dropping of the duplicate RREQs tends to have one RREQ dominating the path-finding process and when multiple route replies follow the reverse paths back to the source, they tend to converge when getting closer to the source. SMR modified the propagation rule of RREQs as follows. Instead of dropping every duplicate RREQ, intermediate nodes forward a duplicate packet if that RREQ packet comes from a link other than the link from which the first RREQ is received, and whose hop count is not larger than that of the first received RREQ. SMR disables route reply from intermediate nodes. Only the destination sends out route replies. The destination replies to the first received RREQ as it is the minimum delay path. It then waits for a certain duration of time to receive more RREQs and learns all possible routes and select the routes that are maximally disjoint to the route that has already been replied. The destination sends one reply to each selected path.

Diversity injection technique and split multipath routing accumulate the path information while propagating the route request messages. The disjointedness of the paths as well as the loop-freedom of each path is therefore not difficult to maintain. *Ad hoc On-demand Multipath Distance Vector Routing*

(AOMDV) is an on-demand multipath routing protocol that is based on distance vector routing and the multiple paths are computed distributively and independently at each hop [32]. AOMDV again makes use of the information in the duplicate RREQ messages. AOMDV propagates the RREQ messages the same way as the basic AODV: only the first received RREQ is further rebroadcast. For the duplicate RREQs, instead of simply ignoring them, AOMDV examines the path information contained in the message for potential alternate reverse path which preserves loop-freedom and link-disjointed-ness among other paths back to the source. For each new alternate path found, the intermediate node generates a RREP message and sends it back to the source along the reverse path if it knows a forward path that has not been used in any previous RREPs for this RREQ. The destination node replies to every RREQ it receives. To ensure the loop-freedom, AOMDV uses the *destination sequence numbers* the same way as AODV to indicate the freshness of the routes. In addition, AOMDV uses the notion of *advertised hop count* to maintain multiple paths for the same sequence number. The advertised hop count is set to the hop count of the longest path available at the time when a node first advertises a path for the destination. It is reset on each new sequence number and remains unchanged until the sequence number changes. A node forms an alternate path through a neighbor only if that neighbor has a smaller advertised hop count than the node itself. Another criterion for the loop-freedom guarantee is that, besides the next hop information, the route table contains the last hop information for each path. Paths have distinct first (i.e., next) hops as well as distinct last hops are disjoint.

3.2.2 Node-Disjoint Paths

Node-disjoint paths are of particular interest in many application scenarios because of the independence and resilience they provide. A number of node-disjoint path-finding algorithms have been proposed in the literature.

AODV-Multipath (AODVM) is one multipath routing protocol that aims to find node-disjoint paths [46]. It is based on AODV. The propagation of RREQs follows the same rule as the basic AODV except that the intermediate nodes are disallowed to send route replies back to the source. Although not further propagated, duplicate RREQs received at an intermediate node are processed for possible alternate paths back to the source. Every node maintains a *RREQ table* that keeps track of all the neighbors from which a RREQ is received and the corresponding cost (hop count) back to the source. When a RREQ reaches the destination, a RREP message is generated and sent back to the last hop node from which the destination received the RREQ. The RREP packet

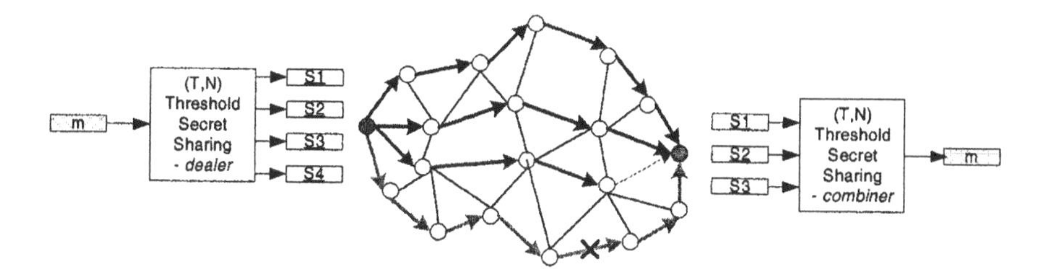

Figure 4: Basic idea of SPREAD.

contains an additional field *last_hop_ID* to indicate the last hop node (i.e., the neighbor of the destination). The RREP message may not follow the exact reverse path. Instead, an intermediate node determines which next hop node the RREP should be sent to based on the information saved in the RREQ table. When an intermediate node receives a RREP, it finds from its RREQ table a shortest path back to the source and sends the RREP to the corresponding next hop node. In AODVM, in order to ensure that a node does not participate in multiple paths, when a node overhears its neighbor's transmission of a RREP message, the node deletes the entry corresponding to that neighbor in its RREQ table so that it won't attempt to use that neighbor for another RREP. If an intermediate node when receiving a RREP message cannot forward it further (i.e., its RREQ table is already empty), it generates a RDER (Route Discovery ERror) message and sends it back to the node from which it receives the RREP. The neighbor, upon receiving the RDER message will try to forward the RREP to a different neighbor. The number of RDERs that a particular RREP can experience is limited so that unnecessary endless attempts can be avoided.

A node-disjoint multipath routing protocol has been described in [13] for WSNs. As we have mentioned in Section 3.1, with the basic directed diffusion scheme, a node is able to determine locally its most preferred neighbor in the direction of the intended sensor node. With a primary path reinforcement message, the sink is able to find the primary path to the node. In the proposed algorithm, once the primary path is settled, the sink sends an alternate path reinforcement in order to find a node-disjoint path. Different from the partially disjoint paths algorithm (i.e., braided multipath routing), where each node on the primary path initiates an alternate path reinforcement, in order to find node-disjoint paths, only the sink initiates the alternate path reinforcement to its next preferred neighbor. That neighbor further propagates the alternate path reinforcement to its most preferred neighbor in the direction of the intended sensor node. If the node that receives the alternate path reinforcement happens to be already on the primary path, it sends a *negative reinforcement* back to the previous node; the previous node then tries its next preferred neighbor. Otherwise the node continues the propagation of the alternate path reinforcement to its most preferred neighbor and so on. This mechanism allows each alternate path reinforcement sent by the sink to find a node-disjoint path between the sink and the intended sensor node. The mechanism can be extended to construct multiple node-disjoint paths by sending out multiple alternate path reinforcement messages, each separated from the next by a small delay.

The multipath routing protocols proposed for our SPREAD scheme are also node-disjoint and are presented in the following section.

4 The SPREAD Scheme

4.1 SPREAD Overview

The idea of SPREAD was first proposed in [22] and then was studied as a complementary mechanism to enhance the secure data delivery service in a MANET in [24] and for secure and reliable data collection service in WSNs in [26]. The basic idea and operation of SPREAD is illustrated in Figure 4. A secret message m is transformed into multiple shares, $S_1, S_2, \cdots$, by a secret sharing scheme, and then delivered to the destination via multiple independent paths. Due to the salient features of secret sharing and the distributed fashion of multipath delivery, the SPREAD has been shown to be more resistent to the node compromise/failure problem, namely, even if a small number of paths/nodes/shares are compromised/lost, the message as a whole is not compromised/lost.

A number of coding schemes can be used to split the traffic for multipath routing in order to enhance reliability. Examples include the well-known Reed-Solomon codes, diversity coding, multiple description coding, etc. In our SPREAD scheme, we used the threshold secret sharing scheme for its add-on security property. A (T, N) threshold secret sharing scheme could transform a secret into N pieces, called *shares* or *shadows*. The nice property of the N shares is that form any less than T shares one cannot learn anything about the secret, whereas with an effective algorithm, one can reconstruct the system secret from any T out of N shares. The generation of the shares is very simple: by evaluating a polynomial of degree $(T-1)$,

$$f(x) = (a_0 + a_1 x + \cdots + a_{T-1} x^{T-1}) \bmod p$$

at point $x = i$ to obtain the ith share:

$$S_i = f(i)$$

where $a_0, a_1, a_2, \ldots, a_{T-1}$ are secret bits and p is a large prime number greater than any of the coefficients and can be made public.

According to the fundamental theorem of algebra, T values of a polynomial of degree $(T - 1)$ can completely determine the polynomial (i.e., all its coefficients), whereas any fewer values cannot determine the polynomial (at least computationally difficult). Thus, any T shares can reconstruct the original secret bits, but any fewer shares cannot. Efficient $(O(T \log^2 T))$ algorithms have been developed for polynomial evaluation and interpolation [7]. Moreover, depending on the number of paths available, the (T, N) value in our SPREAD will not be large. Even the straightforward quadratic algorithms are fast enough for practical implementation.

A challenging job in any multipath routing approach is the efficient and effective multipath routing protocols. In [24] we proposed a multipath finding technique to find multiple node disjoint paths between a single source–destination pair. In fact, most of the current multipath routing protocols fall into this category. In response to the communication pattern in a WSN, we also proposed a novel N-to-1 multipath discovery protocol [26]. Instead of finding multiple paths between a specific source and a specific destination, the N-to-1 multipath discovery protocol takes advantage of the flooding in a typical route discovery procedure and is able to find multiple node-disjoint paths from every sensor node to the common destination (i.e., the sink node) simultaneously in one route discovery. In the rest of the section, we introduce the two routing protocols for SPREAD.

4.2 End-to-End Multipath Routing

Most of the proposed multipath routing protocols are on-demand and they work by broadcasting the route request messages throughout the network and then gathering the replies from the destination following slightly different rules (see Section 3). Although those routing protocols are able to find multiple node-disjoint paths, the path sets found directly through them are short in terms of number of paths and might not be optimal for a particular application as the path selection is usually based on the hop count or propagation delay, not necessarily the desired property such as security in our SPREAD scheme. We proposed a different approach in [24] to find multiple node-disjoint paths. Our approach has two major components. One is the "link cache" organization we studied in [23]. The link cache can take advantage of many multipath routing techniques, such as diversity injection, split multipath routing, etc., as mentioned in the previous section, to help collect diversified path information. The other component is the multipath-finding algorithm that is used to find the maximal number of node-disjoint paths with the desired property. Once the paths are selected, our multipath routing protocol depends on the source routing mechanism to route the packets along the designated multiple paths.

In DSR and other DSR-like on-demand routing protocols, the route replies back to the source contain the complete node list from the source to the destination. By caching each of these paths separately, a "path cache" organization can be formed. This type of cache organization has been widely used in the proposed protocols. Instead of using the returned paths directly, we adopted a "link cache" organization in [23] where each path returned to the source is decomposed into individual links and represented in a unified graph data structure. We also proposed an adaptive link lifetime prediction and stale link removal scheme to work with the link cache. A link cache organization provides the source node a partial view of the network topology, similar to a link-state type of routing protocol. Using such a link cache organization allows us to further optimize the multipath selection based on other cost metrics. Typically by reorganizing the links, more disjoint paths can be found and paths can be selected according to some desired property, such as security or energy consumption, besides propagation delay or hop count. In addition, with a link cache, although we rely on an underlying routing protocol to provide a node with a partial view of network topology, the optimization of the path set can be done solely based on the discovered partial network topology, which is independent of the underlying routing protocols.

For the optimal selection of paths, we proposed a security-related link cost function such that paths can be selected according to their security properties (i.e., the probability that the path might be compromised). Assume that a node, say, n_i, is compromised with probability q_i (a number determined from certain measurements, say, from an intrusion detection device). Assume that the overhearing does not resulting in message compromise (e.g., by link-layer encryption or directional antenna); then the probability that a path from a source s to a destination t, consisting of intermediate nodes, $n_1, n_2, \ldots, n_l$, is compromised, is given by

$$p = 1 - (1 - q_1)(1 - q_2) \cdots (1 - q_l).$$

We define the link cost between n_i and n_j as

$$c_{ij} = - \log \sqrt{(1 - q_i)(1 - q_j)};$$

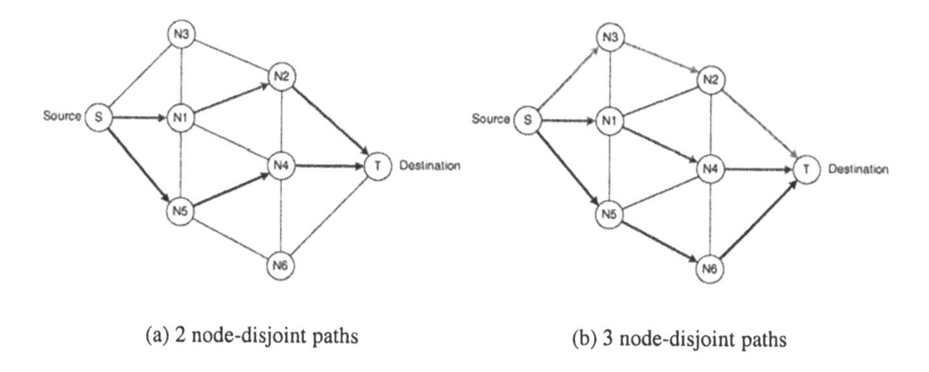

(a) 2 node-disjoint paths

(b) 3 node-disjoint paths

Figure 5: Maximal path finding algorithm.

then we have the path cost

$$\sum c_{ij} = -\log(1 - q_1)(1 - q_2) \cdots (1 - q_l) = -\log(1 - p),$$

and hence, minimizing the path cost with link metrics c_{ij} is equivalent to minimizing the path compromise probability p. If we use this link metric, then we can find multiple paths with a certain bound for path compromise probability.

The next component of our end-to-end SPREAD routing protocol is a maximal node-disjoint path finding algorithm to discover the maximal number of secure paths. The maximal path finding algorithm is an iterative procedure. The most secure path is found first and added to the path set. In each iteration, the number of paths in the set is augmented by one. Details of the algorithm can be found in [24]. Basically each time a new path is added to the selected path set, a graph transformation is performed, which involves a vertex splitting of the nodes on the selected paths (except the source and destination node). Then, the modified Dijsktra algorithm [4] is executed to find the most secure path in the transformed graph. Then, the split nodes are transformed back to the original one, any interlacing edges are erased, and the remaining edges are grouped to form the new path set. This approach has the advantage that it can find maximal node-disjoint paths, which is solely determined by the topology. The order of the selected path does not affect the final number of paths found. As an example shown in Figure 5, after finding the first two node-disjoint paths, the algorithm is able to regroup the links and form a path set consisting of three paths instead of two.

4.3 N-to-1 Multipath Routing

Although end-to-end connections are the most common communication pattern in many networks, a typical task of a WSN is data collection where the base station broadcasts the request for the data of interest and every sensor node (or nodes that have the data of interest) sends its readings back to the base station. Therefore, a routing protocol that is able to efficiently disseminate information from the sink node to the many sensor nodes and find paths from each of the many sensor nodes back to the common sink becomes more desirable in a WSN. For this purpose, Berkeley's TinyOS sensor platform utilizes a flooding-based beaconing protocol [20]. The base station periodically broadcasts a route update. Each sensor node when receiving the update for the first time rebroadcasts the update and marks the node from which it receives the update as its parent. The algorithm continues recursively till every node in the network has rebroadcast the update once and finds its parent. What follows is that all the nodes forward the packets they received or generated to the parent until the packets reach the base station.

As illustrated in Figure 6(a), the beaconing protocol essentially constructs a breadth-first spanning tree rooted at a base station. It finds every sensor node a single path back to the base station efficiently, however, reliability and also security suffer from the single-path routing. The failure of a single node or link will disrupt the data flow from the node itself and all its children. Similarly, the compromise of a single node will cause the information leakage from the node and all its children. In [26] we extend our SPREAD idea into the data collection service in WSNs and present a multipath discovery algorithm that is able to find multiple node-disjoint paths from every sensor node to the base station efficiently. Then each sensor node can follow the SPREAD idea: splitting the data into multiple shares and routing them to the sink node using the multiple node-disjoint paths.

The proposed multipath discovery algorithm consists of two phases. The mechanism used in phase one, termed *branch aware flooding*, takes advantage of the simple flooding technique. Without introducing additional routing messages, the mechanism is able to find a certain number of node-disjoint paths, depending on the density of the network topology.

The general idea of the branch aware flooding is as follows. A simple flooding such as the beaconing protocol essentially constructs a breadth-first spanning tree rooted at a base station. The route update initialized at the base station is first propagated to the immediate neighbors of the base station, e.g., nodes a, b, c, and d as shown in Figure 6(b). Then from each of the immediate neighbors, it will be further propagated and each forms a branch of the tree. The

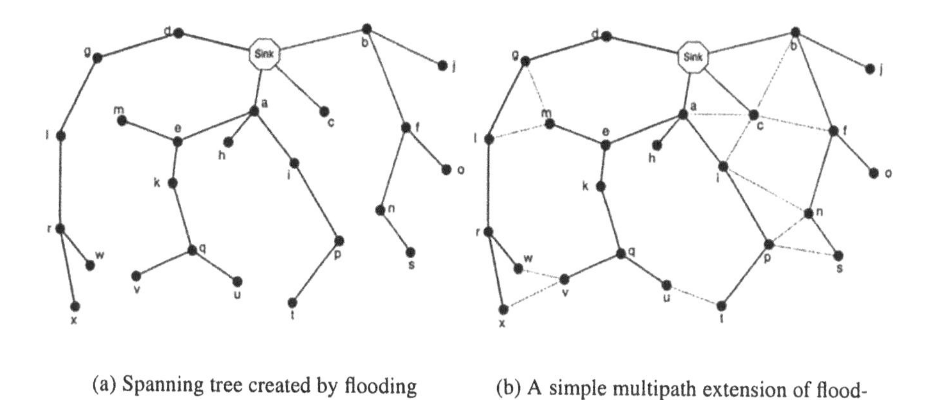

(a) Spanning tree created by flooding

(b) A simple multipath extension of flooding

Figure 6: N-to-1 multipath routing.

number of branches the tree has depends on the number of immediate neighbors the base station has (e.g., four branches in the example where different branches are distinguished by different colors). The maximum number of node-disjoint paths from any node to the base station is thus bounded by the number of the immediate neighbors of the base station. Notice that although each node has a *primary* (in most cases also the shortest) path to the base station by following its tree links up, a link between two nodes that belong to two different branches provides each node an alternate disjoint path to the base station through the other. For example, as shown in Figure 6(b), although node w has the primary path $(w - r - l - g - d - Sink)$ back to the base station, it learns another alternate path $(w - v - q - k - e - a - Sink)$ from node v which is not in the same branch as w when w overhears v's broadcast.

Because each node needs to broadcast the route update message once anyway, the branch aware flooding is able to find a certain number of node-disjoint paths without any additional routing messages. However, the limitation of this method is that it only finds disjoint paths at the nodes where there are direct links to other branches. In the same example, notice that if w further propagates the disjoint paths it learned to its parent or siblings/cousins (but not necessarily the children), its parent or siblings/cousins might learn a new disjoint path as well. For example, node r has the primary path $(r - l - g - d - Sink)$. If it hears a disjoint path $(w - v - q - k - e - a - Sink)$ from w and it does not yet know a path through branch a, it learns a new disjoint path $(r - w - v - q - k - e - a - Sink)$. Therefore, in order to maximize the number of disjoint paths each node may have, a second phase of the multipath routing algorithm is designed which is to further propagate the disjoint paths found in the first phase. The tradeoff of the second phase is that it finds more disjoint paths with additional routing messages. Details of the N-to-1 multipath routing protocol can be found in [26].

4.4 SPREAD Summary

Multipath routing has been a promising technique in MANETs and WSNs to deal with unreliable data communications. It is also a feasible technique due to the dense deployment of nodes in the MANETs/WSNs. SPREAD is an innovative scheme that combines multipath routing and secret sharing techniques to address both reliability and security issues. The end-to-end SPREAD scheme adopts concurrent multipath routing and provides more secure data transmission when messages are transmitted across the insecure network. A certain amount of redundancy can be added without affecting security through the optimal share allocations onto each selected path (see [24] for details on optimal share allocation schemes). The N-to-1 SPREAD scheme is distinguished from all previous work in that the multipath discovery protocol is receiver-initiated (in contrast to the common source-initiated route discovery) and the protocol is efficient in that it finds multiple paths from every sensor node to the base station simultaneously in one route discovery procedure, which fits the special communication pattern (i.e., multiple senders to a single receiver) in the WSN very well. For data delivery, the N-to-1 SPREAD adopts concurrent multipath routing which spreads traffic onto multiple disjoint paths simultaneously between the sensor node and the sink. In addition, taking advantage of the multiple paths available at each node, it also adopts the per-hop multipath packet salvaging technique which uses the multipath alternately and helps to improve the reliability of each packet delivery/path significantly. By the combination of the two, the overall scheme improves both security and reliability with no or very little information redundancy.

A few remarks are necessary here. First, the SPREAD scheme considers security and reliability when messages are transmitted across the network, assuming the source and destination are trusted. It improves the security and reliability of end-to-end message delivery in the sense that it is resilient to a certain number of compromised/faulty nodes but it does not improve the security of each individual node. Secondly, the SPREAD scheme cannot address security, i.e., confidentiality, alone; it only statistically enhances such service. For example, it is still possible for adversaries to compromise all the shares, e.g., by collusion. Finally, SPREAD can be made adaptive in the sense that the source node could make the final decision as to whether a message is delivered at a certain time instance according to the security level and the availability of multiple paths. Moreover, the chosen set of multiple paths may be changed from time to time to avoid any potential capture of those multiple shares by adversaries.

5 Conclusion

Multipath routing has been a promising technique in MANETs and WSNs. It has been shown through both theoretical analysis and simulation results that multipath routing provides many performance benefits, including improved fault tolerance, security, and reliability, improved routing efficiency and reduced routing overhead, more balanced traffic load and energy consumption, reduced end-to-end latency, and aggregated network bandwidth, etc. Significant research efforts have been made and are continuously being made in developing multipath routing protocols and multipath packet forwarding techniques in order to achieve the above-mentioned performance gains effectively and efficiently. Nevertheless, many issues that are directly related to the application of multipath routing remain untouched, such as the integration of the multipath routing into the current single-path routing paradigm, the synchronization of the packets among the multiple paths, and the interfaces of multipath routing protocols to other layers of protocol in the network protocol stack, etc.

Due to space limitations, we are only able to introduce the basic concept of multipath routing, highlight the fundamental techniques used to find the multiple paths, and outline the essential idea of what and why it can help in performance. For detailed algorithms/protocols as well as performance evaluations, interested readers are referred to respective publications.

References

[1] E. Ayanoglu, I. Chih-Lin, R. D. Gitlin, J. E. Mazo, "Diversity coding for transparent self-healing and fault-tolerant communication networks," *IEEE Transaction on Communications*, 41(11):1677–1686, November 1993.

[2] A. F. Akyildiz, W. Su, Y. Sankarasubramainiam, E. Cayirci, "A survey on sensor networks," *IEEE Communications Magazine*, August 2002.

[3] A. Banerjea, "On the use of dispersity routing for fault tolerant realtime channels," *European Transactions on Telecommunications*, 8(4):393–407, July/August 1997.

[4] R. Bhandari, Survivable Networks: Algorithms for diverse routing, *Kluwer Academic*, 1999.

[5] J. Broch, D. Maltz, D. Johnson, Y-C. Hu, J. Jetcheva, "A performance comparison of multi-hop wireless ad hoc network routing protocol," *The 4th Annual ACM/IEEE International Conference on Mobile Computing and Networking (MobiCom'98)*, pp. 85–97, Dallas, TX, October 1998.

[6] J. Chen, "New Approaches to Routing for Large-Scale Data Networks," Ph.D Dissertation, Rice University, 1999.

[7] T. Cormen, C. Leiserson, R. Rivest, *Introduction to Algorithms*, MIT Press, 1990.

[8] H. Chan, A. Perrig, D. Song, "Random key predistribution schemes for sensor networks," *IEEE Symposium on Security and Privacy (SP'03)*, Oakland, CA, May 2003.

[9] I. Cidon, R. Rom, Y. Shavitt, "Analysis of multi-path routing," *IEEE/ACM Transactions on Networking*, 7(6):885–896, December 1999.

[10] J. Chang, L. Tassiulas, "Energy conserving routing in wireless ad hoc networks," *Proceedings INFOCOM 2000*, March 2000.

[11] S.K. Das, A. Mukherjee, et al., "An adaptive framework for QoS routing through multiple paths in ad hoc wireless newtorks," *J. Parallel Distributed Computing*, 63(2003)141–153.

[12] S. De, C. Qiao, H. Wu, "Meshed multipath routing: An efficient strategy in sensor networks," *IEEE Wireless Communications and Networking Conference (WCNC'03)*, New Orleans, LA, March 2003.

[13] D. Ganesan, R. Govindan, S. Shenker, D. Estrin, "Highly-resilient, energy-efficient multipath routing in wireless sensor networks," *Mobile Computing and Communication Review*, 5(4):10–24, October 2001.

[14] E. Gustafsson, G. Karlsson, "A literature survey on traffic dispersion," *IEEE Networks*, 11(2):28–36, March/April 1997.

[15] N. Gogate, S. S. Panwar, "Supporting video/image applications in a mobile multihop radio environment using route diversity," *IEEE International Conference on Communications (ICC'99)*, Vancouver, Canada, June 1999.

[16] C. Intanagonwiwat, R. Govindan, D. Estrin, J. Heidemann, "Directed diffusion for wireless sensor networks," *IEEE/ACM Transactions on Networking*, 11(1):2–16, February 2003.

[17] D. B. Johnson, D. A. Maltz, Y-C. Hu, J. G. Jetcheva, "The dynamic source routing protocol for mobile ad hoc networks," IETF Internet Draft, draft-ietf-manet-dsr-06.txt, November 2001.

[18] P. Johansson, T. Larsson, N. Hedman, B. Mielczarek, M. Degermark, "Scenario-based performance analysis of routing protocols for mobile ad hoc networks," *The 5th Annual ACM/IEEE International Conference on Mobile Computing and Networking (MobiCom'99)*, pp.195–206, Seattle, WA, August 1999.

[19] S. Jain, Yi Lv, S. R. Das, "Exploiting path diversity in the link layer in wireless ad hoc networks," Technical Report, SUNY at Stony Brook, CS department, WINGS Lab, July 2003.

[20] C. Karlof, D. Wagner, "Secure routing in wireless sensor networks: Attacks and countermeasures," *Elsevier's AdHoc Networks Journal, Special Issue on Sensor Network Applications and Protocols*, 1(2–3):293–315, September 2003.

[21] J. Kong, P. Zerfos, H. Luo, S. Lu, L. Zhang, "Providing robust and ubiquitous security support for MANET," *Proceedings of the 9th IEEE International Conference on Network Protocols(ICNP)*, pp. 251–260, 2001.

[22] W. Lou, Y. Fang, "A multipath routing approach for secure data delivery," *IEEE Military Communications Conference (MILCOM 2001)*, Mclean, VA, October 2001.

[23] W. Lou, Y. Fang, "Predictive caching strategy for on-demand routing protocols in ad hoc networks," *Wireless Networks*, vol. 8, issue 6, pp. 671–679, November 2002.

[24] W. Lou, W. Liu, Y. Fang, "SPREAD: Enhancing data confidentiality in mobile ad hoc networks," *IEEE INFOCOM 2004*, HongKong, March 2004.

[25] W. Liu, Y. Zhang, W. Lou, Y. Fang, "Scalable and robust data dissemination in wireless sensor networks," *IEEE GLOBECOM 2004*, Dallas, TX, December 2004.

[26] W. Lou, "An efficient N-to-1 multipath routing protocol in wireless sensor networks," *2nd IEEE International Conference on Mobile Ad-hoc and Sensor Systems (MASS 2005)*, Washington DC, November 2005.

[27] S.-J. Lee, M. Gerla, "AODV-BR: backup routing in ad hoc networks," *IEEE Wireless Communications and Networking Conference (WCNC'00)*, September 2000.

[28] S.-J. Lee, M. Gerla, "Split multipath routing with maximally disjoint paths in ad hoc networks", *International Conference on Communications (ICC'01)*, Helsinki, Finland, June 2001.

[29] N. F. Maxemchuk, "Dispersity routing," *International Conference on Communications (ICC '75)*, pp. 41.10–41.13, San Francisco, CA, June 1975.

[30] N. F. Maxemchuk, "Dispersity routing in high speed networks," *Computer Networks and ISDN Systems*, 25(6):645–661, 1993.

[31] N. F. Maxemchuk, "Dispersity routing on ATM networks," *IEEE INFOCOM'93*, vol.1, pp.347–57, San Francisco, CA, Mar 1993.

[32] M. K. Marina, S. R. Das, "On-demand multipath distance vector routing in ad hoc networks," *9th International Conference on Network Protocols*, Riverside, CA, November, 2001.

[33] A. Nasipuri, R. Castaneda, S. R. Das, "Performance of multipath routing for on-demand protocols in mobile ad hoc networks," *Mobile Networks and Applications*, 6(4):339–349, 2001.

[34] V. D. Park, M. S. Corson, "A highly adaptive distributed routing algorithm for mobile wireless networks," *IEEE INFOCOM'97*, pp. 1405–1413, Kobe, Japan, April 1997.

[35] M. R. Pearlman, Z. J. Haas, "Improving the performance of query-based routing protocols through diversity injection," *IEEE Wireless Communications and Networking Conference (WCNC'99)*, New Orleans, LA, September 1999.

[36] M.R. Pearlman, Z.J. Haas, P. Sholander, S. S. Tabrizi, "On the impact of alternate path routing for load balancing in mobile ad hoc networks," *The ACM Symposium on Mobile Ad Hoc Networking and Computing (MobiHOC'00)*, Boston, MA, August 2000.

[37] P. Papadimitratos, Z.J. Haas, E. G. Sirer, "Path set selection in mobile ad hoc networks," *The ACM Symposium on Mobile Ad Hoc Networking and Computing (MobiHoc'2002)*, EPFL Lausanne, Switzerland, June 2002.

[38] E. M. Royer, C-K Toh, "A review of current routing protocols for ad hoc mobile wireless networks," *IEEE Personal Communications*, 6(2):46–55, April 1999.

[39] D. Sidhu, S. Abdallah, R. Nair, "Congestion control in high speed networks via alternative path routing," *Journal of High Speed Networks*, 2(2):129–144, 1992.

[40] N. Taft-Plotkin, B. Bellur, R. Ogier, "Quality-of-Service Routing Using Maximally Disjoint Paths," *1999 Seventh International Workshop on Quality of Service*, London, UK, June 1999.

[41] A. Tsirigos, Z. J. Haas, "Multipath routing in the presence of frequent topological changes," *IEEE Communication Magazine*, 39(11):132-138, November 2001.

[42] A. Shamir, "How to share a secret," *Communications of the ACM*, 22(11):612–613, November 1979.

[43] G. J. Simmons, "An introduction to shared secret and/or shared control schemes and the application," *Contemporary Cryptology: The Science of Information Integrity*, IEEE Press, pp. 441–497, 1992.

[44] S. Singh, M. Woo, C.S. Raghavendra, "Power aware routing in mobile ad hoc networks," *Proceedings of the Fourth Annual ACM/IEEE International Conference on Mobile Computing and Networking (MobiCom '98)*, 1998.

[45] K. Wu, J. Harms, "Performance study of a multipath routing method for wireless mobile ad hoc networks," *9th International Symposium on Modeling, Analysis and Simulation of Computer and Telecommunication System (MASCOTS'01)*, Cincinnati, Ohio, August 2001.

[46] Z. Ye, S. V. Krishnamurthy, S. K. Tripathi, "A framework for reliable routing in mobile ad hoc networks," *IEEE INFOCOM 2003*, San Francisco, CA, March 2003.

[47] J. Yang, S. Papavassiliou, "Improving network security by multipath traffic dispersion," *IEEE Military Communications Conference (Milcom'01)*, McLean, VA, October 2001.

[48] L. Zhou, Z. J. Haas, "Securing ad hoc networks," *IEEE Network Magazine*, 13(6):24–30, November/December 1999.

[49] S. Zhu, S. Xu, S. Setia, S. Jajodia, "Establishing pairwise keys for secure communication in ad hoc networks: A probabilistic approach," *11th IEEE International Conference on Network Protocols (ICNP'03)*, Atlanta, GA, November 2003.

Chapter 5

Ad Hoc Networks: Optimization Problems and Solution Methods

Carlos A.S. Oliveira
School of Industrial Engineering and Management
Oklahoma State University, Stillwater, OK 74078
E-mail: coliv@okstate.edu

Panos M. Pardalos
Department of Industrial and Systems Engineering
University of Florida, Gainesville, FL 32611
E-mail: pardalos@ufl.edu

1 Introduction

Ad hoc networks have become a very active topic of research in the last years, due to the recent development of improved wireless systems and protocols. An increasing number of devices such as PDAs, mobile phones, and laptops are connected to wireless systems, forming ad hoc networks. In an ad hoc network, nodes act as clients and servers, and no fixed topology is enforced. The flexibility of ad hoc networks also results in added complexity in areas such as routing, reliability, and security, for example. Many of the related design goals can be posed as combinatorial optimization problems and therefore would benefit from a detailed study of their combinatorial structure.

Ad hoc networks have many applications. These include traditional domains where the lack of structure is well recognized, such as in military and rescue operations. However, new opportunities for applications of ad hoc networks have been made available by the introduction of new devices that can be interconnected using a wireless medium. Examples are the new generation of equipment capable of using wireless connection protocols, such as bluetooth, embedded into new laptops, PDAs, cellular phones, etc. In this type of urban application, devices may be able to share resources and information without the necessity of a previous network infrastructure [15].

Such wireless networks have become ubiquitous in the last few years, and sophisticated techniques need to be developed for the best use of their resources. Combinatorial problems arise in different areas such as message delivering, power control, resource management and discovery, reliability, and position measurement. All these problems can benefit from a better understanding of their combinatorial properties, leading to the development of new methods for computing optimal or near-optimal solutions.

Ad hoc networks are also viewed nowadays as an essential infrastructure for the development of sensor network systems [37]. In such systems, a large number of sensors are integrated wirelessly and used to collect information about a resource, activity, or event. Sensors are a key technology for enabling better iteration with physical and computational resources without human intervention, and have been the target of much research effort in the last few years [5, 37, 45].

Algorithms for ad hoc networks have stringent requirements compared to other distributed systems. First, they must be fast and adapted to devices with a small amount of computational power. This is an important requirement, because nodes in a sensor network must be as cheap as possible (for example, systems of one cubic millimeter in size are currently being developed with this objective [45]). Even in more conventional systems, power and computational resources must be conserved. Second, ad hoc networks must explore locality among neighboring elements [30]. This becomes necessary, because it is difficult to execute tasks where the majority of the network must be contacted due to the spatial mobility of nodes.

In this chapter we survey some of the most important optimization problems in ad hoc networks. We aim to provide information specifically about techniques used to solve some of these problems, including mathematical programming formulations, heuristics, and combinatorial and approximation algorithms. In Section 2 we discuss the routing problem in ad hoc networks. This problem has several aspects that can be attacked using optimization, and therefore different techniques are analyzed. In Section 3 we present problems occurring in power control. Section 4 is dedicated to methods for minimization of errors in position measurements. Finally, concluding remarks are presented in Section 5.

2 Routing

Routing is the essential task of transferring packets of information from their sources to destinations. Traditional systems handle this task by maintaining tables with information about all possible routes for each packet. This approach allows the utilization of efficient (polynomial time) routing algorithms [6, 13], but it becomes too expensive for ad hoc networks [22].

In an ad hoc network, routing is complicated by the lack of topological structure. Any node can connect or disconnect from the network at any time. Therefore, issues such as connectedness, error recovery, and survivability, become harder to solve in this scenario [29].

Route discovery and maintenance are two of the main problems faced in the implementation of an ad hoc routing protocol [22]. The discovery phase consists of finding nodes that can be reached from the current position. A routing protocol for this problem must be able to contact surrounding nodes and probe them for information about the remaining parts of the network. The maintenance phase of the process is dedicated to keeping updated information about routes in the event of changes of position, disconnection, and connection of new devices. For more information about this topic, Ramanathan and Steenstrup [38] provided a survey of protocols where routing in ad hoc networks is discussed in terms of their advantages and disadvantages. The main characteristics of the discussed routing protocols were divided according to features such as location tracking, route selection, and forwarding.

We consider in this section the problem of gathering and updating routing information. This problem has been considered using several protocols [26, 43, 31]. Some other approaches for this problem, such as the construction of a virtual backbone, explore the combinatorial structure of the resulting network. This and other techniques are described in the remainder of this section.

2.1 Using a Virtual Backbone to Improve Routing

The simplest way of updating routing information in an ad hoc network is to send data about the neighborhood of nodes through all available links. This simple technique is called *flooding*. The drawbacks of flooding are the excessive amount of data sent through the network, therefore degrading its available bandwidth, the lack of guarantee that the packet will reach a certain node, etc. [21, 33]. To avoid some of these problems, other alternatives for keeping routing information have been proposed, including the creation of a virtual backbone.

A *virtual backbone* is a subset of a network used to connect all client nodes. This is a concept frequently used in wired network systems to simplify and improve the efficiency of routing. In this case, defining the structure of a suitable backbone is one of the subproblems that must be solved with the objective of providing optimal routing between clients.

For ad hoc networks, the concept of *virtual backbone* was proposed by Das and Bharghavan [12] as a way of reducing the amount of information that must be kept by routing algorithms, because only a small subset of the total network must be maintained as a backbone. The idea of using a virtual backbone for ad hoc systems had been proposed earlier by Ephremides et al. [14] and by Gerla and Tsai [17], but only with Das and Bharghavan [12] was the objective of optimizing the size of the backbone explicitly studied. The importance of this goal for ad hoc systems is clear, because a small backbone is easier to maintain and update in case of position changes.

Das and Bharghavan [12] proposed a distributed algorithm to solve the routing problem, based on a virtual backbone technique and using minimum connected dominating sets (MCDS). The general routing algorithm is the following.

Algorithm Route1:

1. Find a backbone B, using an algorithm for MCDS.
2. For each node $v \notin B$, collect information about its neighborhood.
3. Broadcast the topology information to all nodes in B.
4. Determine routes in B using some shortest path algorithm.
5. For each node $v \notin B$, send routing information to v.
6. Update routing information periodically.

The main objective of algorithm *Route1* is to reduce the number of nodes for which routing information must be stored. This is achieved by initially computing a virtual backbone B (the algorithm required for this

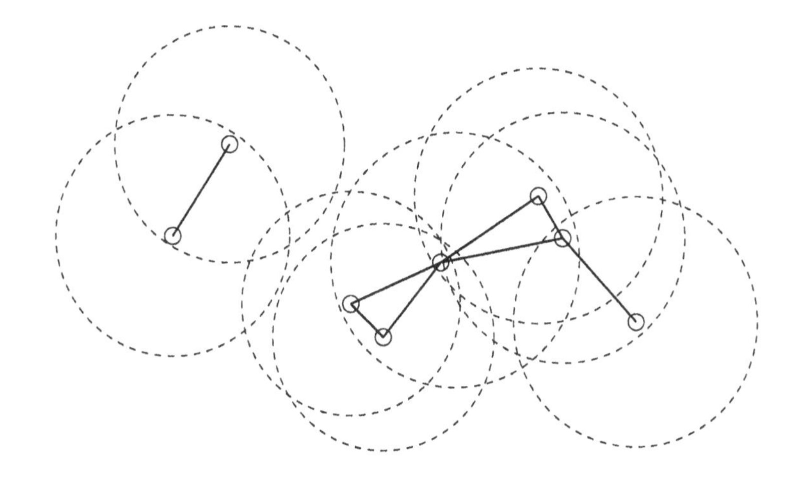

Figure 1: An example of unit graph.

computation is discussed in the remainder of this section). The nodes in the backbone are then capable of directly collecting information about connectedness for neighbor nodes not in B, and sending this information to other backbone nodes. In this way, information about the complete network is updated, and the number of necessary message exchanges is reduced. The following result can be proved for the algorithm above [12].

Theorem 1 *Algorithm* Route1 *computes routes among all nodes in the given network in time $O(\Delta(n + |C| + \Delta))$, where Δ is the maximum number of neighbors for any node, and $|C|$ is the size of the connected dominating set. The number of required messages is $O(n^2/\Delta + n|C| + m + n \log n)$.*

2.2 Unit Disk Graphs

A *unit disk graph* is a graph $G = (V, E)$ with a distance function $d : E \to \mathbb{R}$ and the following property: there is a link between nodes i and j if and only if the distance $d(i, j)$ is at most one unit. Unit disk graphs are frequently applied as a model for ad hoc networks. In this case, the radius of the region covered by the signal of each mobile unit is scaled to be equal to one, and therefore the unit disk graph represents the set of communication links in the network.

Unit disk graphs have many useful geometric properties. In fact, any property of Euclidean graphs is valid for unit disk graphs, because the former is a generalization of the latter. For example, it is known that planar graphs have a structure that allows easier approximation, even for problems NP-hard to approximate in general, such as the traveling salesman problem [3]. General approximation algorithms for problems in planar graphs exist for problems such as maximum independent set, partition into triangles, minimum vertex cover, and minimum dominating set. Baker [4] has shown that for these problems there are polynomial algorithms with constant approximation guarantee. In fact, in most cases the approximation algorithms run in linear time.

One of the useful properties of unit graphs is the relationship between sizes of their independent sets and dominating sets. This property is based on the trade-off between the number of nodes that can be located in a specific region and the number of links that will result from this co-location. A basic lemma proved by Alzoubi et al. [2] and used in the analysis of many approximation algorithms for the connected dominating set problem is the following.

Lemma 2 *Let I be a maximal independent set on a unit disk graph G, and C an optimal connected dominating set on G. Then, $|I| \leq 4|C| + 1$.*

2.3 Minimum Size Virtual Backbone Problem

The general problem of defining a minimum size virtual backbone for ad hoc networks can be modeled as a minimum connected dominating set (MCDS), which is a modified version of the dominating set (DS) problem. In the DS problem, given a graph $G = (V, E)$ with n nodes and m edges, the objective is to compute a set $B \subseteq V$ with the following property: for all $v \in V$, either $v \in B$ or there is a node $w \in B$ such that $(v, w) \in E$ (i.e., v is connected to w). The MCDS problem requires the computation of a minimum-sized DS that is connected; i.e., there is a path between each pair of nodes in B. Solving the minimum DS problem, as well as the MCDS, is known to be NP-hard [16].

Concerning strategies for approximation, some researchers explored the planarity of unit disk graphs to provide algorithms with quality guarantee. For example, Guha and Khuller [18] proposed two approximation algorithms for the MCDS problem. The first algorithm had approximation guarantee $H(\Delta)$, where H is the harmonic function and Δ is the maximum number of neighbors for any node in the network (also known as the maximum degree of G). The second algorithm had an approximation guarantee of $\log \Delta + 3$.

Das and Bharghavan [12] introduced the idea of a minimum size virtual backbone for ad hoc networks, and proposed the use of connected dominating sets to compute such virtual backbones. They discussed two distributed algorithms for the problem, based on existing techniques for CDS computation. In the first algorithm, a dominating set is initially found, and a minimum spanning tree algorithm is used to connect the nodes into a connected dominating set.

The first algorithm (DB1) of Das and Bharghavan [12] can be divided into two stages. In the first stage, the objective is to mark some nodes in the network that will become part of the dominating set. They are selected according to their *effective degree*, i.e., the number of neighbor nodes that have not been previously marked. The *neighborhood in two-hops* $N_2(v)$ is the set of neighbors of a node v together with their own neighbors. If node u has effective degree greater than all elements in $N_2(u)$, then it is marked as an element of the dominating set (ties can be broken by looking at node identifiers). The second step consists of linking the nodes selected in the first step (call this set S), and forming a spanning tree. Edges receive a weight depending on the number of nodes not in S they are linking (1 or 2). Preference is given to edges with smaller costs. The following result summarizes the approximation guarantee of the algorithm [12].

Theorem 3 *Algorithm DB1 is an $H(n)$ approximation algorithm for the minimum connected dominating set problem, where $H(n) = \sum_{i=1}^{\Delta} 1/i$.*

The second algorithm proposed by Das and Bharghavan [12] (DB2) is similar, but this time connectedness is not guaranteed by using a spanning tree algorithm, but rather by growing the dominating set found until a connected set is achieved. A connected component of the current dominating set is called a *fragment*. An *extension* of the fragment is a path with one or two edges, where one or two nodes (respectively) are not in the current fragment. The *effective combined degree* is the number of neighbors of nodes in the extension (counted only once). The algorithm proceeds by adding at each step the extension with the largest effective combined degree. This gives the following result [12].

Theorem 4 *Algorithm DB2 is a $2H(\Delta)$ approximation algorithm for the minimum connected dominating set problem.*

Wu and Li [47] also proposed a distributed algorithm for computing a virtual backbone for ad hoc networks (we call it WL). Their algorithm was designed to be simple, even though it does not provide an approximation guarantee, such as other algorithms for this problem. The basic strategy is to apply two phases, where initially a connected dominating set is found using only local information about the network. In this way, the amount of messages necessary to determine the solution will be reduced, and therefore the algorithm will be more efficient. To do this, the algorithm simply exchanges degree information with all its neighbors. Nodes will decide if they need to be included in the dominating set according to simple comparisons with the connectivity information returned by their neighbors. Then, the number of nodes in the resulting connected dominating set is decreased by the application of rules that guarantee connectedness and dominance.

In the second phase of the WL algorithm, the dominating set is improved by using information about the shortest paths between pairs of nodes. For the final solution returned by the algorithm, it can be shown that all nodes in the resulting connected dominating set appear in the shortest path between some pair of nodes. This property can be used to prove the following result [47].

Theorem 5 *Algorithm WL has total time complexity $O(\Delta^2)$, and message complexity $O(n\Delta)$, with message length $O(\Delta)$.*

2.4 Improved Approximation for the MCDS

Improved approximation algorithms for computing the MCDS were proposed by Cheng and Du [11]. Their work is an extension of previous algorithms proposed by Das and Bharghavan [12], applying the approximation bound proved by Alzoubi et al. [2], and the algorithms of Wu et al. [46], and Wu and Li [47]. Two algorithms are discussed, the first one returning solutions with emphasis on the cost requirements, and the second with greater emphasis on degree constraints of the problem.

The first algorithm by Cheng and Du [11] (which we call CD1) uses colors to define nodes according to different phases of the optimization process. The distributed algorithm starts by setting all nodes to receive color white. Nodes are colored with gray whenever they change to a *dominatee* status. The third possible status represents the fact that a node has a white color but has at least one neighbor colored gray. Finally, the nodes set to the *dominator* state have color black, and this represents that they will be part of the final connected dominating set. Changes between the four states described are allowed by a set of rules described

Ref.	Approximation	Message	Time	Msg. len.
[12]-1	$(2\log\Delta+3)OPT$	$O(n\|C\|+m+\log n)$	$O((n+\|C\|)\Delta)$	$O(\Delta)$
[12]-2	$(2\log\Delta+2)OPT$	$O(n\|C\|)$	$O(\|C\|(\Delta+\|C\|))$	$O(\Delta)$
[47]	N/A	$O(n\Delta)$	$O(\Delta^2)$	$O(\Delta)$
[2]	$8OPT+1$	$O(n\log n)$	$O(n\Delta)$	$O(1)$
[11]-1	$8OPT+1$	$O(n)$	$O(n\Delta)$	$O(1)$
[11]-2	$8OPT$	$O(n)$	$O(n\Delta)$	$O(1)$

Table 1: Comparison among distributed algorithms for computing a virtual backbone (from [11])

in the algorithm. The objective of the algorithm is, starting from some node of the network, to grow a tree made of nodes that reaches the fourth of the described states. The following result summarizes the properties of the algorithm [11].

Theorem 6 *Algorithm CD1 has linear message complexity, time complexity $O(n\Delta)$, where Δ is the maximum degree in the graph, and gives a solution with approximation guarantee $8OPT+1$, where OPT is the size of an optimal MCDS.*

The second algorithm by Cheng and Du [11] (we call it CD2) is a two-phase strategy for the MCDS. It also employs more explicitly the idea of Lemma 2 in order to find an approximate solution to the MCDS. In the first step of the algorithm, a set of messages is used to select nodes forming a maximal independent set I, using an election process similar to the one used in the first algorithm. Then, using the property established in Lemma 2, the second phase of the distributed algorithm connects the nodes in the independent set I in order to build a connected dominating set. The algorithm will ultimately find a set of paths in the network, from the elements in I to each of the remaining nodes. The union of all these paths will compose the final MCDS. It is possible to prove the following result [11].

Theorem 7 *The set S generated by algorithm CD2 is a connected dominating set, and its size is such that $|S| \leq 8OPT$, where OPT is the size of the optimum solution.*

A list of distributed approximation algorithms with their guarantees and respective time complexities is given in Table 1.

2.5 One-Phase Algorithm for Virtual Backbone Computation

A basic problem of the algorithms presented above is that they operate according to phases, where generally a dominating set is initially found, and then connectedness is guaranteed by adding some edges to the original construction. However, from the point of view of the network operation this is suboptimal, because nodes will not be able to communicate during the whole process of selecting the virtual backbone. If the ad hoc network has too many nodes, this becomes a bottleneck for any routing protocol, because the number of changes in the backbone tends to increase, as well as the computing time for a new backbone.

A new algorithm was proposed by Butenko et al. [10] to avoid this problem. The main observation made in this algorithm is that the set of all nodes in a connected network is a trivial connected dominating set. Thus, an algorithm may be composed of just one phase, where initially all nodes are part of the CDS, and elements are removed sequentially until a minimal CDS is found. Using this strategy, connection of the candidate solution is guaranteed at all stages and therefore there is no need to wait until the algorithm is finished to make use of the intermediate solutions.

The decisions of the algorithm are based on the number of neighbors of each node, and if the node is required to maintain the connection of the dominating set. The detailed steps are listed in Figure 2. Initially, nodes are picked according to their degrees, with preferences for nodes with high degrees. In a distributed version of the algorithm, we can assume that a type of election algorithm is run to define the node with high connectivity that will be selected first. Nodes are classified for the purpose of this algorithm into *fixed*, and *non fixed*. A fixed node is defined as one that will be part of the final connected dominating set, either because it has high connectivity or because it is necessary to keep connectivity to the rest of the graph. A non fixed node is one that is currently a candidate for removal.

All nodes are initially defined as non fixed. Each iteration of the algorithm will successively define a fixed node and make all of its neighbors candidates to be removed from the CDS. To determine if a node can be really removed from the current CDS, a distributed connectivity algorithm (such as, for example, distributed depth-first search) must be run. This will determine if the graph remains connected after the removal of a specific node. If the graph is no longer connected, then the candidate node must be fixed as an element of the CDS. Otherwise it is

---

Algorithm Compute CDS:
$D \leftarrow V$ /* all vertices are in the initial solution */
Set all vertices $v \in V$ to non-fixed
while there is $v \in D$ s.t. v is non-fixed **do**
    $F \leftarrow \{v \mid v \text{ is not fixed}\}$
    $u \leftarrow v \in F$ s.t. $\delta(v)$ is minimum
    if $\delta(u) > 1$ and $D \setminus u$ is not connected **then**
        Set u to fixed
    else
        $D \leftarrow D \setminus \{u\}$
        for all $s \in N(u)$ **do** $\delta(s) \leftarrow \delta(s) - 1$
        if there is no $(u, v) \in E$ s.t. v is fixed **then**
            $w \leftarrow v \in D$ s.t. $(u, v) \in E$ and $\delta(v)$ is maximum
            Set w to fixed
        end-if
    end-if
end-while
return D

Figure 2: Computing a connected dominating set.

finally removed from further consideration.

The algorithm presented above was implemented and compared to other methods proposed in the literature (particularly [2] and [9]). The results for several instances of the problem have shown that the algorithm gives solutions which are at least as good as other techniques. It has also the advantage that the intermediate solutions are all feasible for the problem and can be used to avoid the setup time of previous methods.

3 Power Control

Power control is one of the most important issues concerning the efficient operation of ad hoc networks. The main objective in this area is to allow proper operation with the minimum amount of power consumption, satisfying the design goal of ad hoc networks that need to employ cheap clients. This is even more important for sensor networks, where applications may require very small devices (on the scale of millimeters) and their time of operation may span months or even years [20, 45].

Several researchers have considered the problem of power control for ad hoc networks [5, 20, 23, 32, 35, 42]. Many of the initiatives in this area are related to the development of new protocols. Narayanaswamy et al. [32] describe such a protocol, and list a number of features that are affected by power control issues, including bidirectional links, network capacity, power routes, and contention at the media access control layer.

Another protocol for power control in ad hoc networks was presented by Kawadia and Kumar [23]. Their objective is to improve power utilization when nodes are dispersed. Thus, the problem becomes not only to optimize power consumption for individual nodes, but also to organize them into clusters. As a result, three protocols exploring different aspects of the task were proposed and analyzed through simulation. Still another protocol for power control was described in [20].

Singh et al. [42] proposed metrics for the determination of power-aware routes in ad hoc networks. In these metrics, the objective is to minimize power consumption by the optimization of some objective over the paths used by network routes. The first metric, called *energy consumed by packet*, optimizes the following function

$$\min \sum_{i=1}^{k-1} T(n_i, n_{i+1}),$$

where $n_1, \ldots, n_k$ is a set of nodes defining a path from n_1 (the source) to n_k (the destination), and $T(n_i, n_{i+1})$ is the energy consumed to transfer the packet between nodes n_i and n_{i+1}.

The *maximize time to network partition* metric has the objective of load balancing the power consumption among the nodes that separate the source from the destination. It is known from the max-flow min-cut theorem that this partition exists. The objective therefore is to use equally all nodes in the partition in order to minimize power consumption. The metric *minimize cost/packet* considers the nodes with their total power, instead of considering just the amount of power for a specific task. Thus, the objective in this case is to maximize the remaining power of the whole network at any given moment. This objective can be achieved by considering the function $f_i(x_i)$ for each node i, where f_i is a function of the cost of node i and x_i is the amount of power spent up to the current time. An example of a cost function is

$$f_j(x_i) = \frac{1}{1 - g(x_i)},$$

where $0 \leq g(x) \leq 1$ is the remaining fraction of life time and x_i is the voltage measured in node i.

The metric *maximum node cost* tries to minimize the maximum cost spent by each node to deliver a message. Finally, the last metric studied in [42] is used to minimize the variance in power levels of nodes used to send a packet in the network. Simulation results described in [42] showed that ad hoc network protocols designed according to these metrics provided improved results in terms of power consumption when compared to measures in existing routing protocols.

Finally, a clustering technique for power control optimization was recently presented by Bandyopadhyay and Coyle [5]. Their objective is to reduce power consumption on sensor networks by the careful clustering of nodes, when transmitting information collected by the whole set of sensors. The formation of such hierarchical structure would allow a simplification of the transmission process, reducing the amount of resources needed to transfer all data collected by the sensor network. The techniques used in this strategy, however, are limited to the situation where a sample is required from each sensor. Although useful, this is difficult to generalize to other situations occurring in ad hoc networks.

3.1 Minimizing Total Power Consumption

A combinatorial optimization approach for power consumption on ad hoc networks was presented in [35]. In this paper, we considered power optimization for the whole network as the goal to be attained. The advantage of such a global approach is the saving provided by a careful planning of message exchanging. The conditions necessary for the use of such a methodology are the central coordination of nodes, as occurs for example in sensor networks or in military operations.

The problem can be formulated as follows. Initially, let the wireless network be represented by a graph $G = (V, E)$, where $V = \{v_1, \ldots, v_n\}$ is the set of wireless units. For each wireless unit $v \in V$, define $p(v) \in \mathbb{R}$ as the *power level* at which v operates. The set of edges $E(G)$ is determined by the current power level of each node. For example, if unit v operates at power level α, then it can reach the set $N(\alpha, v)$ (also called the *transmission neighborhood*) of wireless units around v. This set has $\delta(\alpha, v) = |N(\alpha, v)|$ elements. Although $N(\alpha, v)$ increases as a function of the coverage of node v, it changes according to discrete steps. Thus, only a discrete number of power levels need to be considered at each node.

We assume also that packets are routed from sources $s_1, \ldots, s_n$ to destinations $d_1, \ldots, d_n$. Our objective is to find the minimum power level necessary to send d units of data from source nodes to destination nodes. The resulting problem is combinatorial because, as we have seen, just a finite number of different configurations of power levels exist. Let these levels be called $l_1, \ldots, l_{n_i}$, for unit i.

An integer programming formulation for the global power minimization problem can be described in the following way. Let x_{il} be a binary variable defined as $x_{il} = 1$ if and only if node i is at the l-th power level. Binary variable w_e^j is defined as $w_e^j = 1$ if and only if edge e is part of the j-th path. Finally, define the binary variable y_e, for $e \in V \times V$ as $y_e = 1$ if and only if $e = (i,j)$ and node j can be reached by i. Then the objective is

$$\min \sum_{i=1}^{n} \sum_{l=1}^{L} p_l x_{il},$$

subject to

$$\sum_{l=1}^{L} x_{il} = 1 \qquad \text{for } i \in \{1, \ldots, n\} \tag{1}$$

$$d(i,j) y_e \leq \sum_{l=1}^{L} p_l x_{il} \qquad \text{for all } (i,j) = e \in E \tag{2}$$

$$A w^j \leq b \qquad \text{for } j \in \{1, \ldots, k\} \tag{3}$$

$$w_e^j \leq y_e \qquad \text{for all } e \in E,\ j \in \{1, \ldots, k\} \tag{4}$$

$$x_{ij} \in \{0,1\} \qquad \text{for } i \in \{1, \ldots, n\} \text{ and } j \in \{1, \ldots, k\} \tag{5}$$

$$w^i \in \{0,1\}^E \qquad \text{for } i \in \{1, \ldots, k\} \tag{6}$$

$$y_e \in \{0,1\} \qquad \text{for all } e \in V \times V, \tag{7}$$

where $A = (a)_{ij}$ is an $n \times n^2$ matrix giving the node-arc incidence relations for a complete graph. Also, b^i, for $i \in \{1, \ldots, k\}$, is a requirements vector; i.e., it has entries $b_j^i = 1$ for $j = s_i$, $b_j^i = -1$ for $j = d_i$ and $b_j^i = 0$ otherwise.

In the objective function, the distance reached by a power level is given by constants p_l, and reaching distance is proportional to power, thus minimizing the sum of $p_l x_{il}$ for all $v_i \in V$ will minimize total power consumption. Constraint (1) ensures that for each node exactly one power level is selected. Then, constraint (2) defines the relationship between variables x_i and y_e. Variable y_e will be 1 if and only if a power level with enough capacity is selected for node i such that $e = (i, j)$.

Constraint (3) ensures that, for all source-destination pairs, there is a feasible path leading from s_i to d_i, for $i \in \{1, \ldots, k\}$. This guarantees the delivery of the existing data from source s_i to destination d_i. We say that these constraints are the *flow constraints*. Inequality (4) is then used to ensure that the data flow described in the previous constraint is feasible with respect to the existing edges. This means that data can be sent through an edge if and only if it really exists in the graph induced by the power levels assigned to each node. Finally, constraints (5) to (7) give domain requirements for the used variables.

The problem described above can be approximately solved using a linear relaxation of the formulation, as described in [35]. Another way to find near-optimal solutions for the problem consists of applying a variable neighborhood search (VNS) algorithm. The VNS is a metaheuristic for combinatorial optimization that has been applied with success to many problems [19]. Its main design goal is to employ different neighborhood structures that can explored using local search methods.

In [35], a VNS algorithm was applied to the power minimization problem. The resulting solution generates a configuration of power levels that can be used to optimize the amount of power necessary for the interchange of messages among the specified sources and destinations.

4 Relative Network Localization Problem

Sensor networks are composed of a potentially large number of nodes organized as a wireless system, with the objective of performing sensing, control, and surveillance activities [25]. The design (or lack of design) of sensor networks introduces many optimization problems, related to power, control, and coordination, for example. One of the problems occurring in sensor networks is the determination of physical location of sensors from measurements made from neighbor nodes.

A simple solution for the location problem would be the use of a geographical location system such as the global positioning system (GPS). However, it is widely accepted that this is not a scalable solution, because it can become very expensive to implement for a large number of nodes [36]. A more realistic solution consists of using measurements taken from neighboring nodes to determine their own positions. The difficulty with this technique, however, is that due to interference and other physical phenomena, distance measurements are not reliable and must be checked against other measurements in order to reduce errors that may have been introduced.

The problem of minimizing errors in the measurement of locations for ad hoc nodes has been recently studied using mathematical programming techniques [34, 39, 40, 41]. The essence of the problem is given by a set of relations over distance and angle measures, which result in a non convex quadratic programming formulation. It is assumed in these formulations that a small set of fixed nodes, or *anchor points*, is available with the purpose of defining fixed coordinates, and for validation of the measurements.

To present the formulation of the minimum measurement error problem, let a_k, for $k \in \{1, \dots, m\}$, represent the anchor points, and x_i, for $i \in \{1, \dots, n\}$ represent positions of sensors. Suppose that, given the set of nodes V, there is a set $N_e \subseteq V \times V$ such that for $(i,j) \in N_e$ the distances measured between sensors i and j are given by $d(i,j)$. Similarly, the distances measured from an anchor point a_k and sensor i, for $(k,i) \in N_e$, are given by $d(k,i)$. Suppose also there is a set $N_l \subseteq V \times V$ such that for each $(i,j) \in N_l$ a lower bound l_{ij} is known. A set N_u also exists, with corresponding upper bounds u_{ij}, for $(i,j) \in N_u$. Then, we can formulate the problem as finding a set of positions x_i, for $i \in \{1, \dots, n\}$ such that

$$
\begin{aligned}
|x_i - x_j|^2 &= d(i,j)^2, & |a_k - x_j|^2 &= d(k,j)^2, && \text{for all } (i,j),\ (k,j) \in N_e\\
|x_i - x_j|^2 &\geq l_{ij}^2, & |a_k - x_j|^2 &\geq l_{ij}^2, && \text{for all } (i,j),\ (k,j) \in N_l\\
|x_i - x_j|^2 &\leq u_{ij}^2, & |a_k - x_j|^2 &\leq u_{ij}^2, && \text{for all } (i,j),\ (k,j) \in N_u.
\end{aligned}
$$

In practice, it is not possible to find exact values for each x_i due to noise or limitations in the measurement capacity of each sensor. Thus, it makes sense to minimize the error of these measurements, which results in a *least squares problem* [44]. The formulation then becomes:

$$\min \sum_{(i,j)\in N_e} |\,|x_i - x_j|^2 - d(i,j)^2| + \sum_{(k,j)\in N_e} |\,|a_k - x_j|^2 - d(k,j)^2|$$
$$+ \sum_{(i,j)\in N_l} \max\{-|\,|x_i - x_j|^2 - l_{ij}^2|, 0\} + \sum_{(k,j)\in N_l} |\,|a_k - x_j|^2 - l_{ij}^2|$$
$$+ \sum_{(i,j)\in N_u} \max\{|\,|x_i - x_j|^2 - u_{ij}^2|, 0\} + \sum_{(k,j)\in N_u} |\,|a_k - x_j|^2 - u_{ij}^2|.$$

4.1 The SDP Approach

The basic difficulty with the previous formulation is that it results in a non convex problem, and therefore it can be quite hard to find its optimum solution. In order to simplify the measurement problem and to provide efficient algorithms for its computation, researchers have proposed relaxations of the original quadratic formulation. Among the proposed techniques, the one of most interest consists of relaxing the formulation to a semidefinite programming problem. This technique has been initially proposed by Alfakih, Khandani, and Wolkowicz [1] for general minimization of errors in location problems (see also the work of Laurent [24]).

Semidefinite programming is a convex programming problem where the feasible solution set is optimized over the cone of semidefinite matrices. Remember that linear programming is basically optimization over the cone of positive values in $\mathbb{R}^n$. Thus, there is some similarity between LP and SDP, which has been frequently explored by researchers in order to provide efficient algorithms for SDP. Most algorithms for SDP rely on interior point methods for calculating optimum solution within a established error ϵ.

A general semidefinite programming problem can be represented as

$$\begin{aligned} &\min \quad Tr(X, C) \\ &\text{subject to} \\ &\quad Tr(A_i, X) = b_i, \qquad \text{for } i \in \{1, \ldots, m\} \\ &\quad X \succeq 0, \end{aligned}$$

where X is a real symmetric $n \times n$ matrix, and $Tr(A, B)$ is the inner product of the matrices A and B given by

$$Tr(A, B) = \sum_{i=1}^{n} \sum_{j=1}^{n} A_{ij} B_{ij}.$$

Note that in the formulation above, $X \succeq 0$ means that X is positive semidefinite, i.e., for any $y \in R^n$, $y^T X y \geq 0$.

In order to find a SDP formulation of the error minimization problem, we perform the following modifications in the previous model: let N_a be the set of pairs $(i, k) \in V \times V$ such that $d(i, k)$ is known for sensor i and anchor point k. Let N_x be the set of pairs $(i, j) \in V \times V$ such that $d(i, j)$ is known for sensors i and j. Then, we consider the problem of finding measurements $x_i \in \mathbb{R}^2$ such that

$$|x_i - x_j|^2 = d(i,j)^2 \qquad \text{for } (i,j) \in N_x \tag{8}$$

$$|a_k - x_j|^2 = d(k,j)^2 \qquad \text{for } (i,j) \in N_a.$$

Let $X = [x_1, \ldots, x_n]$ be a $2 \times n$ matrix giving a solution for the minimization problem above. Let e_{ij} be a vector in $\mathbb{R}^n$ with 1 in position i, -1 in position j, and 0 in all other positions. It can be shown [27] that

$$|x_i - x_j|^2 = e_{ij}^T X^T X e_{ij}, \quad \text{and}$$

$$|a_k - x_j|^2 = [a_k|e_j]^T [I|X]^T [I|X][a_k|e_j].$$

Thus, problem (8) becomes: find matrices $Y \in \mathbb{R}^{n \times n}$ and $X \in \mathbb{R}^{2 \times n}$ such that

$$e_{ij}^T Y e_{ij} = d(i,j)^2 \qquad \text{for } (i,j) \in N_x,$$

$$[a_k|e_j]^T \begin{bmatrix} I & X \\ X^T & Y \end{bmatrix} [a_k|e_j] = d(k,j)^2 \qquad \text{for } (i,j) \in N_a,$$

$$Y = X^T X.$$

To find a semidefinite programming relaxation of the above problem, we simply rewrite the last constraint as

$$Y \succeq X^T X,$$

which means that $Y - X^T X \succeq 0$. Then, to rewrite the problem above as a standard SDP problem, one can use the fact that $Y = X^T X$ is equivalent to

$$\begin{bmatrix} I & X \\ X^T & Y \end{bmatrix} \succeq 0.$$

Thus the problem reduces to finding a vector $Z \in \mathbb{R}^{(n+2) \times (n+2)}$ such that

$$[0|e_{ij}]^T Z [0|e_{ij}] = d(i,j)^2 \qquad \text{for } (i,j) \in N_x$$

$$[a_k|e_j]^T Z[a_k|e_j] = d(k,j)^2 \qquad \text{for } (i,j) \in N_a$$

$$Z \succeq 0.$$

Advanced techniques available for semidefinite programming can be used to solve this problem. It has been observed, for example, that interior point algorithms can solve this formulation in time proportional to $O(n^3)$ [8]. Other approaches such as distributed computing [7] and gradient-based methods [27] have been applied with success to improve the computational efficiency of the existing algorithms. For additional information about theoretical issues related to the semidefinite formulation, one can check the paper by Man-Cho So and Ye [28].

5 Concluding Remarks

Ad hoc networks are an area of telecommunications that has grown in popularity due to its wide applicability. It also presents some interesting and difficult optimization problems. In this chapter we presented some of the optimization issues related to ad hoc networks, as well as techniques employed in their solution. This is an active area of research, and certainly will see many developments in the near future in terms of improved formulations and algorithms.

6 Acknowledgments

This research was partially supported by NSF and AirForce grants.

References

[1] A. Y. Alfakih, A. Khandani, and H. Wolkowicz. Solving Euclidean distance matrix completion problems via semidefinite programming. *Comput. Optim. Appl.*, 12:13–30, 1999.

[2] K. M. Alzoubi, P.-J. Wan, and O. Frieder. Distributed heuristics for connected dominating set in wireless ad hoc networks. *IEEE ComSoc/KICS J. Communication Networks*, 4(1):22–29, 2002.

[3] S. Arora. Approximation schemes for NP-hard geometric optimization problems: A survey. *Math. Program.*, 97:1436–4646, 2003.

[4] B. S. Baker. Approximation algorithms for NP-complete problems on planar graphs. *Journal of the ACM (JACM)*, 41(1):153–180, 1994.

[5] S. Bandyopadhyay and E. J. Coyle. Minimizing communication costs in hierarchically-clustered networks of wireless sensors. *Comput. Networks*, 44(1):1–16, 2004.

[6] R. Bellman. *Dynamic Programming.* Princeton University Press, Princeton, NJ, 1957.

[7] P. Biswas and Y. Ye. A distributed method for solving semidefinite programs arising from ad hoc wireless sensor network localization. Technical report, Dept. of Management Science and Engineering, Stanford University, 2003.

[8] P. Biswas and Y. Ye. Semidefinite programming for ad hoc wireless sensor network localization. In *Proceedings of the Third Tnternational Symposium on Information Processing in Sensor Networks*, pages 46–54. ACM Press, 2004.

[9] S. Butenko, X. Cheng, D.-Z. Du, and P. M. Pardalos. On the construction of virtual backbone for ad hoc wireless network. In S. Butenko, R. Murphey, and P. M. Pardalos, editors, *Cooperative Control: Models, Applications and Algorithms*, pages 43–54. Kluwer Academic, 2002.

[10] S. Butenko, X. Cheng, C. A. Oliveira, and P. M. Pardalos. A new algorithm for connected dominating sets on ad hoc networks. In S. Butenko, R. Murphey, and P. Pardalos, editors, *Recent Developments in Cooperative Control and Optimization*, pages 61–73. Kluwer Academic, 2003.

[11] X. Cheng and D. Du. Virtual backbone-based routing in multihop ad hoc wireless networks. Technical report, Department of Computer Science and Engineering, University of Minnesota, 2002.

[12] B. Das and V. Bharghavan. Routing in ad-hoc networks using minimum connected dominating sets. In *IEEE International Conference on Communications*, 1997.

[13] E. W. Dijkstra. A note on two problems in connexion with graphs. *Numer. Math.*, 1:269–271, 1959.

[14] A. Ephremides, J. Wieselthier, and D. Backer. A design concept to reliable mobile radio networks with frequency hopping signaling. *Proceedings of the IEEE*, 75(1):56–73, 1987.

[15] M. Frodigh, P. Johansson, and P. Larsson. Wireless ad hoc networking: The art of networking without a network. *Ericsson Rev.*, 4:248–263, 2000.

[16] M. R. Garey and D. S. Johnson. *Computers and Intractability, A Guide to the Theory of NP-completeness.* W. H. Freeman, 1979.

[17] M. Gerla and J.-C. Tsai. Multicluster, mobile, multimedia radio network. *ACM J. Wireless Networks*, 1(3):255–265, 1995.

[18] S. Guha and S. Khuller. Approximation algorithms for connected dominating sets. *Algorithmica*, 20(4):374–387, 1998.

[19] P. Hansen and N. Mladenovic. An introduction to VNS. In S. Voss, S. Martello, I. Osman, and C. Roucairol, editors, *Meta-Heuristics: Advances and Trends in Local Search Paradigms for Optimization.* Kluwer Academic, 1998.

[20] W. R. Heinzelman, A. Chandrakasan, and H. Balakrishnan. Energy-efficient communication protocol for wireless microsensor networks. In *Proceedings of the 33rd Hawaii International Conference on System Sciences, Volume 8*, page 8020. IEEE Computer Society, 2000.

[21] P. Johansson, T. Larsson, N. Hedman, B. Mielczarek, and M. Degermark. Scenario based performance analysis of routing protocols for mobile ad hoc networks. In *Proceedings of IEEE MOBICOM*, pages 195–206, Seattle, 1999.

[22] D. B. Johnson. Routing in ad hoc networks of mobile hosts. In *Proceedings of the IEEE Workshop on Mobile Computing Systems and Applications.* IEEE, 1994.

[23] V. Kawadia and P. Kumar. Power control and clustering in ad hoc networks. In *Proceedings of IEEE INFOCOM*, 2003.

[24] M. Laurent. Matrix completion problems. In *The Encyclopedia of Optimization*, volume 3, pages 221–229. Kluwer Academic, 2001.

[25] X.-Y. Li, P.-J. Wan, and O. Frieder. Coverage in wireless ad hoc sensor networks. *IEEE Trans. Comput.*, 52:753–763, 2003.

[26] B. Liang and Z. Haas. Virtual backbone generation and maintenance in ad hoc network mobility management. In *Proceedings of IEEE INFOCOM*, pages 1293–1302, 2000.

[27] T.-C. Liang, T.-C. Wang, and Y. Ye. A gradient search method to round the semidefinite programming relaxation solution for ad hoc wireless sensor network localization. Manuscript available at http://www.stanford.edu/~yyye, 2004.

[28] A. Man-Cho So and Y. Ye. Theory of semidefinite programming for sensor network localization. In *Proceedings of the sixteenth annual ACM-SIAM symposium on Discrete algorithms*, pages 405–414, Philadelphia, PA, USA, 2005. Society for Industrial and Applied Mathematics.

[29] H. Matsuo and K. Mori. Accelerated ants routing in dynamic networks. In *Proceedings of the International Conference On Software Engineering, Artificial Intelligence, Networking and Parallel/Distributed Computing*, pages 333–339, 2001.

[30] S. Meguerdichian, S. Slijepcevic, V. Karayan, and M. Potkonjak. Localized algorithms in wireless ad-hoc networks: location discovery and sensor exposure. In *Proceedings of the 2nd ACM International Symposium on Mobile ad Hoc Networking & Computing*, pages 106–116. ACM Press, 2001.

[31] M. Min, F. Wang, D.-Z. Du, and P. M. Pardalos. A reliable virtual backbone scheme in mobile ad-hoc networks. In *Proceedings of the 1st IEEE International Conference on Mobile Ad hoc and Sensor Systems*, pages 60–69, 2004.

[32] S. Narayanaswamy, V. Kawadia, R. S. Sreenivas, and P. R. Kumar. Power control in ad hoc networks: Theory, architecture, algorithm and implementation of the COMPOW protocol. In *European Wireless Conference*, 2002.

[33] S.-Y. Ni, Y.-C. Tseng, Y.-S. Chen, and J.-P. Sheu. The broadcast storm problem in a mobile ad hoc network. In *Proceedings of IEEE MOBICOM*, pages 151–162, Seattle, 1999.

[34] D. Niculescu and B. Nath. Ad hoc positioning system (APS). *IEEE GLOBECOM*, 1:2926–2931, 2001.

[35] C. A. Oliveira and P. M. Pardalos. A distributed optimization algorithm for power control in wireless ad hoc networks. In *Proceedings of the 18th International Parallel and Distributed Processing Symposium (IPDPS'04)*, volume 7, page 177. IEEE Computer Society, 2004.

[36] N. Patwari and Y. Wang. Relative location in wireless networks. In *Proceedings of the IEEE Spring VTC*, pages 1149–1153, 2001.

[37] G. J. Pottie and W. J. Kaiser. Wireless integrated network sensors. *Commun. ACM*, 43(5):51–58, 2000.

[38] S. Ramanathan and M. Steenstrup. A survey of routing techniques for mobile communications networks. *Baltzer/ACM Mobile Networks and Applications*, 1:89–104, 1996.

[39] C. Savarese, J. Rabay, and K. Langendoen. Robust positioning algorithms for distributed ad-hoc wireless sensor networks. In *USENIX Technical Annual Conference*, 2002.

[40] A. Savvides, C.-C. Han, and M. B. Srivastava. Dynamic fine-grained localization in ad-hoc networks of sensors. *Mobile Comput. Netw.*, pages 166–179, 2001.

[41] A. Savvides, H. Park, and M. B. Srivastava. The bits and flops of the n-hop multilateration primitive for node localization problems. In *Proceedings of the 1st ACM International Workshop on Wireless Sensor Networks and Applications*, pages 112–121. ACM Press, 2002.

[42] S. Singh, M. Woo, and C. Raghavendra. Power aware routing in mobile ad hoc networks. In *Proceedings of IEEE MOBICOM*, 1998.

[43] P. Sinha, R. Sivakumar, and V. Bharghavan. Enhancing ad hoc routing with dynamic virtual infrastructures. In *Proceedings of IEEE INFOCOM*, volume 3, pages 1763–1772, 2001.

[44] I. Söderkvist. On algorithms for generalized least squares problems with ill-conditioned covariance matrices. *Computational Statistics*, 11:303–313, 1996.

[45] B. Warneke, M. Last, B. Liebowitz, and K. S. J. Pister. Smart dust: Communicating with a cubic-millimeter computer. *Computer*, 34(1):44–51, 2001.

[46] J. Wu, F. Dai, M. Gao, and I. Stojmenovic. On calculating power-aware connected dominating sets for efficient routing in ad hoc wireless networks. *IEEE/KICS J. Commun. Netw.*, 4(1):59–70, 2002.

[47] J. Wu and H. Li. On calculating connected dominating set for efficient routing in ad hoc wireless networks. In *Proceedings of the 3rd International Workshop on Discrete Algorithms and Methods for Mobile Computing and Communications*, pages 7–14, 1999.

Chapter 6

Stochastic Programming in Allocation Policies for Heterogeneous Wireless Networks

Abd-Elhamid M. Taha
Department of Electrical and Computer Engineering
Queen's University, Kingston, Canada, K7L 3N6
E-mail: taha@ece.queensu.ca

Hossam S. Hassanein
School of Computing
Queen's University, Kingston, Canada, K7L 3N6
E-mail: hossam@cs.queensu.ca

Hussein T. Mouftah
School of Information Technology and Engineering
Ottawa University, Ottawa, Canada, K1N 6N5
E-mail: mouftah@site.uottawa.ca

1 Introduction

Since its introduction, Linear Programming (LP) has become an inherent planning tool, with applications in nearly all specialization fields. In the specific field of telecommunications, it has been used in temporal resource assignments, physical and virtual facility allocations, traffic flow management and many other various problems. The attractiveness of LP comes from being an easily comprehensible planning tool. Its malleability feature enables practitioners to tailor the formulation according to their specific problems. The advances in solving algorithms, with highly efficient algorithms for specific cases, also add to the motivations of its utility. Moreover, a rather interesting fact is that there are many situations where the formulation can be made and applied without requiring a rigorous understanding of LP's theoretical background.

LP is a mathematical programming technique that is concerned with scenarios in which the elements in question exhibit linear relationships with the constraining resources while yielding a linear cost function. The underlying assumption is that the entities involved in the programming are deterministic. This can mean, for example, that the costs, abundance, etc., of such entities are controllable by the planner. In another instance, the entities' costs can be ascertained to be stable (constant) in the time spans under consideration. However, while there exist a plethora of scenarios where these certainty assumptions hold, there are situations where uncontrollable elements must be involved in the decision-making process. There are also situations where in the time spans being considered constancy assumptions do not hold. It is in these scenarios where the deterministic LP model fails, and herein lies the motivation for a mathematical programming technique that can accommodate uncertainty.

Stochastic Programming (SP) is aimed at scenarios where the elements in question are of uncertain instantaneous value, relative to the time span under consideration, but remain a known, or can be modeled as exhibiting, a quantifiable probabilistic behavior. In other words, the theory behind SP is categorically not aimed at situations of total uncertainty where neither the elements' instantaneous value nor their probabilistic behavior can be determined. It should also be recognized that SP is not about merely substituting averages, or similar statistical properties, into a deterministic mathematical programming model. As is clarified further on, this misconception that a more robust solution is attained by substituting averages renders outcome unrealistic, infeasible or both.

In the field of telecommunications, there are many instances where SP can be applied. Consider, for example, the error and delay introduced in optical data networks at the optical/electrical interfaces. SP can be used to accommodate such irregularities into a more robust channel (wavelength) assignment scheme. Another example is where service providers cannot ascertain the demand at any time instance. Faced with options that, parted, either lead to underutilization or high communication overhead, service providers can utilize SP in inducing a proactive element in the allocation policies that, together with an active negotiation mechanism, can embrace demand unevenness.

In this chapter the application of Stochastic Integer LP (SILP) in the area of future wireless networks is explored. It was perhaps in 1996 that visions for what are now called Heterogeneous Wireless Networks (HWN) were first conceived [12]. In an HWN, a mobile will be able to roam freely, changing association from one type of wireless access network to another. For example, a user may seamlessly switch between a WLAN hotspot in a coffee shop and a cellular network as she leaves, or vice versa. A user might also opt to switch association if she moves within the coverage of an access network that better suits her needs. On the network side, overlaid networks can be used in instances of load balancing and congestion control. Hence, contrary to horizontal handoffs which occur between homogenous systems and are mostly based on signal strength, vertical handoffs, i.e., switching associations between different wireless network paradigms, can be triggered by the conditions or requirements of either the user or the network. Consequently, irregularities are introduced that cannot be accommodated by the traditional resource management techniques, which mainly relied on deterministic allocation models, and hence the need for more robust schemes.

The remainder of this chapter is as follows. In Section 2, after reviewing the basics for the LP formulation, the general model for the SLP formulation is introduced together with some of its different forms. Section 3 discusses how SLP can be used to provide robust allocation policies in a HWN. A note on algorithmic complexity is made in Section 4. Section 5 concludes.

2 A Preliminary on SLP

In a sense, mathematical programming is concerned with finding the maximum or minimum of a certain criterion function of some decision elements, values of which are within a state space bounded by constraints made on the elements in question. In finding the minimum or maximum, the planner is said to be optimizing the criterion function. Generally, a mathematical program takes on the form

$$\max \left\{ c^T x | Ax = b, x \in X \right\}$$

where c^T are the criterion coefficients (costs, revenues, preference, etc.) of the elements in question, x; A are the coefficients associated with x

in the constraints bound by b, and $X \subset \mathrm{R}^n$ is the state space created by the constraints. In the literature, the terms decision variables and objective function correctly refer to x and $c^T x$.

When the objective function and the constraints are both linear, the mathematical program is said to be an LP. Once correctly formulated, methods such as the Simplex Method or the more recent Interior Point algorithm can be used to find the optimal solution. Currently, there are many commercial software packages that can be used in solving numerous types of mathematical programs, including LP.

In integer linear programming, a constraint of integrality is imposed on some or all elements of X, transforming the program into a mixed-integer or integer linear program, respectively. Naturally, as discussed below, this condition does affect the complexity of the solving algorithm.

There are many examples of how LP can be applied in telecommunications. The famous shortest-path algorithm utilized in routing is essentially a linear program [1]. In Mobile IP, a node maintains its IP connectivity to its original home network using registrations through forwarding agents at foreign networks. To enhance and speed registration procedures, the notion of agent hierarchy in foreign networks was introduced. Kamal and El-Rewini address the problem of locating these forwarding agents in a foreign network using integer linear programs [11]. Another instance of utilizing integer programming in a facility allocation problem is the problem addressed by Lee et al. where they approach the issue of access point placement in designing a wireless LAN [14]. The objective of the work was to maximize the coverage as much as possible while minimizing the interference that results from the channel assignment to each access point. A classic example of the employment of LP in resource management can be found in the work by Herzberg and Pitsillides [7]. A more recent application is made by Kumar and Venkataram in [13]. In this work, an admission control based on LP is designed with artificial neural networks (ANNs) and used in solving the resulting LP formulation. The application of ANNs in solving LP has recently attracted some interest as a promising efficient solution [19].

An LP formulation of the above form usually implies that the problem is only comprised of deterministic parameters, specified in coefficients in the objective function and conditions, respectively c^T and A, and the conditions' constraining values, b. Although many practical decision problems can be solved using linear programming, e.g., where parameters involved are constant over sufficiently long periods, there are problems where solutions are required with some parameters possessing a considerable degree of uncertainty.

The objective of Stochastic Programming (SP) is to deal with situations in which the parameters involved in the optimization cannot be ascertained at any time instance, but are known, or can be assumed, to exhibit a quantifiable probabilistic behavior. In other words, SP is not concerned with the cases of total uncertainty where even the probabilistic behavior of the parameters involved cannot be determined.

At the extreme, SP accommodates c^T, A and b simultaneously exhibiting randomness. Limiting the discussions to situations where uncertainty only exists in the right-hand side of the constraints, i.e.,

$$\max \left\{ c^T x | Ax = b, Dx = \xi, x \in X \right\}$$

where ξ is random, solving such a problem reduces to changing the formulation into a deterministic LP equivalent by enumerating the possible outcome scenarios while associating each scenario with its probability. One way of achieving this is by setting a penalty for not satisfying the constraint posed by each scenario. For example, suppose that there are K possible scenarios and that the probability associated with the kth scenario k, is denoted p_k. Let q_k be the penalty per unit of not satisfying the kth constraint, i.e., per unit difference between Tx and ξ_k. With this in mind, the formulation of the deterministic equivalent problem becomes

$$\max \left\{ c^T x + \sum_{k=1}^{K} p_k (q_k)^T y_k | Ax = b, D_k x + y_k = \xi_k, k = 1 : K, x \in X \right\}$$

Here, y_k represents the slackness due to randomness, i.e., not satisfying the condition caused by the probabilistic nature of ξ_k. For more tolerance, SLP also provides a venue for types of problems where stochastic constraints need not be absolutely held, and it is acceptable that these constraints hold with prescribed probabilities. Such constraints, called chance or probabilistic constraints, are added to the formulation in the form

$$P \left(D_k x + y_k \leq \xi_k \right) \geq p, p \in (0, 1)$$

The utilization of SP within the general context of networking has been made before. For example, in [6] Heckmann et al. compared different SP formulations in allocating bandwidth proactively for a given link. It was also employed by Liu in [15] as means of managing stochastic traffic flows. Another place where SP was applied in network design can be found in the work by Riis and Andersen [16]. A comprehensive overview of the application of SP and other optimization under uncertainty techniques is provided by Gaivoronski in [5].

For an excellent overview of SLP, please refer to Sen and Higle's introductory tutorial [17]. For a more rigorous read, refer to *Stochastic Programming* [10] by Kall and Wallace.

3 Robust Allocation Policies in HWNs

Serious efforts are currently underway to hasten the realization of Heterogeneous Wireless Networks (HWNs) or Fourth Generation (4G) wireless networks. Echoing the long-made efforts in Europe, Japan and the United States, the IEEE 802.21 standardization group was initiated in March 2004 to study how intertechnology handoffs can be made [8]. Commercially, WLAN hotspots[1] (L-spots) are being implemented nearly everywhere. Until now, however, the ability to maintain active connections while toggling the association from one network to another (e.g., from 802.11b to CDMA2000) has not been commercialized. The heterogeneity of the various access technologies in addition to the heterogeneity of different networks' architectural structures, all add to the difficulties of implementation. On the network side, functionalities such as admission control, resource and mobility management are required to operate in a dynamic and agile manner. On the user side, stringent constraints are made mainly due to the limited capabilities of any mobile user.

Although HWNs have long been investigated, the issue of joint resource management has not been previously addressed in a direct manner. Several proposals have been made on other various aspects of seamless integration. For example, an IP-based architecture for integrating UMTS and 802.11 WLAN networks is detailed by Jaseemuddin in [9].

[1] The term hotspot can be found in the literature and media to mean two separate things. One is strictly within the context of cellular networks and indicates the situation where there are more users in a certain cell than its design capacity. The other is when a service provider implements WLAN (or Wi-Fi) coverage, whether within a cellular coverage or not. For distinction, the term L-spot is used for the latter.

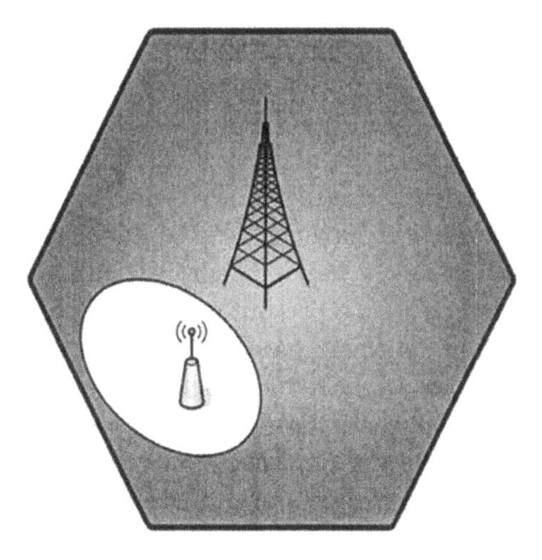

Figure 1: A WLAN overlaid within the coverage of a cell.

Tsao and Lin discuss and compare different non-IP intertechnology connectivity mechanisms in [18]. Performance analysis of handoff procedures in heterogeneous networks can be found in [3] and [2].

In its simplest definition, resource management is concerned with optimally assigning the available resources to the current demand. In order to conceptualize the general model at hand, consider the scenario illustrated in Figure 1, where an L-spot has been overlaid within the coverage of a cellular network. (For the purpose of this study, it is assumed that within the coverage overlap there are no possibilities for horizontal handoffs.) Hence, in the general population, three types of mobile users can be identified: single-mode cellular users, single-mode WLAN users, and dual-mode users, i.e., users who are capable of associating with both types of networks, whether simultaneously or one network at any given time. On the other hand, single-mode users are only capable of connecting to one type of network at any given time.

The existence of dual mode has certain attractive features from both the users' and the networks' sides. Recall that, by design, cellular networks are aimed at users with high mobility and low connection rates and WLAN networks are aimed at users with low mobility and high connection rates. When a dual-mode user, connected to the cellular network, comes within the coverage of an L-spot, she may opt to toggle the association for, say, better services. On the networks' side, when either network experiences congestion, it could force certain users to undergo vertical handoff until the congestion is relieved.

From the above, the motivation for this work can hence be observed. When roaming in homogeneous networks, handoffs are limited to decisions based only on the signal strength received at the user's terminal. In heterogeneous networks, handoffs can be also be triggered due to signal strength conditions. However, handoffs can further be triggered by network conditions, either within a precautionary load balancing procedure or during a congestion relief mechanism. As aforementioned, vertical handoffs can be initiated by user requirements, reacting to the awareness of different access networks with capabilities or, even, with different economic options. These added, and possibly more frequent, triggers to vertical handoffs result in irregularities in demand, posing challenging problems on how resource management in such environments is to be provided. Traditional resource management schemes relied on deterministic allocation models where stabilities in demand can be safely assumed. Hence, such traditional schemes cannot accommodate the irregularities to be expected in HWNs.

For illustration, an exemplary scenario is described where a single data service of fixed bandwidth requirements is provided in both networks. The goal here is accommodating the maximum number of users in both networks residing within the L-spot area while minimizing the costs of resource underutilization and demand rejection. The accommodation preference of each user type is signified through some return value per unit allocation. In this manner, maximizing the allocation concurrently maximizes the service provider's profits. In what follows, an optimization model for demand known with certainty is outlined in order to understand the main operational considerations. Next, the formulation for resource optimization with probabilistic demands is presented. Following that, a numerical example is detailed to observe the main characteristics of a solution obtained under uncertainty. Without loss of generality, all users are assumed to be dual mode users.

Single Common Service-Deterministic Demand (SCS-DD)

Suppose that there exists a single basic service that is commonly provided in both networks. Also, let the bandwidth requirements of this single common service (SCS), denoted Q, be fixed. Refer to the larger network, i.e., the cellular network, as N_1 and the network providing the L-spot as N_2. The maximum capacities for N_1 and N_2 are B_1 and B_2, respectively.

The term maximum capacity of one network refers to the capacity that can be provided for respective users such that at least Q can be provided for each user. For N_1, maximum capacity further implies that is the capacity allocation by N_1 for users residing within the area covered by the L-spot but requesting services from N_1. These capacities represent the capacities available at the time of request arrival. Hence, the maximum number of users that can be supported by N_1 and N_2 is B_1/Q and B_2/Q, respectively.

In the following, index ij is used to distinguish between entities related to different types of users. Entities indexed with $i = j$ are related to users to be admitted in one network, and entities indexed with $i \neq j$ are related to users changing networks. Let D_{ij}, R_{ij} and A_{ij} respectively refer to the demand, rejection (unsatisfied demand) and allocation for users ij, such that

$$R_{ij} = D_{ij} - A_{ij}$$

When the total allocation in a certain network is less than the available resources, i.e.,

$$\sum_i A_{ij} = A_j < \frac{B_j}{Q}$$

then network j is underutilized by U_j, defined as

$$U_j = \frac{B_j}{Q} - A_j$$

An interconnection between the two networks will be dedicated for exchanging control messages and relaying in-flight packets for users undergoing vertical handoff. This interconnection is assumed to follow a certain delay-throughput profile. Let B_v be the upper bound of the number of users above where this exchange cannot be performed with an acceptable delay. Also, let U_v be the underutilization associated with this interconnection. The following equality can then be established.

$$\sum_{i \neq j} A_{ij} + U_v = B_v$$

The constituents comprising the return function of the management operation can now be detailed. Let profit per allocated ij user be x_{ij}. Also, let the costs of unit underutilization for networks j and the interconnection be y_j and y_v, respectively. Finally, denote the cost per user rejection by z_{ij}. With this in mind, the total return function can be stated as

$$\Pi_{SCS-DD} = \sum_{\forall i,j} x_{ij} \cdot A_{ij} - \sum_{s=\{j,v\}} y_s \cdot U_s - \sum_{\forall i,j} z_{ij} \cdot R_{ij}$$

The manner in which various profits and costs are valuated is beyond the scope of this work. However, it should be noted that although in such valuations there is an inherent emphasis on stability, the network economics may very well depend even on the hour-to-hour network dynamics. There are also other aspects that are to be considered. For example, the cost of utilizing (or underutilizing) the interconnection is bound to vary with the bandwidth abundance.

Considering the above, the ILP can be stated as follows.

$$\begin{array}{lll} & \text{Program SCS-DD} & \\ \text{Max} & \Pi_{SCS-DD} & \\ \text{S.t.} & A_j + U_j = \frac{B_j}{Q} & \forall j \\ & \sum_{i \neq j} A_{ij} + U_v = B_v & \\ & A_{ij} + R_{ij} = D_{ij} & \forall i,j \\ & \text{All variables positive, integers} & \end{array}$$

The formulation for the Program SCS-DD signifies the inherent characteristic of any allocation policy mechanism. Simply stated, the allocation for different types of demands is made by weighing the demand against the resources. As such, this formulation could be extended to implement different types of policies. For example, although the allocations for A_{ij} with $i \neq j$ could serve as what is commonly called *guard bandwidth*, fixed proportions of the resources can be further allocated to accommodate operational discrepancies, be they on the part of the network or the demand.

The more important facet of the formulation is that it could be used to study the behavioral aspects of such allocation policies. For example, several interconnection modes have been proposed for vertical system handoffs [18]. The formulation could be used in evaluating which interconnection method, occasionally called coupling, would be more desirable from either the network's or the demand's points of view.

Single Common Service-Probabilistic Demand (SCS-PD)

In this subsection, we make use of the notation used in introducing stochastic programming. We assume that there is a set of scenarios, S, each of which describes an instance where each demand takes on a specific value with a probability known a priori. In other words, in every scenario $s \in S$ the demand $D_{ij}(s)$ takes on specific values with a predetermined probability, the probability that the current $D_{ij}(s)$ is a specific value, i.e., $P(D_{ij}(s) = D) = p_{ij}(s)$.

The demand uncertainty can be imposed on Program SCS-DD through the allocation-rejection-demand constraints, where the penalty can be applied to the rejection. In this manner, the penalty (cost) of unit rejection is $z_{ij}(s)$. As such, the return function to be maximized becomes

$$\Pi_{SCS-PD} = \sum_{\forall i,j} x_{ij} \cdot A_{ij} - \sum_{c=\{j,v\}} y_c \cdot U_c - \sum_{\forall i,j} \sum_{s \in S} p_{ij}(s) \cdot z_{ij}(s) \cdot R_{ij}(s)$$

With this, the SILP can be stated as follows.

$$\begin{array}{lll}
 & \text{Program SCS-PD} & \\
\text{Max} & \Pi_{SCS-PD} & \\
\text{S.t.} & A_j + U_j = \frac{B_j}{Q} & \forall j \\
 & \sum_{i \neq j} A_{ij} + U_v = B_v & \\
 & A_{ij} + R_{ij}(s) = D_{ij}(s) & \forall i, j, s \in S \\
 & \text{All variables positive, integers} &
\end{array}$$

Rightfully, in order to reduce the problem dimensionality, one can consider only the scenarios associated with significant probabilities. Given a probability significance threshold, p_t, an added condition can be that only scenarios with $p_{ij}(s) \geq p_t$, $\forall i, j$, be considered. Naturally, the exact value of pt remains a design factor that can be subjected to optimization.

Although this formulation is aimed at probabilistic demands in a given time period, the formulation is also extendable to other types of uncertainties. In certain situations the returns and costs can be uncertain. For example, when a cell undergoes overload, i.e., becomes a hotspot, it may borrow bandwidth from neighboring cells that belong to different service providers. In this case, the cost of service is coupled with the anticipated demand. This is one example where operational dynamics affect economic considerations.

A Numerical Example

In order to compare Programs SCS-DD and PD, the load on N_1, notated L_1, is given sampler distributions. In other words, the total demand on N_1 is given values ranging from slight underloading to slight overloading, with each value associated with a predefined probability. Denote the optimal value of the objective function in a program with Φ. The percentage deviation of the Φ_{PD} and the average Φ_{DD} relative to the optimal scenario for each demand instance is computed. The values used in comparison are given in Table 1.

Table 1: Values used in comparison

Symbol	Value	Symbol	Value
B_1	50	y_1	3
B_2	200	y_2	3
B_v	50	y_v	1.5
Q	1	z_{11}	2
x_{11}	3	z_{21}	3
x_{21}	5	z_{12}	5
x_{12}	5	z_{22}	4
x_{22}	3	L_2	0.7

The load on N_2, notated L_2, will be fixed at 70% of the resources; i.e., $0.7 \cdot 200 = 140$. Different demands will take faxed proportions of the load: In N_1, D_{11} and D_{21} will comprise 70% and 30% of load, respectively; in N_2, D_{22} and D_{21} will also respectively comprise 70% and 30% proportions of L_2.

For completion of comparison, the values of $z_{ij}(s)$ in Program SCS-PD will take values equal to those used in Φ_{DD}; i.e., equal penalties will be made each i j pair.

Table 2 shows five sample distributions. In each distribution, there are five scenarios, i.e., $S = [1, 5]$, each with a predetermined probability. For clarification and simplification of notation, the probability of each load instance is as follows.

$$P(L_1 = L) = P(D_{11}(s) = 0.7 \cdot L) = P(D_{21}(s) = 0.7 \cdot L)$$

In other words, in each distribution

$$p_{11}(s) = p_{21}(s) = p_1(s)$$

for all scenarios.

Table 2: Load distributions for each of the five scenarios

Distribution	1	2	3	4	5
$P(L_1 = 0.7)$	3/15	1/15	5/15	4/15	2/15
$P(L_1 = 0.85)$	3/15	2/15	4/15	3/15	3/15
$P(L_1 = 1)$	3/15	3/15	3/15	1/15	4/15
$P(L_1 = 1.05)$	3/15	4/15	2/15	3/15	3/15
$P(L_1 = 1.1)$	3/15	5/15	1/15	4/15	2/15

The average optimal solution for each distribution, Φ_{Avg}, is calculated in the following manner. Let Φ_s be the optimal return for load L_1 in scenario s; then the average optimal solution for each distribution is

$$\Phi_{Avg} = \sum_{s \in [1,5]} p_1(s) \cdot \Phi_s$$

Let Φ_{PD} be the optimal solution for Program SCS-PD in a given distribution. The deviation, Δ_s, of Φ_{PD} from Φ_s is hence

$$\Delta_s = \left| \frac{\Phi_s - \Phi_{PD}}{\Phi_s} \right|$$

The average deviation of Φ_{PD}, notated Δ_{PD}, follows.

$$\Delta_{PD} = \sum_s p_1(s) \cdot \Delta_s$$

In a similar fashion Δ_{Avg}, the average deviation of Φ_{Avg}, can be calculated.

$$\Delta_{Avg} = \sum_s p_1(s) \cdot \Delta_s$$

However, for Δ_{Avg} the value of Δ_s is calculated differently than for Δ_{PD}.

$$\Delta_s = \left| \frac{\Phi_s - \Phi_{Avg}}{\Phi_s} \right|$$

The outcomes and deviations of the above settings are presented in Table 3. For all the distributions, it is apparent that Δ_{PD} is consistently less than Δ_{Avg}. This indicates that the solution of Program SCS-PD is closer to the different optimal outcomes of Program SCS-DD. Table 3 also shows that returns of Program SCS-DD are always higher than those of Program SCS-PD. Although this might tempt one to criticize the stochastic formulation, there is an important fact that must not be overlooked. In any given distribution, a solution of any of the scenarios bears a significant probability of being infeasible for other scenarios. This is contrary to the solution of the stochastic program, or rather the specific formulation that is used in this work, where any solution is feasible in all scenarios and hence, the robustness.

Table 3: Outcomes and deviations of Programs SCS-DD and PD

Distribution	Φ_{Avg}	Φ_{PD}	Δ_{Avg} (%)	Δ_{PD} (%)
1	417.1000	349.9000	26.6230	15.8192
2	428.1000	342.2333	25.2030	19.8926
3	406.1000	357.5667	28.5581	13.0213
4	414.9000	351.4333	27.0100	14.9350
5	419.3000	348.3667	26.2359	16.6989

4 Notes on Computational Complexity

In SP, the number of scenarios signifies the enumeration of all possible combinations of all the probabilistic elements involved. Needless to say, this may lead to computationally unbearable situations. As such, it was suggested above that when considering the different scenarios, one should only consider scenarios with somewhat relative significance, i.e., associated with probabilities above a certain threshold probability, p_t. This is one way of reducing the problem dimensionality. Another method is used in scenario generation — that is, when dealing with continuous distribution, sampling is sometimes used in order to linearize the formulation. However, in making such sampling a quantization error is introduced with a value that is directly proportional to the sampling interval. In the above two methods, the tradeoff between the solution goodness and the processing required should not be overlooked.

The introduction of integer constraints in the LP formulations affects the solution space and, consequently, complicates the solution method utilized. One of the methods used in solving ILPs is the branch-and-bound algorithm. In a very loose description, the algorithm is a re-iterated LP algorithm with each iteration having further constraints until a solution with integer values for the constrained elements resides on a corner of the solution space; i.e., an optimal integer solution is found. For an accessible reference on integer programming, please refer to Wolsey's *Integer Programming* [20].

As set forth, to accommodate uncertainty in LP, SLP considers all possible instances, each of which has its unique probability. In SILP, the computational complexity of considering the possible scenarios is added to the already complex ILP solution procedures. Although researchers have always been aware of the possible complexities in solving SP problems, only recently has the complexity of certain forms, such as the one discussed and applied herein, has been computed [4]. The basic form introduced in Section 2, sometimes called the two-stage stochastic recourse, has been shown to be "*either easy or NP-hard, depending only on the absence or presence, respectively, of integer decision variables.*"

5 Conclusions

Linear programming, albeit an essential planning utility, relies on the assumptions that the parameters involved are deterministic. Consequently, the theory of stochastic programming was developed for situations when the parameters involved are known to exhibit a quantifiable probabilistic behavior. In this chapter, the motivations for this development were presented, and a brief overview of one of its basic forms, namely stochastic linear programming, was made. More importantly, an application of stochastic integer linear programming in future wireless networks was presented. Such networks are characterized by various challenges that cannot be accommodated by traditional resource management mechanisms that mainly rely on deterministic optimization models.

In order to evaluate the proposed proactive resource allocation mechanism in a comprehensive manner, further measures need to be taken.

First of all, a behavioral model that captures the characteristics, limitations and requirements of both the users and the networks involved needs to be devised. This model would account for users' mobility, their request rates for various services and the type of services they request based on their awareness of the available access networks. On the network side, the model would mimic the networks' reactions and decisions under diversified loading scenarios. Second, probabilistic demand distributions will be extracted from such model. Once extracted, the distributions will be used in generating the allocation policies using the proposed SILP formulation. Once the allocations are acquired, a simulation environment will be created to evaluate the goodness of the allocations against active conditions.

References

[1] D. Bertsimas and J. Tsitsiklis *Introduction to Linear Optimization*, (Massachusetts, Athena Scientific, 1997).

[2] H. Bing, C. He and L. Jiang, Performance Analysis of Vertical Handover in a UMTS-WLAN Integrated Network, in *Proceedings of IEEE Symposium on Personal, Indoor and Mobile Radio Communications*, September 2003, Volume 1, pp. 187–191.

[3] F. Du, L. M. Ni and A.-H. Esfahanian, HOPOVER: A New Handoff Protocol for Overlay Networks, in *Proceedings of IEEE International Conference on Communications*, May 2002, Volume 5, pp. 3234–3239.

[4] M. Dyer and L. Stougie, Computational Complexity of Stochastic Programming Problems, SPOR-Report 2003/20, in *Reports in Statistics, Probability and Operations Research*, (Department of Mathematics and Computing Science, Eindhoven University of Technology, September 2003).

[5] A. Gaivoronski, Stochastic Optimization Problems in Telecommunications, in *Proceedings of the Applied Mathematical Programming and Modelling Conference*, Varenna, Italy, June 2002.

[6] O. Heckmann, J. Schmitt and R. Stenmetz, Robust Bandwidth Allocation Strategies, in *Proceedings of the IEEE International Workshop on Quality of Service*, May 2002, pp. 138–147.

[7] M. Herzberg and A. Pitsillides, A Hierarchical Approach for the Bandwidth Allocation, Management and Control in B-ISDN, in *Proceedings of the International Conference on Communications*, May 1993, Volume 3, pp. 1320–1324.

[8] http://www.ieee802.org/21/index.html.

[9] M. Jaseemuddin, An Architecture for Integrating UMTS and 802.11 WLAN Networks, in *Proceedings of the IEEE International Symposium on Computers and Communications*, July 2003, Volume 2, pp. 716–723.

[10] P. Kall and S. W. Wllace, *Stochastic Programming* (New York, Wiley, 1994).

[11] A. Kamal and H. El-Rewini, On the Optimal Selection of Proxy Agents in Mobile Network Backbones, in *Proceedings of the International Conference on Parallel Processing*, September 2001, pp. 103–110.

[12] R. H. Katz and E. A. Brewer, The Case for Wireless Overlay Networks, in *Proceedings of the SPIE Multimedia and Networking Conference (MMNC'96)*, San Jose, CA, January 29–30, 1996.

[13] B. P. V. Kumar and Pallapa Venkataram, A LP-RR Principle-Based Admission Control for a Mobile Network, *IEEE Transactions on Systems, Man and Cybernetics*, November 2002, Volume 32, Issue 4, pp. 293–306.

[14] Y. Lee, K. Kim and Yanghee Choi, Optimization of AP Placement and Channel Assignment in Wireless LANs, in *Proceedings of the 27th Annual IEEE Conference on Local Computer Networks*, November 2002, pp. 831–836.

[15] X. Liu, Network Optimization with Stochastic Traffic Flows, *International Journal of Network Management*, July/August 2002, Volume 12, Issue 4, pp. 225–234.

[16] M. Riis and K. A. Andersen, Capacitated Network Design with Uncertain Demand, *INFORMS Journal on Computing*, Summer 2002, Volume 14, Issue 3, pp. 247–260.

[17] S. Sen and J. L. Higle, An Introductory Tutorial on Stochastic Linear Programming Models, *Interfaces*, February 1999, Volume 29, Issue 2, pp. 33–61.

[18] S.-L. Tsao and C.-C. Lin, Design and Evaluation of UMTS-WLAN Interworking Strategies, IEEE Vehicular Technology Conference, Volume 2, pp. 777–781, September 2002.

[19] M. I. Velazco, A. R. L. Oliveira, and C. Lyra, Neural Networks Give a Warm Start to Linear Optimization Problems, in *Proceedings of the International Joint Conference on Neural Networks*, May 2002, Volume 2, pp. 1871–1876.

[20] L. Wolsey, *Integer Programming* (New York, Wiley, 1998).

Chapter 7

Selecting Working Sensors in Wireless Sensor Networks

Haining Chen
Center For Advanced Computer Studies
University of Louisiana at Lafayette, P.O. Box 44330, Lafayette, LA 70504
E-mail: hxc5633@cacs.louisiana.edu

Hongyi Wu
Center For Advanced Computer Studies
University of Louisiana at Lafayette, P.O. Box 44330, Lafayette, LA 70504
E-mail: wu@cacs.louisiana.edu

1 Introduction of Wireless Sensor Networks

The recent advances in integrated circuits, wireless technologies, and micromechanisms enables the manufacture of low-cost, low-power, and multifunctional wireless sensors [1]. A typical wireless sensor consists of several units for sensing, data processing, storage, communication, and/or localization. All of these units are integrated into one tiny sensor with limited size and power supply. Each sensor has one or more than one sensing ability, such as the sensing of temperature, moisture, specific sound signals, and/or light. After the target information is collected, the sensor can perform data processing, e.g., by filtering out irrelevant information or averaging the accumulated data over a required period, and then report the sensing data to a special sensor, which is referred to as a *sink* sensor. For wireless sensors, the data transmission and reception are achieved through radio or infrared interface.

Given the limited sensing, data processing, and transmission power of individual sensors, the sensor network is formed with the collaboration of a large number of low-cost sensors, in order to carry out various tasks, such as civilian surveillance, military command and control, meteorological phenomenon monitoring, etc. The functioning of the sensor network depends on the collective efforts of all the sensors in the network. For instance, the forwarding of the collected information may involve many sensors along one or several routing paths.

Because the sensors are powered by batteries with limited lifetime, it is a challenging design issue in a sensor network to improve energy efficiency and prolong the network working lifetime. In this chapter, we address the issue of selecting working sensors in large-scale distributed wireless sensor networks and introduce several state-of-the-art solutions to this problem.

2 Motivation and Problem Formulation

We first formulate the optimization problem discussed in this chapter: in a large-scale distributed wireless sensor network, given the limited lifetime of each individual sensor, how to select a minimum number of working sensors among all the sensors while maintaining system coverage and/or network connectivity, in order to prolong the total lifetime of the entire sensor network.

The motivation of working sensor selection originates from the discrepancy between the requirement of prolonging the total lifetime of the sensor network and the power limitation of each individual sensor. With strained power limitation, individual sensors cannot last for a long duration. An individual sensor can be in one of the two states: working state or sleeping state. In the working state, the sensor may be transmitting data, receiving data, processing data, and/or sensing data, and all these activities consume a certain amount of power. In the sleeping state, the sensor usually shuts down most circuits and only wakes up once in a while to receive possible control messages and thus consumes very little power. These control messages are important because they will be used by the sensor to switch itself between the working state and the sleeping state.

Redundancy usually exists in a large-scale sensor network. Because there is an oversupply of sensors in the network, it is not necessary to keep all of them in the working state. The basic idea for prolonging network lifetime is to keep only a subset of sensors in the working state at any given time instance, while all other sensors stay in the sleeping state and thus consume no or very little power. If one or more sensors in the working state is running out of power, other sleeping sensor(s) may replace the depleted sensor(s) and turn itself (or themselves) into the working state so that the whole sensor network can still function properly. In this way, the total lifetime of the network is prolonged. The solutions we introduce in this chapter all share the same basic idea described above, yet are different in the way to select the working sensors out of a group of sensors and to switch from the sleeping state to the working state.

To evaluate the effectiveness of the solutions, two important criteria are usually checked against the results of the working sensor selection process. One is coverage requirement; the other is connectivity requirement. Coverage requirement means that the selected working sensors provide full coverage of the area of concern, or if full coverage cannot be met, a coverage percentage threshold will be set as the criterion. Connectivity requirement means that the selected working sensors interconnect themselves so that at least one communication path exists between them, which makes it possible to pass the collected information to the sink or to forward control messages to the selected working sensors.

Besides coverage and connectivity, there are some other design issues that need to be considered, e.g., whether the solution is distributed or centralized, whether the location information of each sensor is needed, and whether a one-shot or multishot scenario is considered. If the solution only deals with the selection of a subset of sensors, we call it one-shot scenario oriented; otherwise, if the solution deals with not only selecting a subset of sensors but also replacing the sensors after their expiration, it is multishot scenario oriented. Another design issue is the signaling overhead. The signaling overhead should be very low so that running the working sensor selection algorithm will not consume a significant amount of power. Scalability is also an important design issue, which means that the proposed solution can scale to a sensor network of large size without incurring any difficulty, and that the network lifetime increases linearly with the number of sensors deployed.

It is worth mentioning that the working node selection problem under the constraints of the coverage and/or connectivity requirement is NP-

Table 1: Design features of the four solutions to the working sensor selection problem

Solutions	PEAS	Connected Sensor Cover	OGDC	Grid-Based Selection
coverage requirement	no	yes	yes	yes
connectivity requirement	yes (conditional)	yes	yes (conditional)	no
distributed or centralized	distributed	both	distributed	both
location info. needed	no	yes	yes	yes
one-shot or multishot	multishot	one-shot	multishot	multishot
signaling overhead	low	medium	medium	low
scalability performance	good	not good	good	good

hard, as is revealed in the following detailed discussions of each solution. The performance of the solutions can be evaluated by the number of working sensors selected, and if the multishot scenario is supported, the other metric can be the total lifetime of the sensor network.

3 Current Solutions to the Working Sensor Selection Problem

Four representative approaches to solve the working sensor selection problem are discussed in this chapter, with their principles illustrated and performances evaluated. These four approaches are named Probing Environment and Adaptive Sleeping (PEAS), Connected Sensor Cover, Optimal Geographical Density Control (OGDC), and Grid-Based Selection. Table 1 summarizes their main features, and the details are elaborated next.

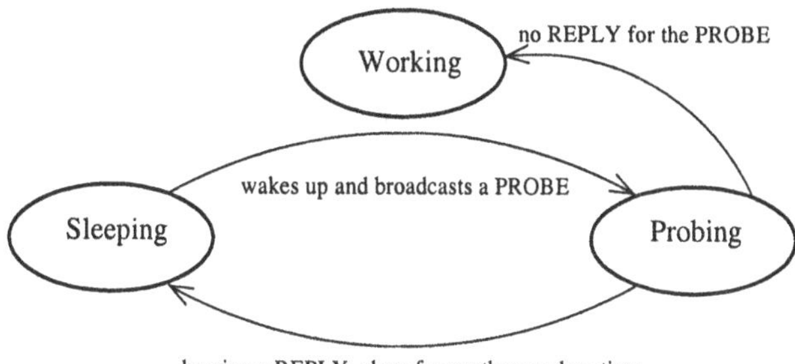

Figure 1: State transition process of every sensor in PEAS.

3.1 Probing Environment and Adaptive Sleeping (PEAS)

Principle. Probing Environment and Adaptive Sleeping (PEAS) is a distributed working sensor selection algorithm proposed in [2]. PEAS does not need sensors' location information and it is multishot scenario oriented. PEAS does not address the coverage requirement. Although PEAS does not guarantee the connectivity requirement, it provides asymptotic connectivity among working sensors under several assumptions.

The basic idea of PEAS is described as follows. Each sensor wakes up from the sleeping state to probe its surrounding environment periodically and makes a decision whether to switch to working state. If there is no working sensor within a certain range, this wake-up sensor switches to the working state; otherwise, this sensor goes back to the sleeping state. Figure 1 illustrates the state transition process in PEAS. Two messages are used to determine whether there is a neighboring sensor nearby, the PROBE message sent by the wake-up probing sensor, and the REPLY message sent by the working sensor(s) upon receiving a PROBE message.

Each individual sensor independently makes its own decision on staying in the working state or not by following the state transition process shown in Figure 1. One advantage of PEAS is that no neighboring sensors' state information needs to be kept by each sensor, which makes this approach simple enough to be implemented in sensors with limited computing ability. Another advantage lies in the fact that PEAS does not depend on the remaining power of each sensor, which makes PEAS appealing to the scenario where sensor failure rate is high.

One important design issue in PEAS is to decide when each sensor should wake up and start probing. Intuitively, the more frequently the sleeping sensors wake up, the better coverage PEAS can achieve. But at the same time, it also results in more signaling overhead and more power consumption. In order to address the tradeoff between coverage and overhead, PEAS takes the advantage of the unique characteristic of the Poisson distribution, and lets the wake-up interval time of each sensor obey the exponential probability density function $f(t_s) = \lambda e^{-\lambda t_s}$, where t_s is the sleeping interval time of each sensor, and variable λ is the *probing rate* of each sensor. In PEAS, the probing sensor adaptively adjusts its own probing rate λ, in order to converge the aggregated probing rate perceived by the working sensor to a desired value λ_d, which is specified by the application.

The process of adaptively adjusting the probing rate is illustrated below. After the number of received PROBE messages reaches k (a constant value large enough, usually set to 32 based on experiments), the working sensor calculates the aggregated probing rate as $\hat{\lambda} = \frac{k}{\Delta t}$, where Δt is the time duration to receive k PROBE messages. Then $\hat{\lambda}$ is included in the REPLY message sent back to the probing sensor. A new probing rate will then be calculated by each probing sensor as $\lambda^{new} = \lambda \frac{\lambda_d}{\hat{\lambda}}$, where λ is the current probing rate. It is known that the sum of the n Poisson distribution is still the Poisson distribution, but with a different parameter $\bar{\lambda} = \sum_{i=1}^{n} \lambda_i$, where λ_i is the probing rate of the ith individual probing sensor surrounding the working sensor. Thus the new aggregated $\bar{\lambda}^{new} = \sum_{i=1}^{n} \lambda_i^{new} = \sum_{i=1}^{n} \lambda_i \frac{\lambda_d}{\hat{\lambda}} = \bar{\lambda} \frac{\lambda_d}{\hat{\lambda}}$. When constant k is large enough, ideally, the measured value of the aggregated probing rate $\hat{\lambda}$ should equal the theoretical value $\bar{\lambda}$. Then we have $\bar{\lambda}^{new} = \bar{\lambda} \frac{\lambda_d}{\hat{\lambda}} \approx \bar{\lambda} \frac{\lambda_d}{\bar{\lambda}} = \lambda_d$, which means that the aggregated probing rate perceived by the working sensor converges to the desired λ_d.

Because the aggregated probing rate at each working sensor can reach the desired value λ_d, $\frac{1}{\lambda_d}$ is the average wake-up time perceived by the working sensor for all the sleeping sensors surrounding it, and $\frac{1}{\lambda_d}$ is also the coverage gap in the time domain. In PEAS, the connectivity requirement is asymptotically met by making two assumptions. The concerned area is divided into square cells, and the border of the square cell is R_p, which equals the probing radius of a sensor. Let R_t denote sensor transmission range. The first assumption is that a sufficient amount of sensors is distributed in the concerned area such that there is at least one sensor in each cell. The second assumption is that $R_t \geq (\sqrt{5}+1)R_p$. Based on these two assumptions, the connectivity requirement is met by proving that the largest distance between any two adjacent working sensors is $(\sqrt{5}+1)R_p$.

Working Scenario. Sensors in PEAS function independently. They wake up and probe, and then either serve as working sensors, or stay as sleeping sensors. The PEAS protocol is simple and the signaling overhead is low. PEAS keeps any two working sensors at least R_p away from each other due to the probing process, which prevents the overlapping of selected working sensors and thus achieves power efficiency. Because the sensors wake up in a distributed manner, the coverage gap is less than that in a synchronized wake-up schedule. The wake-up time interval of each sleeping sensor is randomized according to a Poisson distribution, and the aggregated probing rate will converge to λ_d at the working sensor. Frequent wake-up shortens the coverage gap and helps to counter the sensor failure problem at the expense of more power consumption. So an application-specific λ_d must be carefully chosen in order to maintain adequate working sensors without consuming excessive power. Because the coverage gap in the time domain is inevitable, PEAS does not guarantee coverage.

The performance of PEAS is evaluated by two metrics: the coverage lifetime and the data delivery time. The coverage lifetime represents how long PEAS can maintain certain coverage of the area. The data delivery time deals with the connectivity requirement by measuring how long successful data delivery routes can last. An arbitrary data forwarding protocol is used for the data delivery. The simulation results show a steady linear growth of the coverage lifetime and data delivery time with the increased deployment of sensors, proving the effectiveness of PEAS. Also sshown is that even with the presence of a high failure rate of the sensors, PEAS demonstrates robustness against sensor failures.

3.2 Connected Sensor Cover

Principle. The Connected Sensor Cover approach is proposed in [3]. This approach provides both centralized and distributed algorithms. It is one-shot scenario oriented and needs sensors' location information. It addresses and satisfies both the coverage and connectivity requirements.

The motivation of proposing Connected Sensor Cover is to select a minimum number of connected sensors to perform the query task in a sensor network. Certain sensing information must be forwarded to the sink sensor in the query task, therefore it is necessary for the selected sensors in the query task to be connected to each other. The coverage requirement needs to be satisfied because the query task is to monitor the entire network. The minimum number of working sensors selected is also desirable in order to achieve power efficiency. The Connected Sensor Cover problem is NP-hard, because it can be polynomially mapped to the problem of covering points using line segments, which is proven to be NP-hard [8].

The basic idea of Connected Sensor Cover is to start from an original sensor (typically the initiator of the query task), and to include more sensors into the Connected Sensor Cover set at each stage of the algorithm, as long as there exists a communication path between the already selected sensors and the later included sensors. Thus the connectivity requirement is always guaranteed throughout the sensor selection process. The expanding process continues until the coverage requirement is satisfied. The essential part of this algorithm is to select as few sensors as possible in order to reduce power consumption. The strategy taken is to select such a communication path that covers as many uncovered regions as possible with as few unselected sensors in this path as possible.

For each query, the Connected Sensor Cover algorithm needs to be run if there is no Connected Sensor Cover set available. After one such set has been elected, it can serve as long as the selected sensors have enough remaining power. The Connected Sensor Cover algorithm focuses on how to select such a sensor set. It does not deal with the replacement of depleted sensors. Thus it is one-shot scenario oriented.

Working Scenario. Two algorithms are introduced in the Connected Sensor Cover approach: a centralized algorithm and a distributed algorithm. We first discuss the centralized algorithm. Denote M to be the Connected Sensor Cover set, C to be the set of candidate sensors that are not included in M and whose coverage area intersects with the coverage area of M, and P to be the set of candidate paths. Each candidate path connects some sensors in M with a corresponding candidate sensor in C, and other candidate sensors and unselected sensors may also exist along this candidate path. Initially, M is empty and the originator of the query task will be added into M. In each stage, all the candidate sensors and their candidate paths are evaluated and the most beneficial candidate path is chosen. Then all the unselected sensors in this chosen path are added into M. Here the benefit of a candidate path is calculated as the uncovered area of this candidate path divided by the number of sensors that are in this candidate path but not in M. The idea behind this is to choose a new path with the highest gain in each stage, where the highest gain is indicated by the largest uncovered area per unselected sensor. The above process repeats until the coverage requirement is met.

Figure 2 is an example of candidate sensors and candidate paths. Because P_2 is the most beneficial path, all the unselected sensors in P_2, namely, C_2, I_2 and C_1, are merged into M.

Note that the existence of a candidate path for a candidate sensor is not always guaranteed. For example, if the density of the sensors is not high enough, it is possible that no candidate path can be found before the coverage requirement is satisfied.

When it comes to selecting a minimum number of working sensors, the optimality of this approach is proven to be within $O(r \log n)$ factor of the optimal result, where r is the link radius of the sensor network defined as the maximum number of hops between two sensors whose sensing regions intersect, and n is the total number of sensors in the query region. Here the optimal result refers to the case when only the coverage requirement is considered.

A weighted version of the Connected Sensor Cover algorithm is also proposed by taking into consideration the remaining power of sensors, and favoring those sensors with more remaining power. Thus the stability of the selected sensors is improved. In the weighted version, higher weight is assigned to sensors with less energy and the calculation of the candidate path's benefit is redefined as dividing the uncovered area by the sum of weights in the candidate path, which will put the path with higher weight into an unfavorable situation.

It is natural to expand the centralized algorithm to a distributed one. In the distributed algorithm, similarly, M is empty initially, and a randomly chosen sensor is added to M. Then this most recently added sensor broadcasts the *Candidate Path Search (CPS)* message within $2r$ hops, where r is the network link radius mentioned before. Here $2r$ is used instead of r as the broadcast distance in order to let more sensors receive the *CPS* message. The candidate sensors receiving this *CPS* message will respond with a *Candidate Path Response (CPR)* message back to the originating sensor. After the originating sensor gathers the information of all candidate paths, it selects the most beneficial one. Then a *NewC* message will be unicast from the originating sensor to the new candidate sensor in this chosen path, which will serve as the originating sensor of the *CPS* message for another round of finding the most beneficial path. The above process repeats until the coverage requirement is met. One important issue in distributed algorithms is to keep the signaling overhead small to save energy. The power consumed in the distributed algorithm is related to the chosen value of r, which determines how far away the *CPS* message can travel in the network. Obviously, there is a tradeoff between signaling overhead and optimality of the result. If r is set to a large value, more sensors will be involved in the distributed algorithm and thus more energy will be consumed, although a better result can be achieved with fewer working sensors. On the other hand, if r is set to a small value, less signaling overhead is expected at the expense of a less optimal result with more working sensors. Several strategies are adopted to reduce the signaling overhead, such as reusing the previously calculated results of the candidate paths, eliminating unnecessary broadcasting of *CPS* message, etc.

Simulations are carried out for both centralized and distributed algorithms. The results show that, if the density of the sensor is high enough, only a small portion of the sensors is needed to construct a Connected Sensor Cover set with a small amount of communication overhead. It also shows that the distributed algorithm yields very close results as the centralized algorithm does.

3.3 Optimal Graphical Density Control (OGDC)

Principle. Optimal Geographic Density Control (OGDC) [4] is a distributed multishot scenario oriented working sensor selection algorithm, aiming at satisfying both the coverage and the connectivity requirement. It needs location information.

Several assumptions are made to simplify the working sensor selection problem. First, the power consumption of each individual sensor is assumed to be proportional to its sensing range. The sensor's sensing range is considered to be a circle centered at itself. Thus reducing the overlap of the sensors' sensing ranges is equivalent to reducing the power consumption of the network. Second, all the sensors are supposed to have the same sensing range, and thus selecting a minimum number of working sensors is equivalent to minimizing the overlap of the coverage areas of all selected sensors. Third, each sensor's transmission range is assumed to be at least twice its sensing range. With this assumption, it is proven in [4] that fulfilling the coverage requirement guarantees connectivity at the same time. Therefore, OGDC concentrates on how to satisfy the coverage requirement, while minimizing the number of working sensors.

Two rules are developed to minimize the overlap of the working sensors' coverage areas thus resulting in the minimum number of working sensors. Rule 1 is that the distance of any two neighboring sensors is

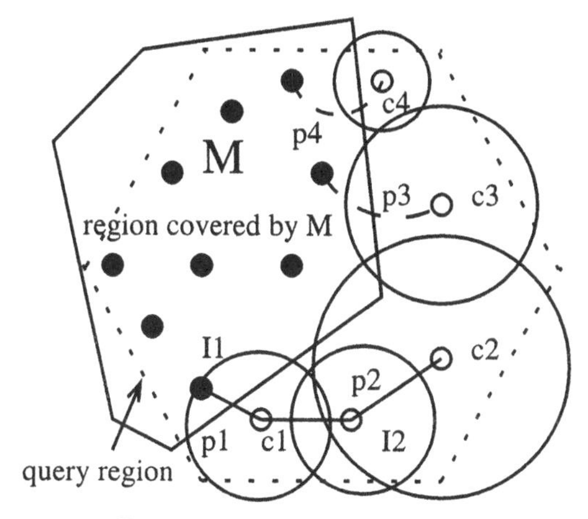

Candidate Sensors: c1, c2, c3, c4

Candidate Paths: p1, p2, p3, p4

p1 = < c1, I1 >; p2 = < c2, I2, c1, I1>

Figure 2: Candidate sensors, C_1, C_2, C_3, C_4 and associated candidate paths, P_1, P_2, P_3, P_4.

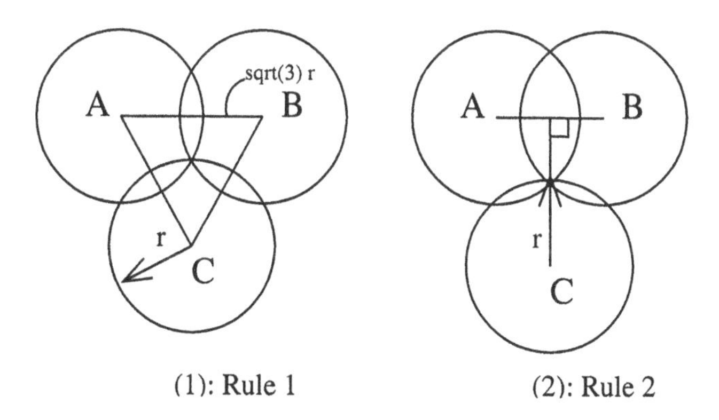

Figure 3: Two rules in OGDC used for selecting the minimum number of working sensors. (1) Rule 1: the positions of any three adjacent sensors A, B and C are adjustable, and (2) Rule 2: two adjacent sensors A and B are fixed in position and the third sensor C is adjustable in position.

$\sqrt{3}\, r$, where r is the sensing radius of a sensor. This rule applies to the scenario when the positions of any three adjacent sensors are adjustable. Rule 2 applies to the scenario where two adjacent sensors are fixed in position, and the position of the third sensor is adjustable. According to rule 2, the third sensor should be placed at a position where only one of the two cross-points of two fixed sensors is covered by the third sensor, and the distance from the center of the third sensor to the covered cross-point is r, and also the line connecting the center of the third sensor and the covered cross-point is vertical to the line connecting the centers of the two fixed sensors. These two rules are illustrated in Figure 3.

Working Scenario. The one-shot working scenario of running OGDC is to apply the above two rules to all the sensors. At first, one or more sensors is selected as starting sensor(s), such as sensor A in Figure 3. Then sensor B is selected if the distance between sensor A and B is (approximately) $\sqrt{3}\, r$, according to Rule 1. Then the third sensor C according to either Rule 1 or Rule 2 is selected. If the desired sensor cannot be found at the optimal position determined by Rule 1 or Rule 2, the sensor nearest to the optimal position will be selected, instead. The above process repeats until every sensor in the network knows its own state for this round of working sensor selection, which is either selected or unselected, corresponding to the working sensor and sleeping sensor, respectively.

For the multishot working scenario, a time interval is chosen to run the OGDC algorithm periodically, and there is a power threshold requirement for each sensor in order to make the sensor eligible to be selected as the working sensor. This power threshold is set to a value high enough to let the selected sensor sustain itself almost throughout the entire time interval for this round of working sensor selection.

The discussion so far is based on the assumption that all sensors have the same sensing range. The OGDC approach is extended in [5] to handle sensors with different sensing ranges, where the basic idea and objective remain the same (i.e., to minimize the overlapped sensing area in order to achieve power efficiency). The optimal solution reached in [5] is similar to that in [4], with adjacent sensors placed in certain positions corresponding to each other. Thus the distributed algorithm can start with one or more initial sensor(s), and other sensors can be selected subsequently by following the optimal solution.

In [4], the performance of the OGDC approach is compared with that of the PEAS approach based on the same power consumption model. The simulation results show that OGDC is better than PEAS in terms of the total lifetime of the sensor network and the number of working sensors selected in each round. One advantage of the OGDC approach is the linear growth of the sensor network's lifetime with the total number of sensors deployed in the area. This is a nice property that makes this approach scalable. One shortcoming of OGDC is that it uses a fixed time interval to initiate a new round of the working sensor selection process, which may not always guarantee full coverage at the end of each round because some selected sensors may be running out of power before the next time interval arrives for another round of working sensor selection.

3.4 Grid-Based Selection

Principle. The Grid-Based Selection approach is proposed in [6] and extended to the multishot scenario in [7]. This approach provides both centralized and distributed algorithms. It only considers network coverage in order to decouple the coverage requirement from the connectivity requirement. In addition, the algorithms need the sensors' location information.

The basic idea is to establish a virtual grid on the sensing area, and let the intersection points of the grid represent the coverage area. Thus each sensor's coverage area is replaced by a set of intersection points inside its own sensing range, which may be in any arbitrary shape. To simplify the following discussion, we assume the sensor's sensing range is a circle with a radius of r centered at the sensor itself. Therefore, the coverage requirement is converted into a set cover problem, where a minimum number of sensors are selected to include all the intersection points of the grid in the concerned area.

Note that the connectivity requirement is always associated with the specific routing protocol used. For example, higher connectivity is needed if the network employs a multipath routing protocol. Here we decouple the coverage requirement from the connectivity requirement, so that the result of the working sensor selection is independent of the routing protocol chosen.

In the centralized algorithm, all sensors' location and coverage information is forwarded to a control sensor that runs the working sensor selection algorithm and passes the results to all the sensors, indicating whether they should stay in the working state or in the sleeping state. In the distributed algorithm, a cluster formation algorithm [9] is employed to form clusters in the sensor network. Each cluster has a cluster head connected to every sensor in this cluster. The cluster head runs the Grid-

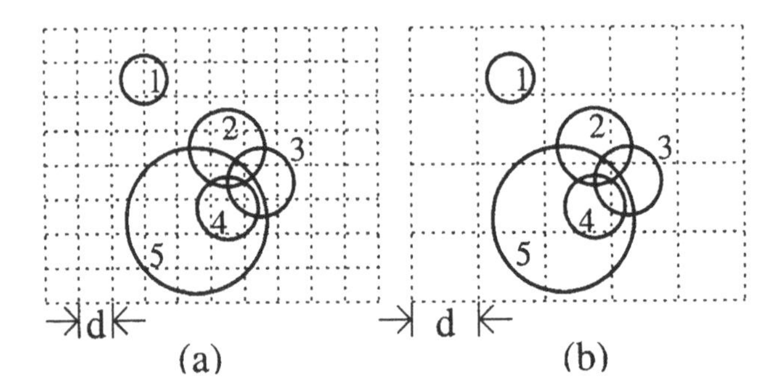

Figure 4: Sampling with different grid distances under (a) dense grid and (b) coarse grid.

Based Selection algorithm to select a working sensor within this cluster. Thus the time complexity of the distributed algorithm is reduced, which makes this approach scalable to sensor networks of large size.

Besides the size of the sensor network, grid density is another important factor that affects the time complexity of the algorithm. A higher grid density means higher resolution of the coverage area, which results in more accurate solutions at the expense of higher computing complexity, because more intersection points are involved in the computation. Figure 4 gives an example of the impact of different grid densities on working sensor selection. Sensor 2 and sensor 3 have different coverage areas (in terms of the covered intersection points) under a dense grid (see Figure 4(a)); in the meantime, sensor 2 and sensor 3 are represented by the same set of covered intersection points under a coarse grid (see Figure 4(b)). A certain grid density corresponds to a certain error rate. The selection of grid density is determined by the application according to the tolerable error rate.

When the selected sensors expire, the coverage area will decrease accordingly. If the coverage area drops below a threshold, the Grid-Based Selection approach will respond immediately by selecting new working sensors to compensate for the reduced coverage area. The advantage of responding immediately to the expiration of any working sensors is that the sensor network can maintain coverage without leaving any coverage gap in the time domain, however, such coverage gap is almost inevitable if a fixed time interval is set to invoke each round of working sensor selection process.

Working Scenario. In the one-shot scenario, the working sensor selection process is converted to the setcover problem, which is proven to be NP-hard [10]. To solve the setcover problem, a simple greedy algorithm and an integer linear programming model are developed. Simulation results show that the integer linear programming model yields more optimal results but is not scalable, whereas the greedy algorithm is scalable and achieves near optimal results.

In the multishot scenario, the dynamic working sensor selection process aims at prolonging the lifetime of the entire network. The working sensor selection problem in the multishot scenario is also NP-hard, because it can be polynomially reducible to the Set-K Cover problem that is proven to be NP-hard [11], by assuming that all the selected working sensors in each round have the same power consumption rate and observing that they will expire at the same time. Four approaches are proposed in [7]. They are classified by whether sensors used in the previous round can be reused (denoted by retrospective and nonretrospective approach), and whether local repair is performed. Here local repair refers to replacing the depleting working sensor with some nearby local sleeping sensor(s), which can compensate for the loss of the coverage area caused by the depleting sensor.

Metrics used to evaluate the performance of the approach include the number of working sensors selected, the signaling overhead involved and the total lifetime achieved. The performance of the Grid-Based Selection approach is compared with that of OGDC under the same power consumption model. It is shown that the Grid-Based Selection approach yields a less optimal result than OGDC does in terms of number of working sensors selected and the total network lifetime, whereas OGDC incurs more computing process in each sensor. Moreover, OGDC requires a certain density of the sensors because otherwise the desired sensors cannot be found at the optimal location required by OGDC. In addition, OGDC does not guarantee full coverage due to the use of a fixed time interval for the multishot scenario of working sensor selection. In the multishot scenario of the Grid-Based Selection approach, simulation results in [7] indicate that the local repair process achieves local optimality and the retrospective approach achieves global optimality. Thus local repair works well in the nonretrospective approach but not in the retrospective approach. The drawback of the retrospective approach is that it usually invokes the working sensor selection algorithm more times than the nonretrospective approach does because some previously used sensors may not last long before depleting themselves, thus the retrospective approach involves more signaling overhead than the nonretrospective approach.

4 Conclusions

We have discussed the optimization problem of selecting working sensors in a wireless sensor network. The goal is to prolong the network lifetime, given the fact that each individual sensor is limited by its power and computing ability. The basic idea is to select a subset of sensors functioning in the working state, while other unselected sensors stay in the sleeping state and consume very little power.

The working sensor selection problem is proven to be NP-hard. In the one-shot scenario, it deals with the optimal solution of selecting one subset of sensors to satisfy the coverage and/or connectivity requirement, whereas in the multishot scenario, it aims at maximizing the total lifetime of the network.

Four representative approaches are discussed in this chapter, with their principles and working scenarios illustrated. Through simulation results, these four approaches all demonstrate their effectiveness in selecting working sensors. The differences among these four approaches can be summarized as the different tradeoffs between several design issues, such as coverage, connectivity, signaling overhead, etc.

In general, the PEAS approach aims at the simplicity of the algorithm that can be implemented on sensors with low computing ability. But it does not guarantee coverage and it can guarantee connectivity only conditionally. The Connected Sensor Cover approach focuses on meeting the coverage and connectivity requirement but it involves heavy computing at each sensor, which affects the scalability of this approach. OGDC emphasizes the minimum number of selected working sensors. It can achieve a near optimal result, although it requires the density of the sensors to be sufficiently high. It aims at satisfying the coverage requirement. At the same time, it meets the connectivity requirement only if the coverage requirement has been satisfied and a specific assumption is made about the relationship between the transmission radius and the sensing radius of the sensors. The Grid-Based Selection approach only addresses the coverage requirement. It forms clusters to scale to large sensor networks. Though with slightly less optimal results compared to OGDC, the Grid-Based Selection approach works well in the network with lower sensor density. In a nutshell, every solution has its own fitness for different application scenarios. A suitable solution should be chosen according to the requirements of the sensor network and the characteristics of the sensors.

Acknowledgment

This research is in part supported by NSF CAREER Award, NSF CNS-0347686.

References

[1] I. F. Akyildiz, W. Su, Y. Sankara and E. Cayirci, "A Survey on Sensor Networks," in *IEEE Communication Magazine*, August 2002.

[2] F. Ye, G. Zhong, J. Cheng, S. Lu and L. Zhang, "PEAS: A Robust Energy Conserving Protocol for Long-lived Sensor Networks," in *Proceedings of the 23rd International Conference on Distributed Computing Systems (ICDCS'03)*, May 2003.

[3] H. Gupta, S. R. Das and Q. Gu, "Connected Sensor Cover: Self-Organization of Sensor Networks for Efficient Query Execution," in *Proceedings of ACM MobilHoc 2003*, June 2003.

[4] H. Zhang and J. C. Hou, "Maintaining Sensing Coverage and Connectivity in Large Sensor Networks," in *NSF International Workshop on Theoretical and Algorithmic Aspects of Sensor, Ad Hoc Wireless, and Peer-to-Peer Networks*, February 2004.

[5] H. Zhang and J. C. Hou, "Maintaining Sensing Coverage and Connectivity in Large Sensor Networks," in *Wireless Ad Hoc and Sensor Networks: An International Journal*, Vol. 1, No. 1-2, pp. 89–123, January 2005.

[6] H. Chen, H. Wu and N-F. Tzeng, "Grid-based Approach for Working Node Selection in Wireless Sensor Networks," in *Proceedings of International Conference on Communications (ICC'04)*, June 2004.

[7] H. Chen, H. Wu and N-F. Tzeng, "Grid-based Dynamic Working Node Selection Algorithms," Technical Report, University of Louisiana at Lafayette, June 2004.

[8] V. S. A. Kumar, S. Arya and H. Ramesh, "Hardness of Set Cover with Intersection 1," in *Automata, Languages and Programming*, 2000.

[9] A. D. Amis, R. Prakash, T. H. P. Vuong and D. T. Huynh, "Max-Min D-Cluster Formation in Wireless Ad Hoc Networks," in *Proceedings of INFOCOM 2000*, March 2000.

[10] M. R. Garey and D. S. Johnson, *Computers and Intractability: A guide to the Theory of NP-Completeness*, W. H. Freeman, 1979.

[11] S. Slijepcevic and M. Potkonjak, "Power Efficient Organization of Wireless Networks," in *Proceedings of IEEE International Conference on Communications*, June 2001.

Chapter 8

Quality of Service Provisioning for Adaptive Multimedia in Mobile/Wireless Networks

Yang Xiao
Department of Computer Science
The University of Memphis, Memphis, TN 38152 USA
E-mail: `yangxiao@ieee.org`

1 Introduction

During the last few years there has been a great demand for multimedia applications with Quality of Service (QoS) in wireless/mobile networks. QoS provisioning for multimedia applications has become more important than before partially due to the service users' requirements. The key factor in QoS provisioning is Call Admission Control (CAC) [10], which can efficiently utilize the system resources and satisfy QoS parameters' requirements. CAC has become vital for multimedia services in a wireless/mobile network's ability to guarantee QoS requirements partially due to the network's limited capacity.

In wireless/mobile networks, many CAC schemes in wireless/mobile networks have been proposed in the literature [1-5,7-10,13,16-17]. We can classify CAC schemes into three categories: the classical non multimedia CAC schemes [2,3], the non adaptive multimedia CAC schemes [1,8,9], and the adaptive multimedia CAC schemes [4,5,10,16,17]. In the three kinds of CAC schemes, different kinds of CAC schemes work on different situations. The classical non multimedia CAC schemes are for non multimedia traffic, e.g., traffic of cellular phone calls. The non adaptive multimedia CAC schemes are for multimedia traffic, but this multimedia traffic is not adaptive. However, the adaptive multimedia CAC schemes are for adaptive multimedia traffic.

The classical non multimedia CAC problems typically try to find a better CAC scheme handling with handoff calls and new calls. There are three approaches for the classical non multimedia CAC schemes: the complete sharing approach, the complete partitioning approach, and the threshold approach [16,17]. The Fractional Guard Channel, a special kind of the threshold approach, is the optimal approach [2,3]. The optimization is in the sense of optimizing the revenue while satisfying the QoS requirement, which is an upper bound of handoff dropping probability. For the non adaptive multimedia CAC schemes, Choi et al. [7] present an optimal centralized CAC scheme; Kwon et al. [8] and Yoon et al. [9] propose an optimal distributed CAC scheme. Most recently, many researchers focus on adaptive multimedia services [4,5,10,16-22]. Kwon et al. [5] investigate the CAC scheme for one class of adaptive multimedia. Kwon et al. [4] seek CAC and bandwidth reallocation algorithms for multi-classes of adaptive multimedia by using a graph resource reallocation approach with a greedy algorithm, which gives a suboptimal or near optimal solution.

In order to provide QoS provisioning, a good service classification is important. In this chapter, we introduce a QoS provisioning framework for multimedia traffic in wireless/mobile networks. We classify services into three categories: Bandwidth Guaranteed (BG) service, Bandwidth Not Guaranteed (BNG) service, and Best Effort (BE) service. BG service is for non adaptive multimedia traffic or non multimedia traffic; BNG service is for adaptive multimedia traffic; whereas BE service is for computer data or multimedia traffic that can be suspended and reactivated. Both the classical non multimedia CAC schemes and the non adaptive multimedia CAC schemes are for the BG service, and the adaptive multimedia CAC schemes are for the BNG service or the BE service. Three traffic descriptors are defined here: Fixed Bandwidth (FB), Minimum Bandwidth (MB), and Upper Bandwidth Vector (UFV). For each of the three categories, traffic descriptors and QoS parameters are defined and specified. Furthermore, we abstract the three categories into a general abstract traffic model. Based on the abstract general model, we present, analytically, an optimal Call Admission Control (CAC) scheme for all service categories that guarantees the QoS parameters' requirements and traffic descriptors, and maximizes the revenue. The proposed CAC scheme adapts well to all service categories, achieves optimal revenue, and guarantees the QoS parameters' requirements and the traffic descriptors.

Our adaptive multimedia traffic (BNG and BE services) adopts the layered coding approach, which was originally introduced in wired networks, where the multimedia receivers can selectively choose a subset of the hierarchical coding depending on capacity [4-5,10,18-22]. When a sender transmits the layered coding of multimedia stream to a receiver, the receiver will receive the whole multimedia stream or a subset of it, depending on the cell load condition. If a cell is overloaded (congestion), the multimedia stream is filtered at the Base Station based on CAC decisions, and the receiver can only receive a subset of the multimedia stream; otherwise, the receiver will receive the whole multimedia stream. The layered coding multimedia play an important role because the non layered coding multimedia (BG service), in fact, is a special case of layered coding multimedia (BNG service). We show that the non layered coding (BG service) optimal CAC is a special case of our optimal CAC scheme (BNG service), and the bandwidth of a call in the cell is not adjustable in the BG service and the bandwidth reallocation actions are null actions (non adaptive).

The rest of this chapter is organized as follows. Service categories and QoS are presented in Section 2. Section 3 describes our traffic model for service categories. Section 4 introduces the background and the ideas for CAC-BRA. Section 5 presents our CAC-BRA for one class case and shows how it maximizes the revenue/utilization and guarantees traffic descriptors and QoS parameters' requirements. In Section 6, we present our CAC-BRA for the multi-class general case. In Section 7 we discuss the complexity issue of our methods in the real system. Finally, we conclude this chapter in Section 8.

2 Service Categories and QoS

2.1 Service Categories

In order to efficiently support a multi-application environment, we classify the services into three categories: Bandwidth Guaranteed (BG) service, Bandwidth Not Guaranteed (BNG) service, and Best Effort (BE) service. Figure 1 shows the three service categories. BG service is analogous to Constant Bit Rate (CBR) service in ATM networks, and it guarantees the required bandwidth. BG service can be used in non adaptive multimedia applications with strict bandwidth requirements or non multimedia applications, where the bandwidth of an ongoing call is fixed during its lifetime.

BNG service is analogous to Available Bit Rate (ABR) service in ATM networks, and it only guarantees a minimum amount of bandwidth. BNG service can be used in adaptive multimedia applications, where the bandwidth of an ongoing call is time-varying during its lifetime.

BE service, however, is analogous to Unspecified Bit Rate (UBR) service in ATM networks, and it does not guarantee any amount of bandwidth. Here, we make an assumption that a call in this service can be suspended and be reactivated because the call's bandwidth could be zero during its lifetime. The bandwidth of an ongoing call in BE service is also time-varying during its lifetime. BE service can be used for non real-time multimedia or computer data: emails, faxes, etc.

If a user does not want his call's bandwidth adaptive, he can request BG service instead of BNG service.

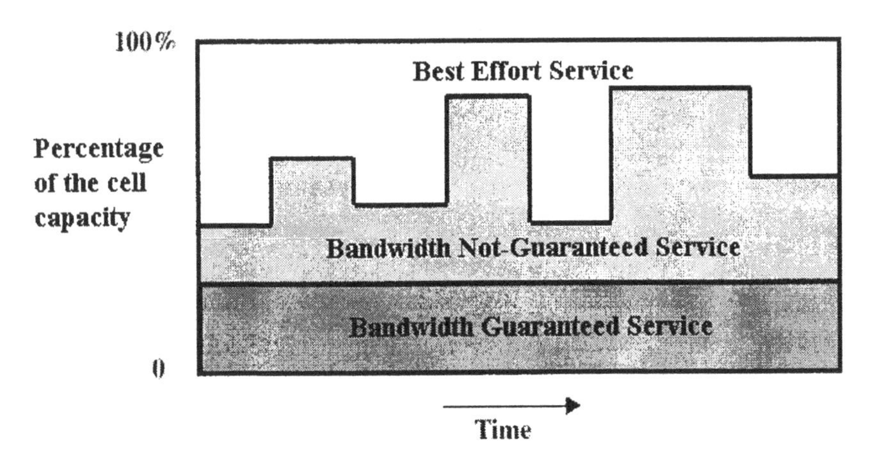

Figure 1: The categories of service.

2.2 Traffic Descriptors and QoS Parameters

When a call requests a service category, it can specify traffic descriptors to express the bandwidth requirements. Three traffic descriptors are defined here: Fixed Bandwidth (FB), Minimum Bandwidth (MB), and Upper Bandwidth Vector (UFV). MB is denoted by b_1, where $b_1 \geq 0$. UFV is denoted $(b_2, ..., b_i, ..., b_K)$, where $b_1 < b_j < b_{j+1}$ for all $j = 2, ..., K-1$, and $K > 1$. MB and UFV will form a Bandwidth Vector (BV) denoted by $(b_1, b_2, ..., b_i, ..., b_K)$.

The most important QoS parameter in wireless/mobile networks is forced termination probability, the probability of terminating an ongoing call before the completion of service [4,5,10]. In our chapter, the forced termination probability does not include handoff case. In other words, it is the probability to be terminated in a cell before it handoffs. Another important QoS parameter in wireless/mobile networks is Handoff Dropping Probability, the probability of dropping a handoff call. These two QoS parameters are defined to quantify QoS: Forced Termination Probability (FTP) and Handoff Dropping Probability (HDP).

We provide the QoS parameters and traffic descriptors for each service category in Figure 2. For each of the service categories, FTP and HDP must be specified. FB must be specified for only BG service; MB must be specified for only BNG service; UBV must specified for both BE service and BNG service.

Categories / Attributes	Bandwidth Guaranteed (BG)	Bandwidth Non-Guaranteed (BNG)	Best Effort (BE)
Fixed Bandwidth (*FB*)	Specified	N/A	N/A
Minimum Bandwidth (*MB*)	N/A	Specified	N/A
Upper Bandwidth Vector (*UBV*)	N/A	Specified	Specified
Handoff Dropping Probability (*HDP*)	Specified	Specified	Specified
Forced Termination Probability (*FTP*)*	Specified	Specified	Specified

* Forced termination probability does not count the handoff case here, i.e. the probability to be terminated in a cell before it handoffs.

Figure 2: QoS and traffic descriptors per category.

In Section 3, we present the traffic model for each service category and specify QoS parameters and traffic descriptors. In Section 5 and Section 6, we present our CAC scheme to optimize the revenue for service providers, and to satisfy QoS parameters' requirements and traffic descriptors for service users.

3 Traffic Model for Service Categories

Here, we try to abstract the three categories into a general traffic model. Our traffic model basically adopts the expressed method of the layered coding approach introduced in [4-5,10,18-22]. The layered (hierarchical) coding multimedia stream was originally introduced in wired networks, where the multimedia receivers can selectively choose a subset of the hierarchical coding depending on its capacity [4-5,10,16-22].

In our case, any traffic class can be expressed as $\{b_1, b_2, ..., b_i, ..., b_K\}$, where $b_j < b_{j+1}$ for all $j = 1, 2, ..., K - 1$, and K is the number of multimedia layers, and b_i is the i-th layer's bandwidth. If $K = 1$ and $b_1 > 0$, we obtain $\{b_1\}$, which is used to describe BG service traffic (traffic descriptor: $FB = b_1$). If $K > 1$ and $b_1 > 0$, we obtain $\{b_1, b_2, ..., b_i, ..., b_K\}$, which is used to describe BNG service traffic (traffic descriptors: $MB = b_1$ and $UBV = (b_2, ..., b_i, ..., b_K)$). If $K > 1$ and $b_1 = 0$, we obtain $\{0, b_2, ..., b_i, ..., b_K\}$, which is used to describe BE service traffic (traffic descriptor: $UBV = (b_2, ..., b_i, ..., b_K)$).

An example of this adaptive multimedia model is a video stream encoded with MPEG-2 I-frames only, MPEG-2 I+P-frames, and MPEG-2 whole (I+B+P) frames. Here, K equals 3; the traffic is $\{b_1, b_2, b_3\}$; b_1 stands for the bandwidth for MPEG-2 I-frames' encoding; b_2 stands for the bandwidth for MPEG-2 I+P-frames' encoding; b_3 stands for the bandwidth for MPEG-2 whole frames' encoding. If the available bandwidth in a cell is enough, the receiver will receive MPEG-2 whole encoding (b_3). Otherwise, the receiver may receive MPEG-2 I+P-frames' encoding (b_2), or MPEG-2 I-frames' encoding (b_1).

BG service traffic is a non adaptive traffic and the bandwidth is fixed. BNG and BE service traffics are adaptive traffics. An adaptive multimedia call can dynamically change its bandwidth depending on the network situation during its lifetime. If a user does not want his call's bandwidth adaptive, he can request BG service instead of BNG service.

BG traffic is a special case of BNG service traffic. In other words, the non layered coding traffic model (BG) is a special case of the layered coding traffic model (BNG). Moreover, the layered coding traffic may include some real layered coding traffic (BNG) and some non layered coding traffic (BG).

To generalize the traffic model, our traffic includes BG, BNG and BE service traffics. For each of the service traffics, there are multiple classes. We assume m classes of users in total: $\{1, 2, ..., m\}$. A user in class-i requires $b_{i,j}$ unites of bandwidth or channels, where $b_{i,j} \in \{b_{i,1}, b_{i,2}, ..., b_{i,j}, ..., b_{i,K_i}\}$ for $1 \leq i \leq m$, and $b_{i,j} < bi, j+1$ for all $1 \leq j \leq K_i - 1, b_{i,1} \geq 0$ and $K_i \geq 1$, where K_i is the number of multimedia layers for class-i users, and $b_{i,j}$ is the j-th multimedia layer's bandwidth for class-i users. We know that some of the classes among $\{1, 2, ..., m\}$ may belong to BG service, some of the classes may belong to BNG service, and some of the classes may belong to BE service. For example, there are 7 classes in a traffic shown as follows.

$$
\begin{array}{l}
\{ \\
\text{Class1}: \{b1\}, \\
\text{Class2}: \{b1, b2, b3\}, \\
\text{Class3}: \{b2\}, \\
\text{Class4}: \{b3\}, \\
\text{Class5}: \{c1, c2, c3, c4\}, \\
\text{Class6}: \{d1, d2, d3, d4, d5\}, \\
\text{Class7}: \{0, f2\} \\
\}
\end{array}
$$

The 1st, 3rd and 4th classes belong to BG service; the 2nd, 5th and 6th classes belong to BNG service; and the 7th class belongs to BE service.

If $K_i = 1$, we have the traffic descriptor, $FB = b_{i,1}$. If $K_i > 1$, we have the traffic descriptors: $MB = b_{i,1}$, $UBV = (b_{i,2}, ..., b_{i,j}, ..., b_{i,K_i})$. For each of the three categories, we require the QoS parameter, $FTP = 0$. In other words, we do not allow forced termination of an existing call in the cell. The arrival calls (i.e., new arrival calls and handoff arrival calls) could be blocked. Moreover, the QoS parameter HDP must be specified for each of the three categories.

Because the BG service is a special case of the BNG service, in Section 5 and Section 6, we do not make a distinction between them when we are working on CAC. In other words, we work on the abstract traffic model introduced above. In the next section, we first introduce some background for CAC-BRA, and then we present its ideas and assumptions.

4 Background and the Ideas

4.1 Semi-Markov Decision Process

The Semi-Markov Decision Process (SMDP) is a dynamic system satisfying Markovian properties. This dynamic system at random points in time is observed and classified into one of a finite number of states, and a decision from a finite decision space, which may depend on the current state, has to be made, and costs are incurred due to the decision made [6]. Markovian properties means that if at a decision epoch the action is chosen in the current state, then both the costs incurred and the time until, and the state at, the next decision epoch depend only on the current state and the chosen action and are thus independent of the past history of the system.

4.2 Definitions and Notations in our SMDP approach

Let us define the following definitions and notations that are used in Section 5 and Section 6.

$\mathbf{x}$ denotes the state of the system that reflects the bandwidth usage in the cell. $\mathbf{S}$ denotes the state space that is a set of all valid states. The decision epochs are the times that immediately follow the arrival and departure events. The decision is made before, rather than after, the occurrence of an event [2,7,8]. A decision epoch is a vector: $\mathbf{v}=(\mathbf{x}, \mathbf{a})$, where $\mathbf{x}$ is the current state, and the variable $\mathbf{a}$ represents the actions corresponding to the event types. In other words, the corresponding action will be chosen if the coming event meets an event type. An element location in vector $\mathbf{a}$ stands for the event type. An action is denoted by $\mathbf{a}$. $\mathbf{\Lambda_x}$ denotes the action space that is a set of chosen actions. Let $\tau(\mathbf{x}, \mathbf{a})$ be the expected sojourn time in present state $\mathbf{x}$ when action $\mathbf{a}$ is chosen, where $\mathbf{x} \in S$ and $\mathbf{a} \in \mathbf{\Lambda_x}$. Let $P(\mathbf{y} \mid \mathbf{x}, \mathbf{a})$ be the transition probability that at the next decision epoch the system will be in state $\mathbf{y}$ if the present state is $\mathbf{x}$ and the action $\mathbf{a}$ is taken, where $\mathbf{x} \in S$ and $\mathbf{a} \in \mathbf{\Lambda_x}$. Let $r(\mathbf{x}, \mathbf{a})$ be the revenue rate [7,8], The revenue rate or the utilization is a reward function. Even though there are other factors that influence revenue rate including bandwidth usage, in this study, we assume that revenue is only related to bandwidth usage.

4.3 Optimal Solution Framework

Here, we summarize a framework to get the optimal solution by the Semi-Markov decision process. The SMDP framework includes six steps. Firstly, a model should satisfy SMDP definition/properties, or we can make reasonable assumptions to let it satisfy SMDP definition/properties. Secondly, we need to properly define the system states ($\mathbf{x}$), the State Space ($\mathbf{S}$), the decision epochs, the actions ($\mathbf{a}$), and the Action Space ($\mathbf{\Lambda}$). Thirdly, we need to formulate the Expected Sojourn Time analytically, when in present state $\mathbf{x}$ and the action $\mathbf{a}$ is taken. Fourthly, we need to formulate the transition probability analytically, which is the probability that at the next decision epoch the system will be in state $\mathbf{y}$ if the present state is $\mathbf{x}$ and the action $\mathbf{a}$ is taken. Fifthly, we need to define a cost function or a reward function. Sixthly, we need to formulate linear programming formulas [6] with the cost/reward function. Finally, we solve the linear programming formulas. The solution was proved mathematically to be the optimal solution in [6].

In Section 5 and Section 6, we use this framework to model call admission control and bandwidth reallocation for both the one-class case and a multi-class case. In fact, we spend quite a lot of time in modeling the actions. The reward function is revenue or utilization. The QoS parameters *HDPs* are also formulated as constraints in the linear programming formulas.

4.4 Ideas

Our goal is to maximize the revenue and at the same time satisfy QoS parameters' requirements, which are upper bounds of Handoff Dropping Probabilities. In the meantime, we need to accept as many calls as possible. Our proposed approach (CAC-BRA) models the call admission control (CAC) and Bandwidth Reallocation Algorithm (BRA) at the same time by using SMDP. Whenever there is an arrival or a departure, the CAC-BRA must make some decisions. The decisions include whether we accept the call, how to reallocate the bandwidths of existing calls to accommodate the newly arrived call if it is an arrival, and how to reallocate the bandwidths of existing calls if a call departs.

For BNG service and BE service, if a cell is under loaded, CAC-BRA tries to allocate as much bandwidth as possible to every call in the cell, where the maximum bandwidth to each call is b_K for one-class case or b_{i,K_i} for a call in class-i for the multi-class case. However, if the cell is over-loaded, some of the calls in the cell might receive a bandwidth lower than b_K for the one-class case or b_{i,K_i} for a call in class-i for the multi-class case. CAC-BRA decides on not only whether an arrival will be accepted, but also which call will be changed to how much bandwidth. In other words, if a new call or a handoff call arrives, some of the calls already in the cell might be forced to lower their bandwidth (minimum is b_1 for the one-class case or $b_{i,1}$ for a call in class-i for the multi-class case) to accommodate the newly arrived call. Our adaptive multimedia framework tries to, at least, allocate a basic bandwidth for every call. On the other hand, if a call completes its job or handoffs to other cells, some of the existing calls in the cell might increase their bandwidth (maximum is b_K for the one-class case or b_{i,K_i} for a call in class-i for the multi-class case). We can use a long couch with a fixed capacity in a meeting room as an analogy. Assume that this couch is the only place where people can sit. If there are more people coming to the meeting room, everyone will sit closer to each other so that the newcomers can be fitted in. If somebody goes away, people on the couch get more space.

As the user moves from one cell to another, the call needs to be allocated some bandwidth/channels in the destination cell. We call this event a handover/handoff event. The probability of a handoff call being dropped is called handoff dropping probability.

In fact, we do not allow the forced termination of existing calls in the cell. The arrival calls (i.e., new arrival calls and handoff arrival calls) could be blocked. Our goal is to maximize the revenue, and at the same time to satisfy the QoS parameters' requirements, which are upper bounds of handoff blocking probabilities.

Because BG service is a special case of BNG service, CAC-BRA for BG service is a special case of CAC-BRA for BNG service, when the bandwidth of a call in the cell is not adjustable in BG service and the bandwidth reallocation actions are null actions (non adaptive).

4.5 Assumptions

We assume that service requests of each class arrive according to a Poisson distribution. Each class user needs a service time requirement (call holding time) that is exponentially distributed. All these distributions are independent of each other. If the current state is given, which is introduced in Sections 5.2 and 6.2, the next state is independent of all past history and is determined by arrival events, departure events, and CAC-BRA decisions. Arrival events include new call arrival events and handoff arrival events, whereas service departures include calls' completions and handoffs to other cells. Therefore, we can use the semi-Markov decision process [6] to model this system. The solution was proved mathematically to be the optimal solution in [6]. Here, we only consider the fixed capacity in a cell. The fixed total number of channels is C.

Next, we present an optimal distributed CAC for a single class of adaptive multimedia services in wireless/mobile networks.

5 CAC-BRA for the One-Class Case

In this section, we propose an optimal call admission control framework with bandwidth reallocation algorithm (CAC-BRA) for a single class of adaptive multimedia services in wireless/mobile networks, where the bandwidth of an ongoing call is time-varying during its lifetime. The optimization is in the sense of optimizing the revenue and satisfying QoS parameters' requirements. We adopt semi-Markov Decision Process (SMDP) approach to model call admission control and bandwidth reallocation algorithm at the same time. The Interior-Point Method in Linear Programming is used to solve the optimal decision problem. We claim that CAC-BRA is a distributed algorithm since it can run on every cell with the same program in a distributed manner.

5.1 One Class Traffic Model

A multimedia call can dynamically change its bandwidth depending on the network situation during its lifetime. In order to simplify the problem, we assume only one class of users in this section. Any customer uses one bandwidth among $\{b_1, b_2, ..., b_i, ..., b_K\}$ [4-5,10,18-22] where $b_j < b_{j+1}$ for $j = 1, 2, ..., K-1$. If $K = 1$, we have non layered coding traffic.

λ_n is the new call arrival rate for customers. λ_h is the handoff call arrival rate for customers. μ is the service rate for customers. h is the rate of handoff to other cells.

5.2 One Class Modeling

Let us define the following parameters. The state of the system is $\mathbf{x} = (x_1, x_2, ..., x_i, ..., x_K)$, where x_i stands for the number of users who are using bandwidth b_i in the system for all $1 \leq i \leq K$. The state space is: $\mathbf{S} = \{\mathbf{x} : x_i \geq 0 \text{ for all } 1 \leq i \leq K; \sum_{i=1}^{K} b_i x_i \leq C\}$. A decision epoch is a vector, $\mathbf{v}$=($\mathbf{x}$, $\mathbf{a}$). An action is denoted by: $\mathbf{a} = (a_1, a_2, \mathbf{d_{1,x,f_1}}, \mathbf{d_{2,x,f_2}}, \mathbf{d_{3,x,f_3}})$, where $a_i(i = 1, 2)$ is the CAC action, $\mathbf{d_{i,x,f_i}}(i = 1, 2, 3)$ is the bandwidth reallocation algorithm (BRA) action, and $a_i \in \{0, 1\}$. a_1 stands for the action for a new call, and a_2 stands for the action for a handoff call. If there is an accepted arrival, the meanings of $a_i(i = 1, 2)$ are: 0 stands for rejection, and 1 stands for acceptance. $\mathbf{d_{1,x,f_1}}$ is the BRA action if there is a new call arrival and the arrival is accepted to have bandwidth b_{f_1}; $\mathbf{d_{2,x,f_2}}$ is the BRA action if there is a handoff call arrival and the arrival is accepted to have bandwidth b_{f_2}; whereas $\mathbf{d_{3,x,f_3}}$ is the BRA action if there is a departure from the calls who are using bandwidth b_{f_3}. We have the following equations for $\mathbf{d_{i,x,f_i}}(i = 1, 2)$, where aw stands for available bandwidth.

$$\mathbf{d_{i,x,f_i}} = \begin{cases} (1-a_i) * \begin{bmatrix} \mathbf{A_1} \\ \mathbf{A_2} \end{bmatrix} + a_i * \begin{bmatrix} \mathbf{A_1} \\ \mathbf{B_2} \end{bmatrix}, & \text{if ab} \geq b_{\mathbf{f_i}} \\ (1-a_i) * \begin{bmatrix} \mathbf{A_1} \\ \mathbf{A_2} \end{bmatrix} + a_i * \begin{bmatrix} \mathbf{B_1} \\ \mathbf{B_2} \end{bmatrix}, & \text{otherwise} \end{cases} \quad (1)$$

$$\mathbf{A_1} = \begin{pmatrix} x_1 & 0 & \dots & 0 \\ 0 & x_2 & \dots & 0 \\ & \dots & & \\ 0 & 0 & \dots & x_K \end{pmatrix}$$

$$\mathbf{B_1} = \begin{pmatrix} y_{1,1} & 0 & \dots & 0 & \dots & 0 \\ y_{2,1} & y_{2,2} & \dots & 0 & \dots & 0 \\ & \dots & & & & \\ y_{j,1} & y_{j,2} & \dots & y_{j,j} & \dots & 0 \\ & \dots & & & & \\ y_{K,1} & y_{K,2} & \dots & y_{K,j} & \dots & y_{K,K} \end{pmatrix}$$

$$\mathbf{A_2} = \begin{pmatrix} 0 & 0 & \dots & 0 \end{pmatrix},$$

$$\mathbf{B_2} = \begin{pmatrix} 0 & \dots & 0 & 1 & 0 & \dots & 0 \end{pmatrix}$$

The bottom row of $((1-a_i) \times \mathbf{A_2} + a_i \times \mathbf{B_2})$ of the matrix $\mathbf{d_{i,x,f_i}}$ stands for the arrival. The value '1' at the f_i -th position of $\mathbf{B_2}$ denotes that the arrival has bandwidth, where $1 \leq f_i \leq K$. We can interpret the rows of $\mathbf{B_1}$ as follows. Among $x_j (1 \leq j \leq K)$ number of calls, who are using bandwidth b_j, there are $y_{j,1}, y_{j,2}, ..., y_{j.j-1}$ number of calls decreasing their bandwidths from b_j to $b_1, b_2, ..., b_{j-1}$, respectively, and there are $y_{j,j}$ number of calls staying their bandwidths, b_j, if the arrival is accepted. We have $y_{k,j} \geq 0 (1 \leq k \leq K \wedge 1 \leq j \leq k)$. Moreover, we have

$$\sum_{j=1}^{k} y_{k,j} = x_k, \text{ for } 1 \leq k \leq K \quad (2)$$

On the other hand, the sum of the column-j of the matrix $\mathbf{d_{i,x,f_i}}$ stands for the new number of calls using bandwidth b_j. Therefore, after the bandwidth reallocation, the new state $(\mathbf{y_i})$ is the result of summing all the columns of $\mathbf{d_{i,x,f_i}}$.

$$\begin{aligned}\mathbf{y_i} &= \sum_{\text{by columns}} \mathbf{d_{i,x,f_i}} \\ &= \begin{cases} (1-a_i)*\mathbf{x}+a_i*(\mathbf{x}+\mathbf{B_2}), & \text{if ab } \geq b_{\mathbf{f_i}} \\ (1-a_i)*\mathbf{x}+a_i*(\mathbf{u}+\mathbf{B_2}), & \text{otherwise} \end{cases}, \text{for} i=1,2\end{aligned} \tag{3}$$

$$\mathbf{u} = (\sum_{j=1}^{K} x_{j,1}, \sum_{j=2}^{K} x_{j,2}, ..., \sum_{j=k}^{K} x_{j,k}, ..., \sum_{j=K}^{K} x_{j,K})$$

For $\mathbf{d_{3,x,f_3}}$, if there are no calls in the cell, there are no departure events. Therefore, we have the following equation for $\mathbf{d_{3,x,f_3}}$.

$$\mathbf{d_{3,x,f_3}} = \begin{cases} \text{zero matrix } ((K+1)\times K) \text{ , if } \sum_{k=1}^{K} x_k = 0 \\ (1-1)*\begin{bmatrix}\mathbf{A_1}\\ \mathbf{C_2}\end{bmatrix} + 1*\begin{bmatrix}\mathbf{D_1}\\ \mathbf{D_2}\end{bmatrix} = \begin{bmatrix}\mathbf{D_1}\\ \mathbf{D_2}\end{bmatrix}, \text{ otherwise}\end{cases} \tag{4}$$

$$\mathbf{D_1} = \begin{pmatrix} y_{1,1} & y_{1,2} & \cdots & y_{1,j} & \cdots & y_{1,K} \\ 0 & y_{2,2} & \cdots & y_{2,j} & \cdots & y_{2,K} \\ & \cdots & & & & \\ 0 & 0 & \cdots & y_{j,j} & \cdots & y_{j,K} \\ & \cdots & & & & \\ 0 & 0 & \cdots & 0 & \cdots & y_{K,K} \end{pmatrix}$$

$$\mathbf{C_2} = \begin{pmatrix} 0 & ... & 0 & -1 & 0 & ... & 0 \end{pmatrix}, \mathbf{D_2} = \begin{pmatrix} 0 & ... & 0 \end{pmatrix}$$

The bottom row ($\mathbf{D_2}$) of $\mathbf{d_{3,x,f_3}}$ is used for the departure. The value $'-1'$ at the f_3 -th position $\mathbf{C_2}$ stands for the departure that has bandwidth b_{f_3}, where $1 \leq f_3 \leq K$. Therefore, we have $x_{f_3} > 0$, and $y_{i,j} \geq 0 (1 \leq i \leq K \wedge i \leq j \leq K)$. We can interpret the rows of $\mathbf{d_{3,x,f_3}}$ except for the bottom row ($\mathbf{D_2}$) as follows. Among x_j (for $1 \leq j \leq K$ and $j \neq f_3$) or $(x_j - 1)$ (for $j = f_3$) number of calls, that are using bandwidth b_j, there are $y_{j,j+1}, y_{j,j+2}, ..., y_{j,K}$ number of calls increasing their bandwidths from b_j to $b_{j+1}, b_{j+2}, ..., b_K$, respectively, and there are $y_{j,j}$ number of calls staying their bandwidths b_j, if the arrival is accepted. Therefore, for $1 \leq k \leq K$ and $k \neq f_3$, we have

$$\sum_{j=k}^{K} y_{k,j} = x_k \tag{5}$$

$$\sum_{j=f_3}^{K} y_{f_3,j} = x_{f_3} - 1 \tag{6}$$

Similar to the previous discussion, the sum of the column-j of the matrix $\mathbf{d_{3,x,f_3}}$ stands for the new number of calls using bandwidth b_j. Therefore, after the bandwidth reallocation, the new state $(\mathbf{y_3})$ is the result of summing all the columns of $\mathbf{d_{3,x,f_3}}$:

$$\mathbf{y_3} = \sum_{\text{by columns}} \mathbf{d_{3,x,f_3}} = \begin{cases} \mathbf{D_2}, \text{if } \sum_{i=1}^{K} x_i = 0 \\ (\sum_{j=1}^{1} y_{j,1}, \sum_{j=1}^{2} y_{j,2}, \ldots, \sum_{j=1}^{K} y_{j,k}, \ldots, \sum_{j=1}^{K} y_{j,K}), \text{ otherwise} \end{cases} \tag{7}$$

If $K = 1$, there is no bandwidth reallocation happening, and it is the BG service. Therefore, we claim that the optimal CAC for BG service is a special case of the optimal CAC-BRA for BNG service, when the bandwidth of a user in the cell is not adjustable in BG service and the BRA actions will be null actions (non adaptive).

We can easily obtain the action space $(\mathbf{\Lambda_x})$, the expected sojourn time $(\tau(\mathbf{x}, \mathbf{a}))$, and the transition probability $(P(\mathbf{y} \mid \mathbf{x}, \mathbf{a}))$ as follows.

$$\begin{array}{l} \mathbf{\Lambda_x} = \{\mathbf{a} \mid a_1, a_2 \in \{0, 1\}; \mathbf{a} \text{ is constrained by (1)-(7);} \\ a_i = 0 \text{ if } \mathbf{y_i} \in \overline{S} \text{ for } i = 1, 2; \mathbf{y_3} \in S\} \end{array} \tag{8}$$

$$\tau(\mathbf{x}, \mathbf{a}) = \frac{1}{\lambda_n a_1 + \lambda_h a_2 + (\sum_{i=1}^{K} x_i)\mu + (\sum_{i=1}^{K} x_i) h} \tag{9}$$

$$P(\mathbf{y}|\mathbf{x},\mathbf{a}) = \begin{cases} (\lambda_n a_1 + \lambda_h a_2)\tau(\mathbf{x},\mathbf{a}), \\ \text{if } \mathbf{y} = \mathbf{y_i}\ (i = 1,2) \text{ defined in (3) and } \sum_{i=1}^{K} y_i = \sum_{i=1}^{K} x_i + 1 \\ (\sum_{i=1}^{K} x_i)(\mu + h)\tau(\mathbf{x},\mathbf{a}), \\ \text{if } \mathbf{y} = \mathbf{y_3} \text{ defined in (7) and } \sum_{i=1}^{K} y_i = \sum_{i=1}^{K} x_i - 1 \\ 0, \text{otherwise} \end{cases} \tag{10}$$

Let r_i be the revenue rate for b_i, and $r(\mathbf{x},\mathbf{a})$ be the revenue rate [7, 8].

$$r(\mathbf{x},\mathbf{a}) = \sum_{i=1}^{K} r_i x_i \tag{11}$$

$$Utilization = \sum_{i=1}^{K} b_i x_i \tag{12}$$

Even though there are other factors that influence revenue rate including bandwidth usage, in this study, we assume that revenue is only related to bandwidth usage.

5.3 Linear Programming Formulas for One Class CAC-BRA

The linear programming associated with the SMDP for maximum revenue is given below with decision variables $\pi_{\mathbf{xa}}, \mathbf{x} \in S, \mathbf{a} \in \mathbf{\Lambda_x}$.

Maximize $\sum_{\mathbf{x}\in S}\sum_{\mathbf{a}\in\Lambda_x} r(\mathbf{x},\mathbf{a})\tau(\mathbf{x},\mathbf{a})\pi_{\mathbf{xa}}$ such that

$$\sum_{\mathbf{x}\in S}\sum_{\mathbf{a}\in\Lambda_x} \tau(\mathbf{x},\mathbf{a})\pi_{\mathbf{xa}} = 1 \tag{13}$$

$$\sum_{\mathbf{a}\in\Lambda_x} \pi_{\mathbf{ya}} = \sum_{\mathbf{x}\in S}\sum_{\mathbf{a}\in\Lambda_x} P(\mathbf{y}|\mathbf{x},\mathbf{a})\,\pi_{\mathbf{xa}} \text{ for } \mathbf{y} \in \mathbf{S} \tag{14}$$

$$\pi_{\mathbf{xa}} \geq 0, \mathbf{x} \in \mathbf{S}, \mathbf{a} \in \mathbf{\Lambda_x}, \tag{15}$$

$$\sum_{\mathbf{x}\in S}\sum_{\mathbf{a}\in\Lambda_x} (1 - a_2)\tau(\mathbf{x},\mathbf{a})\pi_{\mathbf{xa}} \leq P_{BH} \tag{16}$$

The term $\tau(\mathbf{x}, \mathbf{a})\pi_{\mathbf{xa}}$ can be interpreted as the long run fraction of the decision epochs at which the system is in state $\mathbf{x}$, and the action $\mathbf{a}$ is chosen. We consider the QoS parameter's requirement, an upper bound of handoff dropping probability, in the last equation, where P_{BH} is the upper bound of handoff dropping probability. Based on the above linear programming equations, we can obtain optimal CAC-BRA decisions by a linear programming algorithm known as the Interior Point Methods [14,15]. The Interior Point Methods are large-scale linear programming methods. They can be applied to the cases of larger K and m. We discuss the complexity issue in Section 7.

6 CAC-BRA for General Case

In this section, we propose CAC-BRA for multi-classes of adaptive multimedia services in wireless/mobile networks. CAC-BRA optimizes revenue for service providers and satisfies QoS parameters' requirements for service users. It adopts the Semi-Markov decision process to model both call admission control and the bandwidth reallocation algorithm. Because the non adaptive multimedia traffic (BG service) is a special case of adaptive multimedia traffic (BNG service), the non adaptive optimal CAC scheme is a special case of our optimal CAC-BRA scheme. Furthermore, the Interior Point Method in linear programming is used to solve the optimal decision problem.

6.1 General Case Traffic Model

Now, we work on the abstract general traffic model. A multimedia call can dynamically change its bandwidth according to the network situation during its lifetime. We assume m classes of users: $1, 2, ..., m$. A user in class-i requires bi, j unit of bandwidth or channels, where $b_{i,j} \in \{b_{i,1}, b_{i,2}, ..., b_{i,j}, ..., b_{i,K_i}\}$ [4-5,10,18-22] for $1 \leq i \leq m$, and $b_{i,j} < b_{i,j+1}$ for all $1 \leq j \leq K_i - 1$, and $K_i \geq 1$, where K_i is the number of multimedia layers for class-i users, and $b_{i,j}$ is the j-th layer's bandwidth for class-i users. Section 5 is a special case of this section when $m = 1$ and $K_1 = K$.

$\lambda_{i,n}$ is the new call arrival rate for class-i customers. $\lambda_{i,h}$ is the handoff call arrival rate for class-i customers. μ_i is the service rate for class-i customers. h_i is the rate of handoff to other cells for class-i customers.

6.2 General Case Modeling

Let us define the following parameters. The state of the system is $\mathbf{x} = (\mathbf{x_1}, \mathbf{x_2}, ..., \mathbf{x_i}, ..., \mathbf{x_m})$, where $\mathbf{x_i} = (x_{i,1}, x_{i,2}, ..., x_{i,j}, ..., x_{i,K_i})$ for $1 \le i \le m$ and $1 \le j \le K_i$. $x_{i,j}$ stands for the number of class-i users whose bandwidth is $b_{i,j}$ in the cell. The state space is: $\mathbf{S} = \{\mathbf{x}$: $x_{i,j} \ge 0$ for all $1 \le i \le m$ and $1 \le j \le K_i$; $\sum_{i=1}^{m} \sum_{j=1}^{K_i} b_{i,j} x_{i,j} \le C\}$. A decision epoch is a vector, $\mathbf{v}$=($\mathbf{x}$, $\mathbf{a}$). An action is denoted by:

$\mathbf{a} = (a_1, a_2, ..., a_m, a_{m+1}, \ldots a_{2m}, \mathbf{d_{1,x,f_1}}, \mathbf{d_{2,x,f_2}}, ..., \mathbf{d_{2m,x,f_{2m}}}, \mathbf{d_{2m+1,x,s,f_{2m+1}}}$, where a_i $(1 \le i \le 2m)$ is the CAC action, $\mathbf{d_{i,x,f_i}}$ $(i = 1, 2, ..., 2m)$ and $\mathbf{d_{2m+1,x,s,f_{2m+1}}}$ are the BRA actions, and $a_i \in \{0, 1\}$ for $1 \le i \le 2m$. a_i $(1 \le i \le m)$ stands for the action for a new arrival call in class-i, and a_{i+m} $(1 \le i \le m)$ stands for the action for a handoff arrival call in class-i. If there is an accepted arrival, the meanings for a_i $(1 \le i \le 2m)$ are: 0 stands for rejection, and 1 stands for acceptance. $\mathbf{d_{i,x,f_i}}$ $(1 \le i \le m)$ is the BRA action if it is a new call arrival for class-i and the arrival is accepted to have bandwidth b_{i,f_i}; $\mathbf{d_{i+m,x,f_{i+m}}}$ $(1 \le i \le m)$ is the BRA action if it is a handoff call arrival for class-i and the arrival is accepted to have bandwidth b_{i,f_i}; whereas $\mathbf{d_{2m+1,x,s,f_{2m+1}}}$ is the BRA action if there is a departure from the calls in class-s whose bandwidth is $b_{s,f_{2m+1}}$ for $1 \le s \le m$ and $1 \le f_{2m+1} \le K_s$. Let's define g as follows, and $I(e)$ be the indication function that returns 1 if the expression e is true, and returns 0 if the expression e is false. We will have,

$$g = i \times I(i \le m) + (i - m) \times I(i > m) \tag{17}$$

$$\mathbf{d_{i,x,f_i}} = \begin{cases} (1 - a_i) * \begin{bmatrix} \mathbf{E_1} \\ \mathbf{E_2} \end{bmatrix} + a_i * \begin{bmatrix} \mathbf{E_1} \\ \mathbf{F_2} \end{bmatrix}, & \text{if ab } \ge b_{g,f_g} \\ (1 - a_i) * \begin{bmatrix} \mathbf{E_1} \\ \mathbf{E_2} \end{bmatrix} + a_i * \begin{bmatrix} \mathbf{F_1} \\ \mathbf{F_2} \end{bmatrix}, & \text{otherwise} \end{cases}, \tag{18}$$

for $1 \le i \le 2m$

$$\mathbf{E_1} = \begin{pmatrix} x_{g,1} & 0 & \dots & 0 \\ 0 & x_{g,2} & \dots & 0 \\ & & \dots & \\ 0 & 0 & \dots & x_{g,K_g} \end{pmatrix},$$

$$\mathbf{F_1} = \begin{pmatrix} y_{g,1,1} & 0 & \dots & 0 & \dots & 0 \\ y_{g,2,1} & y_{g,2,2} & \dots & 0 & \dots & 0 \\ & & \dots & & \dots & \\ y_{g,j,1} & y_{g,j,2} & \dots & y_{g,j,j} & \dots & 0 \\ & & \dots & & \dots & \\ y_{g,K_g,1} & y_{g,K_g,2} & \dots & y_{g,K_g,j} & \dots & y_{g,K_g,K_g} \end{pmatrix}$$

$$\mathbf{E_2} = \begin{pmatrix} 0 & 0 & \dots & 0 \end{pmatrix},$$

$$\mathbf{F_2} = \begin{pmatrix} 0 & \dots & 0 & 1 & 0 & \dots & 0 \end{pmatrix}$$

The bottom row $((1-a_i) \times \mathbf{E_2} + a_i \times \mathbf{F_2})$ of the matrix $\mathbf{d_{i,x,f_i}}$ denotes the arrival. The value $'1'$ on the f_i -th position of $\mathbf{F_2}$ stands for that the arrival is allocated to bandwidth b_{g,f_g}, where $1 \leq f_g \leq K_g$. We can interpret the rows of $\mathbf{F_1}$ as follows. Among $x_{g,j}$ number of calls ($1 \leq j \leq K_g$), whose bandwidth is $b_{g,j}$, $y_{g,j,1}, y_{g,j,2}, ..., y_{g,j,j-1}$ number of calls decrease their bandwidths from $b_{g,j}$ to $b_{g,1}, b_{g,2}, ..., b_{g,j-1}$, respectively, and $y_{g,j,j}$ number of calls stay their bandwidths $b_{g,j}$, if the arrival is accepted. We have $y_{g,j,k} \geq 0$ ($1 \leq i \leq 2m$, $1 \leq j \leq K_g$ and $1 \leq k \leq j$) and g is defined in (17). Moreover, we have,

$$\sum_{k=1}^{j} y_{g,j,k} = x_{g,j}, 1 \leq i \leq 2m \wedge 1 \leq j \leq K_g \tag{19}$$

On the other hand, the sum of the column-j of $\mathbf{d_{i,x,f_i}}$ stands for the new number of calls using bandwidth $b_{g,j}$ ($1 \leq j \leq K_g$).Therefore, after the bandwidth reallocation, the new state ($\mathbf{y_i}$) is the same for non g-th columns, whereas the g-th column ($\mathbf{z_g}$) is the result of adding all the columns of $\mathbf{d_{i,x,f_i}}$.

$$\mathbf{y_i} = (\mathbf{x_1}, \mathbf{x_2}, ..., \mathbf{x_{g-1}}, \mathbf{z_g}, \mathbf{x_{g+1}}, ..., \mathbf{x_m}),\ i = 1, 2, ..., 2m \tag{20}$$

$$\begin{aligned} \mathbf{z_g} &= \sum_{\text{by columns}} \mathbf{d_{i,x,f_i}} \\ &= \begin{cases} (1-a_i) * \mathbf{x_g} + a_i * (\mathbf{x_g} + \mathbf{F_2}), & \text{if ab } \geq b_{g,f_g} \\ (1-a_i) * \mathbf{x_g} + a_i * (\mathbf{w_g} + \mathbf{F_2}), & \text{otherwise} \end{cases} \\ &\text{for } i = 1, 2...2m \end{aligned} \tag{21}$$

$$\mathbf{w_g} = \left(\sum_{j=1}^{K_g} y_{g,j,1}, \sum_{j=2}^{K_g} y_{g,j,2}, \ldots, \sum_{j=k}^{K_g} y_{g,j,k}, \ldots, \sum_{j=K_g}^{K_g} y_{g,j,K_g} \right) \tag{22}$$

For $\mathbf{d_{2m+1,x,s,f_{2m+1}}}$, if there are no calls in the cell, there are no departure events. Therefore, we have,

$$\mathbf{d_{2m+1,x,s,f_{2m+1}}} = \begin{cases} \text{zero matrix}(\ (K_s+1) \times K_s), \text{ if } \sum\limits_{j=1}^{K_s} x_{s,j} = 0 \\ (1-1) * \begin{bmatrix} \mathbf{G_1} \\ \mathbf{G_2} \end{bmatrix} + 1 * \begin{bmatrix} \mathbf{H_1} \\ \mathbf{H_2} \end{bmatrix} = \begin{bmatrix} \mathbf{H_1} \\ \mathbf{H_2} \end{bmatrix}, \text{ otherwise} \end{cases} \tag{23}$$

$$\mathbf{G_1} = \begin{pmatrix} x_{s,1} & 0 & \ldots & 0 \\ 0 & x_{s,2} & \ldots & 0 \\ & & \ldots & \\ 0 & 0 & \ldots & x_{s,K_s} \end{pmatrix}$$

$$\mathbf{H_1} = \begin{pmatrix} y_{s,1,1} & y_{s,1,2} & \cdots & y_{s,1,j} & \cdots & y_{s,1,K_s} \\ 0 & y_{s,2,2} & \cdots & y_{s,2,j} & \cdots & y_{s,2,K_s} \\ & & \cdots & & \cdots & \\ 0 & 0 & \cdots & y_{s,j,j} & \cdots & y_{s,j,K_s} \\ & & \cdots & & \cdots & \\ 0 & 0 & \ldots & 0 & 0 & y_{s,K_s,K_s} \end{pmatrix}$$

$$\mathbf{G_2} = \begin{pmatrix} 0 & \ldots & 0 & -1 & 0 & \ldots & 0 \end{pmatrix}, \mathbf{H_2} = \begin{pmatrix} 0 & \ldots & 0 \end{pmatrix}$$

The bottom row ($\mathbf{H_2}$) of $\mathbf{d_{2m+1,x,s,f_{2m+1}}}$ is used for the departure. The value $'-1'$ at the f_{2m+1}-th position of $\mathbf{G_2}$ denotes that the departure has bandwidth $b_{s,g}$, where $1 \leq f_{2m+1} \leq K_s$. We have $x_{s,f_{2m+1}} > 0$, and $x_{s,j,n} \geq 0, (1 \leq j \leq K_s$, and $j \leq n \leq K_s)$. We can interpret the rows of $\mathbf{d_{2m+1,x,s,f_{2m+1}}}$ except the bottom row ($\mathbf{H_2}$) as follows. Among $x_{s,j}$, $(1 \leq j \leq K_s \wedge j \neq f_{2m+1})$ or $x_{s,j} - 1$ (for $j = f_{2m+1}$, the departure one) number of calls in class-s whose bandwidths are $b_{s,j}$, $y_{s,j,j+1}, y_{s,j,j+2}, \ldots, y_{s,j,K_s}$ number of calls increase their bandwidths from $b_{s,j}$ to $b_{s,j+1}, b_{s,j+2}, \ldots, b_{s,K_s}$, respectively, and $y_{s,j,j}$ number of calls stay their bandwidths $b_{s,j}$ if the arrival is accepted. Therefore,

$$\sum_{n=j}^{K_s} y_{s,j,n} = x_{s,j} \text{ for } 1 \leq s \leq m, 1 \leq j \leq K_s \text{ and } j \neq f_{2m+1} \tag{24}$$

$$\sum_{n=f_{2m+1}}^{K_s} x_{s,f_{2m+1},n} = x_{s,f_{2m+1}} - 1 \tag{25}$$

Similar to the previous discussion, the sum of column-j $(1 \leq j \leq K_s)$ of the matrix stands for the new number of calls using bandwidth $b_{s,j}$. After the bandwidth reallocation, the new state $(\mathbf{y_{2m+1}})$ is the same for non s-th columns, whereas the s-th column $(\mathbf{p_s})$ is the result of adding all the columns of $\mathbf{d_{2m+1,x,s,f_{2m+1}}}$,

$$\mathbf{y_{2m+1}} = (\mathbf{x_1}, \mathbf{x_2}, ..., \mathbf{x_{s-1}}, \mathbf{p_s}, \mathbf{x_{s+1}}, ..., \mathbf{x_m}) \tag{26}$$

$$\mathbf{p_s} = \sum_{\text{by columns}} \mathbf{d_{2m+1,x,s,f_{2m+1}}} = \begin{cases} \mathbf{H_2}, \text{ if } \sum_{j=1}^{K_s} x_{s,j} = 0 \\ (\sum_{j=1}^{1} x_{s,j,1}, \sum_{j=1}^{2} x_{s,j,2} \cdots \sum_{j=1}^{n} x_{s,j,n} \cdots \sum_{j=1}^{K_s} x_{s,j,K_s}), \text{ otherwise} \end{cases} \tag{27}$$

If $K_i = 1$, the bandwidth reallocation does not exist. It is the BG service case. Moreover, if K_i for all $1 \leq i \leq m$, this is the same case Kwon et al. proposed [8]. The work of Kwon et al. [8] is a special case of our general case. Therefore, we claim that the non layered coding (BG service) optimal CAC is a special case of our optimal General-CAC-BRA (BNG service), when the bandwidth of a user in the cell is not adjustable in BG service and the BRA actions are null actions (non adaptive, BG service case). For example, in such a case, we have the following.

$$\mathbf{e_{i,x,f_i}} = (1 - a_i)\begin{pmatrix} x_i \\ 0 \end{pmatrix} + a_i \begin{pmatrix} x_i \\ 1 \end{pmatrix} \tag{28}$$

We can easily obtain the action space $(\mathbf{\Lambda_x})$, the expected sojourn time $(\tau(\mathbf{x}, \mathbf{a}))$, and the transition probability $(P(\mathbf{y} \mid \mathbf{x}, \mathbf{a}))$ as follows.

$$\begin{aligned} \mathbf{\Lambda_x} = \{\mathbf{a} \mid a_i \in \{0,1\}, (1 \leq i \leq 2m); \mathbf{a} \text{ is constrained by (17)-(27)}; \\ a_i = 0 \text{ if } \mathbf{y_i} \in \overline{S}, (1 \leq i \leq 2m); \mathbf{y_{2m+1}} \in S\} \end{aligned} \tag{29}$$

$$\tau(\mathbf{x},\mathbf{a}) = \frac{1}{\sum_{i=1}^{m} \lambda_{i,n} a_i + \sum_{i=1}^{m} \lambda_{i,h} a_{i+m} + \sum_{i=1}^{m} (\sum_{j=1}^{K_i} x_{i,j})\mu_i + \sum_{i=1}^{m} (\sum_{j=1}^{K_i} x_{i,j} h_i)} \tag{30}$$

$$P(\mathbf{y}|\mathbf{x},\mathbf{a}) = \begin{cases} (\sum_{i=1}^{m} \lambda_{i,n} a_i + \sum_{i=1}^{m} \lambda_{i,h} a_{i+m})\tau(\mathbf{x},\mathbf{a}), \\ \quad \text{if } \mathbf{y} = \mathbf{y}_i, (1 \le i \le 2m) \text{ in (20) and } \sum_{i=1}^{m} \sum_{j=1}^{K_i} y_{i,j} = \sum_{i=1}^{m} \sum_{j=1}^{K_i} x_{i,j} + 1. \\ (\sum_{i=1}^{m} (\sum_{j=1}^{K_i} x_{i,j})\mu_i + \sum_{i=1}^{m} (\sum_{j=1}^{K_i} x_{i,j} h_i)\tau(\mathbf{x},\mathbf{a}), \\ \quad \text{if } \mathbf{y} = \mathbf{y_{2m+1}} \text{ in (26) and } \sum_{i=1}^{m} \sum_{j=1}^{K_i} y_{i,j} = \sum_{i=1}^{m} \sum_{j=1}^{K_i} x_{i,j} - 1. \\ 0, \text{ otherwise} \end{cases} \tag{31}$$

If we let $r_{i,j}$ be the revenue rate for $b_{i,j}$, and $r(\mathbf{x},\mathbf{a})$ be the revenue rate [7,8], we have

$$r(\mathbf{x},\mathbf{a}) = \sum_{i=1}^{m} \sum_{j=1}^{K_i} r_{i,j} x_{i,j} \tag{32}$$

$$Utilization = \sum_{i=1}^{m} \sum_{j=`}^{K_i} b_{i,j} x_{i,j} \tag{33}$$

Even though there are other factors that influence revenue rate including bandwidth usage, in this study we assume that revenue is only related to bandwidth usage.

6.3 Linear Programming Formulas for General CAC-BRA

The linear programming associated with the SMDP for maximum revenue is given below with decision variables $\pi_{\mathbf{xa}}, \mathbf{x} \in S, \mathbf{a} \in \mathbf{\Lambda_x}$:

Maximize $\sum_{\mathbf{x}\in S} \sum_{\mathbf{a}\in\Lambda_x} r(\mathbf{x},\mathbf{a})\tau(\mathbf{x},\mathbf{a})\pi_{\mathbf{xa}}$ such that

$$\sum_{\mathbf{x}\in S} \sum_{\mathbf{a}\in\Lambda_x} \tau(\mathbf{x},\mathbf{a})\pi_{\mathbf{xa}} = 1 \tag{34}$$

$$\sum_{\mathbf{a}\in\Lambda_x} \pi_{\mathbf{ya}} = \sum_{\mathbf{x}\in S} \sum_{\mathbf{a}\in\Lambda_x} P(\mathbf{y}|\mathbf{x},\mathbf{a})\, \pi_{\mathbf{xa}}, \mathbf{y} \in \mathbf{S} \tag{35}$$

$$\pi_{\mathbf{xa}} \geq 0, \mathbf{x} \in S, \mathbf{a} \in \mathbf{\Lambda_x} \tag{36}$$

$$\sum_{\mathbf{x}\in S} \sum_{\mathbf{a}\in\Lambda_v} (1 - a_{i+m})\tau(\mathbf{x},\mathbf{a})\pi_{\mathbf{xa}} \leq P_{i,BH}, (1 \leq i \leq m) \tag{37}$$

The term $\tau(\mathbf{x},\mathbf{a})\pi_{\mathbf{xa}}$ can be interpreted as the long run fraction of the decision epochs at which the system is in state $\mathbf{x}$, and the action $\mathbf{a}$ is chosen. We consider the QoS parameters' requirements, upper bounds of the handoff dropping probabilities in the last equation, where $P_{i,BH}$ is the upper bound of handoff dropping probability for class-i. Based on the above linear programming equations, we can obtain optimal CAC-BRA decisions by a linear programming algorithm known as the Interior-Point Methods [14,15]. They can be applied to the cases of larger K_i and m. We discuss the complexity issue in Section 7.

7 Optimality and Complexity in the Real System

As we discussed in Section 4.2, the SMDP approach will provide an optimal solution. Besides bandwidth usage, there are also other factors that influence revenue. In this study, we assume that revenue is only related to bandwidth usage. Moreover, our proposed approach can get the optimal solution independent of the choice of revenue functions. In other words, no matter how you define a revenue function, as long as it is only related to bandwidth, our proposed approach will give an optimal solution.

Even though we cannot define a deterministic function between revenue and bandwidth usage, we can assume that revenue increases when bandwidth usage increases. Even though the complexity and solution space for the linear programming formulas (13)-(16) or (34)-(37) increase exponentially as K_i or m increases, the large-scale linear programming method based on the Interior Point Methods [14,15] can apply to the cases of larger K_i or m. Moreover, solving the CAC-BRA decisions is an off line procedure so that the decisions are found out before operating CAC [7]. In reality, the values of K_i or m are typically of reasonable size. The disadvantage of this approach in the real system is that the action space is quite large, and it requires some storage space. However, it would be tolerable with reasonable size of K_i or m, and the proposed approach will fit well in real systems.

8 Conclusions

In this chapter, we classify services into three categories: Bandwidth Guaranteed (BG) service, Bandwidth Not Guaranteed (BNG) service and Best Effort (BE) service. For each of the above three categories, traffic descriptors and QoS parameters are defined and specified. We abstract the three categories into a general traffic model. Under such a general abstract traffic model, an optimal CAC scheme that guarantees the QoS parameters' requirements and traffic descriptors, and maximizes the revenue, is presented analytically. The proposed schemes allow us to make decisions on call admission control, as well as bandwidth reallocation at the same time using the semi-Markov decision process approach. The Interior Point Method in linear programming is used to solve the optimal decision problem. With reasonable size of K_i and m, the proposed approach will fit well in the real system.

References

[1] D. Ayyagari and A. Ephremides, *Admission Control with Priorities: Approaches for Multi-rate Wireless Systems*, Mobile Networks and Applications 4, 1999, pp. 209-218.

[2] C. Ho and C. Lea, *Improving Call Admission Policies in Wireless Networks*, Wireless Networks 5, 1999, pp. 257-265.

[3] R. Ramjee, D. Towsley and R. Nagarajan, *On Optimal Call Admission Control in Cellular Networks*, Wireless Networks 3, 1997, pp. 29-41.

[4] T. Kwon, I. Park, Y. Choi, and S. Das, *Bandwidth Adaptation Algorithms with Multi-Objectives for Adaptive Multimedia Services in Wireless/Mobile Networks*, ACM Workshop on Wireless Mobile Multimedia (WOWMOM'99), pp. 51-58, 1999.

[5] T. Kwon, Y. Choi, C. Bisdikian and M. Naghshineh, *Call Admission Control for Adaptive multimedia in Wireless/Mobile Network*, ACM

Workshop on Wireless Mobile Multimedia (WOWMOM'98), pp. 111-116, 1998.

[6] H.C. Tijms, Stochastic Modeling and Analysis: A Computational Approach (Wiley, New York, 1986).

[7] J. Choi, T. Kwon, Y. Choi and M. Naghshineh, *Call Admission Control for Multimedia Service in Mobile Cellular Networks: A Markov Decision Approach*, IEEE ISCC'00, 2000.

[8] T. Kwon, Y. Choi and M. Naghshineh, *Optimal Distributed Call Admission Control for Multimedia Service in Mobile Cellular Network*, Mobile Multimedia Conference (MoMuc'98), Berlin, October, 1998.

[9] D I. Yoon and B. Lee, *A Distributed Dynamic Call Admission Control that Supports Mobility of Wireless Multimedia Users*, ICC'99, 1442-1446, 1999.

[10] T. Kwon, Y. Choi, C. Bisdikian and M. Naghshineh, *QoS Provisioning for Adaptive Multimedia in Wireless/Mobile Networks*, Wireless Networks, Vol. 9 , No. 1 (January 2003), pp. 51-59.

[11] S.M. Ross, Introduction to Probability Methods (Academic Press, 1997).

[12] D. G. Luenberger, Linear and non-linear programming, second edition (Addison Wesley, 1984).

[13] S. Biswas and B. Sengupta, *Call Admissibility for Multirate Traffic in Wireless ATM Networks*, IEEE INFOCOM'97, pp. 650-659, 1997.

[14] Y. Zhang, *Solving Large-Scale Linear Programming by Interior-Point Methods under the MATLAB Environment*, Department of Mathematics and Statistics, University of Maryland, Technical Report TR 96-01, July, 1995.

[15] G. Y. Zhao, *Interior-Point Methods with Decomposition for Solving Linear Programs*, Journal of Optimization Theory and Applications, Vol. 102, No. 1, pp. 169-192, 1999.

[16] M. Naghshineh and M. Schwarz, *Distributed Call Admission Control in Mobile/Wireless Networks*, IEEE PIMRC '95, pp. 289-293, 1995.

[17] C. Yoon and C. Un, *Performance of Personal Portable Radio Telephone Systems with and without Guard Channels*, IEEE Journal on Selected Areas in Communications, Vol. 11, No. 6, pp. 911-917, August, 1993.

[18] M. Naghshineh and M. Willebeek-LeMair, *End-to-End QoS Provisioning in Multimedia Wireless/Mobile Networks Using an Adaptive Framework*, IEEE Communications Magazine, Vol. 35, No. 11, pp. 72-81, November, 1997.

[19] V. Bharghavan, K. Lee, S. Lu, S. Ha, J. Li and D. Dwyer, *The TIMELY Adaptive Resource Management Architecture*, IEEE Personal Communications Magazine, Vol. 5, No. 8, August, 1998.

[20] A. K. Talukdar, B.R. Badrinath and A. Acharya, *Rate Adaptive Schemes in Networks with Mobile Hosts*, ACM/IEEE MobiCom'98, October, 1998.

[21] S. K. Das and S. K. Sen, *Quality-of-Service Degradation Strategies in Multimedia Wireless Networks*, IEEE Vehicular Technology Conference (VTC'98), Ottawa, Canada, May, 1998.

[22] K. Lee, *Adaptive Network Support for Mobile Multimedia*, ACM MobiCom'95, pp. 62-74, 1995.

[23] M. Sidi and D. Starobinski, *New Call Blocking versus Handoff Blocking in Cellular Networks.* Proceedings of IEEE INFOCOM 1996, pp. 35-42, 1996.

[24] H. Takanashi and S.S. Rappaport, *Dynamic Base Station Selection for Personal Communication Systems with Distributed Control Schemes*, Proceedings of Vehicular Technology Conference (1997) pp. 1787-1791.

Chapter 9

MAC-Throughput Analysis of CDMA Wireless Networks Based on a Novel Collision Model

Yunnan Wu
Microsoft Research,
One Microsoft Way, Redmond, WA, 98052, USA
E-mail: yunnanwu@microsoft.com

Xiang-Gen Xia
Dept. of Electrical and Computer Engineering,
University of Delaware,
Newark, DE 19716, USA
E-mail: xxia@ee.udel.edu

Qian Zhang
Dept. of Computer Science,
Hong Kong University of Science and Technology,
Clear Water Bay, Kowloon, Hong Kong
Email: qianzh@cs.ust.hk

Wenwu Zhu
Intel China Research Center Ltd.,
8F, Raycom Infotech Park A,
No. 2, Kexueyuan South Road, Beijing, 100080, China
E-mail: wenwu.zhu@intel.com

Ya-Qin Zhang
Microsoft Corporation,
One Microsoft Way, Redmond, WA 98052, USA
E-mail: yzhang@microsoft.com

1 Introduction

Wireless packet networks have been gaining unprecedented momentum in recent years. Numerous types of wireless consumer devices have sprung up in the market, a representative example being IEEE 802.11 devices. Medium access control (MAC) is a fundamental problem in wireless packet networks, because the wireless medium is shared among multiple users. The medium access problem is stated as follows. Packets arrive at nodes in the network at random time instants, in accordance with the end-to-end communication demand and the routing arrangement. At any time, several nodes, each having some outgoing packets, may choose to transmit packets. The received signal at each intended receiver is corrupted by noise and by mutual interference between the active transmitters. Thus, the medium access control problem involves issues such as interference, noise, and the random packet arrivals. Given the random packet arrivals, random access schemes that alleviate the need to centrally coordinate the users in a wireless network are often preferred because these schemes require minimal exchange of signaling overhead.

The history of random access dates back to 1970, when Abramson proposed the ALOHA scheme [1] for the multiple access channel. The ALOHA scheme is simple: whenever a packet arrives at a transmitter, it will be immediately transmitted; if the transmission were unsuccessful (assuming instantaneous feedback was available), the transmitter will wait a random amount of time and try again. A simple reception model of the physical layer is assumed: if two packets are transmitted in an overlapping interval, a *collision* is said to occur and neither packet will be correctly received. Subsequently, slotted ALOHA was proposed [2] where nodes are synchronized with aligned packet transmission boundaries.

The conventional collision model is restrictive as a characterization of the physical layer. Over the years, advanced communications and signal processing techniques such as spread spectrum and space–time processing have greatly improved the capability of the physical layer. In particular, more than one packet may be successfully received in one time slot, given several concurrent transmissions.

Enhanced physical layer capability would subsequently have an impact on the MAC layer performance. The objective of the current paper is to quantify such effects. We specifically consider the MAC-layer throughput of a DS-CDMA wireless network. With DS-CDMA, each user is assigned a signature sequence of length N (the spreading factor), which is used to modulate the signals from the user. Assigning different spreading sequences for signals from different users is essentially assigning users different directions in the N-dimensional signal space. The correlations of the spreading sequences determine the level of interference among different users. A simple yet very useful scheme of assigning the signature sequences is to let each user pick its signature sequence at random. With random sequence allocation, the signals from multiple active users "point to" randomly chosen directions in the signal space. DS-CDMA with random sequence allocation presents a distributed scheme of coordinating accesses to the medium. In essence, with random sequence allocation, the "communication channels" are automatically shared among the active users. Thus, compared with time-division multiplexing and frequency-division multiplexing, which are based on allocating orthogonal channels, spread spectrum allows the available resource (time, bandwidth) to be more flexibly shared.

Power control presents a significant degree of freedom in the system. One possible formulation of the power control problem is to minimize the total power while providing the required signal-to-interference-and-noise ratios (SINR) for the active links, i.e., transmitter–receiver pairs. Under certain modeling assumptions, this formulation turns out to be a linear optimization. Foschini and Miljanic [3] and Mitra [4] proposed a distributed adaptive algorithm for this power control problem, where the receiver of each link measures and feeds back its locally perceived SINR and the transmitter of the link then updates its transmitting power. If a feasible power allocation exists, then the distributed algorithm converges to the minimum-power solution. If a feasible power allocation does not exist, then the algorithm diverges if there is no power constraint. Several subsequent works [5–7] considered the convergence properties when there are maximum power constraints. In [8], Bambos, Chen, and Pottie proposed a distributed power control algorithm based on admission control. They introduced an active link protection mechanism, which maintains the SINR of active links above the required thresholds, as new links are trying to access the medium. As network congestion builds up, established links sustain their quality, whereas incoming ones may be blocked and rejected.

In this chapter, we define collision as the situation where it is impossible to guarantee the SINR requirement for all concurrent transmissions through power control. Such a collision model presents one concrete example of multipacket reception with spread spectrum. As a modeling simplification, we assume that when a collision occurs, none of the active users can successfully deliver its packet. This assumption may correspond to the divergence of the distributed adaptive algorithm proposed in [3,4]. As mentioned briefly in the previous paragraph, there are distributed power control techniques that can reject certain links and maintain some links at the required SINR, when it is impossible to meet the SINR requirements for all active links. With these techniques, when a collision occurs, the links that can be eventually established may depend on the specific control mechanism, as well as many other details, such as what are the power constraints and which links appear slightly earlier than the others. Because of these complications, it seems not easy to arrive at a clean and yet general model reflecting these advanced techniques that handle the collision. We leave the modeling of these enhancements as a future investigation.

User capacity, optimal power control, and DS-CDMA spreading code design in a CDMA wireless network have been recently considered in [9]–[19] and based on both the MMSE and the matched filter receivers. The user capacity refers to the maximum number of active users in a network such that it is possible to have a power allocation with the SINR at each receiver above a desired value β. In [10], the user capacity is derived for a random DS-CDMA network when $K/N \to \alpha$ as $N \to \infty$, where N is the processing gain (or the CDMA code length) and K is the number of active users. Some related results are also reported in [11]. In [13], the user capacity

$$K < N(1 + \frac{1}{\beta}), \tag{1}$$

and the optimal DS-CDMA spreading code are derived for an arbitrary processing gain N, where β is a desired SINR at the receiver and the optimal (deterministic) CDMA code reaches the user capacity for a given SINR β. These results are generalized from synchronous to asynchronous systems in [12,17]. The user capacity formula (1) holds for a random DS-CDMA network when the CDMA code length N is sufficiently long and for a non-random DS-CDMA network of an arbitrary CDMA code length with an optimized CDMA code. For a high data rate system, the former may not be desirable. For a time-varying channel, the optimal CDMA code has to be updated to achieve the user capacity because the optimal code depends on the SINR, which may have some difficulty in a mobile environment. In this chapter, we consider a DS-CDMA network using random spreading with a given code length. We derive and simulate the collision probabilities based on the single-user matched filter receiver in Section 2.

In Section 3, we study the MAC-layer performance of a wireless network via a queueing analysis. The main objective of this section is to investigate the *stability region*, that is, the set of arrival rates such that the queueing system is stable. We consider a saturated slotted ALOHA system where it is assumed that every user always has a packet to transmit. The "saturation" assumption decouples the analysis among users and hence enables closed form derivation of the multiple-user throughput region. Moreover, the saturated case is the basis for more advanced interacting queueing analysis. We begin the analysis by investigating the MAC-layer throughput region for two transmitter–receiver pairs, assuming a very general reception model at the physical layer. (For convenience in presentation, we use the word "user" to represent a transmitter–receiver pair, when there is no ambiguity; for example, two transmitter–receiver pairs will be called two users.) In this special case, an explicit characterization of the stability region is presented. However, for more than two users, it is difficult to obtain such an explicit characterization. We then examine a symmetric system where the transmission probabilities of all users are identical and the symmetric multipacket reception model proposed by Ghez, Verdú and Schwartz [27] is assumed. The symmetric multipacket reception model is a general mathematical model that encompasses many concrete models of the physical layer reception. The proposed collision model with random spreading and power control can be regarded as a special case of the symmetric multipacket reception model. We then investigate the MAC-layer throughput performance and the overall collision probability under this collision model.

2 Collision Probability Analysis

Although the cellular model is a special case of the ad hoc model, we begin the analysis with the cellular model for ease in illustration and later provide its extension to the ad hoc model. We adopt the notation used in [10, 17] and review some basics. Let $\mathbf{s}_k \in \{-1/\sqrt{N}, 1/\sqrt{N}\}^N$ be the signature sequence or spreading sequence of length N used by the kth user and p_k be the received power of the kth user. Assume there are K active users with K information symbols X_k as i.i.d. random variables of $E[X_k] = 0$ and $E[|X_k|^2] = p_k$, $k = 1, 2, \ldots, K$. Then, the received signal of a symbol-synchronous system is

$$\mathbf{Y} = \sum_{k=1}^{K} X_k \mathbf{s}_k + \mathbf{W}, \tag{2}$$

where $\mathbf{Y} \in \mathcal{R}^N$ is the $N \times 1$ received signal vector and $\mathbf{W}$ is the AWGN noise vector of distribution $N(0, \sigma^2 I)$.

When the single-user matched filter receiver is used, the signal-to-interference-and-noise ratio of the kth user is (see, for example, [10])

$$\mathrm{SINR}_k = \frac{p_k}{\sum_{l \neq k} p_l (\mathbf{s}_k^T \mathbf{s}_l)^2 + \sigma^2}. \tag{3}$$

The common SINR target β for all users is said to be feasible if and only if one can find nonnegative powers $\{p_k\}_{k=1}^K$ such that $\mathrm{SINR}_k \geq \beta$ for all k; i.e.,

$$p_k \geq \beta \left(\sum_{l \neq k} p_l (\mathbf{s}_k^T \mathbf{s}_l)^2 + \sigma^2 \right), \quad k = 1, 2, \ldots, K. \tag{4}$$

Let $\mathbf{A} \triangleq (a_{kl})_{1 \leq k,l \leq K}$ where $a_{kl} = (\mathbf{s}_k^T \mathbf{s}_l)^2$ if $k \neq l$ and $a_{kl} = 0$ if $k = l$, $\mathbf{p} = (p_1, \ldots, p_K)^T$, $\mathbf{1} = (1, \ldots, 1)^T$. Then, (4) can be rewritten as

$$\mathbf{p} \geq \beta(\mathbf{A}\mathbf{p} + \sigma^2 \mathbf{1}), \tag{5}$$

where $\geq$ means the componentwise $\geq$. Let $\rho_A \equiv \max_k |\lambda_k|$, $k = 1, \ldots, K$ be the maximum modulus of the eigenvalues of $\mathbf{A}$, $\{\lambda_k\}$. Because $\mathbf{A}$ is nonnegative, by the Perron–Frobenius theorem (see, e.g., [20]), $\mathbf{A}$ has a real nonnegative eigenvalue, called the Perron–Frobenius eigenvalue, equal to ρ_A. Then the series expansion

$$(\mathbf{I} - \beta\mathbf{A})^{-1} = \mathbf{I} + \beta\mathbf{A} + \beta^2\mathbf{A}^2 + \cdots$$

converges if and only if

$$\rho_A < \frac{1}{\beta}. \tag{6}$$

Furthermore, the above power control problem (4) is feasible if and only if $\rho_A < 1/\beta$, and the componentwise smallest feasible power vector $\mathbf{p}_{opt}$ can be solved from the following linear system

$$\mathbf{p}_{opt} = \beta(\mathbf{A}\mathbf{p}_{opt} + \sigma^2 \mathbf{1}); \tag{7}$$

i.e.,

$$\mathbf{p}_{opt} = \sigma^2\beta(I - \beta\mathbf{A})^{-1}\mathbf{1}. \tag{8}$$

Foschini and Miljanic proposed a distributed power control algorithm that computes the optimal power vector $\mathbf{p}_{opt}$ [3]. The algorithm is based on the following iterative equation

$$\mathbf{p}(j+1) = \beta\mathbf{A}\mathbf{p}(j) + \beta\sigma^2\mathbf{1}, \tag{9}$$

where $\mathbf{p}(j)$ is the power vector in the jth iteration. Equation (9) can be transformed into a different form:

$$p_k(j+1) = \frac{\beta}{\mathrm{SINR}_k(j)} p_k(j), \quad k = 1, \ldots, K, \tag{10}$$

where $p_k(j)$ is the transmitting power of the kth link in the jth iteration, and $\mathrm{SINR}_k(j)$ is the SINR of the kth receiver in the jth iteration. Equation (10) leads to a simple distributed implementation: each transmitter separately increases its power when the current SINR observed at its associated receiver is below β and decreases it otherwise. If $\rho_A < \frac{1}{\beta}$, then the algorithm converges to $\mathbf{p}_{opt}$; if $\rho_A > \frac{1}{\beta}$, then the algorithm diverges.

By assuming that a proper power control can be done, a collision will not occur if and only if there is a proper power vector $\mathbf{p}$ such that a desired SINR β at all receivers is reached; i.e., the power control problem (4) is satisfied (feasible). Therefore, a collision will not occur if and only if $\rho_A < 1/\beta$. This implies that the network (single-cell) collision probability $Pr(collision|K)$ with K active users for a desired SINR β at all the receivers is

$$Pr(collision|K) = Pr\left(\rho_A \geq \frac{1}{\beta}\middle| K \text{ active users}\right), \tag{11}$$

where the probability is taken over all the random binary signature vectors $\mathbf{s}_k$.

Because the probability $Pr\left(\rho_A \geq \frac{1}{\beta}|K \text{ active users}\right)$ deals with the eigenvalues of a $K \times K$ matrix of random variables, it is difficult to have an analytical expression. However, because it only depends on three parameters β, K, and N when all the signature sequences are binary random variables, it can be simulated a priori using the Monte Carlo trials. Nevertheless, this probability can be bounded by using Gerschgorin's theorem; see, for example, [21]:

$$\rho_A < \max_{k=1,\cdots,K} \sum_{l \neq k} (\mathbf{s}_k^T \mathbf{s}_l)^2. \tag{12}$$

Thus,

$$\begin{aligned} &Pr\left(\rho_A \geq \frac{1}{\beta}\middle| K \text{ active users}\right) \\ <&1 - Pr\left(\max_{k=1,\cdots,K}\left(\sum_{l\neq k}(\mathbf{s}_k^T\mathbf{s}_l)^2\right) < \frac{1}{\beta}\right) \\ =&1 - Pr\left(\bigcap_{k=1}^{K}\left(\sum_{l\neq k}(\mathbf{s}_k^T\mathbf{s}_l)^2 < \frac{1}{\beta}\right)\right). \end{aligned} \quad (13)$$

The above collision probability analysis can be generalized to (i) not all users have the same desired SINR β (i.e., the kth user's desired SINR is β_k) and (ii) ad-hoc network. For case (i), we have

$$p_k \geq \beta_k\left(\sum_{l\neq k} p_l(\mathbf{s}_k^T\mathbf{s}_l)^2 + \sigma^2\right), \quad k = 1, 2, \ldots, K. \quad (14)$$

$$\mathbf{p} \geq \mathbf{F}\mathbf{p} + \mathbf{b}, \quad (15)$$

where $f_{kl} = \beta_k(\mathbf{s}_k^T\mathbf{s}_l)^2$ if $k \neq l$ and $f_{kl} = 0$ if $k = l$, $\mathbf{b} = (\beta_1\sigma^2, \cdots, \beta_K\sigma^2)^T$. For case (ii), let us consider K communicating pairs, say (x_k, y_k), $k = 1, 2, \ldots, K$, where x_k represents the transmitter and y_k represents the receiver. Let g_{kl} denote the path gain from x_k to y_l and p_k denote the transmit power at x_k. Then, the power control problem (4) for the ad hoc network becomes

$$p_k g_{kk} \geq \beta\left(\sum_{l\neq k} p_l g_{lk}(\mathbf{s}_k^T\mathbf{s}_l)^2 + \sigma^2\right), \quad k = 1, 2, \ldots, K. \quad (16)$$

Correspondingly, the element a_{kl} in matrix $\mathbf{A}$ in (5) becomes $(\mathbf{s}_k^T\mathbf{s}_l)^2\frac{g_{lk}}{g_{kk}}$ if $k \neq l$. The remaining collision probability analysis (8)–(12) is the same after the matrix $\mathbf{A}$ is replaced.

3 MAC Performance Analysis

Consider a saturated slotted ALOHA system with M transmitter–receiver pairs. For simplicity, we assume the set of transmitters is disjoint from the set of receivers. The following assumptions are common in the literature. Each node has an infinite buffer. Time is slotted, with the slot size being the transmission time of a packet. Packets arrive at each transmitter according to a Bernoulli process (i.e., i.i.d. from slot to slot with mean λ_i), and the arrival processes at different transmitters are independent. Assume instantaneous feedback is available at the end of each slot that informs each active transmitter whether its transmitted packet is successfully received by the receiver. Packets that failed to be received will be put into the head positions of the associated queues, awaiting retransmission at a later stage. In each slot, assume transmitter i always transmits a packet with probability p_i. More specifically, when its buffer is empty, it continues to transmit *dummy* packets. The dummy packets can result in collision, but the successful reception of the dummy packet does not reduce the queue size.

The system can be characterized by an M-dimensional Markov chain where the state vector is the vector of queue lengths. The transition of the M-dimensional Markov chain is determined by the arrival process and the departure process, which are subsequently determined by the reception model at the physical layer. The system is *stable* if all the queue lengths converge in distribution to proper limiting distributions. Simply put, in a stable queueing system, the probability of queue overflow is arbitrarily small. Define

$$\mathbf{\Lambda} \equiv (\lambda_1, \lambda_2, \dots, \lambda_M)^T \tag{17}$$

$$\mathbf{p} \equiv (p_1, p_2, \dots, p_M)^T. \tag{18}$$

The main objective in this section is to investigate the *stability region*, that is, the set of arrival rates $\mathbf{\Lambda}$ such that there exists a set of transmission probabilities $\mathbf{p}$ leading to a stable system.

The assumption that dummy packets are transmitted when queues are empty is termed "saturation". In contrast, a more realistic setup would be to assume that empty queues do not transmit packets. The saturation assumption simplifies the problem by decoupling the queues that would otherwise be interacting with each other. Correspondingly, the M-dimensional Markov chain becomes separable into M one-dimensional Markov chains with the saturation simplification; otherwise, resolving the M-dimensional Markov chain is known to be difficult, even for the conventional collision model. With the saturation simplification, the service time in each queue is a stationary process that depends on the reception model at the physical layer and the transmission probabilities of other transmitters. Furthermore, the analysis of saturated slotted ALOHA is useful in providing performance bounds for more realistic settings of interacting queues, as demonstrated in [22] under the conventional collision model.

3.1 Review of Analysis Under Conventional Collision Model

Before proceeding to the analysis for the advanced physical layer reception model, we review the throughput analysis for saturated slotted ALOHA systems under the conventional collision model, given by Abramson [23]. The service rate of the ith queue is

$$Pr(\text{the user transmits and succeeds}) = p_i \prod_{j:j\neq i} (1-p_j). \tag{19}$$

According to Loynes' theorem [24], if the arrival process and service process of a queue are both stationary, and the average arrival rate is less than the average service rate, then the queue is stable; if the average arrival rate is greater than the average service rate, the queue is unstable. Therefore, the saturated slotted ALOHA system under the conventional collision model with $(\mathbf{\Lambda}, \mathbf{p})$ is stable if

$$\lambda_i < p_i \prod_{j:j\neq i} (1-p_j), \quad \forall i. \tag{20}$$

and unstable if

$$\exists i, \quad \lambda_i > p_i \prod_{j:j\neq i} (1-p_j). \tag{21}$$

Then the stability region is

$$\bigcup_{p_i\in[0,1]} \left\{ \mathbf{\Lambda} \middle| \lambda_i \leq p_i \prod_{j:j\neq i} (1-p_j), \quad \forall i \right\}, \tag{22}$$

which includes the boundary points as a convention. It was shown by Abramson that the outer boundary of the throughput region is the set of all points such that the transmission probabilities sum up to 1.

For the special case with two transmitters, the stability region takes the following parametric form,

$$\bigcup_{\substack{p_1\in[0,1],\\ p_2\in[0,1]}} \left\{ (\lambda_1, \lambda_2) \,|\, \lambda_1 \leq p_1(1-p_2),\ \lambda_2 \leq p_2(1-p_1) \right\}, \tag{23}$$

which can be simplified to

$$\left\{(\lambda_1, \lambda_2) \middle| \sqrt{\lambda_1} + \sqrt{\lambda_2} \leq 1\right\}. \tag{24}$$

In the above, the region in (24) is derived for the saturated ALOHA system. Without the saturation assumption, Tsybakov and Mikhailov [25] showed the stability region remains the same for two interacting queues. However, for general $M > 2$, the stability region for a saturated system may not be the same as that for an unsaturated system. A precise characterization of the latter remains unknown for $M > 3$. To date, only upper and lower bounds are known. This should be indicative of the difficulty in solving interacting queue systems.

3.2 Performance Under General Reception Model

Now we are ready to extend the analysis to the case where the physical layer reception model is more complicated. Let the total number of users be M. Define an indicator vector $\mathbf{u} = (u_1, \ldots, u_M)$ where $u_i = 1$ indicates that the ith user is active and $u_i = 0$ otherwise. Let $|\mathbf{u}|$ denote the number of 1s in the indicator vector.

Table 1: Reception model with two users

$\mathbf{u}$	Pr(user 1 succeeds$\|\mathbf{u}$)	Pr(user 2 succeeds$\|\mathbf{u}$)
(1,1)	α_1	α_2
(1,0)	β_1	0
(0,1)	0	β_2

First, let us begin the analysis with the simplest case: two users. The reception model is characterized in Table 1. Assume $\beta_1 > \alpha_1$ and $\beta_2 > \alpha_2$, which is to say that the presence of interference reduces the probability of success. Proceeding similarly as above, the stability region for the saturated slotted ALOHA system consists of the pairs (λ_1, λ_2) satisfying

$$\lambda_1 \leq p_1 p_2 \alpha_1 + p_1(1 - p_2)\beta_1 \tag{25}$$

$$\lambda_2 \leq p_1 p_2 \alpha_2 + p_2(1 - p_1)\beta_2, \tag{26}$$

for some $p_1, p_2 \in [0, 1]$.

Example 1 *Fig. 1 shows the stability region for slotted ALOHA as in (25) and (26), with* $M = 2$, $\beta_1 = \beta_2 = 1$, *and* $\alpha_1 = \alpha_2 = \alpha$; *the region is obtained by sampling uniformly over* $p_1 \in [0,1]$, $p_2 \in [0,1]$ *and then computing the boundary of the region. The region reflects the throughput tradeoff between the two users.*

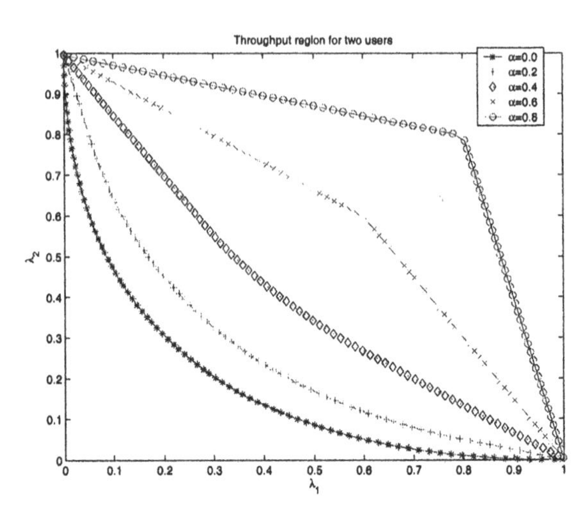

Figure 1: Stability region for slotted ALOHA: $M = 2$, $\beta_1 = \beta_2 = 1$, $\alpha_1 = \alpha_2 = \alpha$.

It can be observed from Fig. 1 that properties of the stability region change with respect to different α, *the probability of collision when both users are active. For* $\alpha > 0.5$, *the region appears to be bounded by two lines.*

We shall derive a more explicit characterization of the stability region $\{(\lambda_1, \lambda_2)\}$, eliminating the two parameters p_1 and p_2. First, perform some change of variables as

$$\tilde{\lambda}_1 \equiv \left(1 - \frac{\alpha_2}{\beta_2}\right) \frac{\lambda_1}{\beta_1} \tag{27}$$

$$\tilde{\lambda}_2 \equiv \left(1 - \frac{\alpha_1}{\beta_1}\right) \frac{\lambda_2}{\beta_2} \tag{28}$$

$$\tilde{p}_1 \equiv \left(1 - \frac{\alpha_2}{\beta_2}\right) p_1 \tag{29}$$

$$\tilde{p}_2 \equiv \left(1 - \frac{\alpha_1}{\beta_1}\right) p_2. \tag{30}$$

After this transformation, the stability region takes the parametric form

$$\bigcup_{\substack{\tilde{p}_1\in[0,1-\alpha_2/\beta_2],\\ \tilde{p}_2\in[0,1-\alpha_1/\beta_1]}} \left\{(\tilde{\lambda}_1,\tilde{\lambda}_2)\,\middle|\,\tilde{\lambda}_1 \le \tilde{p}_1(1-\tilde{p}_2),\ \tilde{\lambda}_2 \le \tilde{p}_2(1-\tilde{p}_1)\right\}, \tag{31}$$

which coincides with (23) except that the union in (23) is taken over $p_1 \in [0,1]$, $p_2 \in [0,1]$, whereas the union in (31) is taken over a smaller region $\tilde{p}_1 \in [0, 1-\alpha_2/\beta_2]$, $\tilde{p}_2 \in [0, 1-\alpha_1/\beta_1]$. Thus, the stability region is certainly contained in

$$\left\{(\tilde{\lambda}_1,\tilde{\lambda}_2)\,\middle|\,\sqrt{\tilde{\lambda}_1}+\sqrt{\tilde{\lambda}_2}\le 1\right\}. \tag{32}$$

The boundary curve, which we denote by C, can be expressed in the parametric form (34).

$$C \equiv \left\{(\tilde{\lambda}_1,\tilde{\lambda}_2)\,\middle|\,\sqrt{\tilde{\lambda}_1}+\sqrt{\tilde{\lambda}_2}= 1\right\} \tag{33}$$

$$= \Big\{(\tilde{\lambda}_1,\tilde{\lambda}_2)\,\Big|\,\tilde{\lambda}_1 = \tilde{p}_1(1-\tilde{p}_2),\ \tilde{\lambda}_2 = \tilde{p}_2(1-\tilde{p}_1),\ \text{for some } \tilde{p}_1,\tilde{p}_2 :\ \tilde{p}_1\in[0,1],\ \tilde{p}_2\in[0,1],\ \tilde{p}_1+\tilde{p}_2=1\Big\}. \tag{34}$$

Because the parameter region $\tilde{p}_1 \in [0,1-\alpha_2/\beta_2]$, $\tilde{p}_2 \in [0,1-\alpha_1/\beta_1]$ is smaller than $p_1 \in [0,1]$, $p_2 \in [0,1]$, we would expect that the boundaries of the parameter region may affect the stability region. This is indeed the case. If we let the first user transmit with probability 1, i.e., $p_1 = 1$ or $\tilde{p}_1 = 1-\alpha_2/\beta_2$, and let p_2 vary in $[0,1]$, the boundary of this region is a line segment L_1 (35) with parametric form (36).

$$L_1 \equiv \left\{(\tilde{\lambda}_1,\tilde{\lambda}_2)\,\middle|\,\frac{\tilde{\lambda}_1}{1-\alpha_2/\beta_2}+\frac{\tilde{\lambda}_2}{\alpha_2/\beta_2}=1,\ \tilde{\lambda}_1\in\left[\left(1-\frac{\alpha_2}{\beta_2}\right)\frac{\alpha_1}{\beta_1},\ 1-\frac{\alpha_2}{\beta_2}\right]\right\} \tag{35}$$

$$= \left\{(\tilde{\lambda}_1,\tilde{\lambda}_2)\,\middle|\,\tilde{\lambda}_1 = \tilde{p}_1(1-\tilde{p}_2),\ \tilde{\lambda}_2 = \tilde{p}_2(1-\tilde{p}_1),\ \text{for some } \tilde{p}_1,\tilde{p}_2 :\ \tilde{p}_1 = 1-\frac{\alpha_2}{\beta_2},\ \tilde{p}_2\in\left[0,1-\frac{\alpha_1}{\beta_1}\right]\right\} \tag{36}$$

By symmetry, if we set $p_2 = 1$ and $p_1 \in [0,1]$, the boundary of this region is a line segment L_2 (37) with parametric form (38).

$$L_2 \equiv \left\{ (\tilde{\lambda}_1, \tilde{\lambda}_2) \middle| \frac{\tilde{\lambda}_1}{\alpha_1/\beta_1} + \frac{\tilde{\lambda}_2}{1 - \alpha_1/\beta_1} = 1, \ \tilde{\lambda}_1 \in \left[0, \ \left(1 - \frac{\alpha_2}{\beta_2}\right) \frac{\alpha_1}{\beta_1}\right] \right\} \tag{37}$$

$$= \Big\{ (\tilde{\lambda}_1, \tilde{\lambda}_2) \Big| \tilde{\lambda}_1 = \tilde{p}_1(1 - \tilde{p}_2), \ \tilde{\lambda}_2 = \tilde{p}_2(1 - \tilde{p}_1), \text{ for some } \tilde{p}_1, \tilde{p}_2 : \ \tilde{p}_1 \in \left[0, 1 - \frac{\alpha_2}{\beta_2}\right], \ \tilde{p}_2 = 1 - \frac{\alpha_1}{\beta_1} \Big\}. \tag{38}$$

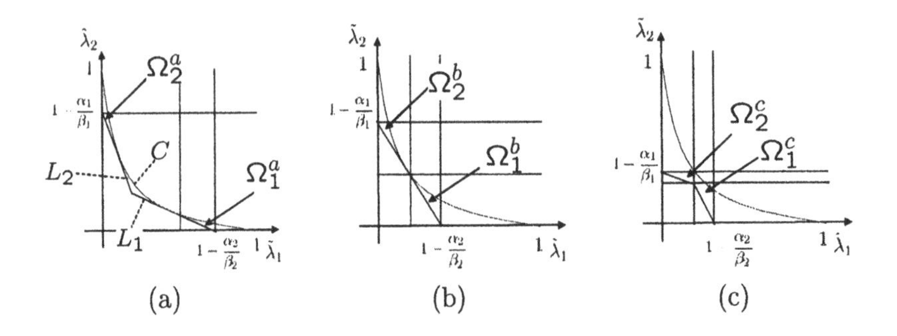

Figure 2: Stability region for saturated slotted ALOHA with two users. Three cases are shown at (a), (b), and (c), respectively.

Now we discuss the relative positions of the three curves, C, L_1, L_2. From (34), (36), (38), the curves, C, L_1, L_2, correspond to $(\tilde{\lambda}_1, \tilde{\lambda}_2) = (\tilde{p}_1(1-\tilde{p}_2), \tilde{p}_2(1-\tilde{p}_1))$ over different domains of $(\tilde{p}_1, \tilde{p}_2)$. Suppose $(\tilde{p}_1, \tilde{p}_2)$ and $(\tilde{p}'_1, \tilde{p}'_2)$ map into the same $(\tilde{\lambda}_1, \tilde{\lambda}_2)$, i.e.,

$$\tilde{\lambda}_1 = \tilde{p}_1(1 - \tilde{p}_2) = \tilde{p}'_1(1 - \tilde{p}'_2), \tag{39}$$

$$\tilde{\lambda}_2 = \tilde{p}_2(1 - \tilde{p}_1) = \tilde{p}'_2(1 - \tilde{p}'_1). \tag{40}$$

Then it can be checked via elementary algebra that either $(\tilde{p}'_1, \tilde{p}'_2) = (\tilde{p}_1, \tilde{p}_2)$ or $(\tilde{p}'_1, \tilde{p}'_2) = (1 - \tilde{p}_2, 1 - \tilde{p}_1)$. This property facilitates examining the relative positions of C, L_1, L_2 using the parametric forms.

The two line segments, L_1 and L_2, intersect at a point

$$\tilde{\lambda}_1 = \left(1 - \frac{\alpha_2}{\beta_2}\right) \frac{\alpha_1}{\beta_1}, \quad \tilde{\lambda}_2 = \left(1 - \frac{\alpha_1}{\beta_1}\right) \frac{\alpha_2}{\beta_2}, \tag{41}$$

corresponding to $\tilde{p}_1 = 1 - \frac{\alpha_2}{\beta_2}$, $\tilde{p}_2 = 1 - \frac{\alpha_1}{\beta_1}$. This intersection point corresponds to the case where both transmitters are active with probability $p_1 = p_2 = 1$.

The relative positions of C, L_1 and L_2 can be classified into three cases, as illustrated in Fig. 2 (a)(b)(c), respectively.

(a) If $\frac{\alpha_1}{\beta_1} + \frac{\alpha_2}{\beta_2} < 1$, then C intersects with L_1 at the point $(\tilde{\lambda}_1, \tilde{\lambda}_2) = (\tilde{p}_1(1-\tilde{p}_2), \tilde{p}_2(1-\tilde{p}_1))$ associated with parameters

$$\tilde{p}_1 = 1 - \frac{\alpha_2}{\beta_2}, \quad \tilde{p}_2 = \frac{\alpha_2}{\beta_2}, \tag{42}$$

and intersects with L_2 at the point $(\tilde{\lambda}_1, \tilde{\lambda}_2) = (\tilde{p}_1(1-\tilde{p}_2), \tilde{p}_2(1-\tilde{p}_1))$ associated with parameters

$$\tilde{p}_1 = \frac{\alpha_1}{\beta_1}, \quad \tilde{p}_2 = 1 - \frac{\alpha_1}{\beta_1}. \tag{43}$$

(b) If $\frac{\alpha_1}{\beta_1} + \frac{\alpha_2}{\beta_2} = 1$, then the two line segments L_1 and L_2 collapse into one line segment connecting $(1 - \frac{\alpha_2}{\beta_2}, 0)$ with $(0, 1 - \frac{\alpha_1}{\beta_1})$. In this case, this line segment $L_1 \cup L_2$ is tangent with C.

(c) If $\frac{\alpha_1}{\beta_1} + \frac{\alpha_2}{\beta_2} > 1$, then $L_1 \cup L_2$ does not intersect with C.

Theorem 3.1 *The stability region for the two-user saturated slotted ALOHA system under the reception model in Table 1 is as follows.*

(a) If $\frac{\alpha_1}{\beta_1} + \frac{\alpha_2}{\beta_2} < 1$, then the region is given by the points $\{(\tilde{\lambda}_1, \tilde{\lambda}_2)\}$ in the first quadrant lying below the boundary curve

$$\begin{cases} \frac{\tilde{\lambda}_1}{\alpha_1/\beta_1} + \frac{\tilde{\lambda}_2}{1-\alpha_1/\beta_1} = 1, & \tilde{\lambda}_1 \in \left[0, \left(\frac{\alpha_1}{\beta_1}\right)^2\right] \\ \sqrt{\tilde{\lambda}_1} + \sqrt{\tilde{\lambda}_2} = 1, & \tilde{\lambda}_1 \in \left[\left(\frac{\alpha_1}{\beta_1}\right)^2, \left(1 - \frac{\alpha_2}{\beta_2}\right)^2\right] \\ \frac{\tilde{\lambda}_1}{1-\alpha_2/\beta_2} + \frac{\tilde{\lambda}_2}{\alpha_2/\beta_2} = 1, & \tilde{\lambda}_1 \in \left[\left(1 - \frac{\alpha_2}{\beta_2}\right)^2, 1 - \frac{\alpha_2}{\beta_2}\right]. \end{cases} \tag{44}$$

Note that the three segments correspond to L_1, C, and L_2, respectively.

(b) If $\frac{\alpha_1}{\beta_1} + \frac{\alpha_2}{\beta_2} = 1$, then the region is given by the points $\{(\tilde{\lambda}_1, \tilde{\lambda}_2)\}$ in the first quadrant lying below $L_1 \cup L_2$, which in this case collapses into a single line segment

$$\frac{\tilde{\lambda}_1}{1 - \frac{\alpha_2}{\beta_2}} + \frac{\tilde{\lambda}_2}{1 - \frac{\alpha_1}{\beta_1}} = 1. \tag{45}$$

(c) If $\frac{\alpha_1}{\beta_1} + \frac{\alpha_2}{\beta_2} > 1$, then the region is given by the points $\{(\tilde{\lambda}_1, \tilde{\lambda}_2)\}$ in the first quadrant lying below $L_1 \cup L_2$, which is

$$\begin{cases} \frac{\tilde{\lambda}_1}{\alpha_1/\beta_1} + \frac{\tilde{\lambda}_2}{1-\alpha_1/\beta_1} = 1, & \tilde{\lambda}_1 \in \left[0, \left(1 - \frac{\alpha_2}{\beta_2}\right) \frac{\alpha_1}{\beta_1}\right] \\ \frac{\tilde{\lambda}_1}{1-\alpha_2/\beta_2} + \frac{\tilde{\lambda}_2}{\alpha_2/\beta_2} = 1, & \tilde{\lambda}_1 \in \left[\left(1 - \frac{\alpha_2}{\beta_2}\right) \frac{\alpha_1}{\beta_1}, 1 - \frac{\alpha_2}{\beta_2}\right]. \end{cases} \tag{46}$$

The three cases are illustrated in in Fig. 2(a)(b)(c), respectively.

Proof. It is easy to check that each point on the boundary of the region given in Theorem 3.1 falls into the stability region. Thus the region specified in Theorem 3.1 is contained in the stability region.

It remains to prove that none of the points falling outside the region specified in Theorem 3.1 is achievable. For ease of explanation, we define several regions for the three cases, as illustrated in Fig. 2. For case (a), define region Ω_1^a as the intersection of the following three regions

$$\sqrt{\tilde{\lambda}_1} + \sqrt{\tilde{\lambda}_2} \leq 1, \tag{47}$$

$$\frac{\tilde{\lambda}_1}{1 - \alpha_2/\beta_2} + \frac{\tilde{\lambda}_2}{\alpha_2/\beta_2} > 1, \tag{48}$$

$$\tilde{\lambda}_1 \in \left[\left(1 - \frac{\alpha_2}{\beta_2}\right)^2, 1 - \frac{\alpha_2}{\beta_2}\right]. \tag{49}$$

For case (c), define region Ω_1^c as the intersection of

$$\sqrt{\tilde{\lambda}_1} + \sqrt{\tilde{\lambda}_2} \leq 1, \tag{50}$$

$$\frac{\tilde{\lambda}_1}{1 - \alpha_2/\beta_2} + \frac{\tilde{\lambda}_2}{\alpha_2/\beta_2} > 1, \tag{51}$$

$$\tilde{\lambda}_1 \in \left[\left(1 - \frac{\alpha_2}{\beta_2}\right)\frac{\alpha_1}{\beta_1}, 1 - \frac{\alpha_2}{\beta_2}\right]. \tag{52}$$

For case (b), define region Ω_1^b as Ω_1^c. As shown in Fig. 2, we also define Ω_2^a, Ω_2^b, Ω_2^c as the symmetric counterparts.

Clearly, none of the points above C is within the stability region. Furthermore, $\tilde{\lambda}_1 \leq 1 - \frac{\alpha_2}{\beta_2}$ and $\tilde{\lambda}_2 \leq 1 - \frac{\alpha_1}{\beta_1}$. Thus we just need to show that none of the points in Ω_1^w, Ω_2^w are achievable for each case $w \in \{a, b, c\}$. Due to symmetry, we focus on establishing that Ω_1^w is not achievable. This is done by contraction. Suppose there is an achievable point $(\tilde{\lambda}_1, \tilde{\lambda}_2)$ in Ω_1^w. Then there must exist a pair $(\tilde{p}_1, \tilde{p}_2)$ with $(\tilde{p}_1(1 -$

$\tilde{p}_2), \tilde{p}_1(1-\tilde{p}_2)) \in \Omega_1^w$. From the definition of Ω_1^w, there must exist $(\tilde{p}_1', \tilde{p}_2')$ such that

$$\tilde{p}_1' = 1 - \alpha_2/\beta_2, \tag{53}$$

$$\tilde{p}_1(1 - \tilde{p}_2) = \tilde{p}_1'(1 - \tilde{p}_2'), \tag{54}$$

$$\tilde{p}_2(1 - \tilde{p}_1) > \tilde{p}_2'(1 - \tilde{p}_1'). \tag{55}$$

Thus, $\tilde{p}_1' > \tilde{p}_1$ and $\tilde{p}_2' > \tilde{p}_2$. From (54), we have

$$\frac{\tilde{p}_1}{\tilde{p}_1'} = \frac{1 - \tilde{p}_2'}{1 - \tilde{p}_2} \Leftrightarrow \frac{\tilde{p}_1' - \tilde{p}_1}{\tilde{p}_2' - \tilde{p}_2} = \frac{\tilde{p}_1'}{1 - \tilde{p}_2}. \tag{56}$$

From (55), we have

$$\frac{1 - \tilde{p}_1}{1 - \tilde{p}_1'} > \frac{\tilde{p}_2'}{\tilde{p}_2} \Rightarrow \frac{\tilde{p}_1' - \tilde{p}_1}{\tilde{p}_2' - \tilde{p}_2} > \frac{1 - \tilde{p}_1'}{\tilde{p}_2}. \tag{57}$$

Thus,

$$\frac{\tilde{p}_1'}{1 - \tilde{p}_2} > \frac{1 - \tilde{p}_1'}{\tilde{p}_2} \Rightarrow \tilde{p}_1' + \tilde{p}_2 > 1. \tag{58}$$

Therefore,

$$\tilde{p}_2' > \tilde{p}_2 > \alpha_2/\beta_2. \tag{59}$$

For case (a), this contradicts the fact that in region Ω_1^a, $\tilde{p}_2' \leq \alpha_2/\beta_2$. For cases (b) and (c), a contradiction is: $\tilde{p}_2' > \alpha_2/\beta_2 \geq 1 - \alpha_1/\beta_1$. Hence the proof. □

Recently, in [26], Naware, Mergen, and Tong presented the two-user stability region for the case where the saturation assumption is removed. In [26], the authors first derived the stability region $\{(\lambda_1, \lambda_2)\}$ for fixed (p_1, p_2), whereas here the stability region $\{(\lambda_1, \lambda_2)\}$ for fixed (p_1, p_2) can be obtained as the rectangle (25) and (26). Without the saturation assumption, the stability region for fixed (p_1, p_2) is larger. However, it is interesting to note that the stability region as the union of $\{(\lambda_1, \lambda_2)\}$ achievable under all (p_1, p_2), is the same both with and without assuming saturated queues. The derivations for the stability region of $\{(\lambda_1, \lambda_2)\}$ here are different from those of [26] because the stability region $\{(\lambda_1, \lambda_2)\}$ for fixed (p_1, p_2) is different. The work [26] also identified an important observation: in the slotted ALOHA system without the saturation assumption, when $\alpha_1/\beta_1 + \alpha_2/\beta_2 \geq 1$ (cases (b) and (c)), the stability region coincides with the throughput region that can be supported by central scheduling, and it can be achieved by having both users transmitting with probability one.

In the following, we investigate the general case with multiple users, i.e., $M > 2$. Extending previous discussions for $M = 2$, according to Loynes' theorem, the stability region for the saturated slotted ALOHA system comprises the points $\mathbf{\Lambda}$ that satisfy

$$\lambda_i \leq \sum_{\mathbf{u}} Pr(\mathbf{u}) \cdot Pr(i\text{th user succeeds}|\mathbf{u}), \quad \forall i = 1, \ldots, M, \tag{60}$$

where $Pr(\mathbf{u})$ is the probability of state $\mathbf{u}$,

$$Pr(\mathbf{u}) = \prod_{j=1}^{M} (u_j + (1 - 2u_j)(1 - p_j)). \tag{61}$$

By adding up all the expressions in (60), the sum throughput can be obtained as

$$\begin{aligned} T_{sum} &= \sum_{i=1}^{M} \sum_{\mathbf{u}} Pr(\mathbf{u}) \cdot Pr(i\text{th user succeeds}|\mathbf{u}) \\ &= \sum_{\mathbf{u}} Pr(\mathbf{u}) \cdot E(\text{number of successful transmissions}|\mathbf{u}). \end{aligned} \tag{62}$$

3.3 Performance Under Symmetric Multipacket Reception Model

It is difficult to further simplify (60) for the general case. In the literature, Ghez, Verdú, and Schwartz [27] proposed a symmetric multipacket reception model, as the simplification of the general (hence asymmetric) reception model such as in Table 1. The reception model at the physical layer is characterized by a matrix $\mathbf{C}$ whose entry at (n, k), $C_{n,k}$, represents the probability that k packets are successfully received when n packets are simultaneously transmitted. Symmetry among the users is assumed. Thus if n packets are transmitted, the success probability of any given packet out of the n packets is

$$\sum_{k=1}^{n} C_{n,k} \frac{k}{n}. \tag{63}$$

Now we consider a symmetric system where all transmission probabilities p_i take the same value p and examine the throughput performance under the symmetric multipacket reception model. In this case, the maximum achievable throughput is the same for all transmitters. We define the symmetric throughput, T_{sym}, as the sum of the commonly achievable throughput in this symmetric system. From (62), we have

$$T_{sym} = \sum_{n=1}^{M} \binom{M}{n} p^n (1-p)^{M-n} \sum_{k=1}^{n} kC_{n,k}. \tag{64}$$

Because the expression is a polynomial, standard numerical optimization techniques can be used to obtain the optimal p in $[0,1]$ that maximizes T_{sym}. As the number of users M goes to $+\infty$, the binomial distribution converges to a Poisson distribution with $\lambda = Mp$. Thus, with an infinite population, the symmetric throughput is

$$T_{sym}^{\infty} = \sum_{n=1}^{+\infty} e^{-\lambda} \frac{\lambda^n}{n!} \sum_{k=1}^{n} kC_{n,k}. \tag{65}$$

3.4 Performance Under Collision Model in Section 2

The collision model discussed in Section 2 can be viewed as a special case of the multipacket reception model, where $C_{n,n} = 1 - Pr(collision|n)$ and $C_{n,k} = 0$, $k < n$. Thus

$$T_{sym} = \sum_{n=1}^{M} \binom{M}{n} p^n (1-p)^{M-n} (1 - Pr(collision|n))n \tag{66}$$

$$T_{sym}^{\infty} = \sum_{n=1}^{+\infty} e^{-\lambda} \frac{\lambda^n}{n!} (1 - Pr(collision|n))n. \tag{67}$$

Besides the throughput, we are interested in the overall collision probability averaged over all the states. Similar to the throughput analysis, we have:

$$Pr(collision) = \sum_{n=1}^{M} \binom{M}{n} p^n (1-p)^{M-n} Pr(collision|n) \tag{68}$$

and for an infinite population

$$Pr(collision) = \sum_{n=1}^{\infty} e^{-\lambda} \frac{\lambda^n}{n!} Pr(collision|n). \tag{69}$$

4 Numerical Examples

In this section we present some numerical examples. The processing gain N is set to be 8. The potential number of users M is 32. We test for two values for the SINR requirement: $\beta = 5$ dB and $\beta = 10$ dB.

Fig. 3 shows the Monte Carlo simulation results (10,000 trials) of the collision probabilities in (11) and its Gerschgorin bound in (13), where the x-axis is K, and $\beta = 5$ dB and $\beta = 10$ dB, and $N = 8$. Notice that the user capacities in (1) for $N = 8$ and $\beta = 5$ dB and $\beta = 10$ dB are 10 and 8, respectively. One can see that these capacities cannot be achieved if random CDMA binary spreading sequences are used.

Fig. 4 gives the symmetric throughput, T_{sym} as in (66), under different transmission probabilities p for the cellular with $\beta = 5$ dB and $\beta = 10$ dB, respectively. The vertical lines intersecting the curves correspond to the values $\frac{1}{M} \arg\max_n (1 - Pr(collision|n))\, n$. It is observed that this choice of p is not always the optimal but is near optimal. This can be explained via the structure of (66). The probability mass function of the binomial distribution has a bell shape with the maximum attained around the mean Mp. Therefore a high sum throughput can often be achieved by adjusting p so that the peak of this bell shape matches the peak of $(1 - Pr(collision|n))\, n$. Fig. 5 shows the overall collision probabilities in (68).

5 Conclusion

In this chapter, we introduced a collision model of a DS-CDMA network using random spreading sequences, where a collision is said to occur if there is no proper power control scheme for achieving a desired SINR at receivers. We derived and simulated the collision probabilities, assuming matched filter receivers.

A queueing analysis with the saturated slotted ALOHA model was conducted to study the MAC-layer throughput performance for systems equipped with enhanced packet reception capabilities, an example being the proposed collision model. Under a general physical reception model, the stability region for saturated slotted ALOHA for two transmitter–receiver pairs is explicitly characterized. Then we discussed throughput under a symmetric multipacket reception model, as a special case of the general reception model, and under the proposed collision model, as a special case of the symmetric multipacket reception model.

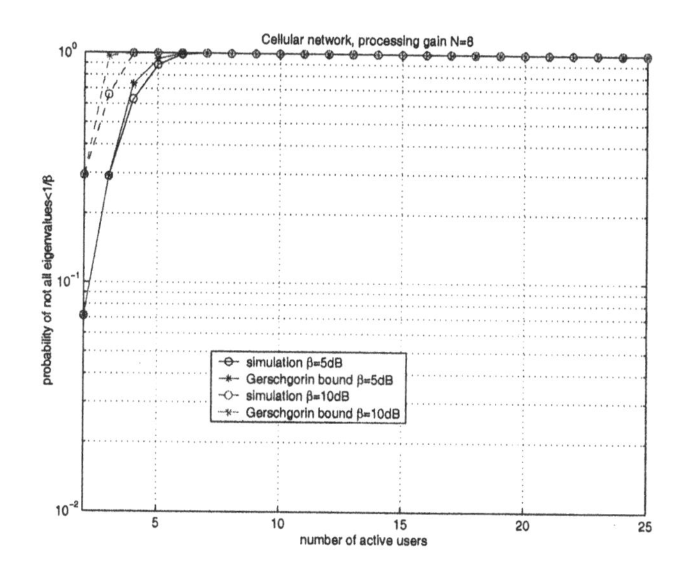

Figure 3: Collision probability for the cellular model, $N = 8$.

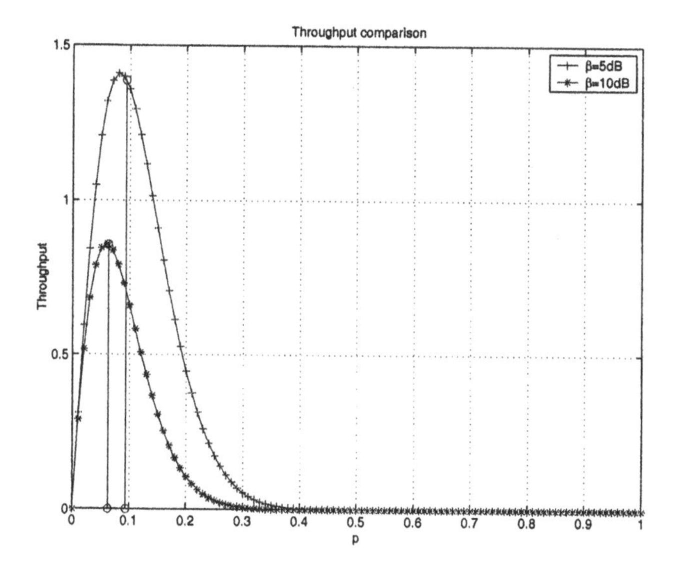

Figure 4: Throughput comparison.

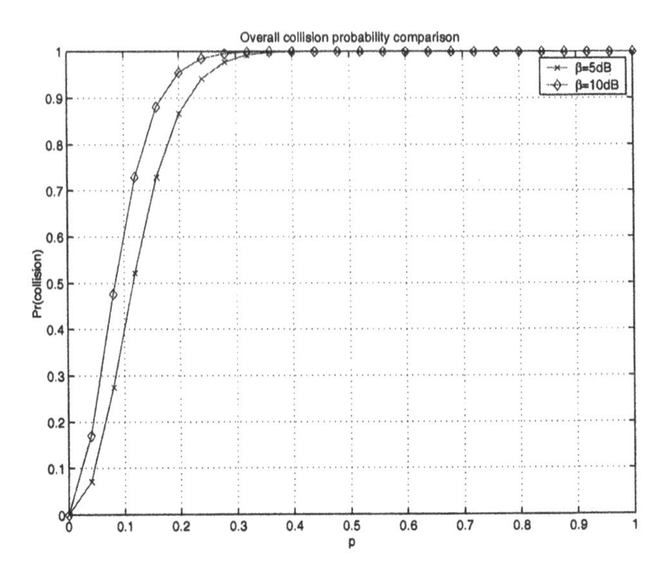

Figure 5: Overall collision probability comparison.

Acknowledgement

Xia's work was partially supported by the Air Force Office of Scientific Research (AFOSR) under Grant No.FA9550-05-1-0161 and the National Science Foundation under Grants CCR-0097240 and CCR-0325180.

References

[1] N. Abramson, "The Aloha system – another alternative for computer communications," in *AFIPS Conference Proceedings*, vol. 36, pp. 295–298, 1970.

[2] L. Roberts, "Aloha packet sytem with and without slots and capture," Stanford Research Institute, Advanced Research Projects Agency, Network Information Center, Tech. Rep. ASS Note 8, 1972.

[3] G. J. Foschini and Z. Miljanic, "A simple distributed autonomous power control algorithm and its convergence," *IEEE Trans. Veh. Tech.*, vol. 42, pp. 641–646, Apr. 1993.

[4] D. Mitra, "An asynchronous distributed algorithm for power control in cellular radio systems," in Proc. 4th WINLAB Workshop, Rutgers University, New Brunswick, NJ, 1993.

[5] R. Yates, "A framework for uplink power control in cellular radio systems," *IEEE J. Select. Areas Commun.*, vol. 13, pp. 1341–1347, July 1995.

[6] S. Grandhi, R. Yates, and J. Zander, "Constrained power control," *Wireless Commun.*, vol. 2, no. 3, Aug. 1995.

[7] M. Andersin, Z. Rosberg, and J. Zander, "Distributed discrete power control in cellular PCS," in *Proc. Int'l Symp. on Personal, Indoor, and Mobile Radio Comm. (PIMRC)*, Toronto, Sept. 1995.

[8] N. Bambos, S. C. Chen, and G. J. Pottie, "Channel access algorithms with active link protection for wireless communication networks with power control," *IEEE/ACM Trans. Networking*, vol. 8, no. 5, pp. 583–597, Oct. 2000.

[9] W. M. Jang, B. R. Vojvic, and R. L. Pickholtz, "Joint transmitter-receiver optimization in synchronous multiuser communications over multipath channels," *IEEE Trans. on Communications*, vol. 46, pp. 269–278, Feb. 1998.

[10] D. N. C. Tse and S. V. Hanly, "Linear multiuser receivers: effective interference, effective bandwidth and user capacity," *IEEE Trans. on Information Theory*, vol. 45, pp. 641–657, March 1999.

[11] S. Verdú and S. Shamai, "Spectral efficiency of CDMA with random spreading," *IEEE Trans. on Information Theory*, vol. 45, pp. 622–640, March 1999.

[12] Kiran and D. Tse, "Effective interference and effective bandwidth of linear multiuser receivers in asynchronous CDMA systems," *IEEE Trans. on Information Theory*, vol. 46, pp. 1426–1447, July 2000.

[13] P. Viswanath, V. Anantharam, and D. N. C. Tse, "Optimal sequences, power control, and user capacity of synchronous CDMA systems with linear MMSE multiuser receivers," *IEEE Trans. on Information Theory*, vol. 45, pp. 1968–1983, Sept. 1999.

[14] P. Viswanath and V. Anantharam, "Optimal sequences and sum capacity of synchronous CDMA systems," *IEEE Trans. on Information Theory*, vol. 45, pp. 1984–1991, Sept. 1999.

[15] M. Rupf and J. L. Massey, "Optimum sequence multisets for synchronous code-division multiple-acess channels," *IEEE Trans. on Information Theory*, vol. 40, pp. 1261–1266, July 1994.

[16] S. Ulukus and R. D. Yates, "Adaptive power control and MMSE interference suppression," *Wireless Networks*, vol. 4, pp. 489–496, 1999.

[17] S. Ulukus and R. D. Yates, "Optimum signature sequence sets for asynchronous CDMA systems," in *38th Annual Allerton Conf. on Communications, Control and Computing*, Oct. 2000.

[18] T. F. Wong and T. M. Lok, "Transmitter adaptation in multicode DS-CDMA systems," *IEEE J. Selected Areas in Communications*, vol. 19, pp. 69–82, Jan. 2001.

[19] S. V. Hanly and D. N. C. Tse, "Resource pooling and effective bandwidths in CDMA networks with multiuser receivers and spatial diversity," *IEEE Trans. on Information Theory*, vol. 47, pp. 1328–1351, May 2001.

[20] E. Seneta, *Non-negative Matrices and Markov Chains*, (Springer Verlag, New York, 2nd ed. 1981).

[21] J. H. Wilkinson, *The Algebraic Eigenvalue Problem*, (Clarendon Press, Oxford, 1965).

[22] W. Luo and A. Ephremides, "Stability of N interacting queues in random-access systems," *IEEE Trans. on Information Theory*, vol. 45, no. 5, pp. 1579–1587, Jul. 1999.

[23] N. Abramson, "Packet switching with satellites," in *AFIPS Conf. Proc.*, vol. 42, pp. 695–702, 1973.

[24] R. Loynes, "The stability of a queue with non-independent inter-arrival and service times," *Proc. Camb. Philos. Soc.*, vol. 58, pp. 497–520, 1962.

[25] B. Tsybakov and W. Mikhailov, "Ergodicity of slotted ALOHA systems," *Probl. Inform. Transmission*, vol. 15, pp. 301–312, Oct./Dec. 1979.

[26] V. Naware, G. Mergen, and L. Tong, "Stability and delay of finite user slotted ALOHA with multipacket reception," *IEEE Trans. on Information Theory*, vol. 51, no. 7, pp. 2636–2656, July 2005.

[27] S. Ghez, S. Verdú, and S. Schwartz, "Stability properties of slotted ALOHA with multipacket reception capability," *IEEE Trans. on Automatic Control*, vol. 33, pp. 640–649, July 1988.

Chapter 10

Information-Directed Routing in Sensor Networks Using Real-Time Reinforcement Learning

Ying Zhang
Palo Alto Research Center
3333 Coyote Hill Road, Palo Alto, CA 94304
E-mail: yzhang@parc.com

Juan Liu
Palo Alto Research Center
3333 Coyote Hill Road, Palo Alto, CA 94304
E-mail: Juan.Liu@parc.com

Feng Zhao
Microsoft Research
One Microsoft Way, Redmond, WA 98052
E-mail: zhao@microsoft.com

1 Introduction

The primary task of a sensor network is sensing, that is, to collect information from a physical environment in order to answer a set of user queries or support other decision-making functions. Typical high-level information processing tasks for a sensor network include detection, tracking, or classification of physical phenomena of interest such as people, vehicles, fires, or seismic events. Routing in a sensor network is not just about getting data from one point to another in the network. It must be optimized with respect to both data transport and information gathering. In other words, the routing structure must match the way the physical information is generated and aggregated by the network. For example, tracking a moving signal source may require the routing algorithm to combine information sequentially along a path, whereas querying the average temperature over an extended region may use a tree structure to aggregate the data from the region.

A broad class of sensor network problems can be characterized as collaborative signal and information processing problems. In such problems, a number of sensor nodes may possess useful information for a sensing task. The goal is to define and manage dynamic groups of such nodes, maximizing information extracted while keeping resource usage to a minimum. A number of approaches along this line have been reported in the literature (see, for example, [1, 2, 3]). As these approaches have demonstrated, a routing decision each local node makes during the information gathering process depends on the data generation model of the signal sources. This blurring of the abstraction barrier between applications and data transport is one of the characteristics of sensor networks. The key is for routing algorithms to handle and exploit constraints from data generation in a principled way. One would like to find a path that is not only efficient in terms of communication cost, but also aggregates as much information as possible along the path, so that the user can have a good estimate or description about the phenomenon of interest at the query or the destination node. In this sense, routing is more than a message-transporting mechanism, but also contributes to the successive message refinement.

Routing for ad hoc networks is a well-studied problem. Examples include OLSR (Optimized Link State Routing) [4], DSDV (Destination-Sequenced Distance Vector) [5], and AODV (Ad hoc On-demand Distance Vector routing) [6] protocols. Of recent interest is the topic of energy-aware routing; methods have been proposed, for example, in [7, 8], to plan paths minimizing the chance of node energy depletion. Geographic routing has also been developed. GPSR [9] routes data around a network hole, using a stateless protocol over a planar subgraph of the network topology. Geocasting [10] routes data to a geographically defined region. However, none of the above ad hoc routing algorithms considers information gathering and aggregation while routing data to a node or region, which is a major concern for sensor networks.

Routing protocols that explicitly explore data aggregation have also been developed. For example, directed diffusion [11] is a type of publish-

and-subscribe that sets up network paths between data source nodes and data sink nodes. It floods the network with data interest, and uses network parameters such as latency to autonomously reinforce good paths. However, directed diffusion does not integrate collaborative signal processing, so the paths generated can be inefficient for target tracking.

Quality of Service (QoS) routing strategies for mobile ad hoc networks have also been proposed [12, 13], where routing objectives and constraints can be specified. However, the types of objectives and constraints are limited to be *additive* (i.e., $\min \sum c_i$, e.g., delay) or *concave* (i.e., $\max \min b_i$, e.g., bandwidth), [1] to which, as we show later in this chapter, collaborative signal processing does not belong. Also in most cases, those strategies are not suitable to highly dynamic networks, because they first establish a route between the source and the sink and then follow up with a route maintenance phase if the route is broken. Message-initiated Constraint-Based Routing (MCBR) [14] developed a set of QoS-aware meta-routing strategies based on real-time reinforcement learning [15]. A couple of MCBR meta-strategies have been implemented in Berkeley motes, a widely used platform for sensor networks, and have shown good performance both in simulation and in real hardware.

In this chapter, we apply real-time reinforcement learning to Information Directed Routing (IDR) in sensor networks, where learning is used to locate the target or the destination. Combining with information utilities, this method effectively generates paths which co-optimize for both communication metrics (e.g., small number of hops) and information metrics (e.g., tracking errors), even in the existence of network holes and unpredictable moving targets. Simulation results show that such a strategy is effective for both common query routing scenarios: routing a user query from an arbitrary node to the vicinity of signal sources and back, or to a prespecified destination, maximizing information accumulated along the path.

The rest of the chapter is organized as follows. Section 2 and Section 3 summarize the general problem of information-directed routing, with target tracking as a canonical problem for IDR [16, 17]. Information models, routing protocols, and performance metrics are also discussed, and the existing solutions are presented. Section 4 develops new near-optimal solutions to information-directed routing for the two common

[1] Note that the definition of concave here is not related to the convexity in nonlinear optimization.

information extraction scenarios, using real-time reinforcement learning techniques. Section 5 presents simulation results, demonstrating the benefit of using real-time reinforcement learning for information-directed routing. Section 6 concludes the chapter with discussions and future work along this direction.

2 Information-Directed Routing

As we have discussed in Section 1, routing in sensor networks is often coupled with sensing applications. Here we consider two source-initiated on-demand routing scenarios.

- **Routing query to a high-activity region:** This scenario is illustrated in Figure 1(a). The user issues a query from an arbitrary peripheral sensor node, which we call a query proxy node, requesting the sensor network to collect information about a phenomenon of interest. The query proxy has to figure out where such information can be collected and routes the query toward the high information content region. This differs from routing in communication networks where the destination is often known a priori to the sender. Here, the destination is unknown and is dynamically determined by the routing state and physical phenomenon.

- **Routing from a proxy to an exit, collecting information along path:** This scenario is pictured in Figure 1(b). The user, for example, an police officer, may issue a query to a query proxy, asking the sensor network to collect information and report to an exit node or a base station, for example, a police station, where the information can be extracted for further processing. In this scenario, the query proxy and the destination node may be far away from the high information content region. A path taking a detour toward the high information region shall be preferable to the shortest path from the query proxy to the destination.

We formulate the information-directed routing problem as an optimization problem. We use a graph $G = (V, E)$ to describe the sensor network structure. V is a collection of vertices corresponding to sensor nodes. E is a collection of edges corresponding to internode connectivity. Associated with an edge between two nodes v_i and v_j is a communication

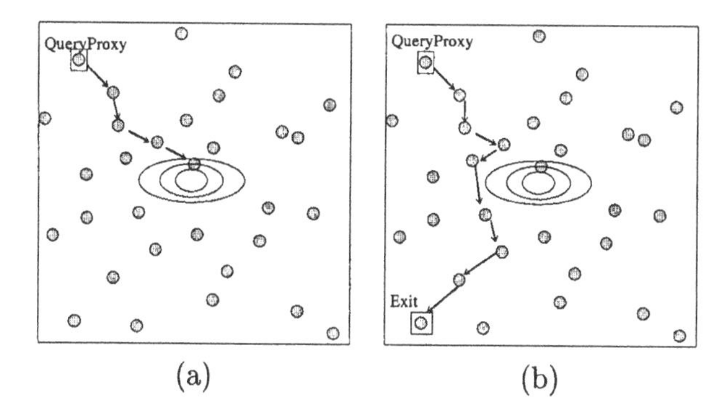

Figure 1: Routing scenarios: (a) Routing from a query proxy to the high activity region and back. (b) Routing from a query proxy to an exit or destination node. The co-centric ellipses represent isocontours of an information field, which is maximal at the center.

cost c_{v_i,v_j}. The information-directed routing problem can be formulated as finding a path $\{v_1, v_2, \cdots, v_T\}$ minimizing the total cost

$$E = \sum_i c_{v_i,v_{i+1}} - \gamma I(v_1, v_2, \cdots, v_T). \tag{1}$$

The first term measures the communication cost. The second is the negative information $-I(v_1, v_2, \cdots, v_T)$, representing the total contribution from the sensors v_1, v_2, $\cdots$, v_T. Under this formulation, routing is to find a path with maximum information gain at moderate communication cost. The regularization parameter γ controls the balance. In a network where shortest path is desired, γ is set to zero. For applications where information aggregation is of primary concern and communication cost is relatively low, γ should be set to high.

When the information gain is additive, i.e.,

$$I(v_1, v_2, \cdots, v_T) = I(v_1) + I(v_2) + \cdots + I(v_T), \tag{2}$$

the path-finding problem can be simplified considerably. It can be converted to the equivalent problem of finding the shortest path in a modified graph G'. G' has the same set of vertices and edges as G, but has a modified cost $c'_{v_i,v_j} = c_{v_i,v_j} - \gamma I(v_j)$ associated with each edge.

Strictly speaking, the additive condition (2) may not always hold in sensor network applications. The phenomenon of interest may have evolving *state*, and the information that a sensor contributes may be state-dependent. For instance, in the target-tracking example discussed later in Section 3, the state is the probability distribution (belief) of target location. Suppose sensors v_1 and v_2 have information value I_{v_1} and I_{v_2} with respect to the current belief state. After applying sensor v_1's measurement, the belief state has changed; hence I_{v_2} is obsolete and needs to be re-computed based on the new state. The information contribution I_{v_1} and I_{v_2} cannot be added together to account for the total contribution. The state-dependency property adds difficulty to routing, making it a combinatorial problem. Our strategy is to use heuristics to approximate the real cost (1). Though information is not strictly additive, the sum of individual information $\sum I(v_k)$ can often be considered as a reasonable approximation of $I(v_1, \cdots, v_T)$ in cases where the belief state varies slowly.

3 IDR in Target-Tracking Applications

As an example of data generation processes in a sensor network, consider tracking a point signal source, or target, in a 2-D region. The goal of tracking is to estimate target location $x^{(t)}$ based on a set of measurements $\overline{z^{(t)}} = \{z^{(0)}, z^{(1)}, \cdots, z^{(t)}\}$, indexed by time t, and collected by a set of nodes. To accomplish this, we use a statistical framework of sequential Bayesian filtering, a generalization of the well-known Kalman filtering [3]. For space reasons, we only briefly review the key concepts here. At time t, one has some rough prior knowledge about where the target is, usually in the form of a probability density function $p(x^{(t)}|\overline{z^{(t)}})$ (called belief). At time $t+1$, a new measurement $z^{(t+1)}$ is collected. Sequential Bayesian filtering incorporates the measurement and updates the belief to $p(x^{(t+1)}|\overline{z^{(t+1)}})$ via Bayesian inference:

$$p(x^{(t+1)}|\overline{z^{(t+1)}}) \propto p(z^{(t+1)}|x^{(t+1)}) \cdot \int p(x^{(t+1)}|x^{(t)}) \cdot p(x^{(t)}|\overline{z^{(t)}})\, dx^{(t)}. \quad (3)$$

The integral represents how prior belief is propagated to the current time step via the target dynamics $p(x^{(t+1)}|x^{(t)})$. The updated belief is the posterior of target location after observing all the measurements up to time $t+1$. The method repeats as time advances. For more details about sequential Bayesian filtering in target tracking, please refer to previous work [3].

In sequential Bayesian filtering, sensor information is aggregated incrementally. Sensors along a routing path can contribute to target tracking via their measurements. To illustrate the aggregation of information,

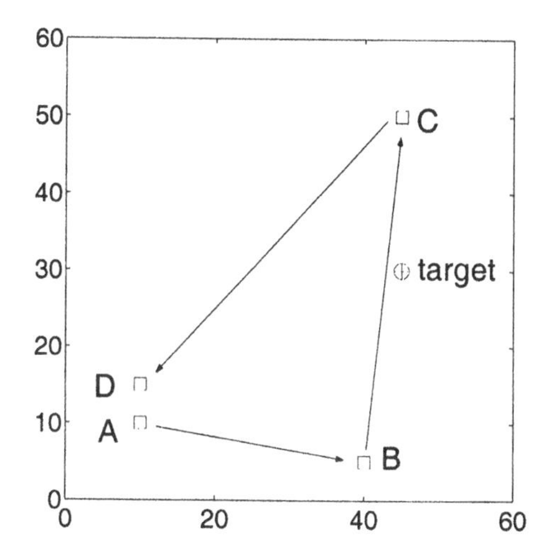

Figure 2: A sample sensor network layout: sensors are marked by squares, with labels A, B, C, and D. Arrows represent the order in which sensor data is to be combined. The target is marked by "$\oplus$".

we consider a simple sensor network example consisting of four sensors, A, B, C, and D, as shown in Figure 2. Belief about the target location is shown in Figure 3 using grayscale grids. The brighter grid means that the target is more likely to be at the grid location. We assume a very weak initial belief, uniform over the entire sensor field, knowing only that the target is somewhere in the region. Figures 3(a)–(d) show how information about the target location is updated as sensor data is combined in the order of $A \to B \to C \to D$. At each step, the active sensor node, marked with a diamond, applies its measurement to update the belief. The localization accuracy is improved over time: the belief becomes more compact and its centroid moves closer to the true target location.

Routing, in this target-tracking setting, can be guided by the estimation result. The two IDR scenarios introduced in Section 2 become the following.

- **Target-Tracking Scenario**: Find where the target is, and route query to the vicinity of the target.
- **Target-Query Scenario**: Route query from a proxy to an exit, collecting information about the target location.

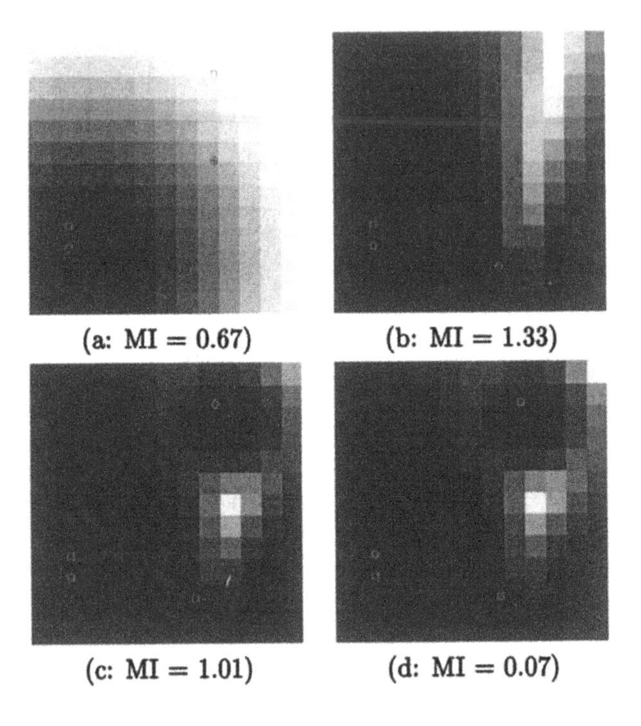

Figure 3: Progressive update of believe states of the target position, as sensor data is aggregated along the path $ABCD$. Figures (a)–(d) plot the resulting belief after each update. MI indicates information that is described in the next section obtained by Eq. (5).

3.1 Information Modeling

To quantify the contribution of individual sensors, we consider mutual information [3], a measure with a root in information theory and commonly used for characterizing the performance of data compression, classification, and estimation algorithms. As becomes clear shortly, this measure of information contribution can be estimated *without* having to first communicate the sensor data. The mutual information between two random variables U and V with a joint probability density function $p(u, v)$ is defined as $MI(U; V)$ which is

$$E_{p(u,v)}\left[\log \frac{p(u, v)}{p(u)p(v)}\right] = D(p(u|v)||p(u)), \tag{4}$$

where $D(\cdot||\cdot)$ is the Kullback–Leibler divergence [18] between two distributions. Under the sequential Bayesian filtering method in Eq. (3), the information contribution of sensor k with measurement $z_k^{(t+1)}$ is

$$I_{MI,k} = MI(X^{(t+1)}; Z_k^{(t+1)}|\overline{Z^{(t)}} = \overline{z^{(t)}}). \tag{5}$$

Intuitively, it indicates how much information $z_k^{(t+1)}$ conveys about the target location $x^{(t+1)}$ given the current belief. Other information metrics, such as the Mahalanobis distance [16], have also been proposed. They are computationally simpler to evaluate and are often good approximations to the mutual information under certain assumptions.

It is worth pointing out that information is an expected quantity rather than an observation. For example, in Eq. (5), the mutual information $I_{MI,k}$ is an expectation over all possible measurements $z_k^{(t+1)} \in \mathcal{R}$, and hence can be computed before $z_k^{(t+1)}$ is actually observed. In a sensor network, a sensor may have local knowledge about its neighborhood, such as the location and sensing modality of neighboring nodes. Based on such knowledge alone, the sensor can compute the information contribution from each of its neighbors. It is unnecessary for the neighboring nodes to take measurements and communicate back to the sensor. In our previous work [3], we provide a detailed algorithm describing how mutual information is evaluated based on knowledge local to the leader sensor. With little modification, this evaluation method can be extended to other information metrics.

In general, the information contribution of each sensor is state dependent. The information metric $I_{MI,k}$ of (5) depends on the belief state

$p(x^{(t)}|\overline{z^{(t)}})$. Revisiting the sensor network example in Figure 3, we compute the information contribution for each sensor. Note that sensors A and D are very similar and physically close by. Despite such similarity, the information values differ significantly (0.67 for A and 0.07 for D). Visually, as can be observed from Figure 3, sensor A brings significant changes to the initial uniform belief. In contrast, sensor D hardly causes any changes. The reason for the difference is that A applies to a uniform belief state, whereas D applies to a compact belief as shown in Figure 3(c).

State-dependency is an important property of sensor data aggregation, regardless of specific choices of information metric. Intuitively, how much new information a sensor can bring depends on what is already known. Note that in sensor networks, sensor measurements are often correlated. Hence a sensor's measurement is not "entirely new"; it could be just repeating what its neighbors have already reported. In the example above, sensor D is highly redundant with sensor A. Such redundancy shows up in the belief state, and thus should be discounted.

3.2 Protocols

We assume each node is aware of its own position, for example, using a GPS device or other location services. Each node also has knowledge about its local (one-hop) neighborhood, including node positions and sensor modalities, link quality, and one-hop communication cost. Such knowledge can be established through local message exchange between neighbors during network initialization and discovery. With these assumptions, the routing protocols described here can be regarded as a form of source-initiated on-demand routing.

We assume that a sensor field of size $X \times Y$ is discretized to $dX \times dY$ cells where d is the density. A belief state is a distribution $p : dX \times dY \rightarrow [0, 1]$ with $\sum p = 1$. We assume that every packet includes the current belief state about the target and the time when this belief was calculated. A packet sent from the source node combines the initial belief state (a uniform distribution) and its local sensor reading to obtain the current belief state. A forwarded packet uses the received belief state and the assumption of the maximum speed of the target to calculate its predicted belief state, then combines with its local sensor reading to obtain the current belief state. In other words, the Bayesian filter (3) is applied to each packet to update the belief state at each node.

For the target-tracking routing scenario, a forwarded packet is considered to be "arrived" at its destination if (1) the sensor reading is large and the increased utility is small, or (2) the maximum number of hops is reached. In this case, the final belief state at the "destination" will be sent back to the query (source) node.

For the target-query routing scenario, the destination is a known base station, addressed by its ID, or any sensor node, addressed by its attributes (e.g., location). The belief state at the destination will be used to estimate the target's location.

3.3 Performance Metrics

There are various types of performance metrics for ad hoc routing in sensor networks [15], including, e.g., latency, throughput, loss/success rate, energy consumption, and lifetime of the network.

To measure the IDR tracking performance, we consider two quantities: (1) the mean-squared error (MSE) $E_{p(x^{(t)}|\overline{z^{(t)}})}||x^{(t)} - x_{true}^{(t)}||^2$, and (2) the size of the belief state.

The MSE describes the tracking accuracy, and the belief size reflects uncertainty in the estimate. In this chapter, the belief size is calculated as follows. Let the number of cells with likelihood value exceeding a threshold be S, the belief size is obtained by $\sqrt{S}/d$ where d is the density of cells.

We can combine these two metrics into one as a measurement ϵ for the quality of tracking; e.g.,

$$\epsilon = \mu + \sigma/k, \tag{6}$$

where μ is the mean error (square root of MSE), σ is the belief size, and k is a constant.

Furthermore, if we assume that the communication cost is identical for every hop (e.g., when all nodes use the same power level to transmit), we can compute the overall cost combining the quality of tracking and communication as

$$c = \lambda h + \epsilon, \tag{7}$$

where h is the number of hops from the source to the destination, λ is a coefficient, and ϵ is defined in Eq. (6).

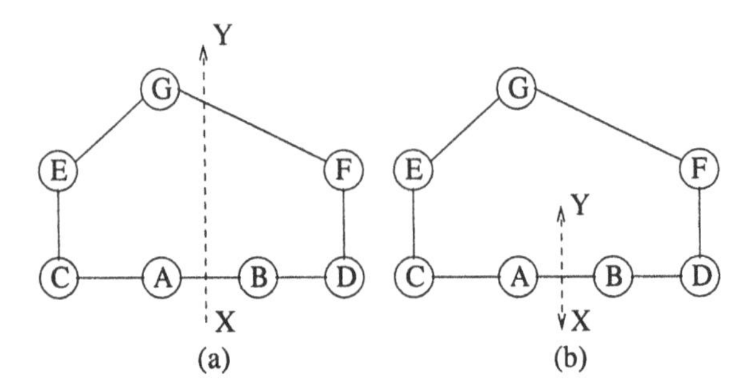

Figure 4: Routing in the presence of sensor holes. A through G are sensor nodes. All edges have unit communication cost. The dashed lines plot target trajectory. In (a), the target is moving from X to Y. In (b), the target is bouncing back and forth between X and Y.

3.4 Existing Solutions

Loosely speaking, there have been two types of approaches developed for IDR. The first type is greedy, routing packets to a neighbor with the highest information utility. An example is the Constrained Anisotropic Diffusion Routing (CADR) described in [1, 16]. This approach works well when the sensor network layout is uniform. For sensor networks with inhomogeneity such as holes, the relay may get trapped due to the greedy nature. Figure 4a provides a simple example. Here we use the inverse of Euclidean distance between a sensor and the target to measure the sensor's information contribution (assuming these information values are given by an "oracle"). The problem with greedy search is independent of the choice of information measure. Consider the case that the target moves from X to Y along a straight line (see Figure 4a). At time $t = 0$, node A is the leader, and can relay the information to its neighbor B or C. The relay goes to B because it has a higher information value. By the same criterion, B then relays back to A. The relay keeps bouncing between A and B, and the target moves away. The path never gets to nodes E, F, or G, who may become informative as the target moves closer to Y. The culprit in this case is the "sensor hole" the target went through. The greedy algorithm fails due to its lack of knowledge beyond the immediate neighborhood.

The second type uses a multiple step look-ahead to get around the problem of search getting trapped at sensor holes. The look-ahead horizon should be large enough and comparable to the diameter of sensor holes, yet not too large to make the communication and computation cost prohibitive. For static sensor networks, the sensors can explore their local area in the network discovery phase and store in cache the information about inhomogeneity. This is done, for example, in [19]. Later in the path planning phase, such information will be helpful in selecting the value for the look-ahead horizon. Our earlier work [17] routes queries to the target vicinity using this look-ahead method. In the simulation study, it has been shown that the algorithm, with a suitable look-ahead horizon, is much more likely to succeed in routing a message around sensor holes (by a factor of 4–11) when compared to a greedy approach. At the same time it produces less error (by a factor of 2–4 in the simulation) in tracking a signal source.

The look-ahead approach has also been successfully used in the second IDR scenario, as studied in [17]. To route to a destination with a query of the target location, a real-time A* (RTA*) search [20] has been used. The baseline A* search is a best-first search, where the merit of a node is assessed as the sum of the actual cost g paid to reach it from the query proxy, and the estimated cost h to pay in order to get to the exit node (often known as the "cost-to-go"). It keeps a moving frontier of $g + h$, and iteratively expands the nodes on the frontier until the exit is reached. The resulting path is guaranteed to be optimal if the estimated h never exceeds the true cost-to-go; i.e., h is admissible. The well-known Dijkstra's algorithm can be considered as a special case of A* search, with $h = 0$, always an underestimate. RTA*, as a real-time variation, guarantees to find a path if it exists, but as a price to pay for real-time operations, the solution may lose the optimality of the baseline A* search and may be suboptimal. The result of RTA* largely depends on the quality of the cost-to-go estimation h: one has to estimate the information contribution of sensors lying ahead, based on the currently available information alone without further querying or communication. The multiple step look-ahead is used for the estimation of h. It has been shown in simulation that the RTA* routing for IDR, compared to the shortest path from source to destination, produces less error in locating a signal source. Thus the RTA* routing is more effective in collecting information.

4 Real-Time Learning for IDR

There are two limitations for the existing solutions. Maintaining an M-hop neighborhood, where M is the look-ahead horizon, can be expensive in communication and nontrivial in computation. For the RTA* algorithm, the location of the destination has to be known to the query node which may not be the case for mobile sensors. To overcome these problems, we propose new IDR solutions based on real-time reinforcement learning.

4.1 Real-Time Reinforcement Learning for Routing

Real-time or agent-centered search has been an active research area in AI for the past decade [21]. Routing with an additive objective can be formulated as a weighted shortest-path problem, which can be solved by dynamic programming off line or real-time reinforcement learning online. Q-learning, a type of reinforcement learning [22], has been applied to telecommunication network routing, i.e., Q-routing [23] and sensor network routing [15]. It has been shown in both simulation and real hardware that this type of routing performs well in dynamic situations, e.g., forwarding packets to a mobile destination.

From a learning perspective, the problem of routing can be considered as designing a policy at each node, so that the overall path from source v_0 to destination v_d — $v_0, v_1, \cdots, v_d$ — is *optimal*, i.e., $\min \Sigma c_{v_i}$, where c_v is the *local* cost function at node v and Σc_{v_i} is the global *additive* objective. Routing for additive objectives is a *weighted* shortest-path problem.

All learning-based strategies typically consist of an *initialization* phase and a *forwarding* phase. The initialization phase normally either establishes a neighborhood by sending "hello packets" from each node, or generates a spanning tree by broadcast from the base station. The forwarding phase is a policy improvement phase, deciding which neighbor to pass the packet to according to its current estimates.

One can define a cost function on each node, called *Q-value*, indicating the minimum cost from this node to the destination. For a distributed sensor network, Q-value is initially unknown, and an initial estimation is made according to the type of message or by an initial flooding from the destination. Furthermore, a node also stores its neighbors' Q-values, *NQ-values*, which are estimated initially according to the neighbors' attributes and updated when packets are received from neighbors.

For wireless networks, value iteration in reinforcement learning can be realized in a cost-effective way [15]. For each packet sent out from a node, the current Q-value of the node with respect to that destination is attached. All the nodes are set to be in *promiscuous* listening mode. In other words, when a packet is sent, all neighbors, not just the designated receiver, will hear the packet and make use of the Q-value sent with the packet. Learning is triggered whenever a node overhears a packet, and its own Q-value is updated by

$$Q_m \leftarrow (1-\alpha)Q_m + \alpha(c_m + \min_n NQ_m(n)) \qquad (8)$$

where $0 < \alpha \leq 1$ is the learning rate, $c_m > 0$ is the local objective or cost, and n is a neighbor of this node.

In a forward search-based routing strategy, a node holding the current packet forwards it to the neighbor for which it has the best NQ-value. For dynamic and asymmetric networks, implicit confirmation is also used [15], which we do not discuss further in this chapter.

4.2 Routing a Query to Track a Target

In target-tracking applications, it is important to be able to initiate a query from an arbitrary entry node to find out the current status of a target, as illustrated in Figure 1(a). Ideally, the entry point node (query proxy node) would like to contact the nodes in the vicinity of the target, or the high information content region. Due to the distributed nature of ad hoc sensor networks, the query proxy may not be aware of existence and whereabouts of the high information content region. Using real-time reinforcement learning, packets discover the whereabouts of the target during routing.

Figure 5 shows the algorithm on each node. When a node receives a packet, it learns its Q-value, which is the number of hops to the sensors at the vicinity of the target in this case. If the packet is addressed to this node, it will either forward the packet, or, when the maximum number of hops is reached, send the packet back to the query node with the belief. The algorithm selects the next node to forward the packet according to the following rule: if the sensor is close to the target, the costs of the neighbors are the combination of the hop counts to the target and the information gains; otherwise, the costs are the hop counts to the target only. In either case, the neighbor with the minimum cost is selected as the next forward node.

```
Q: Q value
NQ: NQ values
R: sensor reading
Rm: threshold for at target
Ru: threshold for near target
Is: information gains
Cs: costs
C0: maximum hops
λ: coefficient
0 < α < 1: learning rate

received (m) at w from node u do
  if R > Rm then
    Q ← 0; //at target
  else // learn hops to target
    NQ(u) ← m.Q;
    Q ← (1-α) Q + α (1+min_v NQ(v));
  end
  if m is addressed to w then
    m.hops ← m.hops +1;
    if m.hops < C0 then forward(m);
    else send m back to the query node;
    end
  end
end

forward (m) at w do
  calculate Is for neighbors according to Eq. (5);
  if R > Ru then
    Cs ← λ NQ - Is; //close to target, use information gain also
  else
    Cs ← NQ; //too far from target, use hop counts only for learning
  end
  v ← argmin_n Cs(n);
  m.Q ← Q;
  send(m) to v;
end
```

Figure 5: Learning-based IDR for tracking a target.

When the target is moving, the nodes that detect the target (when sensor reading R > Rm) will broadcast with their new Q-value, 0. The other sensors that are not at the vicinity of the target will update their Q-value in two cases: (1) received a broadcast packet from the destination node with Q-value 0; or (2) overheard a packet sent by a sensor node. For a network with enough packets, the Q-value field will be established soon, following the motion of the target. For a network without much traffic, extra control packets can be added for learning purposes, e.g., broadcast M-hops from the destination, or broadcast until the Q-value change is small. There is a tradeoff between the energy cost and the tracking performance in this case, which shall be investigated in the future.

With the learning algorithm, we revisit the examples in Figure 4. If the target is traveling in a straight line as in Figure 4a, starting from A, the path will bounce between A and B for a while. But as the target gets far from A and B, the hop counts (Q-value) of A and B will be increased so packets start to forward to C or D. When the target gets close to G, G will be the destination, and the path will extend to G via $ACEG$ or $BDFG$. On the other hand, if the target is traveling as in Figure 4b, then B is always the most informative sensor in A's neighborhood, and vice versa, assuming the sensor readings at both A and B are above the threshold. The learning algorithm selects the path alternating between A and B.

4.3 Routing a Query About a Target to a Destination

In the scenario pictured in Figure 1(b), the goal is to route a query from the query proxy to the exit point and accumulate as much information as possible along the way, so that one can extract a good estimate about the target state at the exit node.

As in [17], for this task, we set the total communication cost (hop counts) close to some prespecified amount C_0. Here the total cost C_0 is treated as a *soft* constraint, which is a "hypothetical" cost that the routing algorithm aims to achieve.[2] The value of C_0 controls the tradeoff between the communication cost and information aggregation.

Figure 6 shows the algorithm on each node. When a node receives a

[2] An alternative formulation of treating C_0 as a hard constraint that must be satisfied strictly. However, finding an optimal path under this hard constraint will require global knowledge about the sensor network and thus is inapplicable to ad hoc sensor networks.

Qd: Q value for destination
NQd: NQ values for destination
Qt: Q value for target
NQt: NQ values for target
δ: estimation error
R, Rm, Is, Cs, C_0, λ, α: same as in Figure 5

received (m) at w from node u **do**
  if R > Rm **then**
    Qt ← 0; m.foundTarget ← true; *//at target*
  else *// learn hops to target*
    NQt(u) ← m.Qt;
    Qt ← (1-α) Qt + α (1+min$_v$ NQt(v));
  end
  if w is destination **then**
    Qd ← 0; *//at destination*
  else *// learn hops to destination*
    NQd(u) ← m.Qd;
    Qd ← (1-α) Qd + α (1+min$_v$ NQd(v));
  end
  if m is addressed to w **then**
    m.hops ← m.hops +1;
    if w is not destination **then forward**(m); **end**
  end
end
forward (m) at w **do**
  L ← C_0 - m.hops; *//hops left*
  if L > (Qd+δ) **then** *//get more information*
    calculate Is for neighbors according to Eq. (5);
    if m.foundTarget **then** Cs ← NQd; **else** Cs ← NQt; **end**
    Cs ← λ Cs - Is;
    v ← argmin$_n$ Cs(n);
  else *//go to destination as soon as possible*
    v ← argmin$_n$ NQd(n);
  end
  m.Qt ← Qt; m.Qd ← Qd;
  send(m) to v;
end

Figure 6: Learning-based IDR for query about a target.

packet, it learns both of its Q-values for the target and for the destination. A packet will be marked if it has visited the sensor nodes at the target. If the packet is addressed to this node and it is not the destination, the packet will be forwarded. The algorithm selects the next node to forward the packet according to the following rule: if the number of hops left exceeds an estimation error δ plus the estimated hop counts of this node to the destination, the neighbor with the smallest hop counts to the destination is selected. Otherwise, the cost will be evaluated as the combination of Q-value and the information gain. In the latter case, before the packet arrives at the target, the Q-value of the target will be used; after the packet arrives at the target, the Q-value of the destination will be used. In both cases, the neighbor with the minimum cost is selected as the next forward node. The estimation error δ represents the estimated maximum difference between the actual hops to the destination and its Q-value to the destination.

This algorithm behaves in a way that packets are first attracted to the target and then attracted to the destination. Parameters C_0, δ and λ can be tuned to alter the path. Low C_0 value favors shorter paths, and high C_0 allows longer paths with more effective information aggregation. The increase of δ will increase the probability that the maximum hop counts do not exceed C_0. Coefficient λ in addition controls the tradeoff between information gains and the communication cost.

For this routing scenario, if the destination is known (e.g. a base station), it is worthwhile to learn the Q-values to the destination by flooding the network during initialization. Even if the destination is moving, the Q-values will be readjusted as the result of learning, similar to what we have discussed in the previous scenario. On the other hand, if the destination is unknown, the first few packets will travel extra hops in order to learn the whereabouts of the destination.

5 Experimental Results

5.1 Simulation Environment

We have simulated the IDR protocols using Prowler [24], a probabilistic wireless network simulator. Prowler provides a radio fading model with packet collisions, static and dynamic asymmetric links, and a CSMA MAC layer.

Two types of sensors are modeled for target tracking: acoustic amplitude sensors and Direction-Of-Arrival (DOA) sensors. The acoustic amplitude sensors output sound amplitude measured at each microphone, and estimate the distance to a target based on the physics of sound attenuation. The DOA sensors are small microphone arrays. Using beamforming techniques, they determine the direction the sound comes from, i.e., the bearing of the target. The detailed description of these two types of sensors can be found in [3].

We simulate a sensor field of dimension 6×15. Sensor layout is generated as follows: first generate a uniform grid of 15 rows and 6 columns evenly covering the region, then perturb the grid points with a uniform noise distribution in $[-0.1, 0.1]$. The resulting sensor layout is plotted in Figure 7a. To test the routing performance in the presence of sensor holes, sensors within the rectangular region centered at (2.5, 4) with size (4, 2) are removed. The resulting sensor network is shown in Figure 7b. The sensor network consists of roughly 90% amplitude sensors and 10% DOA sensors (i.e., the probability of the DOA sensor is 0.1), randomly spread over the sensor region.

The signal strength of the radio is set so that the maximum communication range is about 2 grids. In this simulator, as in real situations, neighborhood is established by communication rather than calculated by distances, and the results of neighborhood vary from run to run. Figure 7 shows snapshots of the communication links.

The routing algorithms are evaluated in terms of the IDR performance metrics presented in Section 2.

5.2 Routing to a Stationary Target

A stationary target is simulated at location (2.5, 7.5). We use the sensor closest to the lower-left corner (0, 0) as the query proxy node. Starting from the proxy node, we would like to estimate the target location and shoot the query toward it. In simulation, we allow a path length of 20 hops and examine the IDR performance at the end of the path. The sensor network is inhomogeneous with a sensor hole, as shown in Figure 7b.

For this routing task, we compare the learning algorithm with the greedy CADR algorithm. Each method is simulated with 100 independent runs, with each run 15 packets, 5 seconds a packet, from the source. The results are summarized and reported in Figure 8. Compared to the greedy CADR algorithm, the learning algorithm significantly improves the tracking performance after the first few packets.

Figure 9 visualizes the paths produced by the greedy CADR and the

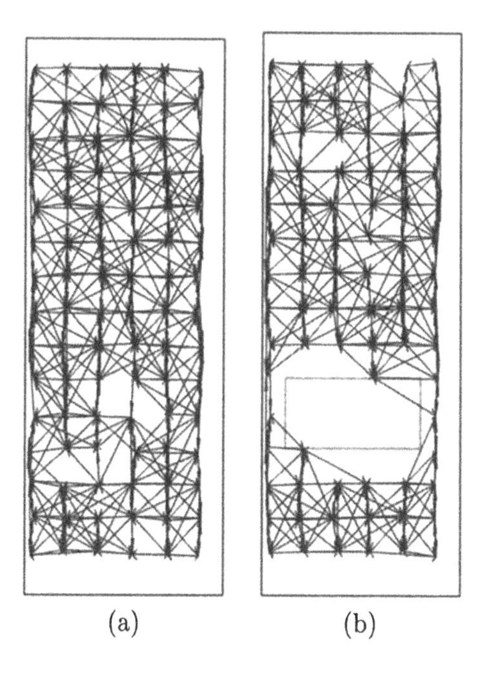

Figure 7: Examples of simulated sensor layout: (a) homogeneous and (b) with a sensor hole.

learning algorithm. The greedy algorithm path is plotted in Figure 9a. The path gets stuck and spends most of the hops bouncing between nodes on the lower side of the sensor hole. For comparison, the learning path of the tenth packet is plotted in Figure 9b. The path manages to get around the sensor hole, and ends at a node around the true target location. The tracking performance is also much improved; the target location estimate is more accurate.

5.3 Routing to a Moving Target

In tracking applications, the target is often nonstationary. In principle, a moving target can be considered as an extension of the stationary target case; the target can be considered as approximately stationary within a short time interval. Here we simulate a target moving along the straight line $x = 2.5$ (the center line of the sensor field along the vertical dimension) with speed $v = 0.1/\mathrm{s}$. The query enters at the node closest to the initial target position (2.5, 0).

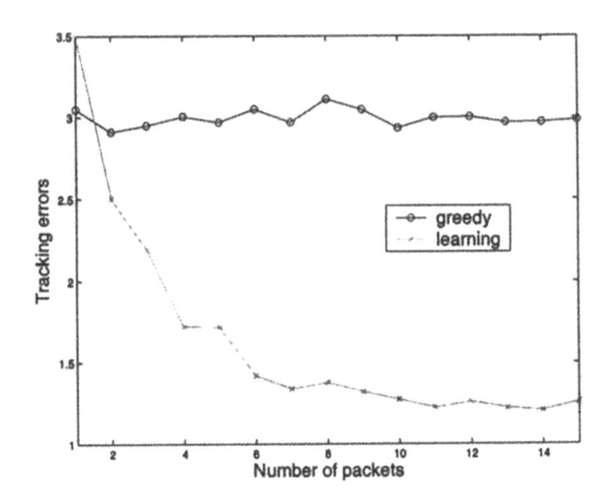

Figure 8: Tracking performance for a static target: greedy CADR algorithm vs. the learning algorithm. Tracking errors are calculated according to Eq. 6 with $k = 3$.

We compare the learning algorithm with the greedy CADR, with 100 independent runs for each. Same as the previous test, each run shoots 15 packets, 5 seconds a packet, from the source. The performance is summarized in Figure 10. Here we observe similar characteristics as in the stationary case. With the greedy algorithm, tracking errors are getting larger when the target moves into the hole and up. With the learning algorithm, the performance gets worse only for the first couple of packets and new paths are learned soon.

5.4 Routing to a Destination

Routing from a query proxy to an exit node or destination is tested with a stationary target at (4, 7.5). The selection of query proxy and exit node can be arbitrary. We select the query proxy node as the node closest to the lower-left corner (0, 0), and the exit node as the node closest to the upper-left corner (0, 14). The tests are performed on a homogeneous sensor network (as in Figure 7a).

Recall from Section 2 that the information-directed routing problem is essentially a tradeoff between the communication expense and information aggregation. We simulated three different situations: (1) shortest

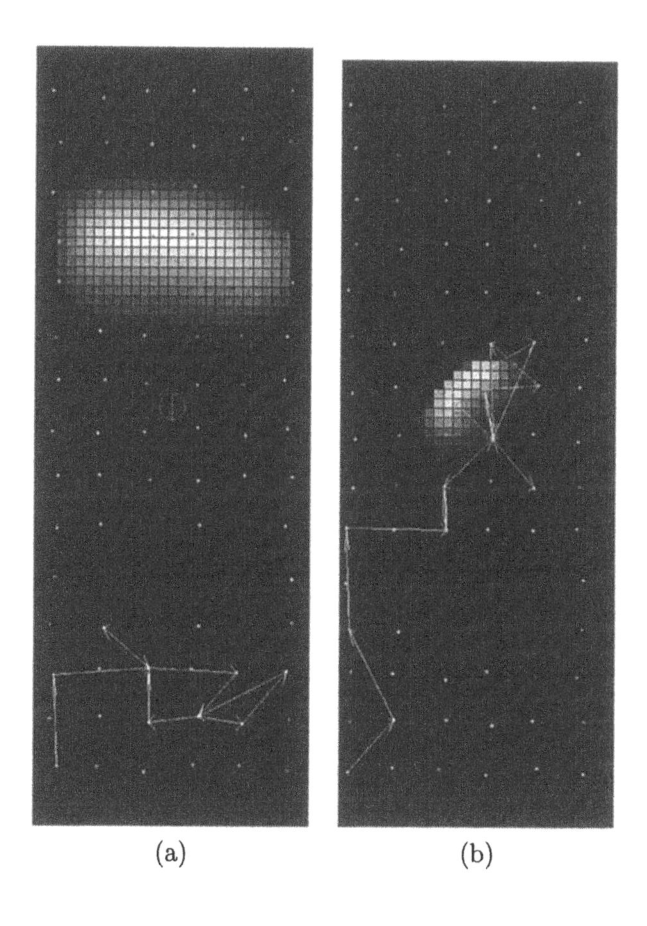

Figure 9: Query routing path produced by (a) the greedy CADR algorithm and (b) tenth packet of the learning algorithm. The target is located at (2.5, 7.5), roughly in the middle of the sensor field. It is marked with "$\oplus$". The selected paths are marked with lines.

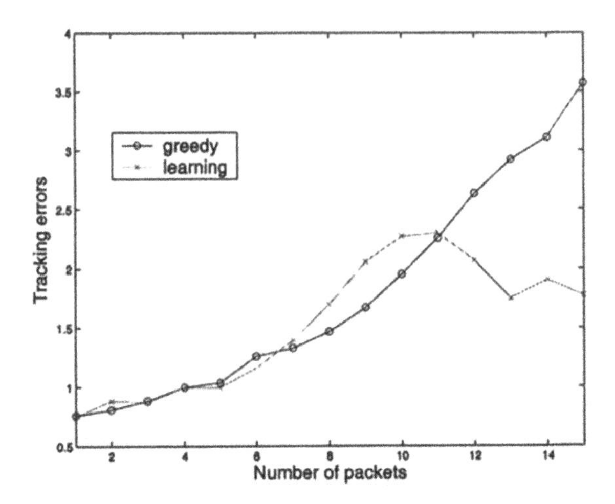

Figure 10: Tracking performance for a moving target: greedy CADR algorithm vs. the learning algorithm. Tracking errors are calculated according to Eq. 6 with $k = 3$.

path to the destination by setting C_0 small; (2) a detoured path (with C_0 20) to obtain the target information when the destination was known a priori; and (3) the same as (2) but the destination was unknown a priori. The second situation simulates the cases when the destination is a base station, and the third situation simulates the cases when the destination is an arbitrary node in the network. For the second situation, the Q-values to the destination are learned during the initialization by flooding from the destination. For the third situation, the Q-values to the destination are learned during routing.

We tested these situations with 100 independent runs for each. Same as the previous test, each run shoots 15 packets, 5 seconds a packet, from the source. The results are compared in terms of tracking errors (Figure 11a), number of hops (Figure 11b) to the destination, and the overall performance (Figure 11c). We can see the learning algorithm works well for tracking whether or not the destination is known a priori. When the destination is unknown, the paths are long initially but gradually reduce to the allowed length C_0 as a result of learning. The learning algorithm with known destination gets the best overall performance, and the learning algorithm with unknown destination improves the performance over time due to shorter path lengths.

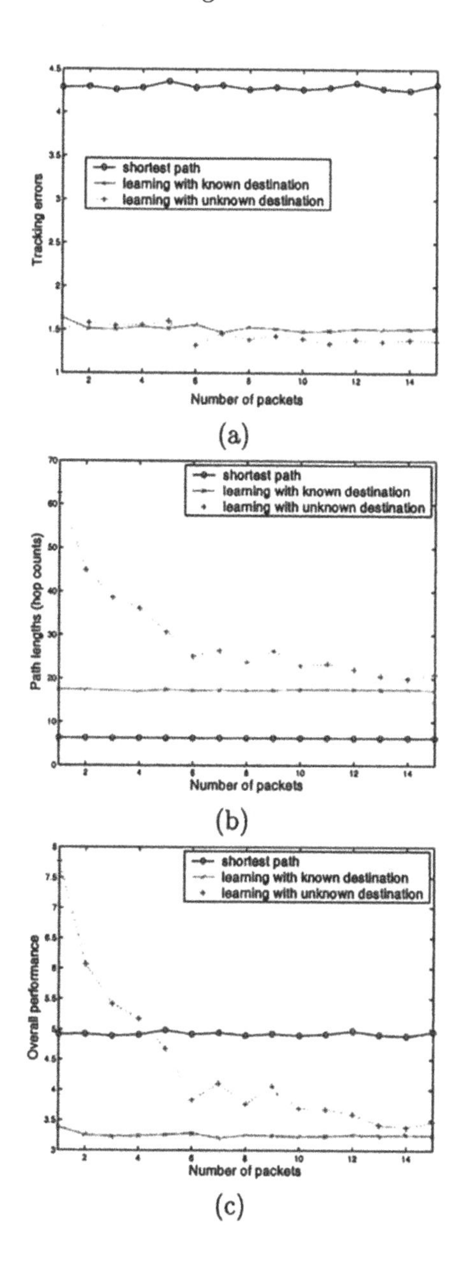

Figure 11: Shortest path vs. the learning algorithm with known and unknown destinations: (a) Tracking errors are calculated according to Eq. 6 with $k = 3$ (b) Path lengths are the hop counts from the source to the destination (c) Overall performance values are obtained according to Eq. 7 with $\lambda = 0.1$. Note that the same λ is used in the algorithm in Figure 6.

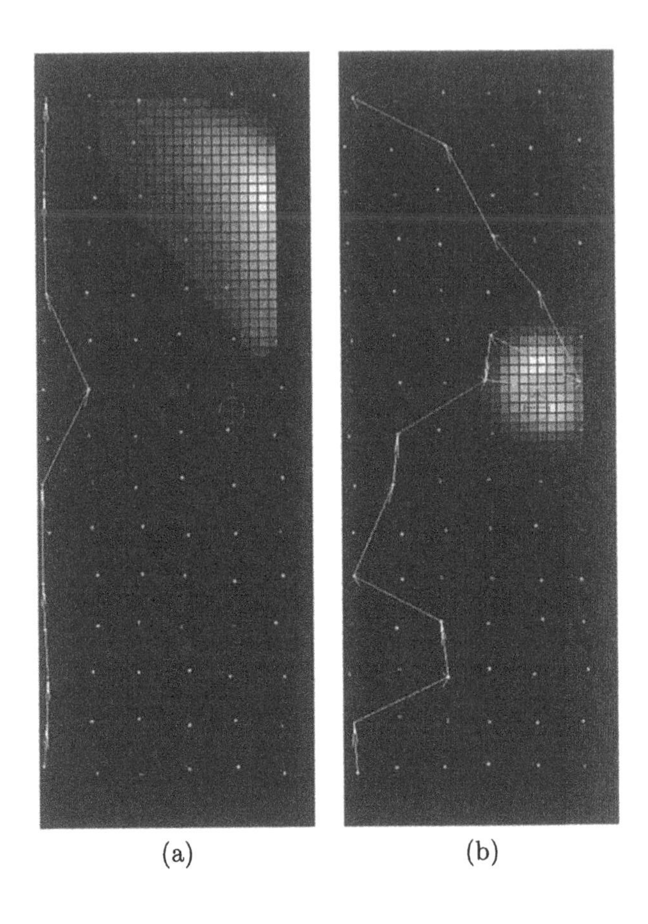

Figure 12: Query routing path produced by (a) the shortest path and (b) the learning algorithm. The target is located at (4, 7.5), roughly in the middle-right side of the sensor field. It is marked with "⊕". The selected paths are marked with lines.

Figure 12 visualizes selected paths with different lengths. The shortest path is shown in Figure 12a. It has 6–7 hops and mostly follows a vertical line from the query proxy to the exit node. The belief state is fairly big, and cannot localize the target. Figure 12b shows a longer path of about 20 hops. Starting from the proxy, the path bends toward the target direction in an attempt to accumulate information. The tracking performance is vastly better than the shortest path.

6 Conclusions

We have demonstrated the benefits of using real-time reinforcement learning for information-directed routing that jointly optimizes for maximal information gain and minimal communication cost. In the simulation study, we have shown that: (1) for tracking targets, the learning algorithm, compared to the previous greedy algorithm, is able to route around the sensor hole after a couple of packets and to track the moving target effectively; and (2) for querying targets, the learning algorithm performs well in both situations with known or unknown destinations.

Unlike look-ahead routing algorithms [17], knowledge about the network structure or the application does not play an important role in learning-based algorithms for IDR. On the other hand, the more the system learns, (e.g., hop counts to the destination), the better the algorithm performs. For a static network, the performance should converge to its optimal. For a dynamic network, if the speed of change is commensurate with the speed of learning, the system should be adaptable to the dynamic situations.

Although we presented IDR usage in the context of localization and tracking problems, the general idea of using information to guide routing applies to other problems as well. For example, in monitoring and detection problems, information may be defined as the reduction of uncertainty in the hypothesis test of target presence. In classification problems, information may relate to how sensor measurement affects the overall classification error. The specific form of information model may vary; the basic structure of the routing algorithms stays the same.

We presented the algorithms for scenarios with a single stimulus. The algorithms can be generalized to handle multiple stimuli, using a spanning tree [25]. Generalizing the algorithms described here to handle dynamically moving stimuli while maintaining good approximations to the optimal routing tree remains a future research topic.

Acknowledgment

This work is funded in part by Defense Advanced Research Project Agency contract # F33615-01-C-1904.

References

[1] F. Zhao, J. Shin, and J. Reich, "Information-driven dynamic sensor collaboration," *IEEE Signal Processing Magazine*, vol. 19, pp. 61–72, Mar. 2002.

[2] R. Brooks, C. Griffin, and D. Friedlander, "Self-organized distributed sensor network entity tracking," *International Journal of High-Performance Computing Applications*, vol. 16, no. 3, 2002.

[3] J. Liu, J. E. Reich, and F. Zhao, "Collaborative in-network processing for target tracking," *EURASIP, Journal on Applied Signal Processing*, vol. 2003, pp. 378–391, Mar. 2003.

[4] T. Clausen, G. Hansen, L. Christensen, and G. Behrmann, "The optimized link state routing protocol, evaluation through experiments and simulation," in *IEEE Symposium on Wireless Personal Mobile Communication*, 2001.

[5] C. E. Perkins and P. Bhagwat, "Highly dynamic destination-sequenced distance-vector routing (DSDV) for mobile computers," *Computer Communications Review*, pp. 234–244, 1994.

[6] C. E. Perkins and E. M. Royer, "Ad-hoc on-demand distance vector routing," in *Proc. 2nd IEEE Workshop on Mobile Computer System and Applications*, pp. 90–100, Feb. 1999.

[7] Q. Li, J. Aslam, and D. Rus, "Online power-aware routing in wireless ad hoc networks," in *Proc. MobiCom* (Rome, Italy), July 2001.

[8] R. C. Shah and J. M. Rabaey, "Energy aware routing for low energy ad hoc sensor networks," in *Proc. IEEE Wireless Communication and Networking Conference* (Orlando, FL), Mar. 2001.

[9] B. Karp and H. T. Kung, "Greedy perimeter stateless routing for wireless networks," in *Proc. of MobiCom* (Boston, MA), Aug. 2000.

[10] Y.-B. Ko and N. H. Vaidya, "Geocasting in mobile ad hoc networks: Location-based multicast algorithms," in *Proc. IEEE Workshop on Mobile Computer Systems and Applications* (New Orleans), Feb. 1999.

[11] C. Intanagonwiwat, R. Govindan, and D. Estrin, "Directed diffusion: A scalable and robust communication paradigm for sensor networks," in *Proc. MobiCOM 2000* (Boston, MA), Aug. 2000.

[12] S. Chen, "Distributed quality-of-service routing in ad hoc networks," *IEEE Journal on Selected Areas in Communications*, vol. 17, no. 8, Aug. 1999.

[13] K. Chen, S. H. Shah and K. Nahrstedt, "Cross-layer design for data accessibility in mobile ad hoc networks," *Wireless Personal Communication*, no. 21, pp. 49–76, 2002.

[14] Y. Zhang and M. Fromherz, "Message-initiated constraint-based routing for wireless ad hoc sensor networks," in *Proc. IEEE Consumer Communication and Networking Conference*, Jan. 2004.

[15] Y. Zhang, M. Fromherz and L. Kuhn, "Smart routing with learning-based QoS-aware meta-strategies," in *Proc. Quality of Service in the Emerging Networking, Lecture Notes in Computer Science*, vol. 3266, pp. 298–307, 2004.

[16] M. Chu, H. Haussecker, and F. Zhao, "Scalable information-driven sensor querying and routing for ad hoc heterogeneous sensor networks," *International Journal of High-Performance Computing Applications*, vol. 16, no. 3, 2002.

[17] J. Liu, F. Zhao and D. Petrovic, "Information-directed routing in ad hoc sensor networks," *IEEE Journal on Selected Areas in Communications*, Special issue on Self-Organizing Distributed Collaborative Sensor Networks, vol. 23, no. 4, pp. 851–861, 2005.

[18] T. M. Cover and J. A. Thomas, *Elements of Information Theory*. New York, John Wiley and Sons, Inc., 1991.

[19] Q. Huang, C. Lu, and G.-C. Roman, "Mobicast: Just-in-time multicast for sensor networks under spatiotemporal constraints," in *Information Processing in Sensor Networks, Proc. of IPSN 2003*, Apr. 2003.

[20] R. Korf, "Real-time heuristic search," *Artificial Intelligence*, vol. 42, pp. 189–211, 1990.

[21] S. Koenig, "Agent-centered search," *AI Magazine*, vol. 22, no. 4, pp. 109–131, 2001.

[22] R. S. Sutton and A. G. Barto, editors, "Reinforcement learning: an introduction," Cambridge, MA, MIT Press, 1998.

[23] J. A. Boyan and M. L. Littman, "Packet routing in dynamically changing networks: A reinforcement learning approach," edited by J. D. Crowan, G. Tesauro, and J. Alspector, in *Advances in Neural Information Processing Systems*, vol. 6, pp. 671–678, San Francisco, Morgan Kaufmann, 1994.

[24] G. Simon, "Probabilistic wireless network simulator," in *http://www.isis.vanderbilt.edu/projects/nest/prowler/.*

[25] E. J. Cockayne and D. G. Schiller, "Computation of Steiner minimal trees," in *Combinatorics (Conference on Combinatorial Mathematics)* (D. J. A. Welsh and D. R. Woodall, eds.), (Southend-on-Sea, Essex, England), pp. 53–71, Institute of Math. and Its Applications, 1972.

Chapter 11

QoS Provisioning Strategies in LEO Satellite Networks

Stephan Olariu
Department of Computer Science
Old Dominion University
Norfolk, VA 23529-0162, USA
E-mail: olariu@cs.odu.edu

1 Introduction

It is increasingly clear that the use of e-commerce is becoming an important new tool for the manufacturing and retail industries and an imperative new development for all industries striving to maintain a competitive edge. Companies can use the mechanisms and technologies offered by e-commerce to put their stores on-line. At the same time, customer mobility has emerged as an important catalyst triggering a new paradigm shift that is redefining the way we conduct business. An important feature of mobile commerce is that the on-line store is accessible 24 hours per day to a potentially huge customer base scattered around the world [2, 10, 17, 18, 20, 26]. By their very nature, mobile commerce applications will rely increasingly on wireless communications. To be cost effective, and thus economically viable, these applications will have to reach a wide consumer base. In turn, this suggests that a wide coverage - indeed, a global one - is a must in mobile commerce. As illustrated in Figure 1, such a global coverage can only be achieved by a combination of terrestrial and satellite networks.

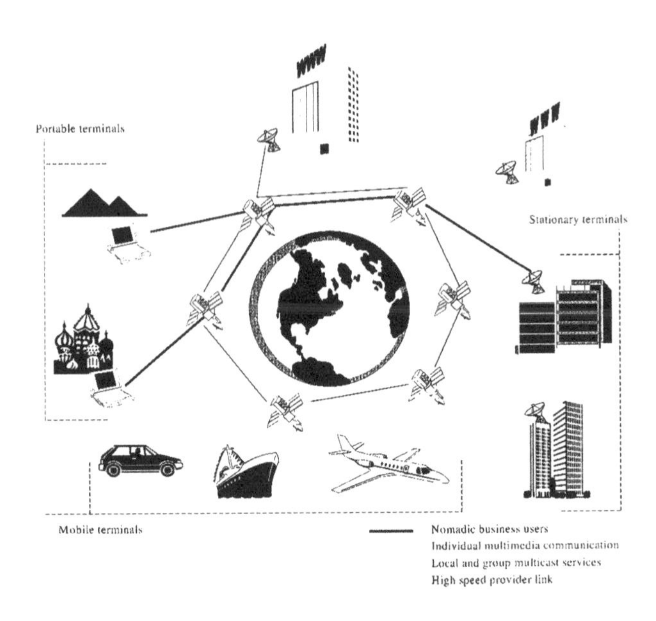

Figure 1: An integrated terrestrial and LEO satellite system.

In response to the increasing demand for truly global coverage needed by Personal Communication Services (PCS), a new generation of mobile satellite networks intended to provide *anytime-anywhere* communication services was proposed in the literature [1, 7, 13, 16, 17, 18, 23, 37]. LEO satellite networks, deployed at altitudes ranging from 500 km to 2000 km, are well suited to handle bursty Internet and multimedia traffic and to offer anytime-anywhere connectivity to Mobile Hosts (MH). LEO satellite networks offer numerous advantages over terrestrial networks including global coverage and low cost-per-minute access to MHs equipped with handheld devices. Because LEO satellite networks are expected to support real-time interactive multimedia traffic they must be able to provide their users with Quality-of-Service (QoS) guarantees including bandwidth, delay, jitter, call dropping, and call blocking probability [5, 9, 11, 23].

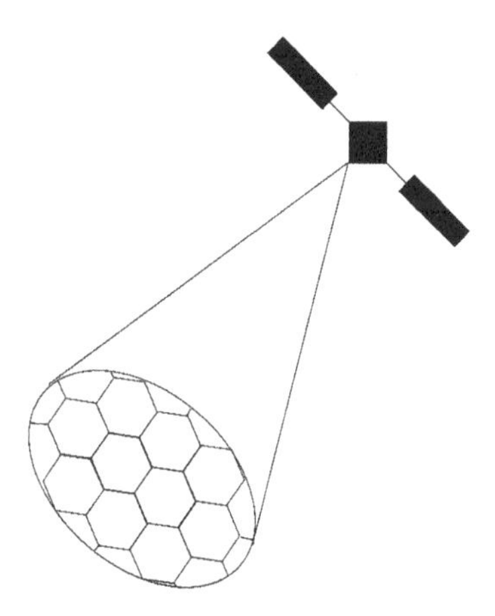

Figure 2: Illustrating a satellite footprint and the corresponding spotbeams.

1.1 LEO Satellite and Main QoS Parameters

Although providing significant advantages over their terrestrial counterparts, LEO satellite networks present protocol designers with an array of daunting challenges, including handoff, mobility, and location management [1, 5, 7, 9, 23]. Because LEO satellites are deployed at low altitude, Kepler's third law implies that these satellites must traverse their orbits at a very high speed. We assume an orbital speed of about 26,000 Km/h. Referring to Figure 2, the coverage area of a satellite - a circular area of the surface of the Earth - is referred to as its *footprint.* For spectral efficiency reasons, the satellite footprint is partitioned into slightly overlapping cells, called *spotbeams.* As their coverage area changes continuously, in order to maintain connectivity, MHs must switch from spotbeam to spotbeam and from satellite to satellite, resulting in frequent intra- and inter-satellite handoffs. In this chapter, we focus on intra-satellite handoffs, referred to, simply, as *handoffs.* A well-known strategy for reducing handoff failure is to reserve bandwidth for the exclusive use of handoff connections [4, 6, 19, 35]. In the fixed reservation approach, a fixed percentage of the available bandwidth in a cell is permanently reserved for handoff. In the predictive reservation strategy, bandwidth is reserved dynamically using a probabilistic approach [8, 15, 21, 27, 32].

Due to the large number of handoffs experienced by a typical connection during its lifetime, resource management and connection admission control are very important tasks if the system is to provide fair bandwidth sharing and QoS guarantees. In particular, a reliable handoff mechanism is needed to maintain connectivity and to minimize service interruption to on-going connections, as MHs roam about the system. In fact, one of the most important QoS parameters for LEO satellite networks is the call dropping probability (CDP), quantifying the likelihood that an on-going connection will be force-terminated due to an unsuccessful handoff attempt. The Call Blocking Probability (CBP) expresses the likelihood that a new call request will not be honored at the time it is placed. The extent to which the existing bandwidth in a spotbeam is efficiently used is known as Bandwidth Utilization (BU). The main goal becomes to provide acceptably low CDP, CBP while, at the same time, maximizing bandwidth utilization [17, 18, 21, 23, 27, 37].

The main goal of this chapter is to survey a number of resource allocation and handoff management strategies for multimedia LEO satellite networks. These protocols espouse a different resource allocation strategy, each being suited for a class of practical applications. We examine these protocols in detail and evaluate and contrast their performance in terms of achieving a number of QoS parameters.

The remainder of the chapter is organized as follows: Section 2 offers a succinct review of MAC protocols for satellite networks; Section 3 reviews the recent literature dealing with admission control and bandwidth allocation strategies for QoS provisioning in multimedia LEO satellite networks; Section 4 discusses the mobility and traffic model assumed in this work. Section 5 looks at SILK, a selective look-ahead call admission and handoff management protocol; Section 6 presents the details of Q-WIN; Section 7 provides the details of OSCAR, an opportunistic call admission and handoff management protocol. Section 8 introduces the simulation model and offers the performance evaluation of SILK, Q-WIN, and OSCAR in terms of CDP, and CBP, and bandwidth utilization. Finally, Section 9 offers concluding remarks and maps out directions for further work.

2 A Survey of MAC Protocols for Satellite Communications

Common intended access to a physically unshareable resource requires arbitration. Because in wireless communications the transmission medium cannot be shared without loss of information, Medium Access Control (MAC) plays the role of an arbiter stipulating, for each of the intended users, the conditions and parameters of exclusive access rights. The major challenge in designing a good MAC protocol is to provide each user with the negotiated QoS parameters while, at the same time, utilizing the network resources as efficiently as possible. Thus, an ideal MAC protocol must strike a balance between the conflicting goals of QoS provisioning, efficiency, simplicity, and service controllability.

MAC protocols have been developed for different environments and applications. A survey of MAC protocols for wireless ATM networks is given in [3, 9, 12, 14, 28, 29, 30, 31, 34]. For a survey of MAC protocols for satellite networks we refer the reader to Peyravi [28, 29, 30]. The MAC protocol classification proposed in [28, 29, 30] is based on the static or dynamic nature of the channel, the centralized or distributed control mechanism for channel assignments, and on the adaptive behavior of the control algorithm.

According to this classification one can define five groups of MAC protocols:

- Fixed assignment protocols,
- Demand assignment protocols,
- Random access protocols,
- Hybrid of random access and reservation protocols, and
- Adaptive protocols.

The *Fixed and Demand Assignment* protocols are contention-free, using either static allocation (fixed assignment) or dynamic allocation (demand assignment). The remaining three groups of protocols are contention-oriented, where collision occurs on either the data channel (random access) or on the signaling channel (hybrid of random access and reservation).

In the *Fixed Bandwidth Assignment* (FBA) schemes [3, 28], a terminal on the Earth is given a fixed number of slots per frame for the duration of the connection. The technique has the advantage of simplicity, but lacks flexibility and reconfigurability. The main disadvantage of this solution is that during idle periods of a connection slots go unused. Fixed allocation of channel bandwidth leads to inefficient use of transponder capacity. Because of the constant assignment of capacity such techniques are not suitable for VBR applications.

In the *Demand Assignment Multiple Access* (DAMA) schemes, the capacity of the up-link channel is dynamically allocated, on demand, in response to station requests based on their queue occupancies. Thus, the time-varying bandwidth requirements of the stations can be accommodated and no bandwidth is wasted. Dynamic allocation using reservation (implicit or explicit) increases transmission throughput. Real-time VBR applications can use this scheme. DAMA does not presuppose any particular physical layer transmission format. It can be implemented using TDMA, FDMA, and CDMA. DAMA protocols consist of a phase for bandwidth request and a phase of data transmission. The main disadvantage of this solution is the long delay experienced by the data in the queues from the time the stations send bandwidth requests to the satellites until the time the data is received at the destination.

In the *Random Access* (RA) schemes [28] the slots of the frame are available to all the MHs. Each station can send its traffic over the randomly selected slot(s) without making any request. There is no attempt to coordinate the ready stations to avoid collision. In case of collision, data is corrupted and has to be retransmitted. The technique is simple to implement and adaptive to varying demand, but very inefficient in bandwidth utilization. The resulting delays are in most of the cases not acceptable for real-time traffic.

The *Hybrid MAC schemes* attempt to combine the good features of different techniques to improve QoS. Hybrid protocols derive their efficiency from the fact that reservation periods are shorter than transmission periods. Examples of Hybrid MAC schemes include (see [28, 29, 30]):

- *Combined Random/Reservation* schemes combine DAMA with RA. The main advantage is the very good bandwidth utilization at high loads and low delay. The disadvantage of the technique is the necessity to monitor the RA part of the channel for possible collisions, which leads to large processing time.

- In *Combined Fixed/Demand Assignment* a fixed amount of up-link bandwidth is always guaranteed to the stations. The remaining bandwidth is assigned by DAMA. The efficiency of the technique depends on the amount of fixed bandwidth allocated to the stations.

- In the *Combined Free/Demand Assignment* (CFDAMA) scheme, the reserved slots are assigned to the stations by DAMA based on their demands. The unused slots are distributed in a round-robin manner based on weighting algorithms [14, 28, 29, 30].

- In *Adaptive Protocols* the number of the contenders is controlled to reduce the collisions, and the channel switches between random access mode and reservation mode depending on the traffic load. Adaptive protocols are a wide area with many possible solutions having different advantages and disadvantages. In view of their complexity, these protocols are not suitable for satellite communications.

3 State of the Art

In wireless mobile networks the radio bandwidth is shared by a large number of MHs. An important property of the network is that a MH would change its access points several times. This fact causes technical problems in which fair sharing of bandwidth between handoff connections and new connections is required. Resource allocation and Call Admission Control (CAC) are very important tasks that need to be performed efficiently in order to achieve high bandwidth utilization and QoS provisioning in terms of low dropping probability for handoffs and reasonably low blocking probability for new connections.

In the remainder of this section we survey a number of call admission algorithms proposed in the literature. We refer the reader to Table 1 for an at-a-glance comparison among several resource allocation strategies and CAC algorithms for LEO satellite networks found in the literature.

Del Re et al. [6] proposed a mobility model and different resource allocation strategies for LEO satellite networks. A queueing strategy of handoff requests is proposed aiming to reduce handoff blocking probability. The dynamic channel allocation scheme is expected to be the best scheme to optimize system performance.

Later, Mertzanis et al. [19] introduced two different mobility models for satellite networks. In the first model, only the motion of the satellite is taken into account, whereas in the second model, other motion components such as the rotation of the Earth and user mobility are considered.

To design a CAC algorithm for mobile satellite systems, the authors introduced a new metric called mobility reservation status, which provides the information about the current bandwidth requirements of all active connections in a specific spotbeam in addition to the possible bandwidth requirements of mobile terminals currently connected to the neighboring spotbeams. A new call request is accepted in the spotbeam where it originated, say m, if there is sufficient available bandwidth in the spotbeam and the mobility reservation status of particular neighboring spotbeams have not exceeded a predetermined threshold TNewCall. If a new call is accepted, the mobility reservation status of a particular number S of spotbeams will be updated. A handoff request is accepted if bandwidth is available in the new spotbeam and also the handoff threshold (THO) is not exceeded. The key idea of the algorithm is to prevent handoff dropping during a call by reserving bandwidth in a particular number S of spotbeams into which the call is likely to wander. The balance between new call blocking and handoff call blocking depends on the selection of predetermined threshold parameters for new and handoff calls. However, during simulation implementation, we found that the scheme has a problem of determining threshold points for the case of LEO satellite networks.

Uzunalioglu [33] proposed a call admission strategy based on the MH location. In his scheme, a new call is accepted only if the handoff call blocking probability of the system is below the target blocking rate at all times. Thus, this strategy ensures that the handoff blocking probability averaged over the contention area is lower than a target handoff blocking probability PQoS (QoS of the contention area). The system always traces the location of all the MHs in each spotbeam and updates the MH's handoff blocking parameters. The algorithm involves high processing overhead to be handled by the satellite, and seems therefore to be unsuitable for high-capacity systems where a satellite footprint consists of many small-sized spotbeams, each having many active MHs.

Cho [4] employs MH location information as the basis for adaptive bandwidth allocation for handoff resource reservation. In a spotbeam, bandwidth reservation for handoff is allocated adaptively by calculating the possible handoffs from neighboring beams. A new call request is accepted if the beam where it originated has enough available bandwidth for new calls. The reservation mechanism provides a low handoff blocking probability compared to the fixed guard channel strategy. However,

Ref.	Multi-service	Reservation strategy	Resource allocation	Necessary info	CAC criteria	QoS
[6]	No	FCA	DCA	Residual time	Local	No
[19]	Yes	Fixed	FCA	Residual time	Local	No
[33]	No	No	FCA	User location	Threshold	Yes
[4]	No	Adaptive	DCA	User location	Local	No
[8]	Yes	Probabilistic	FCA	Residual time	Local Non-local	Yes
[32]	Yes	Predictive	DCA	Residual time	Local Non-local	Yes
[24]	Yes	Adaptive	DCA	User location	Local Non-local	Yes
[25]	Yes	No	DCA	User location	Local Non-local	Yes

Table 1: A synopsis of recent resource allocation and handoff control strategies.

the use of location information in handoff management suffers from the disadvantage of updating locations which results in high processing load to the on-board handoff controller, thereby increasing the complexity of terminals. The method seems suitable only for fixed users.

El-Kadi et al. [8] proposed a probabilistic resource reservation strategy for real-time services. They also introduced a novel call admission algorithm where real-time and non real-time service classes are treated differently. The concept of a sliding window is proposed in order to predict the necessary amount of reserved bandwidth for a new call in its future handoff spotbeams. For real-time services, a new call request is accepted if the spotbeam where it originated has available bandwidth and resource reservation is successful in future handoff spotbeams. For non real-time services, a new call request is accepted if the spotbeam where it originated satisfies its maximum required bandwidth. Handoff requests for real-time traffic are accepted if the minimum bandwidth requirement is satisfied. Non real-time traffic handoff requests are honored if there is some residual bandwidth available in the cell.

Todorova et al. [32] proposed a selective look-ahead strategy, that they call SILK, specifically tailored to meet the QoS needs of multimedia connections where real-time and non real-time service classes are treated differently. The handoff admission policies introduced distinguish between both types of connections. Bandwidth allocation only pertains to real-time connection handoffs. To each accepted connection, bandwidth is allocated in a look-ahead horizon of k cells along its trajectory, where k is referred to as the depth of the look-ahead horizon. The algorithm offers low CDP, providing for reliable handoff of on-going calls and acceptable CDP for new calls.

Recently, Olariu et al. [24] proposed Q-WIN, a novel admission control and handoff management strategy for multimedia LEO satellite networks. A key ingredient in Q-WIN is a novel predictive resource allocation protocol. Extensive simulation results have confirmed that Q-WIN offers low CDP, acceptable CBP for new call attempts, and high BU. Q-WIN involves some processing overhead. However, as the authors showed, this overhead is transparent to the MHs, being absorbed by the on-board processing capabilities of the satellite. Their simulation results show that Q-WIN scales to a large number of MHs.

Quite recently, Olariu et al. [25] proposed OSCAR (OpportuniStic Call Admission pRotocol) which provides a lightweight and robust solution to both call admission and handoff management in LEO satellite networks. Being simple and robust, OSCAR was designed with efficiency and scalability in mind. Along this line of thought, one of the features that sets OSCAR apart from existing protocols is that it avoids the overhead of reserving resources for MHs in a series of future spotbeams. Instead, OSCAR relies on a novel opportunistic bandwidth allocation mechanism that is very simple and efficient. OSCAR borrows some ideas from Q-WIN [24] in that it uses a number of virtual windows to make an opportunistic decision for admission of new users. Extensive simulation results have shown that, without maintaining any queues or expensive reservations, OSCAR achieves results comparable to those of Q-WIN: it offers very low call dropping probability, providing for reliable handoff of on-going calls, good call blocking probability for new call requests, and high bandwidth utilization.

4 Mobility Model and Traffic Parameters

Although several mobility models exist for LEO satellites [1, 13, 22, 23] it is customary to assume a one-dimensional mobility model where the MHs move in straight lines and at a constant speed, essentially the same as the orbital speed of the satellite [6, 7, 22]. For simplicity, all the spotbeams (also referred to as *cells*) are identical in shape and size. Although each spotbeam is circular, we use squares to approximate spotbeams (we note that some authors use regular hexagons instead of squares).

The diameter of a cell is taken to be 425 Km [5, 6, 34]; the time t_s it takes a MH to cross a cell is, roughly, 65 seconds. Referring to Figure 3, the MH remains in the cell where the connection was initiated for t_f time, where t_f is uniformly distributed between 0 and t_s. Thus, t_f is the time until the first handoff request, assuming that the call does not end in the original cell. After the first handoff, a constant time t_s is assumed between subsequent handoff requests until call termination.

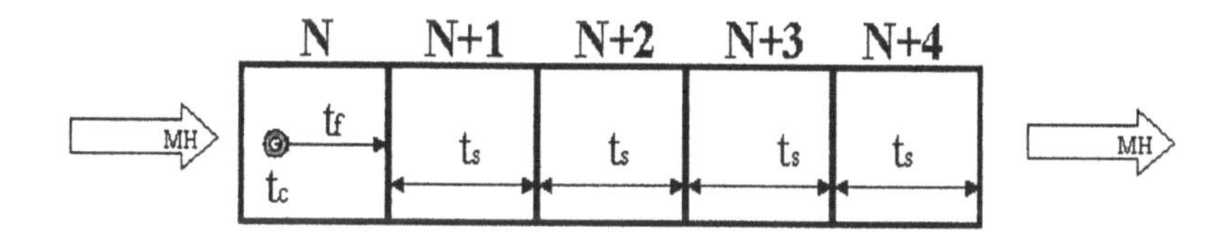

Figure 3: Illustrating some of the mobility and cell parameters.

As illustrated in Figure 3, when a new connection C is requested in cell N, it is associated with a trajectory, consisting of a list $N, N+1, N+2, \ldots, N+k, \ldots$ of cells that the connection may visit during its lifetime. The holding times of the connections are assumed to be exponentially distributed with mean $\frac{1}{\mu}$.

Assume that C was accepted in cell N. After t_f time units, C is about to cross into cell $N+1$. Let p_f be the probability of this first handoff request. El-Kadi et al. [8] have shown that

$$p_f = \Pr[H > t_f] = \frac{1}{t_s} \int_0^{t_s} e^{-\mu t} dt = \frac{1 - e^{-\mu t_s}}{\mu t_s}. \tag{1}$$

Moreover, El-Kadi *et al.* [8] have shown that due to the memoryless property of the exponential distribution the probability of the $(k+1)$-th, $(k \geq 1)$, handoff request is

$$\Pr[H > t_f + k \cdot t_s | H > t_f + (k-1) \cdot t_s] =$$

$$\frac{\Pr[H > t_f + k \cdot t_s]}{\Pr[H > t_f + (k-1) \cdot t_s]} = \frac{e^{-\mu(t_f + k \cdot t_s)}}{e^{-\mu(t_f + (k-1) \cdot t_s)}} = e^{-\mu t_s}$$

which, as expected, is independent of k. Consequently, we write

$$p_s = e^{-\mu t_s} \qquad (2)$$

denote the probability of a *subsequent* handoff request. It is important to note that t_f, t_s, p_f and p_s are mobility parameters that can be easily evaluated by the satellite using its on-board processing capabilities.

The traffic offered to the satellite system is assumed to belong to two classes:

1. Class I traffic - real-time multimedia traffic, such as interactive voice and video applications,

2. Class II traffic - non real-time data traffic, such as email or ftp.

When a mobile host requests a new connection C in a given cell, it provides the following parameters:

- The desired class of traffic for C (either I or II)

- M_C, the desired amount of bandwidth for the connection

- If the request is for a Class I connection the following parameters are also specified:

  - m_C, the minimum acceptable amount of bandwidth, that is the smallest amount of bandwidth that the source requires in order to maintain acceptable quality, e.g., the smallest encoding rate of its codec,
  - θ_C the largest acceptable call dropping probability that the connection can tolerate, and
  - $\frac{1}{\mu_C}$, the mean holding time of C.

5 SILK - A Selective Look-Ahead Bandwidth Allocation Scheme

A well-known strategy for reducing handoff failure is to reserve bandwidth for the exclusive use of handoff connections [7, 8, 34, 36]. In the fixed reservation approach a fixed percentage of the available bandwidth in a cell is permanently reserved for handoff. In the predictive reservation strategy, bandwidth is reserved dynamically using a probabilistic approach [14].

The main goal of this subsection is to spell out the details of SILK, a selective look-ahead bandwidth admission control and handoff management scheme.

5.1 SILK - The Basic Idea

SILK admission policies distinguish between Class I and Class II connections. As in [27], Class I handoffs are admitted only if their minimum bandwidth requirements can be met. However, Class II handoff requests will be accepted as long as there is some residual bandwidth left in the cell. Thus, bandwidth allocation only pertains to Class I handoffs. The key idea of SILK is to allocate to each accepted Class I connection bandwidth in a look-ahead horizon of k cells along its trajectory. Here, k is referred to as the depth of the look-ahead horizon.

The intuition for this concept is provided by the fact that the deeper the horizon, the smaller the likelihood of a handoff failure, and the smaller the CDP. Because at setup time the connection C specifies the CDP it can tolerate, it implicitly specifies the depth k of the corresponding look-ahead horizon. Thus, for each connection C, SILK looks ahead just enough to ensure that the CDP of θ_C can be enforced. This idea is quite different from the scheme of El-Kadi et al. [8] where bandwidth is reserved in *all* the cells on the trajectory, resulting in a potential wasteful allocation of resources. Thus, in SILK, the look-ahead allocation is determined by the negotiated QoS.

Let p_h denote the handoff failure probability of a Class I connection, that is, the probability that a handoff request is denied for lack of resources. Let S_k denote the event that a Class I connection C admitted in cell N goes successfully through k handoffs and will, therefore, show up in cell $N + k$. It is easy to confirm that the probability of S_k is

$$\Pr[S_k] = p_f \cdot (1 - p_h) \cdot [p_s \cdot (1 - p_h)]^{k-1} \quad (3)$$

where

- $p_f \cdot (1 - p_h)$ is the probability that the first handoff request is successful, and
- $[p_s \cdot (1 - p_h)]^{k-1}$ is the probability that all subsequent $k - 1$ handoff requests are also successful.

Likewise, let D_{k+1} be the event that C will be dropped at the next handoff attempt. Thus, we have

$$\begin{aligned} \Pr[D_{k+1}] &= \Pr[S_k] p_s p_h = p_f p_s p_h (1 - p_h)[p_s \cdot (1 - p_h)]^{k-1} \\ &= p_f p_h [p_s \cdot (1 - p_h)]^{k-1} \end{aligned}$$

as $p_s p_h$ is the probability that the connection will attempt but fail to secure the $(k + 1)$-th handoff.
Now, assuming that connection C has negotiated a CDP of θ_C, we insist that

$$\Pr[D_{k+1}] = p_f p_h [p_s \cdot (1 - p_h)]^{k-1} = \theta_C$$

from where we obtain:

$$k = \left\lfloor \frac{\log \frac{\theta_C}{p_f p_h}}{\log p_s (1 - p_h)} \right\rfloor + 1. \tag{4}$$

There are a number of interesting features of equation (4) that we now point out. First, the only *variable* parameter in the equation is p_h. All the others are known beforehand. Todorova *et al.* [32] argued that the satellite maintains p_h as the ratio between the number of unsuccessful handoff attempts and the total number of handoff attempts. Second, since p_h may change with the network conditions, the depth k of the look-ahead horizon will also change accordingly. This interesting feature shows that SILK is indeed adaptive to traffic conditions. Finally, k is dynamically maintained by the satellite either on a per-connection or, better yet, on a per-service class basis, depending on the amount of on-board resources and network traffic.

As it turns out, equation (4) is at the heart of SILK. The details are spelled out as follows:

- In anticipation of its future handoff needs, bandwidth is allocated for connection C in a number k of cells corresponding to the depth of its look-ahead horizon; no allocation is made outside this group of cells.

- For $1 \leq i \leq k$, allocate in cell $N+i$ an amount of bandwidth equal to $B_{N+i} = m_C \Pr[S_i]$.

- This amount of bandwidth will be allocated for connection C during the time interval

$$I_{N+i} = [t_C + t_f + (i-1)t_s, t_C + t_f + it_s]$$

where t_C is the time connection C was admitted into the system.

As pointed out by Todorova et al. [32], SILK is lightweight. Indeed, the mobility parameters t_f and t_s are readily available and the look-ahead horizon k is maintained by the satellite for each service class. Similarly, since the trajectory of connection C is a straight line, the task of computing for every $1 \leq i \leq k$ the amount of bandwidth B_{N+i} to allocate, as well as the time interval I_{N+i} during which B_{N+i} must be available is straightforward and can be easily computed by the satellite using its on-board capabilities.

5.2 SILK - The Call Admission Strategy

Connection admission control is one of the fundamental tasks performed by the satellite network at call setup time in order to determine if the connection request can be accepted into the system without violating prior QoS commitments. The task is non-trivial because the traffic offered to the system is heterogeneous due to new call attempts and handoff requests. SILK's call admission strategy involves two criteria.

- The first call admission criterion, which is local in scope, applies to both Class I and Class II connections, attempting to ensure that the originating cell has sufficient resources to provide the connection with its desired amount of bandwidth.

- The second admission control criterion which is global in scope, applies to Class I connections only, attempting to minimize the chances that, once accepted, the connection will be dropped later due to a lack of bandwidth in some cell into which it may handoff.

Consider a request for a new Class I connection C in cell N at time t_C and let t_f be the estimated residence time of C in cell N. The second criterion was inspired by the sliding window criterion first proposed by El-Kadi et al. [8]. However, unlike [8], SILK only looks at the first k cells on C's trajectory. The connection satisfies the second criterion if all these k cells have sufficient bandwidth to accommodate C, that is for every i, $(1 \leq i \leq k)$, the amount of residual bandwidth in the cell during the time interval I_{N+i} must not be less than B_{N+i}. The motivation for this second criterion is very simple: if the residual bandwidth available in cell $N+i$ is less than the projected bandwidth needs of connection C, it is very likely that C will be dropped. To avoid such a situation, connection C is not admitted into the system. Thus, the second admission criterion acts as an additional safeguard against a Class I connection to be accepted, only to be dropped at some later point.

6 Q-WIN - A Predictive Resource Allocation and Management Protocol

The main goal of this section is to discuss in full detail the Q-WIN protocol proposed in [24]. A key ingredient in Q-WIN is a novel predictive resource allocation protocol. Q-WIN involves some processing overhead. However, as it turns out, this overhead is transparent to the MHs, being absorbed by the on-board processing capabilities of the satellite. Consequently, Q-WIN is expected to scale and to accommodate a large population of MHs.

6.1 Q-WIN - The Data Structures

Recall that, as discussed above, the traffic offered to the satellite system belongs to two classes: Class I traffic, real-time multimedia traffic, such as interactive voice and video applications; and Class II traffic, non real-time data traffic, such as email or ftp. When a MH requests a new connection C in a given cell, it provides the following parameters: the desired class of service, the desired amount M_C of bandwidth for the connection, as well as the minimum acceptable amount of bandwidth, m_C, that the source requires in order to maintain acceptable quality.

A Class I connection C in a generic cell N is said to be:

- **Regular** if C has confirmed bandwidth reservations in cells $N+1$

and $N + 2$. The regular connections in cell N are maintained in the queue $R(N)$.

- **1-short** if C has a confirmed bandwidth reservation in cell $N + 1$ but not in cell $N + 2$. The 1-short connections in cell N are maintained in the queue $S1(N)$.
- **2-short** if C has no confirmed reservation in either of the cells $N+1$ and $N + 2$. The 2-short connections in cell N are maintained in the queue $S2(N)$.

It is important to note that 2-short connections are liable to be dropped at the next handoff attempt, whereas 1-short connections are in no imminent danger of being dropped. The stated goal of Q-WIN's bandwidth allocation scheme is to minimize the likelihood of a dropped connection, by striving to render as many Class I connections as possible. The principal vehicle for achieving this goal is a judicious priority-based bandwidth allocation strategy. Finally, we note that Class II connections in cell N are maintained in a separate queue $Q(N)$.

6.2 Q-WIN - The Call Admission Strategy

Consider a request for a new connection C in cell N. Q-WIN relies in its call admission strategy on two criteria:

- The first call admission criterion, which is local in scope, ensures that the originating cell N has sufficient resources to provide the connection with its desired amount of bandwidth M_C. Both Class I and Class II connections are subject to this first admission criterion. A Class II connection request that satisfies the first admission criterion is accepted into the system and placed into the queue $Q(N)$ of Class II connections currently in cell N. On the other hand, if the first admission criterion is not satisfied, the connection request is immediately rejected.
- The second admission control criterion, which is non-local in scope, applies to Class I connections only, attempting to minimize the chances that, once accepted, the connection will be dropped later due to a lack of bandwidth in some cell into which it may handoff.

Consider a request for a new Class I connection C in cell N at time t_C and let t_f be the estimated residence time of C in N. Referring to

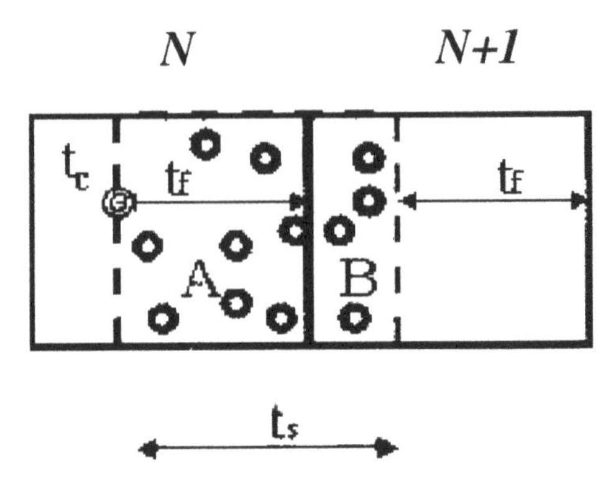

Figure 4: Illustrating the sliding window concept.

Figure 4, the key observation that inspired the second criterion is that when C is about to handoff into cell $N+1$, the connections resident in $N+1$ are likely to be those in region A of cell N and those in region B of cell $N+1$. More precisely, these regions are defined as follows:

- A connection is in region A if at time t_C its residual residence time in cell N is less than or equal to t_f.

- A connection is in region B if at time tC its residual residence time in cell $N+1$ is larger than or equal to t_f.

In general, the satellite does not know the exact position of a new call request in generic cell N. This makes the computation of the bandwidth committed to connection in areas A and B difficult to assess. In what follows we describe a heuristic that attempts to approximate the bandwidth held by the connections in A and B.

For this purpose, we partition the union of cells N and $N+1$ into $m+1$ virtual windows $W_0, W_1, W_2, \ldots, W_m$, each of width t_s. In this sequence, W_0 is the base window, and its left boundary is normalized to 0. For every i, $(1 \leq i \leq m)$,

$$\text{window } W_i \text{ stretches from } \frac{it_s}{m} \text{ to } t_s + \frac{it_s}{m}. \tag{5}$$

In particular, by equation (5), window W_0 coincides with cell N, and window W_m coincides with cell $N+1$. We refer the reader to Figure 5 for an illustration, with $m = 5$. All the virtual windows have the exact shape and size of a cell (though, for clarity purposes, they were drawn differently in Figure 5).

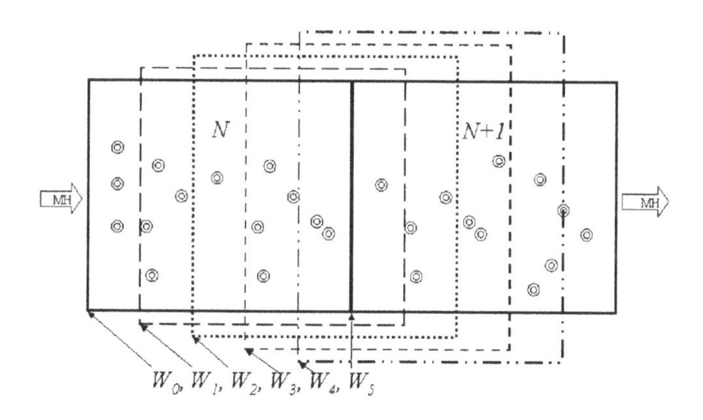

Figure 5: *Illustrating the virtual windows* $W_0, W_1, W_2, \ldots, W_m$.

For later reference, we partition a generic window W_i into a *left* sub-window W_i^N and a *right* sub-window W_i^{N+1} denoting, respectively, the intersection of W_i with cells N and $N+1$. Also, Q-WIN distinguishes between MHs that have experienced a handoff (referred to as *old*) from those that have not (referred to as *new*). As we are about to describe, MHs may or may not be assigned timers. Specifically, each old MH is assigned a timer θ; no timer is assigned to new MHs.

Upon entering a new cell, θ is set to t_s (the time it takes to traverse a cell). Every time unit, θ is decremented by one, making it close to zero by the time it is about to reach the next handoff. For illustration purposes, we note that in Figure 5, since $m = 5$, W_1^N contains the old MHs in cell N with $\theta \leq 65 - \frac{65}{5} = 52$; likewise, W_1^{N+1} contains the old MHs in cell $N+1$ with $\theta \geq 52$. Similarly, W_2^N contains the old MHs in cell N with $\theta \leq 65 - 2 \times \frac{65}{5} = 39$, and so on.

Let B_i and D_i denote, respectively, the total amount of bandwidth in use by the old and new MHs in window W_i. Notice that the amount of bandwidth B_i is easy to compute by the satellite because, by virtue of timers, the position of old MHs, up to the granularity of a virtual window, is known. The location of new MHs (i.e., newly accepted ones that have not yet experienced their first handoff) is unknown. It is, therefore, difficult to determine D_i exactly. However, as Todorova et al. [32] pointed out, it is reasonable to assume that within each of the cells N and $N+1$ these MHs are uniformly distributed. Notice that this does not imply a uniform distribution of new MHs across the union of cells N and $N+1$. Let n_N and n_{N+1} stand, respectively, for the number of new MHs in cells N and $N+1$. As illustrated in Figure 6, the assumption of uniform distribution of new MHs in cell N implies that the expected number of MHs in W_i^N is $n_N\left(1-\frac{i}{m}\right)$. Likewise, since the new MHs are uniformly distributed in cell $N+1$, the expected number of new MHs in W_i^{N+1} is $n_{M+1} \times \frac{i}{m}$. Thus, by a simple computation we obtain the following approximation for D_i

$$D_i = \left(1 - \frac{i}{m}\right) n_N + \frac{i}{m} n_{N+1}. \tag{6}$$

Let M stand for the total bandwidth capacity of a cell. Using B_i and the value of D_i from equation (6), the virtual window W_i determines the residual bandwidth $R_i = M - B_i - D_i$. If, $R_i \geq M_C$, W_i *votes* in favor of accepting the new request C with desired bandwidth M_C; otherwise it votes against its admittance. After counting the votes, if the majority of the virtual windows had voted in favor of admittance, the new connection request is admitted into the system. Otherwise, it is rejected. Once admitted, the desired bandwidth of connection C is reserved in the current cell, and the connection is placed in queue $S2(N)$.

6.3 Q-WIN - The Handoff Management Scheme

The main goal of this subsection is to spell out the details of Q-WINs handoff management scheme. Recall that for each cell the satellite maintains four different queues R, Q, $S1$ and $S2$ as described in Subsection 6.1. The management of these queues is the workhorse of our scheme and is discussed in detail in Subsection 6.4.

Before presenting the details of managing Class I handoff requests, we note that the handoff request of a Class II connection in $Q(N-1)$ attempting to transit into cell N, is accepted as long as cell N has some residual bandwidth available, and placed in the queue $Q(N)$ of Class II connections in cell N. It is important to observe that for Class II connections no minimum bandwidth guarantees are honored. This is

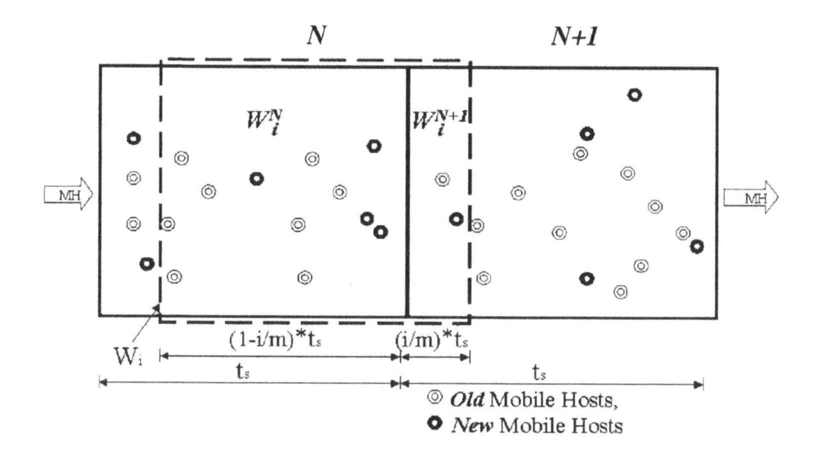

Figure 6: Illustrating the computation of D_i.

consistent with the view that Class II connections do not have hard delay requirements: a lower bandwidth allocation will naturally translate into a longer end-to-end delay.

Now, consider a regular Class I connection C in cell $N - 1$ that performs a handoff into cell N. Recall that, since C is regular, there is bandwidth reserved for C in both cells N and $N + 1$ and, thus, the handoff is guaranteed to be successful. Two cases may arise at this point:

- If a minimum of m_C units of bandwidth can be reserved on C's behalf in cell $N + 2$ all is well, C becomes a regular connection in cell N, and is placed in queue $R(N)$.
- If bandwidth cannot be secured for C in cell $N + 2$, C becomes 1-short in cell N and is placed in queue $S1(N)$.

Next, assume that a 1-short Class I connection C in cell $N - 1$ performs a handoff into cell N. Since C is 1-short in $N - 1$, the handoff is successful, as there exists bandwidth reserved on its behalf in cell N. The following three cases may arise:

- If a minimum of m_C units of bandwidth can be reserved on C's behalf in cells $N + 1$ and $N + 2$, C becomes a regular connection in cell N and is placed in queue $R(N)$.

- If a minimum of m_C units of bandwidth can be reserved on C's behalf in cell $N+1$ but not in cell $N+2$, C becomes 1-short in N and is placed in queue $S1(N)$.
- If bandwidth cannot be secured for C in either of the cells $N+1$ and $N+2$, C becomes 2-short in cell N and is placed in queue $S2(N)$.
- Finally, consider a 2-short Class I connection C in cell $N-1$ that is attempting a handoff into cell N. Because there is no reservation for C in cell N and, as we show, there is not enough bandwidth to satisfy the minimum of m_C units, C will be dropped.

6.4 Q-WIN - The Queue Management

For a generic cell N, the four queues $Q(N)$, $R(N)$, $S1(N)$, and $S2(N)$ are maintained as priority queues. This means that each connection within the queues is processed based on the residual time it will spend in cell N. The priority regimen is suggested by the fact that connections closest (in time) to handoff have to get attention first. It is important to note that, except for the connections that terminate in cell N, all other connections in the cell must occur in exactly one of the queues above. The priority rules introduced between the different types of queues are explained later on.

Consider an arbitrary group of cells $N-2$, $N-1$, and N. Bandwidth in cell N is reclaimed as a result of call terminations occurring in cell N or bandwidth de-allocation due to termination of Class I calls in previous cells. Let $Avail(N)$ denote the amount of bandwidth currently available in cell N.

The processing specific to $R(N-1)$ is as follows: when a connection C in $R(N-1)$ is about to handoff into cell N, the satellite attempts to reserve bandwidth for C in cell $N+1$. If this reservation attempt is successful, at the handoff point, C is removed from $R(N-1)$ and added to $R(N)$, in other words, C is now a regular connection in cell N. If, on the other hand, the reservation attempt described above fails, C is removed from $R(N-1)$ at the handoff point, and added to $S1(N)$, as in cell N, connection C is 1-short.

The processing of queue $Q(N-1)$ is straightforward: when a connection C in $Q(N-1)$ is about to handoff into cell N, the satellite checks if $Avail(N) > 0$, that is, if some bandwidth is available in cell N. If such is the case, C is removed from $Q(N-1)$ and added to $Q(N)$. If

$Avail(N) = 0$, that is, no bandwidth is available in cell N, C is removed from $Q(N-1)$ and dropped.

The *first priority* in queue management is to reserve bandwidth in cell N on behalf of connections in the queue $S2(N-1)$ that are in imminent danger of being dropped, since there is no reserved bandwidth for them in cell N.

The *second priority* in queue management is to reserve bandwidth in cell N on behalf of connections in queue $S1(N-2)$. These connections are in no imminent danger of being dropped, for they have guaranteed bandwidth reserved in cell $N-1$. However, as already mentioned, the philosophy of Q-WIN is to ensure that there are as many regular connections as possible.

The *third priority* goes to new call admission requests. Recall that, as discussed above, a new Class I request C in cell N is accepted into the network only if cell N can honor the desired bandwidth requirement of the connection and, in addition, if the majority of the windows of cell N are in favor of admittance of the MH.

The *fourth priority* goes to Class II handoff connection. Handoff request of a Class II connection in $Q(N-1)$ attempting to transit into cell N, is accepted as long as cell N has some residual bandwidth available (recall that Class II connections have no hard delay requirements and, therefore, are willing and able to tolerate delays), and placed in the queue $Q(N)$ of Class II connections in cell N.

7 OSCAR: An Opportunistic Resource Management Protocol

The main idea behind OSCAR is to propose a multiple virtual window call admission protocol and average line mechanism based dynamic channel reservation for handoff calls for multimedia LEO satellite networks. The essence of this predictive resource allocation protocol is that it achieves results comparable to those of Q-Win by eliminating queues. Even though it adds up more processing time, the overhead of maintaining queues during heavy traffic is avoided and hence makes this algorithm simpler and less dependent on buffers. Moreover, the processing time is transparent to the MH, being absorbed by the on-board processing capabilities of the satellite. Consequently, OSCAR scales to a large number of users. In Section 8 the performance of OSCAR is compared to that of Q-WIN and other recent schemes proposed in the literature. Simulation results show that OSCAR offers low call dropping probability, providing for reliable handoff of on-going calls, and good call blocking probability for new call requests, while maintaining high bandwidth utilization.

Consider a request for a new connection C in cell N. Very much like SILK [32] and Q-WIN [24], OSCAR [25] bases its connection admission control on a novel scheme that combines the following two criteria

- *Local availability:* The first call admission criterion, which is local in scope, ensures that the originating cell N has sufficient resources to provide the connection with its desired amount of bandwidth M_C.
- *Short-term guarantees:* The second admission control criterion, which is non-local in scope, attempts to minimize the chances that, once accepted, the connection will be dropped later due to a lack of bandwidth.

However, unlike both SILK and Q-WIN that either look at a distant horizon or maintain rather complicated data structures, OSCAR looks ahead only one cell. Surprisingly, as simulation results indicate, this short horizon works well when supplemented by an opportunistic bandwidth allocation scheme. OSCAR's second admission criterion relies on a novel idea that is discussed in full detail below.

7.1 OSCAR - Generalities

Because OSCAR inherited many of the design ideas of Q-WIN, the discussion in this subsection parallels the one in Section 6. Consider a request for a new connection C in cell N at time t_C and let t_f be the estimated residence time of C in N. Referring back to Figure 4, the key observation that inspired our second criterion is that when C is about to handoff into cell $N+1$, the connections resident in $N+1$ are likely to be those in region A of cell N and those in region B of cell $N+1$. As pointed out by Olariu et al. [24], the satellite controller does not know the exact position of a new call request in generic cell N. This makes the computation of the bandwidth committed to connection in areas A and B difficult to assess. In what follows we describe a heuristic that attempts to approximate the bandwidth held by the connections in A and B. For this purpose, just as Q-WIN does, OSCAR partitions the union of cells N and $N+1$ into $m+1$ virtual windows $W_0, W_1, W_2, \ldots, W_m$ each of width t_s. We refer the reader to Figure 5 for an illustration. We also find it convenient to borrow the terminology from Section 6.2.

7.2 OSCAR - The Handoff Management Scheme

A well-known strategy for reducing handoff failure is to reserve bandwidth for the exclusive use of handoff connections [4, 6, 19, 23, 27]. In the fixed reservation approach, a fixed percentage of the available bandwidth in a cell is permanently reserved for handoff. It is well known that the fixed reservation strategy is either wasteful of bandwidth or outright ineffective depending on the amount of bandwidth reserved. In the predictive reservation strategy, bandwidth is reserved dynamically using a probabilistic approach [8, 15, 21, 24, 32]. OSCAR implements a novel idea of the predictive strategy combined with an opportunistic handoff management scheme.

In OSCAR handoff calls fall into one of the two types discussed below.

- **Type 1:** those that are still not assigned a timer i.e. these are newly admitted calls that are about to make their first handoff
- **Type 2:** those that are assigned a timer i.e., the calls that have already made one or more handoffs.

It is important to observe that by virtue of our call admission scheme that is looking at both the originating cell and the next one along the MH's path, handoffs of Type 1 succeed with high probability. We are, therefore, only showing how to manage Type 2 handoffs. The details of this scheme are discussed below.

Each cell in the network dynamically reserves a small amount of bandwidth specifically for handoffs of Type 2. When a Type 2 handoff request is made, the algorithm will first try to satisfy the request by allotting the bandwidth from the reserved channel. If the reserved channel is already full, the request will be allotted the bandwidth from the remaining available bandwidth of the cell. Otherwise, the handoff request is dropped.

Let the maximum amount of bandwidth that could be reserved be β_{max} - a small percentage of total available bandwidth. The amount of bandwidth reserved for Type 2 handoffs dynamically varies between 0 and β_{max} depending on the relative position of the *average load line* in the previous neighboring cell.

To explain the concept of average load line, consider a cell N, and refer to Figure 7. Assume that cell $N-1$ contains k Type 2 handoff calls with *residual residence times* in cell $N-1$ denoted by $t_1, t_2, \ldots, t_k$ such that $t_1 \leq t_2 \leq \cdots \leq t_k$ and let the corresponding amounts of bandwidth

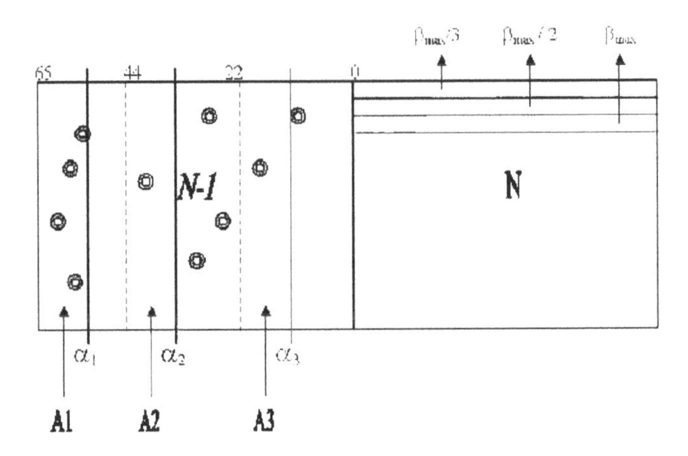

Figure 7: Illustrating the handoff scheme in OSCAR.

allocated to the calls be $b_1, b_2, \ldots, b_k$. Write $B = \sum_{j=1}^{k} b_j$. The average load line L is defined as $L = \frac{t_i + t_{i+1}}{2}$ where i is the smallest subscript for which the inequality (7) below holds.

$$\sum_{j=1}^{k} b_i \geq \left\lceil \frac{B}{2} \right\rceil . \tag{7}$$

We note that form a computational standpoint determining the average load line L is a simple instance of the prefix sums problem and can be handled easily by the satellite.

7.3 OSCAR - The Dynamic Reservation Scheme

The dynamic channel reservation scheme in cell N can be explained as follows. Since cell N knows about its neighbors, it can track all the Type 2 handoff calls in cell $N-1$ as shown. $A1$, $A2$ and $A3$ represent equal-sized areas of a cell $N-1$. The average load line L will always fall into one of these three areas depending upon the distribution of Type 2 handoff calls. In Figure 7, α_1, α_2, α_3 represent the positions of the average line. The bandwidth for Type 2 calls in cell N is reserved depending upon the position of the average line as detailed below:

- If the position of average load line L is at α_1 in area $A1$, then it can be inferred that roughly half of the bandwidth required by Type 2 handoff calls is concentrated in area $A1$. Since L is relatively far from cell N, an amount $\frac{\beta_{max}}{3}$ of bandwidth is reserved for Type 2 handoff calls in cell N as shown, in such a way that more bandwidth is available for other call requests.
- If the average load line L is at α_2 in $A2$, then an amount $\frac{\beta_{max}}{2}$ of bandwidth is reserved in cell N.
- If the average load line L is at α_3 in $A3$, then a maximum bandwidth of β_{max} is reserved for Type 2 calls in cell N.

8 Performance Evaluation

The main goal of this section is to analyze the performance of SILK, Q-WIN, and OSCAR through simulation. We begin by describing the simulation model and then we go on to present the results of our simulation. The simulation model and system parameters are presented in Subsection 8.1. A detailed performance comparison of the three protocols is presented in Subsection 8.2.

8.1 Simulation Model

In order to set the stage for a fair and accurate evaluation and comparison of SILK, Q-WIN, and OSCAR we have developed a detailed simulation model that we describe next. Each spotbeam has a server that manages its bandwidth in accord with the specific policy of each protocol. The system parameters used in the simulation model are presented in Figure 8.

The generic server functions across the three protocols we implemented are:

- To monitor the amount of available bandwidth in each of the spotbeams
- To reserve bandwidth required by individual connections
- To accept or reject new call requests
- To accept or reject Type 1 and Type 2 requests.

Spotbeam parameters						
Radius	Capacity	Speed				
212 km	30000 Kbit	26000 km/h				
Service parameters						
Parameters	**Class 1**			**Class2**		
	Type1	*Type 2*	*Type 3*	*Type 1*	*Type 2*	*Type 3*
Mean duration (s)	180	300	600	30	180	120
Maximum bandwidth (kbps)	30	256	6000	*20*	*512*	10000
Minimum bandwidth (kbps)	30	256	1000	*5*	*64*	1000

Figure 8: Illustrating the simulation parameters.

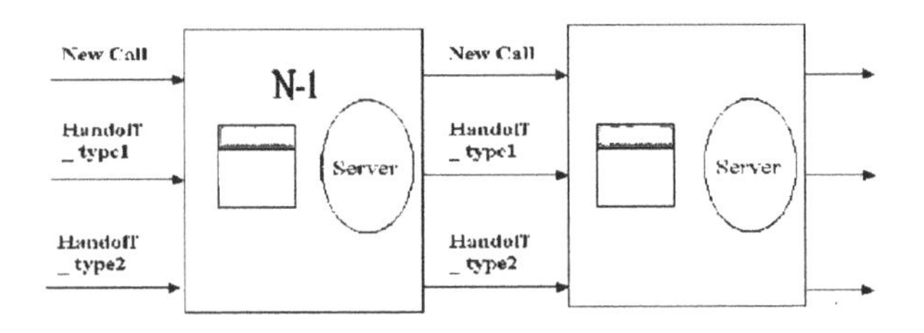

Figure 9: The simulation model.

Referring to Figure 9, the new call arrival rate follows a Poisson distribution and holding times are exponentially distributed. We define three types of services with different QoS requirements and assume equal mean arrival rate for each service type and fixed bandwidth capacity in each spotbeam.

8.2 Simulation Results

The simulation results are shown in Figures 10 - 13. We compare the CDP, CBP, and BU performance of OSCAR with those of Q-WIN [24] and SILK [32].

Figure 10 shows that OSCAR has almost the same CBP as Q-WIN and much better than SILK. A more detailed comparison results for OSCAR and Q-WIN is illustrated in Figure 11.

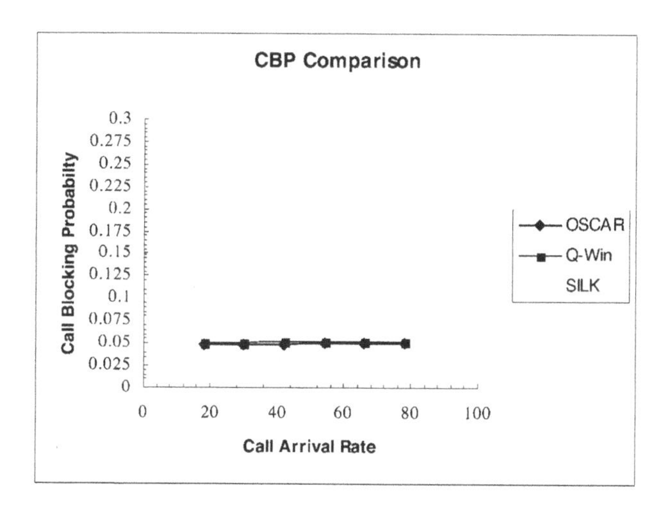

Figure 10: Illustrating call blocking probability.

Figure 12 shows that the CDP of OSCAR is a little higher than Q-WIN but is fair enough to conclude that the handoff management scheme yields almost the same results as SILK and Q-WIN.

It is well known that the goals of keeping the call dropping probability low and that of keeping the bandwidth utilization high are conflicting. It is easy to ensure a low CDP at the expense of bandwidth utilization and similarly, it is easy to ensure a high bandwidth utilization at he expense of call dropping probability. The challenge, of course, is to come up with a handoff management protocol that strikes a sensible balance between the two. As Figure 13 indicates, OSCAR features high bandwidth utilization in addition to keeping the call dropping probability low.

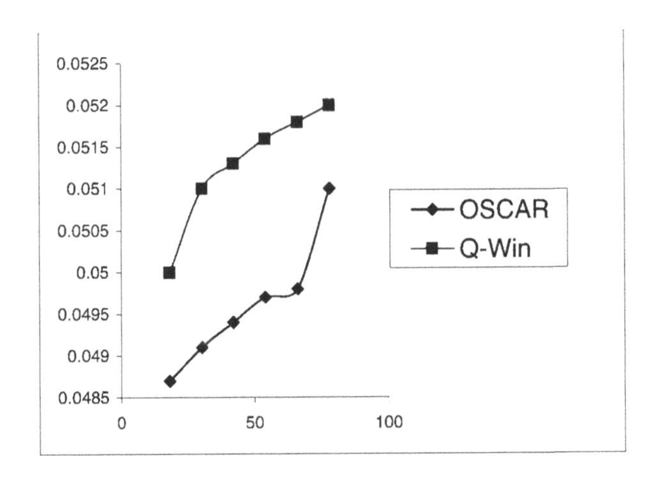

Figure 11: A detailed comparison of CBP in OSCAR and Q-WIN.

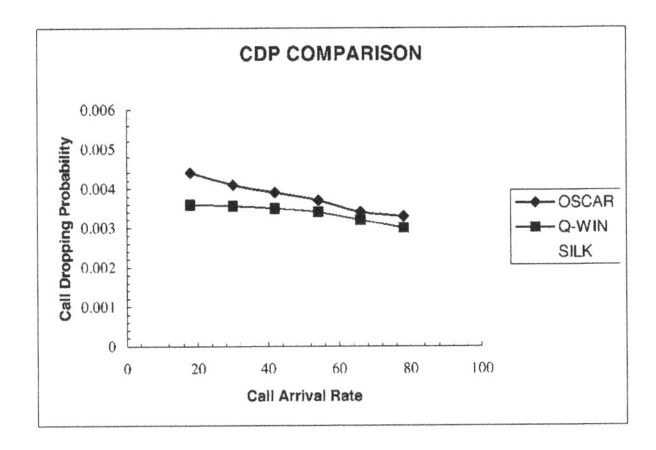

Figure 12: Illustrating call dropping probability.

9 Concluding Remarks

LEO satellites are expected to support multimedia traffic and to provide their users with the appropriate QoS. However, the limited bandwidth

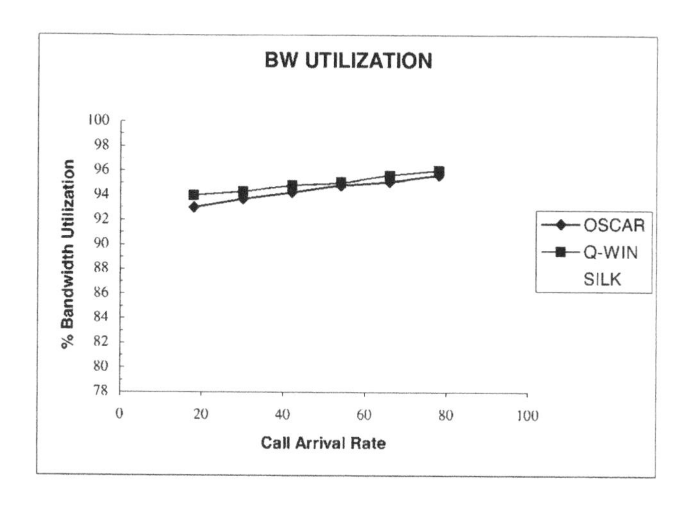

Figure 13: A comparison of bandwidth utilization.

of the satellite channel, satellite rotation around the Earth, and mobility of MHs makes QoS provisioning and mobility management a challenging task. The main goal of this chapter was to survey a number of recently-proposed call admission and handoff management schemes for LEO satellite networks. The schemes that we compared, SILk [32], Q-WIN [24], and OSCAR [25] were specifically tailored to meet the QoS needs of multimedia connections, where real-time and non real-time service classes were differently treated. These protocols feature a different philosophy of admission and bandwidth management. In a sense, they complement each other since the solutions they offer may each appeal to a different set of applications.

We have implemented these protocols and have evaluated their performance by extensive simulation. Our simulation results exposed differences in performance due to design decisions. In summary, these protocols are well suited for QoS provisioning in multimedia LEO satellite networks.

Acknowledgment: This work was supported, in part, by NATO grant PST.CLG979033. The author is grateful to Mona El-Kadi, Hoang Nam Nguyen, Syed Rizvi, Rajendra Shirhatti, Alexander Markhasin, and Petia Todorova for many insightful discussions on QoS provisioning in LEO satellite networks.

References

[1] I. F. Akiyldiz and S. H. Jeong, Satellite ATM networks: A survey, *IEEE Communications*, 35(7), 1997.

[2] S. J. Barnes and S. L. Huff, Rising sun: iMode and the wireless Internet, *Communications of the ACM*, 46(11), 2003, 79-84.

[3] H. C. B. Chan and V. C. M. Leung, Dynamic bandwidth assignment multiple access for efficient ATM-based service integration over LEO satellite systems, *Proc. Northcon98*, Seattle, WA, October 1998.

[4] S. Cho, Adaptive dynamic channel allocation scheme for spotbeam handover in LEO satellite networks, *Proc. IEEE VTC 2000*, 1925-1929.

[5] G. Comparetto and R. Ramirez, Trends in mobile satellite technology, *IEEE Computer*, 1997, 44-52.

[6] E. Del Re, R. Fantacci, and G. Giambene, Efficient dynamic channel allocation techniques with handover queuing for mobile satellite networks, *IEEE Journal of Selected Areas in Communications*, 13(2), 1995, 397-405.

[7] E. Del Re, R. Fantacci, and G. Giambene, Characterization of user mobility in Low Earth Orbiting mobile satellite systems, *Wireless Networks*, 6, 2000, 165-179.

[8] M. El-Kadi, S. Olariu and P. Todorova, Predictive resource allocation in multimedia satellite networks, *Proc. IEEE GLOBECOM*, San Antonio, TX, November 25-29, 2001.

[9] J. Farserotu and R. Prasad, A survey of future broadband multimedia satellite systems, issues and trends, *IEEE Communications*, 6, 2000, 128-133.

[10] I. Foster and R. G. Grossman, Data integration in a bandwidth-rich world, *Communications of the ACM*, 46(11), 2003, 51-57.

[11] E. C. Foudriat, K. Maly and S. Olariu, H3M - A rapidly deployable architecture with QoS provisioning for wireless networks, *Proc. 6th IFIP Conference on Intelligence in Networks*, Vienna, Austria, September 2000.

[12] A. Huang, M-J. Montpetit and G. Kesidis, ATM via satellite: A framework and implementation, *Wireless Networks*, 4, 1998, 141-153.

[13] A. Jamalipour and T. Tung, The role of satellites in global IT: trends and implications, *IEEE Personal Communications*, 8(3), 2001, 5-11.

[14] T. Le-Ngoc and I. M. Jahangir, Performance analysis of CFDAMA-PB protocol for packet satellite communications, *IEEE Transactions on Communications*, 46(9), 1998, 1206-1214.

[15] D. Levine, I. F. Akyildiz and M. Naghshineh, A resource estimation and call admission algorithm for wireless multimedia networks using the shadow cluster concept, *IEEE/ACM Transactions on Networking*, 5, 1997, 1-12.

[16] M. Luglio, Mobile multimedia satellite communications, *IEEE Multimedia*, 6, 1999, 10-14.

[17] A. Markhasin, S. Olariu and P. Todorova, QoS-oriented medium access control for all-IP/ATM mobile commerce applications, in Nan Si Shi (Ed.) *Mobile Commerce Applications*, (Hershey, PA, Idea Group, 2004, pp. 303-331).

[18] A. Markhasin, S. Olariu and P. Todorova, An overview of QoS-oriented MAC protocols for future mobile applications, in M. Kosrow-Pour (Ed.), *Encyclopedia of Information Science and Technology*, (Hershey, PA, Idea Group, 2005).

[19] I. Mertzanis, R. Tafazolli, and B. G. Evans, Connection admission control strategy and routing considerations in multimedia (Non-GEO) satellite networks, *Proc. IEEE VTC*, 1997, 431-436.

[20] Mobile Media Japan, Japanese mobile Internet users: http://www.mobilemediajapan.com/, January 2004.

[21] M. Naghshineh and M. Schwartz, Distributed call admission control in mobile/wireless networks, *IEEE Journal of Selected Areas in Communications*, 14, 1996, 711-717.

[22] H. N. Nguyen, S. Olariu, and P. Todorova, A novel mobility model and resource allocation strategy for multimedia LEO satellite networks, *Proc. IEEE WCNC*, Orlando, Florida, March 2002.

[23] H. N. Nguyen, *Routing and Quality-of-Service in broadband LEO satellite networks*, (Boston, Kluwer Academic, 2002).

[24] S. Olariu, S. A. Rizvi, R. Shirhatti, and P. Todorova, Q-WIN - A new admission and handoff management scheme for multimedia LEO satellite networks, *Telecommunication Systems*, 22(1-4), 2003, 151-168.

[25] S. Olariu, R. Shirhatti, and A. Y. Zomaya, OSCAR: An opportunistic call admission and handoff management scheme for multimedia LEO satellite networks, *Proc. International Conference on Parallel Processing* (ICPP'2004), Montreal, Canada, August 2004.

[26] S. Olariu and P. Todorova, QoS provisioning in LEO Satellites: A resource reservation framework, *IEEE Potentials*, August 2004, 11-18.

[27] C. Oliviera, J. B. Kim, and T. Suda, An adaptive bandwidth reservation scheme for high-speed multimedia wireless networks, *IEEE Journal of Selected Areas in Communications*, 16, 1998, 858-874.

[28] H. Peyravi, Multiple access control protocols for the Mars regional networks: A survey and assessments, Tech. Report, Kent State University, September 1995.

[29] H. Peyravi, Medium access control protocol performance in satellite communications, *IEEE Communications Magazine*, 37(3), 1999, 62 -71.

[30] H. Peyravi and D. Wieser, Simulation modeling and performance evaluation of multiple access protocols, *Simulation*, 72(4), 1999, 221-237.

[31] X. Qiu and V. O. K. Li, Dynamic reservation multiple access (DRMA): A new multiple access scheme for Personal Communication Systems (PCS), *Wireless Networks*, 2, 1996, 117-128.

[32] P. Todorova, S. Olariu, and H. N. Nguyen, A selective look-ahead bandwidth allocation scheme for reliable handoff in multimedia LEO satellite networks, *Proc. ECUMN2002*, Colmar, France, April 2002.

[33] H. Uzunalioglu, A connection admission control algorithm for LEO satellite networks, *Proc. IEEE ICC*, 1999, 1074-1078.

[34] M. Werner, J. Bostic, T. Ors, H. Bischl, and B. Evans, Multiple access for ATM-based satellite networks, *Proc. European Conference on Network and Optical Communications*, NOC98, Manchester, UK, June 1998.

[35] M. Werner, C. Delucchi, H.-J. Vogel, G. Maral and J.-J. De Ridder, ATM-based routing in LEO/MEO Satellite networks with inter-satellite links, *IEEE Journal on Selected Areas in Communications*, 15(2), 1997, 69-82.

[36] J. E. Wieselthier and A. Ephremides, Fixed- and movable-boundary channel-access schemes for integrated voice/data wireless networks, *IEEE Transactions on Communications*, 43(1), 1995, 64-74.

[37] W. K. Wong, H. Zhu and V. C. M. Leung, Soft QoS provisioning using the token bank fair queues scheduling algorithm, *IEEE Wireless Communications*, 10(3), 2003, 8-16.

Chapter 12

Quasi-Optimal Resource Allocation in Multispot MFTDMA Satellite Networks

Sara Alouf
INRIA, the National Institute for Research in Computer Science and Control
06902 Sophia Antipolis, France
E-mail: `Sara.Alouf@sophia.inria.fr`

Eitan Altman
INRIA, the National Institute for Research in Computer Science and Control
06902 Sophia Antipolis, France
E-mail: `Eitan.Altman@sophia.inria.fr`

Jérôme Galtier
INRIA, the National Institute for Research in Computer Science and Control
06902 Sophia Antipolis, France
E-mail: `Jerome.Galtier@sophia.inria.fr`
France Telecom Research and Development,
06921 Sophia Antipolis, France
E-mail: `Jerome.Galtier@rd.francetelecom.com`

Jean-François Lalande
INRIA, the National Institute for Research in Computer Science and Control
06902, Sophia Antipolis, France
E-mail: `Jean-Francois.Lalande@sophia.inria.fr`

Corinne Touati
Institute of Information Sciences and Electronics
University of Tsukuba, Japan
E-mail: `corinne@osdp.is.tsukuba.ac.jp`

1 Introduction

We consider a multispot geostationary satellite system for which a manager assigns satellite uplink MFTDMA (Multifrequency Time-Division Multiple Access) slots to service providers (operators). The service providers themselves operate a park of terminals distributed on the satellite area of cover. Concerning the radio channel, the satellite divides the time and frequency spectrum into time slots. Geographically, the terminals are distributed on zones, themselves being included in *spots*. A spot is an area covered by a satellite beam, as illustrated in Figure 1.

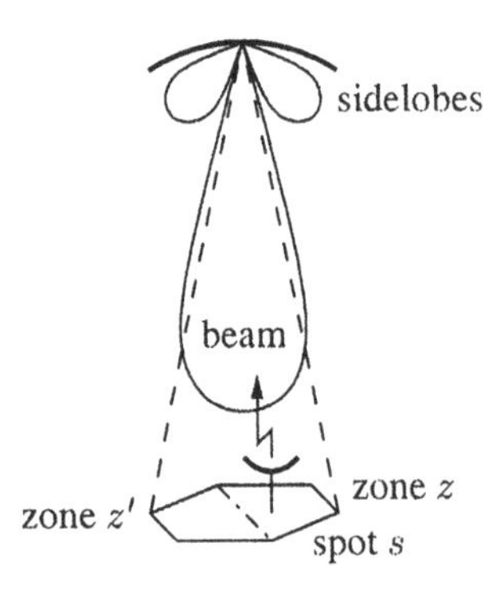

Figure 1: Illustration of a spot and an antenna beam.

Radio interference imposes constraints on the slots that can simultaneously be assigned in different spots that have the same frequency. A slot cannot be assigned simultaneously to more than one zone in a spot. Spots are given colors (bands of frequencies) and spots of different colors do not interfere, but spots of the same color do, and a slot can be assigned to an operator in a given zone only if the interference it experiences with the other active zones is below a given threshold. Slot assignment is static but can be changed once per hour (due to changes in demands, or to changes in atmospheric conditions, on the other hand). Every hour, the demand of the service providers is re-evaluated and a new allocation could be generated. Due to real-time constraints, solutions are needed within a few minutes.

Our goal is to maximize the throughput of the system. A simplified version of our problem can be modeled as a so-called "k-colorable induced subgraph" problem where one considers a graph $G = (V, E)$ consisting of finitely many nodes and directional links [32]. A valid coloring of the graph consists of coloring nodes such that no nodes with a common link have the same color. We look for a subset of nodes $V' \subset V$ and edges $E' \subset E$ such that the induced subgraph is k-colorable; i.e., there is a coloring for the subgraph (V', E') of cardinality at most k. The problem consists in finding such a graph with the maximal number of nodes. This problem turns out to be NP complete [13, GT20]. Actually, our problem is even more complex because arcs have weights (arc weights are related to the amount of interference) and exclusion constraints are more complex. This is in contrast to slot allocation in satellite switched TDMA systems that have polynomial solutions as they correspond to coloring problems with a simple bipartite graph topology [6]. In this chapter, instead of using coloring approaches, we propose to solve the problem differently, using a linear and integer programming approach with column generation.

This work[1] is clearly motivated by the cost of the design of satellite antennas: the cost of an antenna is a strong function of its size, roughly speaking, proportional to the diameter cubed [2]. Larger antennas generate small interference and have better gain, but increase tremendously the cost of the satellite. One of the goals of this approach is to tune precisely the assignment problem given its profile in terms of interference and gain. We show that in return, our program can derive which interference is responsible for (sometimes substantial) loss of capacity for a given demand.

In our experiments to evaluate the proposed approach, we use two series of data corresponding to 8 and 32 spots per color, respectively. We assume that there are three zones per spot, and four types of carriers.[2] Our work is focused on one of the colors of the bandwidth (recall that spots of different colors do not interfere with each other), so that the complete processing phase should use the same program for each color (if necessary in a parallel way). In our experiments, the total number of (time) slots that can be assigned is set to 3456.

We propose in this chapter a linear and integer programming approach that allows us to solve the problem almost optimally. For the 8-spot case, the problem is solved in a minute or so, with a guarantee of consuming at most 1% more bandwidth than the absolute optimum. The dual/primal approach is ex-

---

[1]This work is part of research convention A 56918 between INRIA and ALCATEL SPACE INDUSTRIES (contract number 1 02 E 0306 00 41620 01 2).

[2]Carriers have different bandwidths thus providing different slot durations. The use of a specific carrier by a given terminal is determined by the terminal's transmission capability.

ploited in a master/slave fashion, where the master program is a heuristic that finds noninterfering zones that are directly translated into valid columns for the primal problem handled by the slave program. This approach can output the interfering configurations that limit the optimization up to a certain threshold. This information is extremely important for the design of antennas because it explains the characteristics of the antennas that lead to performance limitation. In other words, our approach identifies the interfering configurations that are crucial to optimization, and this information has to be taken into account when designing antennas. Designers have to make sure that the antennas do not impair such configurations. Last, we show that, in the 32-spot case, our program can output solutions that in practice have good performance.

The structure of the chapter is as follows. Related references are briefly discussed in Section 2. The system model and its constraints are presented in Section 3. The resolution of the time slot allocation problem throughout a simple example is detailed in Section 4, whereas the general solution is detailed in Section 5. Numerical results are presented in Section 6, followed by a concluding section.

2 Related Work

In the vast majority of the cases, the related references that have appeared in the past dealt with simpler models which, in some cases, have been solvable using polynomial algorithms. We wish to mention, however, that problems with similar nature but with simpler structure have also been treated in the context of scheduling in ad hoc networks; see e.g., [14] and the references therein.

In this section, we focus on algorithmic approaches for solving the slot allocation (or "burst scheduling") problem, that have appeared in the literature. Empirical approaches for burst scheduling have been proposed in [21, 22, 29]. Concerning other aspects of TDMA satellites, we direct the reader to the paper [1] which surveys issues such as architecture, synchronization, and some physical layer considerations and discusses papers presenting probabilistic performance evaluation techniques related to TDMA systems.

A few words on the terminology: we use the standard term SS/TDMA for Satellite-Switched Time-Division Multiple Access. We also frequently find in the literature the expression "burst scheduling". The term "burst" does not refer to a burst in the input traffic (the data) but rather to the fact that traffic is not transmitted continuously but in bursts.

The input to the slot assignment problem in TDMA systems is often a traffic matrix whose ijth element — denoted as $\delta_{i,j}$ — describes the amount of traffic to be shipped from zone i to zone j, or equivalently the time to transfer it at a fixed channel rate.

In [19], the author considers n transponders to switch an $n \times n$ demand matrix. Each terminal can either send or receive with one transponder at a time. The author shows that the minimal time to transfer a complete matrix corresponds to

$$\max\left(\max_{1\leq i\leq n}\sum_{1\leq j\leq n}\delta_{i,j},\ \max_{1\leq j\leq n}\sum_{1\leq i\leq n}\delta_{i,j}\right).$$

The author provides an algorithm that achieves this bound, and in order to do so, the frame should be divided into a number of switching modes, that correspond to several assignments of the transponders. This number of switching modes is minimized under the condition that the time to transfer is minimal. He shows that at most $n^2 - 2n + 2$ different switching modes are necessary.

From the algorithmic aspect, this reference can be explained in simple terms. First note that given the maximum row and column sums of the matrix, it can be greedily completed into a matrix with constant row and column sums, simply by marking all the *deficient* rows and columns (i.e., those that do not reach the maximum) and increasing at a step an element sharing a deficient row and a deficient column. Then a maximum bipartite matching (what they refer to as System of Distinct Representatives in the paper; see also the improvement of [28]) will find a switching mode. It results that less than $n^2 + 2n + 2$ steps are necessary because at least one element of the matrix goes to zero at a time.

In [3], the authors extend the results of Inukai [19] in the case where k transponders are present and the demand matrix is $n \times m$. In this case, the minimal time that one could expect to transfer the matrix is equal to:

$$\max\left(\max_{1\leq i\leq n}\sum_{1\leq j\leq m}\delta_{i,j},\ \max_{1\leq j\leq m}\sum_{1\leq i\leq n}\delta_{i,j},\ \sum_{1\leq i\leq n}\sum_{1\leq j\leq m}\delta_{i,j}/k\right).$$

The authors give an algorithm that achieves this bound, and manage to bound the number of switching modes used to $n^2 - n + 1$ if $n = m = k$, and $nm + k + 1$ otherwise. However experimental results suggest that this number is substantially lower than that bound in practice. The algorithmic ingredients are essentially the same as before.

In [16], the authors again consider n transponders to switch an $n \times n$ demand matrix, but this time, interferences are taken into consideration. The interferences are modeled as constraints both on the uplink and downlink of the system, with respective undirected graphs G_U and G_D. The graph G_U associates a vertex with each terminal, and has an edge between terminals u and v if and only if terminals u and v cannot communicate at the same time. G_D is defined similarly. The authors demonstrate the NP-hardness of this problem and propose a solution in the context of polarization, which is the case when two independent channels are used to transmit the traffic. They accordingly propose a two-step algorithm: (i) divide the matrix into two parts using supposedly planarity properties, minimizing the interference using a MAX-CUT algorithm (the algorithm they use is optimal only in the planar case; note, however, that a 0,87-approximation algorithm in polynomial time of this problem in the general case has been since discovered; see [15]), and (ii) in each of the obtained two parts, minimize the number of necessary time slots to transmit without interference, developing various coloring heuristics (e.g., brute force, greedy algorithms) that will help to incrementally construct a suboptimal schedule, selecting a "good" interference-free matrix at each step. Note that this approach fails to give a result on the global optimality of the problem. In fact, only the second part of the algorithm really addresses the problem. Indeed, if a minimum number of time slots can be found in the *general* case, polarization (or other types of separate band assignment, such as frequency division) can be efficiently exploited by splitting the final schedule in two parts (or more, in the case of frequencies) and assigning a part to each polarity (or frequency). Note that our method can be easily adapted to this case and can give general optimality guarantees.

In [23] the problem of finding a solution when n transponders are present and an $n \times n$ demand matrix is given is studied under the particular restriction that only a restricted set of switching matrices can be used. In such a case, of course, the authors notice that linear programming can minimize the total transfer time, which means that the solution of the problem can be found efficiently. Rather, they consider some even more specific conditions on the switching matrices and give in that particular case even faster algorithmic solutions.

The type of problem studied in the previous references and the results therein obtained were later extended in [4–6, 26, 30, 31]. It is important to mention that no interference problem is considered in these papers.

In [5], the authors consider the problem of adding some second-priority traffic to some existing schedule. It is argued to be important when some streaming communications (voice, video) are taken to compute the switching modes, to which some additional data may be assigned. The authors claim a NP-completeness theorem, and give some heuristics to approach the problem. Note, however, that an alternative solution would then be to recompute the switching matrices from scratch and see whether it increases the total communication time.

Additionally, note that a considerable amount of work as been done on this topic; see for instance [7, 12, 17, 20, 25, 27, 33].

3 The Model

In this section, we present the model considered in the rest of the chapter. We introduce the spots configurations in Section 3.1 and the interference model in Section 3.2. Section 3.3 presents some practical details concerning the computation of the carrier-to-interference ratio. Last, informations related to the terminals (capacity of transmission, carrier used, demand) are provided in Section 3.4.

3.1 Spatial Reuse

The total satellite bandwidth is subdivided in several equally-large bandwidths. Each one of these will be assigned a *color*. Every spot is assigned a unique fixed color, implying that all terminals of a spot can transmit within the bandwidth corresponding to the spot's color. Every color may be assigned to several spots. This is the concept of spatial reuse (see for instance [10]). Observe that terminals in different spots of the same color will interfere with each other when using the same frequency band within the spots total bandwidth. Multiple terminals will not be allowed to transmit if the global interference generated is too high, as it will impair the correct reception of the data by the satellite. Color assignment is given as an entry of our problem. Examples of color assignment can be seen in Figure 2(a), resp. Figure 2(b), when 3 colors, resp. 4 colors, are used.

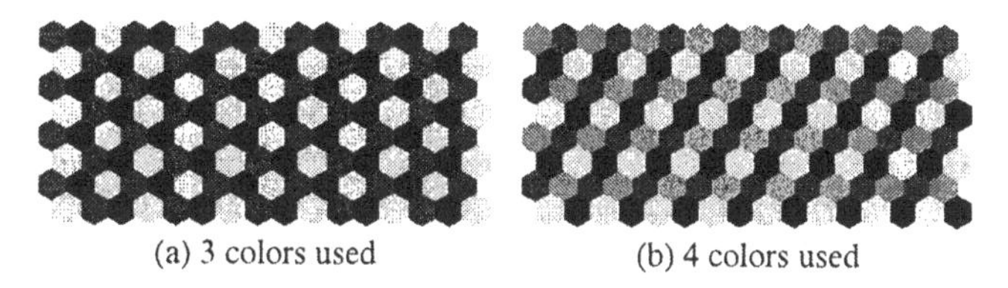

(a) 3 colors used (b) 4 colors used

Figure 2: Spatial distribution of spots and optimal reuse of colors.

Since colors do not overlap in bandwidth, they are completely independent from each other. Hence, resource allocation can be done for each color separately. The original problem has simply to be split in the number of colors used, and each resulting problem can be solved independently from the others. Hereafter, we will consider only the problem of resource allocation within the same color. Without loss of generality, we will consider a spatial reuse of 4 colors. Let N denote the numbers of spots having the same color, and B denote the color bandwidth. We are particularly interested in the case where $N \leq 32$. Figure 3 depicts the spots, configuration within one color when 4 colors are used.

Figure 3: Spatial distribution of spots using the same color (4-color case).

Different spots of the same color are allowed to transmit only if the overall level of interference is acceptable and does not impair the correct reception of the transmitted signals at the satellite. In the following section, we introduce an allocation criterion as a means to check if it is safe to activate one spot or another. This allocation criterion will condition any frequency reuse between spots of the same color.

3.2 Interference Level

To take into account the real conditions of radio propagation, it is necessary to account for the position of the terminals within a given spot. The spot is usually large enough to have different channel conditions in different geographical regions. We therefore divide a spot into a number of *zones* (typically 2 or 3), assuming that each zone exhibits the same propagation conditions in all its area. The radio propagation experienced by a terminal is thus completely characterized by the zone where the terminal is.

If a terminal is transmitting at time t, using carrier f, we say that its zone/spot is active in (t, f). Whenever a zone is active, its transmission will generate interference over all other spots using the same carrier at the same time. Note that this interference will be the same over any zone of a given active spot. The importance of the interference is directly affected by the size of the antennas' sidelobes. Figure 4 illustrates well how a transmission can interfere over others. It is clear from Figure 4 that the interference, generated over spot s' by a terminal in spot s, located in a zone other than zone z, will be different. It should also be clear that the interference generated by an active zone is the same over all zones within the same active spot.

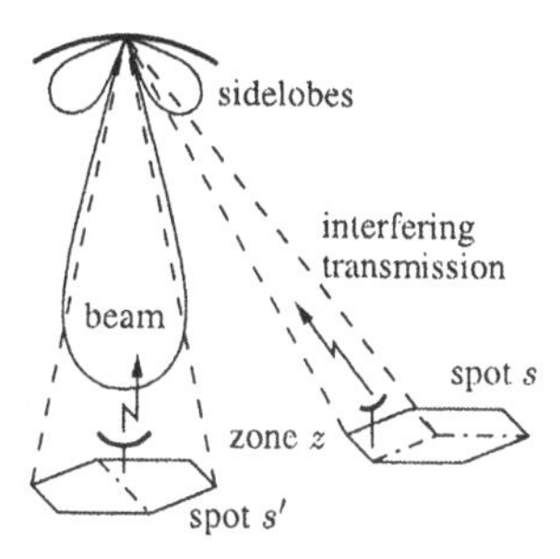

Figure 4: Interference generated by a terminal in zone z over a terminal in spot s'.

Let $G(z)$ denote the minimal antenna gain corresponding to zone z. Let $I(s, z)$ denote the maximal interference generated over spot s by a transmission in zone z. It is the maximal antenna gain in the sidelobes corresponding to zone z, when the main beam is directed to spot s. If zone z belongs to spot s then $I(s, z) = 0$. The received signal at the satellite is useful only if its power amplitude is large enough compared to the power of the interfering signals. In other words, the carrier-to-interference ratio should be beyond a certain threshold σ, otherwise the satellite cannot properly handle the received transmission. Hence, a zone z could be active in (t, f) if and only if the following criterion is satisfied,

$$\frac{C}{I} = \frac{G(z)}{\sum_{z' \text{ active in } (t,f)} I(s(z), z')} \geq \sigma, \tag{1}$$

where $s(z)$ denotes the spot in which zone z is located.

Note that the interference considered in our model is much more realistic than the ones considered in [16]. Indeed, only two terminals could interfere with each other and in this case, only one of these terminals will be allowed to transmit at a given time. Our model is more complex as multiple interfering communications are possible given that the interference threshold is observed.

3.3 Interference Model in Numerical Results

The power of the interfering signal used in (1) depends on the size of the antenna. Small sidelobes lead to weak interference. Unfortunately, we do not have data on the power distribution of the interfering signal over all geographical areas; we therefore assume the following: neighboring spots are the ones generating the highest interference with each other; remote spots still interfere with each other but not as significantly. In the results of Section 6, the values in decibels of the gain $G(z)$ (resp., interference $I(s,z)$) are taken randomly in the interval $[40,41]$ (resp., $[11,15]$) decibels. Thus, we use these different quantities:

$$I_1(z) = \sum_{z' \text{ neighbor, active in } (t,f)} I(s(z),z') \tag{2}$$

$$I_2(z) = \sum_{z' \text{ active in } (t,f)} I(s(z),z') \tag{3}$$

$$I(z) = I_1(z) + (1-\gamma)\left(I_2(z) - I_1(z)\right)$$

where γ is a given weight. Equation (1) is replaced with

$$\frac{C}{I} = \frac{G(z)}{I(z)} \geq \sigma. \tag{4}$$

The term $I_2(z) - I_1(z)$ designates the interference generated by all active zones in nonneighboring spots. Therefore, the interference generated by remote spots is reduced by a factor $1-\gamma$. Observe that taking $\gamma = 0$ is equivalent to considering that all interference is equally important (Eqs. (1) and (4) will be exactly the same), and having $\gamma = 1$ nullifies the effect of transmissions in nonneighboring spots over the zone at hand.

3.4 Types of Terminals and Demand

Terminals have different capabilities of transmission. A given type of terminals will use a unique frequency band. Hereafter, we classify terminals according to their capability of transmission, and use the notation $t_k, k = 1,\ldots,\tau$ to refer to a given type of terminals (τ referring to the number of different types of terminals), the ascending order corresponding to the ascending slot duration. Every type of terminal t_k is assigned a unique bandwidth, denoted by t_k^b. In our problem, the ratio of the bandwidths of any two different types is either an integer or the inverse of an integer and is called the multiplicity. Also, the duration of a slot is dependent on the terminal's type. The idea is to have the same amount of data transmitted in a slot time whichever the type of terminal at hand: for any type t_k, the product of its bandwidth, t_k^b, and its slot duration, denoted by t_k^t, is a constant: $t_k^b t_k^t = \Delta$. Table 1 reports the values used to evaluate our algorithm. Observe that type t_1 is the smallest in time and largest in bandwidth, whereas type t_4 has the longest time slot duration and the narrowest bandwidth. From the table we can write $t_1^b = 2t_2^b = 8t_3^b = 32t_4^b$, or equivalently, $t_4^t = 4t_3^t = 16t_2^t = 32t_1^t$.

Table 1: Test values of terminal types

Type	Maximum number of time slots per frame	Maximum number of carriers per spot bandwidth
t_1	192	18
t_2	96	36
t_3	24	144
t_4	6	576

The individual demands of all terminals in a zone are aggregated according to the type of terminals, and hence, the bandwidth used by every type. Let $d(z, t_k)$ denote the demand in time slots in zone z expressed in time slots of type t_k, for any zone z and any type t_k.

4 Solving a Simple Example

In this section, we consider the simple case where there is only one type of terminal, i.e., all terminals use the same amount of bandwidth to transmit their data. For every carrier, the channel can be accessed simultaneously by multiple terminals/zones according to the Time-Division Multiple Access (TDMA) technique. Solving the resource allocation problem translates then into the following question: which zones are allowed to transmit in a given time slot and using a given carrier?

Consider the example illustrated in Figure 5. There are 3 spots transmitting in the same color, each spot having 2 zones. When active, every zone generates

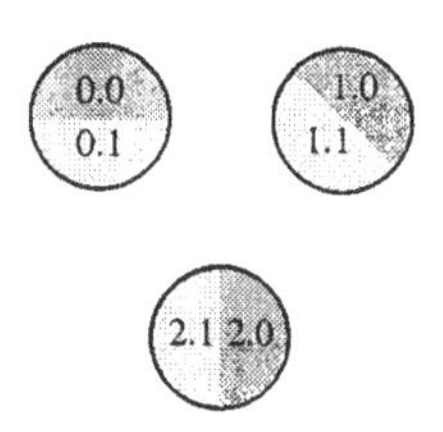

Figure 5: Example with 3 spots.

a certain level of interference over all other spots (gain and interference can be found in Table 2; values are not in dB). Every spot can have either one of its zones active, or be inactive (recall that only one zone in a given spot can be active at a given time). Hence, there are $3^3 = 27$ possibilities in our simple example.

Table 2: Gain and interference of the 6 zones in the example

Zone	Gain	$I(\text{Spot } 0, \cdot)$	$I(\text{Spot } 1, \cdot)$	$I(\text{Spot } 2, \cdot)$
0.0	4	-	5	3
0.1	6	-	5	7
1.0	3	4	-	2
1.1	8	7	-	10
2.0	5	3	7	-
2.1	5	7	3	-

Considering any zone from the example, this zone can be active (*on*) only if its carrier-to-interference ratio is above a certain value. This ratio will naturally depend on whether the other spots are active or not (*on* or *off*). For every zone considered, there are 9 possible situations, as reported in Table 3. Let $\sigma = 0.3$. All of the situations where only two spots are active are valid, because the carrier-to-interference ratio is higher than 0.3 for all zones in every such situation (refer to last column and last row for every zone). Among all $2^3 = 8$ situations where 3 spots are active, only 3 are valid. For instance, if zones 0.0, 1.0, and 2.0 are active, it appears that the carrier-to-interference ratio is above $\sigma = 0.3$ for zones 0.0 and 2.0, but not for zone 1.0. This combination is therefore not valid and should not be used in the allocation procedure. The only 3 combinations with 3 active spots that are valid are illustrated in Figure 6. The reader can check that, for each combination, all zones satisfy the allocation criterion.

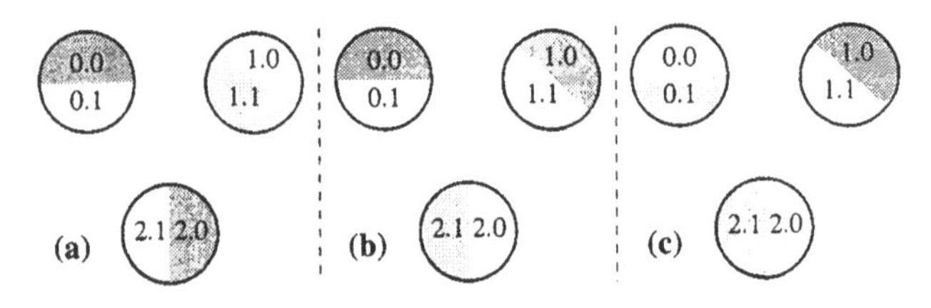

Figure 6: Valid 3-spot combinations for a threshold $\sigma = 0.30$.

Observe that the 3-spot combinations transmit more data, at the same time,

Table 3: Values of the carrier-to-interference ratio for all zones in all situations

C/I for Zone 0.0	Zone 1.0 *on*	Zone 1.1 *on*	Spot 1 *off*
Zone 2.0 *on*	0.57	0.40	1.33
Zone 2.1 *on*	0.36	0.29	0.57
Spot 2 *off*	1.00	0.57	-
C/I for Zone 0.1	Zone 1.0 *on*	Zone 1.1 *on*	Spot 1 *off*
Zone 2.0 *on*	0.86	0.60	2.00
Zone 2.1 *on*	0.55	0.43	0.86
Spot 2 *off*	1.50	0.86	-
C/I for Zone 1.0	Zone 0.0 *on*	Zone 0.1 *on*	Spot 0 *off*
Zone 2.0 *on*	0.25	0.25	0.43
Zone 2.1 *on*	0.38	0.38	1.00
Spot 2 *off*	0.60	0.60	-
C/I for Zone 1.1	Zone 0.0 *on*	Zone 0.1 *on*	Spot 0 *off*
Zone 2.0 *on*	0.67	0.67	1.14
Zone 2.1 *on*	1.00	1.00	2.67
Spot 2 *off*	1.60	1.60	-
C/I for Zone 2.0	Zone 0.0 *on*	Zone 0.1 *on*	Spot 0 *off*
Zone 1.0 *on*	1.00	0.56	2.50
Zone 1.1 *on*	0.38	0.29	0.50
Spot 1 *off*	1.67	0.71	-
C/I for Zone 2.1	Zone 0.0 *on*	Zone 0.1 *on*	Spot 0 *off*
Zone 1.0 *on*	1.00	0.56	2.50
Zone 1.1 *on*	0.38	0.29	0.50
Spot 1 *off*	1.67	0.71	-

than the 2-spot combinations which are less efficient.

4.1 Case of a Simple Demand

Assuming that there is a demand of 100 time slots per *zone*, it is clear that the minimum number of time slots necessary to fulfill the demand is 200, because only one zone per spot can be active at any time. For the first 100 time slots, the combination in Figure 6(a) can be used to satisfy the demand of zones 0.0, 1.1, and 2.0, and for the second 100 time slots, the combination in Figure 6(c) can be used to satisfy the demand of zones 0.1, 1.0, and 2.1, which solves the problem.

4.2 Case of a More Complex Demand

Consider here a demand slightly more complex than in the previous case, as can be seen in Table 4. The demand per *spot* is 200 time slots, as in the previous

Table 4: Demand (in time slots) of the different zones

Zone	0.0	0.1	1.0	1.1	2.0	2.1
Demand	50	150	50	150	150	50

case, but more than 200 time slots are needed to satisfy all zones, because the 3 combinations of Figure 6 cannot be used as efficiently as before. It is clear that the combination in Figure 6(a) can still be used for 50 time slots to satisfy the demand of zone 0.0, and zones 1.1 and 2.0 are left with 100 time slots demand to satisfy. Also, the combination in Figure 6(c) can be used for 50 time slots to satisfy the demand of zones 1.0 and 2.1, and zone 0.1 is left with an unsatisfied demand of 100 time slots. To complete the allocation problem, we can use combinations with only two active zones, allocating 50 time slots to each one of the following combinations: (i) zones 0.1 and 1.1; (ii) zones 0.1 and 2.0; and (iii) zones 1.1 and 2.0. Observe that the allocation procedure consists mainly in allocating 250 time slots to combinations of zones, provided that these combinations are valid.

Looking at Figure 6, we see that combinations (b) and (c) differ only on spot 0. They both include zones 1.0 and 2.1, but although the first combination includes zone 0.0, the second includes zone 0.1. It is therefore possible to merge these combinations into one, composed of *any* zone of spot 0 and zones 1.0 and 2.1. Hereafter, we use the term "family" to refer to such a combination of zones/spots. Observe that it is possible to use a given family when allocating slots, even though not all zones within this family need to be active. This observation will add flexibility to the solution. Using the same amount of time slots as before, that is 250, the allocation to satisfy the demand of Table 4 could now be satisfied as expressed in Table 5. In this solution, zone 0.0 will be assigned 50 extra time slots.

5 Solving the General Case

As seen in the previous section, to solve the allocation problem in the simple case where there is only one type of terminals, we have first computed the carrier-to-interference ratio for all zones that let us identify the valid combinations, or families, of zones that are allowed to transmit simultaneously.

Table 5: A more efficient solution to the example

Number of time slots	Family to use	Active zones
100	Zones 0.0, 1.1, 2.0	Zones 0.0, 1.1, 2.0
50	Spot 0, Zones 1.0, 2.1	Zones 0.1, 1.0, 2.1
50	Spots 0, 1	Zones 0.1, 1.1
50	Spots 0, 2	Zones 0.1, 2.0

Second, we have allocated a certain number of time slots for some families in order to satisfy the demand of all zones. To solve the allocation problem in general (arbitrary number of zones/spots, arbitrary demand, and multiple types of terminals) we have to (i) generate families of spots/zones that are valid (see Section 5.1), (ii) identify the number of time slots of each type to allocate to which families in order to satisfy the demand (see Section 5.3), and (iii) allocate the required number of time slots by placing the carriers in the radio channel and the time slots in the corresponding time frames (see Section 5.2). The details of step (ii) are presented through Sections 5.4 to 5.8. Section 5.9 presents a wrap-up of our approach.

5.1 Solving Interference Problems

Our approach is mainly based on the following key observation: for any time t and any frequency f, there exists at least one family of zones that can be simultaneously active. Let Z denote one such family; we therefore have:

$$\frac{G(z)}{\sum_{z' \in Z} I(s(z), z')} \geq \sigma \quad \forall z \in Z. \tag{5}$$

Naturally, there could be in family Z no more than one (active) zone per spot. This concept of concurrent transmissions is somehow similar to graph coloring [18], where families of independent edges are used to solve the problem. Here, we use families of zones allowed to transmit at the same time (and at the same frequency).

In practice, there are a very large number of families checking this criterion. It is possible to have families that differ only by one spot, according to which zone in the spot is active (see the example in Section 4). As already said, such families can be merged in a single family, keeping in mind that, for that particular spot, several zones could be allowed to be active. This merging will add flexibility to the use of the resulting family. To solve the interference problem, we generate a certain number of families, that are used later on in the time slot allocation procedure. It is crucial to generate in the first place the most efficient families, or in other words, the families having the highest possible number of zones that can be active in (t, f), while presenting the highest flexibility.

5.1.1 Generating Generic Families

The threshold of interference σ is given as an input. If σ is very weak (for instance 10 dB, which is not very realistic), all spots can be active in (t, f). As σ increases, fewer spots can be active simultaneously using the same frequency. The difficulty here is to have the maximum number of active spots/zones for a given σ.

Recall the allocation criterion given in (4). It makes the distinction as to whether the interfering terminal is in a neighboring spot. Terminals in the vicinity are considered to interfere more than remote terminals. It then comes out that inactive spots should be geographically distributed for increased efficiency. We consider situations where only a restricted set of spots are inactive. We call a configuration 6/7 (resp., 5/7, 4/7) when at most 6 (resp., 5, 4) spots over a vicinity of 7 are active. We illustrate in Figure 7 such possible configurations. We translate the illustrated patterns (that have maximality properties on the infinite grid) to obtain a limited but efficient series of families.

Figure 7: Example of configurations 6/7 and 5/7.

It is obvious that there are 7 distinct configurations 6/7 as there are 7 possible positions for the inactive spots in a line. Also, there are $7 \times 3 = 21$ distinct configurations 5/7, because every configuration 6/7 generates 3 possible configurations 5/7 according to whether there are 0, 1, or 2 active spots between two inactive spots in any horizontal line. In a similar way, there are in total $7 \times 5 = 35$ configurations 4/7, as every configuration 6/7 generates 5 possible configurations 4/7.

5.1.2 Status of a Spot

In the previous section, we have introduced efficient spatial configurations of active/inactive spots that homogeneously distribute the inactive spots. We believe that these configurations are more efficient than others as they will allow a larger number of spots to be active given the same threshold σ. Each one of these configurations yields several families of active zones. Indeed, spots are usually divided into few zones (typically 2 or 3), and there are several possibilities for having a spot active. As (i) the power gain depends on the geographical zone within a spot, and (ii) the interference generated over the spot depends on which zones have transmitted the interfering signals, it is quite possible that one zone in a spot does not check the allocation criterion (4) whereas another zone in the very same spot does. Therefore, every spot will be assigned a *status* describing which zones can potentially be active. If a spot s has $nbZones(s)$ zones, then its status takes value in the interval $[0, 2^{nbZones(s)} - 1]$. For instance, the status of a 3-zone spot could take on one of the following values (a 2-zone spot could take on one of the first 4 statuses in the list):

0: The spot is inactive;

1: Zone 0 checks (4), hence it could transmit;

2: Zone 1 checks (4), hence it could transmit;

3: Zones 0 and 1 check (4); hence either one could transmit;

4: Zone 2 checks (4), hence it could transmit;

5: Zones 0 and 2 check (4); hence either one could transmit;

6: Zones 1 and 2 check (4); hence either one could transmit;

7: All zones check (4); hence either one could transmit;

Instead of generating families of zones, we generate families of spots and assign to each spot the convenient status given the allocation threshold σ. Allocating time slots to a 3-zone spot with status 7 would actually be done by allocating the time slots to *either* one of its 3 zones, which increases freedom and improves the efficiency of our approach.

5.1.3 Simplifying the Computation of the Allocation Criterion

At the beginning of Section 5.1, we defined a family of zones Z satisfying (5). In this section, we derive a similar equation for families of spots. Instead of checking the allocation criterion (4) for every zone, we have to check it for every spot. To be able to check if a spot could be active and decide which status it could have, we assign to every spot a gain and an interference over other spots.

The gain of a spot is defined as the minimum value of the gains of its zones that are active (information available from the status of the spot). Let $G(s)$ denote the spot gain; we can write

$$G(s) = \min_{z \text{ in } s, \text{ active}} G(z).$$

The interference generated over spot s by spot s' is defined as the maximum value of the interference generated by all zones of spot s' that could potentially be active. It is denoted as $I(s, s')$. We have

$$I(s, s') = \max_{z' \text{ in } s', \text{ active}} I(s, z').$$

Recall the sums $I_1(z)$ and $I_2(z)$ introduced in (2)–(3). They represent the overall interference generated by active zones in neighboring spots and in all spots, respectively. Let $I_1(s)$ and $I_2(s)$ be their equivalent at the spot level:

$$I_1(s) = \sum_{s' \text{ neighbor, active}} I(s, s'), \qquad I_2(s) = \sum_{s' \text{ active}} I(s, s')$$

Similarly to what we did at the zone level, the total level of interference generated over a spot s is computed as

$$I(s) = \gamma I_1(s) + (1 - \gamma) I_2(s).$$

Thus, a spot is said to be *valid* if it checks the following criterion

$$\frac{G(s)}{I(s)} \geq \sigma. \tag{6}$$

The advantage of using (6) rather than using (4) is clear from the following example. Consider a spot whose status is 7. This means that it has 3 zones that could all be active (of course, not together). To check this hypothesis, one would have to check if each zone satisfies the criterion (4). It is definitely more advantageous to use instead the criterion (6) as the computation time would be greatly reduced. Observe that (6) implies (4). For any active zone z in spot s:

$$\begin{aligned} \frac{G(s)}{I(s)} &= \frac{G(s)}{\gamma I_1(s) + (1 - \gamma) I_2(s)} \\ &\leq \frac{G(z)}{\gamma I_1(z) + (1 - \gamma) I_2(z)} = \frac{G(z)}{I(z)}. \end{aligned}$$

Thus, if a spot with a given status is valid, then all of its zones corresponding to its status are valid.

For maximum flexibility, we would like to have all spots in a family have a status equal to $2^{nbZones(s)} - 1$. To that purpose, we first generate families of spots, all having the highest status, and then test their validity. That can be done by checking the allocation criterion (6) for all spots in a family.

5.1.4 Heuristics for Generating Valid Families

We want to maximize the number of active zones; we start by generating the 7 families 6/7 in which any active spot s has the status $2^{nbZones(s)} - 1$ and inactive ones have status 0. We then successively test the validity of these families and separate them in two pools, one for valid families and the other for nonvalid families. We do the same with families 5/7, 4/7, etc.

To make a non-valid family become valid, some of its active zones should be deactivated. For instance, if a 3-zone spot having status 7 (any one of its 3 zones could be active) is not valid, then we should test the validity of its family when its status is 3, 5, or 6 (zone 2, zone 1, or zone 0 are deactivated). The following heuristic is used.

1. Randomly choose a nonvalid family from the pool of nonvalid families;
2. As long as the family is not valid, do:
   (a) Randomly choose a spot,
   (b) If its status is nonnull and the spot is nonvalid, deactivate at random one of the active zones; keep a record of the spot identifier;
3. Try, for a certain number of times, to reactivate zones that were deactivated in step 2 and test the validity of the resulting family after each try: an amendment is adopted only if the family is valid;
4. Compare the valid family obtained in step 3 with those in the pool of valid families. In case of redundancy, increment a counter of redundancies and reject the family; otherwise, add the family to the pool of valid families. Return to step 1 to generate another family.

This algorithm stops either when the desired number of valid families is reached, or when the counter of redundancies has reached a given maximum value. At this point, we have generated valid families of spots. In every spot s of a valid family, $0, \ldots, nbZones(s)$ zones are candidates in the time slot allocation procedure.

5.2 Placing the Carriers in the Radio Channel

The constraints on the radio channel deal with the spot bandwidth B and the time frame duration T. When planning the allocation of a time slot from a given carrier to a given type of terminal, one schematically uses a rectangle of a *fixed* surface equal to Δ in the time–frequency space (recall Section 3.4). See, for instance, zone 0.1 in Figure 11 in which two different types of terminals are used.

Thus, if the types of terminals are denoted by subscripts from 1 to τ (ordered by decreasing bandwidth), and if x^{t_k} denotes the number of time slots of type t_k used in the spot, we then have:

$$\sum_{k=1}^{\tau} x^{t_k} \leq \frac{BT}{\Delta}. \tag{7}$$

In other words, the maximal surface, in the time–frequency space, that can be allocated to a spot is equal to the product BT, yielding an upper bound equal to BT/Δ on the number of time slots that can be allocated.

The following lemma is used to establish the properties of a filling of time slots.

Lemma 5.2.1 *Let $G = (V, E)$ be a directed graph with $V = \{t_1, \ldots, t_\tau\}$ and $E = \{(t_j, t_k) : j < k\}$. Define $w(t_j, t_k) = w_{(j,k)} = \frac{t_j^b}{t_k^b} - 1$. Then any path in G from t_i to t_j has a weight less than $w_{(i,j)}$. In particular, any path in G from t_1 to t_τ has a weight less than $w_{(1,\tau)}$.*

Proof. Note that G is transitive. Thus, if $(t_i, t_j) \in E$ and $(t_j, t_k) \in E$, then $(t_i, t_k) \in E$. Observe that for any numbers x and y such as $x > 1$ and $y > 1$, we have

$$(x - 1) + (y - 1) = xy - 1 - (x - 1)(y - 1) < xy - 1. \tag{8}$$

The weight of the path $t_i \to t_j \to t_k$ is equal to $w_{(i,j)} + w_{(j,k)} < w_{(i,k)}$ (take $x = t_i^b/t_j^b$ and $y = t_j^b/t_k^b$ in (8)), which concludes the proof. □

Example 5.2.2 *Figure 8 illustrates the graph $G(V, E)$ corresponding to the data in Section 3.4.*

Thereafter, we show that a path in this graph corresponds to losses due to the geometrical structure of the problem. Any change in type during the

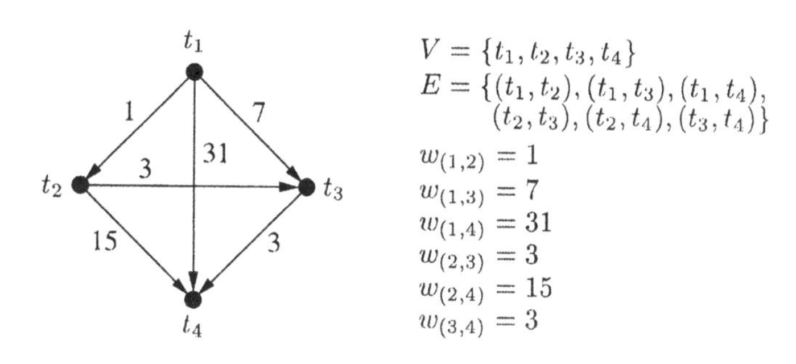

Figure 8: Graph G according to Table 1.

placement process will incur a waste in space in the time–frequency space. Changing from type t_i to type t_j ($j > i$) will cause *at most* an unused space equal to $w_{(i,j)}$. To minimize the space that could be lost, the best thing to do is to place the types monotonically. We have opted to fill the time-frequency space from left to right and top to bottom using the ascending order of types. The maximum number of unused time slots with this policy is given by the weight along a path in G that goes from t_1 to t_τ. We know from Lemma 5.2.1 that this maximum is less than $w_{(1,\tau)}$. To be more precise, this maximum (obtained in the worst case) is exactly the sum of the weights along the path followed in graph G to go from t_1 to t_τ.

Result 5.2.3 *It is feasible to place, in the time–frequency space, x^{t_k} time slots of type t_k, for $k = 1, \ldots, \tau$ if*

$$\sum_{k=1}^{\tau} x^{t_k} \leq \frac{BT}{\Delta} - w_{(1,\tau)}. \tag{9}$$

Equation (9) is therefore a sufficient condition for a placement algorithm.

Proof. We give the sketch of the proof. To prove that Eq. (9) is sufficient to find a placement, we have to devise a placement policy that will succeed in placing all time slots in the space $B \times T$ without wasting more than the theoretical maximum waste $w_{(1,\tau)}$. We start by subdividing the total space $B \times T$ in rectangles whose "frequency" dimension is the maximum bandwidth of a time slot and whose "time" dimension is the maximum duration of a time slot. Such rectangles have a bandwidth t_1^b and a duration t_τ^t, and are referred to as "rectangles $(1, \tau)$".

The rectangles $(1, \tau)$ are filled according to the ascending order of types. To evaluate the "waste" within a single rectangle $(1, \tau)$ induced by this filling policy, two cases have to be considered whether a change of type forces the filling of the next rectangle $(1, \tau)$, or not.

Case 1. If the bottom-right corner of a rectangle $(1, \tau)$ is reached, there will be no waste when switching to the following rectangle. The only waste (if any) will be "internal" to the rectangle $(1, \tau)$ at hand. If this rectangle contains types t_i and t_j, it is allowed to "waste" a space equal to $w_{(i,j)}$. Indeed, it subdivides itself into rectangles (i, j) of types increasing from t_i to t_j. In the worst case, we lose the sum of the weights in a path of the graph of Lemma 5.2.1 going from type t_i to type t_j. Rectangles containing only one type of time slot should not cause any waste in space.

Case 2. If a change in type forces the switching to the following rectangle whereas the bottom-right corner of the current rectangle has not been reached, then there will be an additional waste apart from the internal one computed in Case 1. If the current rectangle $(1, \tau)$ contains types t_i and t_j and the following one has a slot of type t_k in its top-left corner, then the current rectangle could potentially have a waste larger than $w_{(i,j)} + w_{(j,k)}$, but it will be at most equal to $w_{(i,k)}$.

Over all rectangles $(1, \tau)$ in the space $B \times T$, the total waste is equal to the sum of the waste within every rectangle. We know from Lemma 5.2.1 that this sum will be at most equal to $w_{(1,\tau)}$, yielding Result 5.2.3. □

Observe that time slots of types t_1 and t_τ could never be placed simultaneously in a rectangle $(1, \tau)$ as can be seen in the example in Figure 9. The same observation holds for types t_i and t_j and rectangles (i, j) (see rectangle I in Figure 9). Observe also that a rectangle (i, j) can have exactly t_i^b/t_j^b time slots exclusively of type t_i or of type t_j (see rectangles II and III in Figure 9). The surface of a rectangle (i, j) is exactly $(t_i^b/t_j^b)\Delta$. Therefore, the space $B \times T$ could be subdivided into exactly $BT/((t_1^b/t_\tau^b)\Delta)$ rectangles $(1, \tau)$.

Introduce $\delta = w_{(1,\tau)}$. From now on, we consider the following constraint on the number of time slots to be used

$$\sum_{k=1}^{\tau} x^{t_k} \leq \frac{BT}{\Delta} - \delta.$$

We know from Result 5.2.3 that the placement is feasible if this constraint is respected. Observe that for the data in Table 1, this constraint allows us to solve the problem of the placement by sacrificing less than $w_{(1,4)}/3456 = 0.897\%$ of the bandwidth. This ratio depends on the data of the problem and cannot

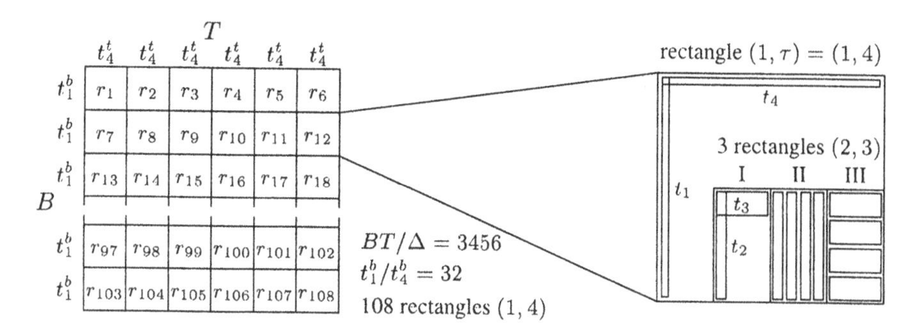

Figure 9: Subdivision of $B \times T$ corresponding to Table 1 and rectangles (i, j).

be guaranteed for any instance of the problem. The only guarantee is that the fraction of lost bandwidth will be less than $w_{(1,\tau)}\Delta/(BT)$. Nevertheless, the given example of Table 1 is representative of the possible instances, and a loss of roughly 1% is definitely satisfying regarding the complexity of the problem. It might be possible to do even better than that by adopting a lower value of δ, assuming that the arrangement will still be feasible. In practice, one can carry out the placement according to many other policies, which may lead to a waste smaller than that of the preceding proof.

It is not trivial to write the placement algorithm. We therefore give just its simplest version[3] in Algorithm 1. Algorithm 1 consists in filling the space from left to right and "jumping" to the order of multiplicity when there is a change in the type. The set R of rectangles divides the space $B \times T$ and is ordered as illustrated in the example of Figure 9.

Figure 10 depicts a sample output of the placement algorithm working with 4 types of terminals, according to the data in Table 1. This placement is obtained when using Algorithm 1. There are 9 time slots of type t_1, 3 time slots of type t_2, 11 time slots of type t_3, and 4 time slots of type t_4 to place in the time–frequency space of size $B \times T$. The orders of multiplicity are 2 between types t_1 and t_2, and 4 between types t_2 and t_3, and between types t_3 and t_4 (see Section 3.4). The rectangles drawn in dotted lines are "lost spaces" whereas the rectangles in continuous features are time slots of different types placed on the time-frequency space. The vertical dotted lines correspond to the "jumps" occurring at the changes in types. The dotted line at number 10 (resp., 16) corresponds to the change from type t_1 to t_2 (resp., from type t_2 to t_3). One time

[3]We consider here the case where only terminals of the same type can transmit together. We see later on in Section 5.3 that it is possible to assign different types of carriers to distinct spots.

Algorithm 1 Placement algorithm.

Input: An ordered heap H of types $\{t_1, \dots, t_1, t_2, \dots, t_2, \dots, t_\tau, \dots, t_\tau\}$. An ordered set R of $(1, \tau)$ rectangles ordered from left to right and top to bottom

Ouput: A placement on $B \times T$

```
1: Let r be the first (1,τ) rectangle in R
2: Set t_old = head(H)
3: while H is not empty do
4:    Dequeue t_k from H
5:    if t_k cannot be placed in r then
6:       Fill empty space in r (if any) with empty types t_old {waste}
7:       Select next rectangle r in R
8:    else
9:       if t_k and t_old are different then
10:         Jump to next multiple of t_k^b/t_old^b filling with empty types t_old
            {waste}
11:      end if
12:   end if
13:   Set t_old = t_k
14:   Put t_k in r with leftmost, topmost policy
15: end while
```

slot of type t_1 and three time slots of type t_2 were lost in these jumps. To place the first time slot of type t_4, a new rectangle had to be used. The empty space in the first one was filled with five empty time slots of type t_3. The placement of Figure 10 incurred a total loss of 1 + 3 + 5 = 9 time slots ($9 < w_{(1,4)} = 31$). The example of Figure 10 illustrates well the two cases considered in the proof of Result 5.2.3: the change from type t_1 to t_2 is considered in Case 1 (internal waste) and the change from type t_3 to t_4 is considered in Case 2.

The lost space in this example may seem very significant compared to the total amount of time slots to be placed (exactly 27), but this is because of the little demand. As explained before, the losses are inevitable when changing types. The advantage of this policy is that all lost spaces are nicely formatted: they can be used if the demand for certain types of slots is larger.

To conclude this section, we give an example of placement in which the maximum waste allowed is attained. It is the case when there are only 2 time slots to place, the first of type t_1 and the second of type t_τ. Two rectangles $(1, \tau)$ are needed here, and each will have exactly one time slot. There will be $w_{(1,\tau)}$ time slots lost in the first rectangle.

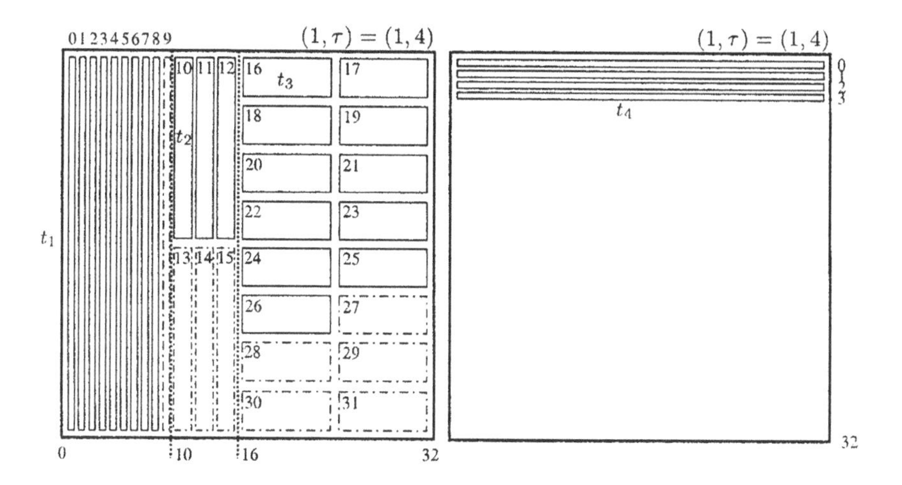

Figure 10: Sample output of the placement algorithm.

5.3 Satisfying the Global Demand

Instead of allocating time slots of a certain type to a spot, we propose to allocate slots to typified families, i.e., simultaneously in all spots. In a typified family, distinct spots can be assigned different types. If family F_i assigns type t_k to spot s, we note $F_i^T(s) = t_k$.

Initially, we consider families with only one type. Thus, for a family F_i, we can choose a type of terminal t_k which is used on all concerned spots (another family $F_{i'}$ would use another type $t_{k'}$). In other words, $\forall s,\ F_i^T(s) = t_k$. Such families are denoted as 1-typified families. We place this 1-typified family, in the time–frequency space, at exactly the same place for all concerned spots, implying that all spots would use the same frequency band. In this way, we are sure that the allocation criterion is respected, because of the definition of a family. Over other frequency bands, another family could be used to satisfy another (or the same) demand.

Figure 11 shows a possible placement of the radio resources. If $F_i^T(s) = t_k$, we note (F_i, t_k) in the rectangle concerned. Thus, this notation is found in all active zones of a family (for instance, zones 0.1 and 2.0 for family F_2). The constraints of capacity on each zone, in terms of bandwidth and time frame are ensured by the constraint of surface of a rectangle (F_i, t_k) on the rectangle $B \times T$.

A family can possibly have several types of terminals according to its different spots. It is the case, for example, for the rectangles (F_3, t_1) in zones 0.2

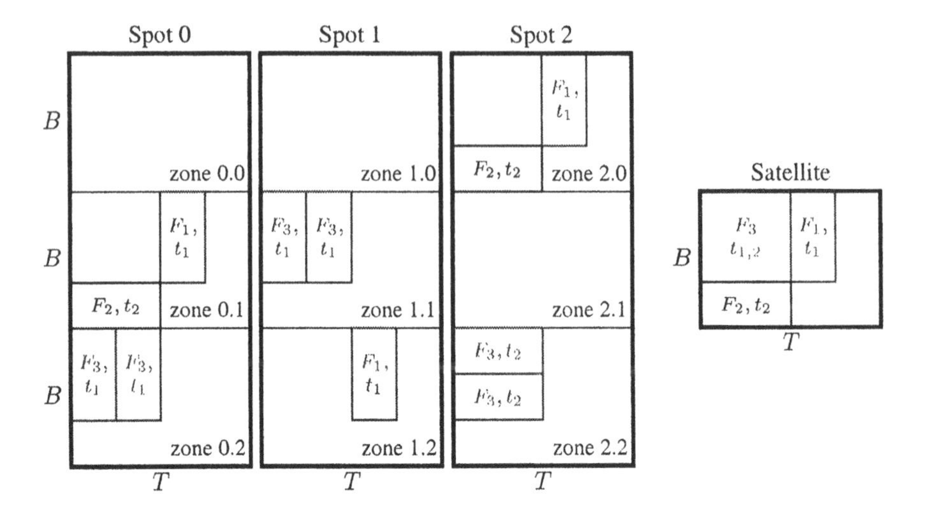

Figure 11: A global example of arranging families.

and 1.0 and (F_3, t_2) in zone 2.2. We say that family F_3 is 2-typified and its type is denoted $t_{1,2}$, as can be seen in the diagram at the right of Figure 11 (satellite point of view). These families have a specific order of multiplicity. If t_k is the type in the family having the larger bandwidth and $t_{k'}$ that with the narrower bandwidth, then the order of multiplicity of the family is

$$F_i^M = \frac{t_k^b}{t_{k'}^b} \in \mathbb{N}^* - \{1\}.$$

5.4 Linear Program $\mathcal{P}$

In this section, we define the linear program used to compute a solution, based on the typified families described earlier. Without loss of generality, we consider the case where each spot has three zones. We model the constraints for satisfying demands with Equations (11) to (13). Equation (10) provides the time–frequency space constraint of Result 5.2.3.

The variables of the linear program, denoted $\mathcal{P}$, are the x_{F_i}, which represent the number of times that the typified families are used. They must be integer variables. Let $\mathcal{I}$ be the current set of typified families used to solve $\mathcal{P}$. Recall that $d(z, t_k)$ is the demand for type t_k, as defined in Section 3.4. Let $F_i^A(z) = on$ denote if zone z could be active, and $F_i^A(z) = off$ otherwise. $\mathcal{P}$ is then defined as min J where

$$J = \sum_{i \in \mathcal{I}} F_i^M x_{F_i} \leq \frac{BT}{\Delta} - \delta \tag{10}$$

$$\forall k \in [1, \tau],\ \forall z \in s,\ \sum_{i \in \Gamma(z,k)} F_i^M x_{F_i} \geq d(z, t_k) \tag{11}$$

$$\forall k \in [1, \tau],\ \forall z, z' \in s,\ \sum_{i \in \Gamma(z,z',k)} F_i^M x_{F_i} \geq d(z, t_k) + d(z', t_k) \tag{12}$$

$$\forall k \in [1, \tau],\ \forall\, s,\ \sum_{i \in \Gamma(z,z',z'',k)} F_i^M x_{F_i} \geq d(z, t_k) + d(z', t_k) + d(z'', t_k) \tag{13}$$

with:

$$\begin{aligned} \Gamma(z, k) &= \{i \in \mathcal{I}/F_i^T(s) = t_k,\ F_i^A(z) = on\} \\ \Gamma(z, z', k) &= \{i \in \mathcal{I}/F_i^T(s) = t_k,\ (F_i^A(z) = on \text{ or } F_i^A(z') = on)\} \\ \Gamma(z, z', z'', k) &= \{i \in \mathcal{I}/F_i^T(s) = t_k,\ \exists z \in s/F_i^A(z) = on\}. \end{aligned}$$

It is obvious that if (10) is not satisfied, no integer solution can be found. Therefore, we choose to consider the occupied surface as the objective function to minimize. Minimizing J results in the maximization of reuse of the resources and thus in the maximization of the system throughput.

Result 5.4.1 *Equations* (11) *to* (13) *guarantee the satisfaction of the demand in type* t_k.

Proof. The satisfaction of the demand in type t_k can be computed on a flow from a source s, while passing by 3 arcs (or $nbZones(s)$, if there are $nbZones(s)$ zones) of respective capacities $d_0 = d(z_0, t_k)$, $d_1 = d(z_1, t_k)$, and $d_2 = d(z_2, t_k)$, as seen in Figure 12. The capacities of the other arcs, denoted by $C[z_0, z_1, z_2]$, $C[z_j, z_{j'}]$ for $j \neq j'$, $\{j, j'\} \subset \{0, 1, 2\}$, and $C[z_j]$, $j \in \{0, 1, 2\}$, are given by:

$$\begin{aligned} C[z_0, z_1, z_2] &= \sum_i x_{F_i}^{t_k} \times U[Fi, \{z_0, z_1, z_2\}] \\ C[z_j, z_{j'}] &= \sum_i x_{F_i}^{t_k} \times U[Fi, \{z_j, z_{j'}\}] \\ C[z_j] &= \sum_i x_{F_i}^{t_k} \times U[Fi, \{z_j\}] \end{aligned}$$

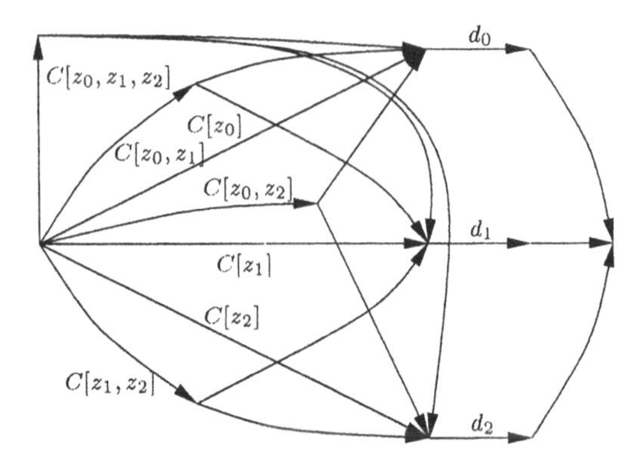

Figure 12: Modeling the constraints of zones as flows.

where $U[Fi, Z]$ is equal to 1 when F_i could activate either one of the zones of the set Z, and to 0 otherwise. The capacities of all other arcs in the figure are assumed infinite. Indeed, by the theorem of Ford and Fulkerson [11] (or in its version of Menger [24]), there is a maximum integer flow from the source to the sink, which is equal to the cardinality of a minimal cut. However, there are 8 cuts of finite size (or $2^{nbZones(s)}$ in the case of $nbZones(s)$ zones), according to the choice of the arcs of capacity d_0, d_1, and d_2. One of these equations is trivial because it stipulates that the flow of the zones must be less than $d_0 + d_1 + d_2$. The 7 others are checked by our linear program. □

5.5 Optimal Typification of Families

There exist τ^N different ways of typifying a given nontypified family F_i. As it is too much to include them all in $\mathcal{P}$ we use the concept of *generation of columns* [9]. A column corresponds to one valid typified family. The optimal float solution is obtained when $\mathcal{I}$ is the set of *all* valid typified families, a set that is too large to be used in practice. Actually, the process initializes $\mathcal{I}$ as the set of homogeneously typified families. However, given a restricted $\mathcal{I}$, dual properties allow us to identify new columns to be added to $\mathcal{I}$ to improve the solution. We show in the following that dual properties characterize nontypified families, which greatly simplifies the problem of identifying an optimal $\mathcal{I}$.

Let $\mathcal{P}$ be rewritten as follows:

$$\text{Minimize} \quad f = c \cdot x$$
$$\text{Such that} \quad \begin{cases} Ax & = & b \\ x & \geq & 0 \end{cases}.$$

Let A_B denote the matrix extracted from the corresponding system of equations, and x_B be the vector of the associated families. Let x_N denote the vector of the other families, and A_N be the corresponding matrix. In the same way, we subdivide c in c_B and c_N. We can write

$$A_B x_B + A_N x_N = b \quad \text{and} \quad f = c_B x_B + c_N x_N.$$

It then follows,

$$\begin{aligned} x_B &= A_B^{-1} b - A_B^{-1} A_N x_N, \\ f &= c_B A_B^{-1} b + (c_N - c_B A_B^{-1} A_N) x_N. \end{aligned}$$

The equations above return a basic solution to the system with $x_N = 0$. The system is optimal if and only if

$$c_N - c_B A_B^{-1} A_N \geq 0.$$

Thus, the system is improvable if and only if a negative coefficient can be found in the above vector. We further decompose A_N by writing $A_N = [A_{\alpha_1} \cdots A_{\alpha_m}]$ where m is the number of columns of A_N, each column corresponding to a family with subscript α_j. In particular, we have $c_{\alpha_j} = F_{\alpha_j}^M$.

Result 5.5.1 *For any nontypified family $\mathcal{F}$, there exist a constant $K_{\mathcal{F}} > 0$ and a function $\kappa_{\mathcal{F}}$, mapping pairs (type, Spot) to positive real numbers, such that for all typified families deriving from $\mathcal{F}$, we have*

$$c_i - c_B A_B^{-1} A_i = F_i^M \left(K_{\mathcal{F}} - \sum_{s \text{ spot}} \kappa_{\mathcal{F}}(F_i^T(s), s) \right).$$

Proof. Observe that, for a given line of A corresponding to a family F_i, denoted as A_i, all coefficients are either 0 or F_i^M. In addition, if F_i and F_j are typified families deriving from the same nontypified family, then $A_i/F_i^M = A_j/F_j^M$. Also, if c_i and c_j are the coefficients of c corresponding to F_i and F_j, then $c_i/F_i^M = c_j/F_j^M$. Observe that a spot s corresponds specifically to certain lines of A, given by $P_s A$ where P_s is the corresponding projection. If F_i and F_j derive from the same nontypified family and $F_i^T(s) = F_j^T(s)$, then $P_s A_i/F_i^M = P_s A_j/F_j^M$. Last, defining the following constants $K_{\mathcal{F}} := c_i/F_i^M$ and $\kappa_{\mathcal{F}}(F_i^T(s), s) := c_B A_B^{-1} P_s A_i/F_i^M$ yields the result. □

The optimal solution of our program is obtained when $\mathcal{I}$ is the set of *all* typified valid families. Because this set is too large to be used for a computation, we simply start with a restricted $\mathcal{I}$ which is progressively augmented to reach the optimum.

Result 5.5.2 *The program $\mathcal{P}$ with the restricted set of families $\mathcal{I}$ is improvable with respect to the set of all valid families if there exists a nontypified family $\mathcal{F}$ such that*

$$K_{\mathcal{F}} - \sum_{s \text{ spot}} \max_{t \text{ } type} \kappa_{\mathcal{F}}(t, s) < 0. \quad (14)$$

If we find one or several nontypified families which show that the system is improvable, we can strictly improve the solution by introducing the corresponding typified families (with the types found by the above maximization) into the linear program. This property considerably reduces the number of searches to be made in order to reach the optimal solution. In practice, as long as it is assumed that the solution is improvable, it will be possible to restrict the search by choosing a type for all spots in a subset of $\{t_1, \ldots, t_T\}$, reducing thereby the coefficient of multiplicity of the derived families and thus, the difficulty of the integrity constraints.

5.6 The Slave Program

Given a set of nontypified valid families, the slave program assigns the types to the families and returns the exact solution of $\mathcal{P}$ among all possible types. At first, the families are 1-typified with all possible types. The solution returns a dual that allows us to derive the improving 2-typified families according to Section 5.5. Then the linear program is solved again and eventually the dual will generate new 2-typified families. The process is iterated until no new 2-typified families are obtained, which means that we have reached the optimal solution given (i) the current set of nontypified families and (ii) the fact that only 2-typified families are used. The same process is done until τ-typified families are considered.

5.7 The Master Program

In this section, we show how we exploit the properties derived in Section 5.5 to find new valid families that will eventually lead to one $\mathcal{I}$ having the optimal solution.

A spot s being either inactive, or either one of its $nbZones(s)$ zones being active, it will have $nbZones(s) + 1$ possible states. Hence, for N spots, all having the same number of zones, there will be $(nbZones(s) + 1)^N$ combinations to test. For instance, there will be $4^8 = 65536$ combinations to test for an 8-spot configuration in which each spot has exactly 3 zones, which is very reasonable. However, when the number of spots increases, it will no longer be

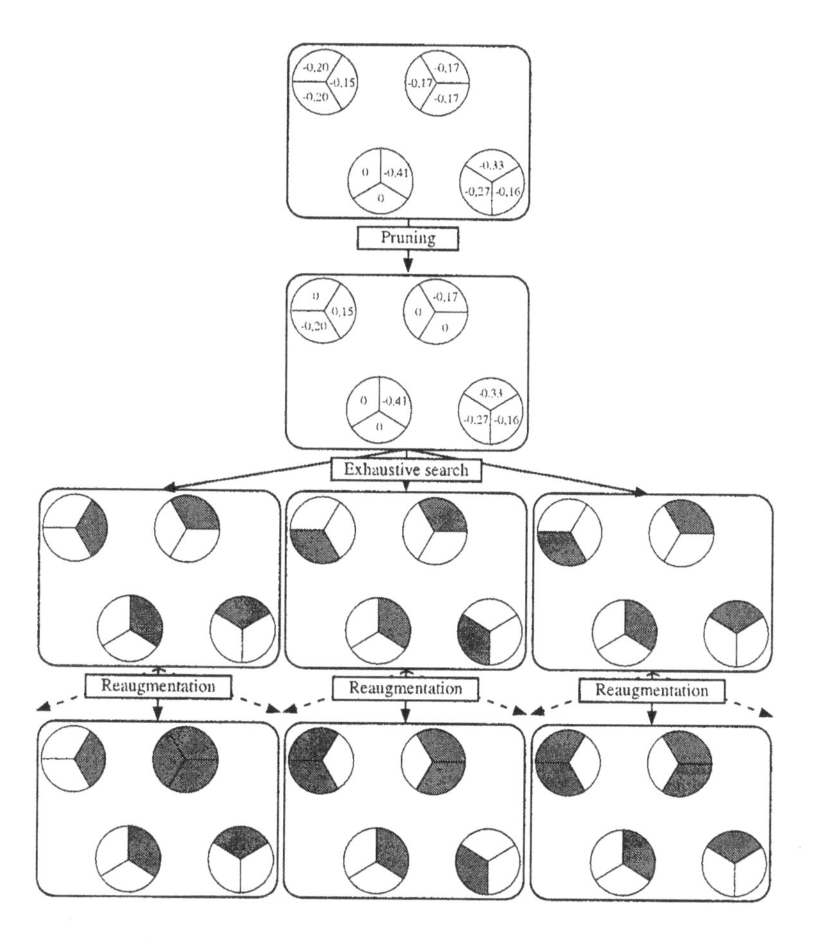

Figure 13: Pruned search of improving families.

reasonable to generate all families, which makes it difficult to find the optimal float solution.

Fortunately, for moderate numbers of spots, we are still able to derive an optimal solution in a relatively small time, thanks to a pruning technique described hereafter.

- A "pruner" selects zones within a spot. If several zones have the same gain, then only one of these is selected for an exhaustive search. This step is called "pruning".

- The p families with the highest "improvement potential" are selected.

These are the ones having the highest sum in (14).

- Every selected family is "reaugmented" whenever possible. In other words, if there are zones satisfying the allocation criterion without invalidating the family, then these are incorporated in the family.

The valid families generated by this technique are added to $\mathcal{I}$ and used in the next iteration to solve the linear program. This methodology is depicted in Figure 13.

5.8 Integer Solution to $\mathcal{P}$

The resolution of the slave programs enables the generation of the columns giving the best floating solution in each case. All these columns are then introduced into a new *integer* linear program, and are candidates to return the best possible *integer* solution. We stress that a solution exists with a number of non-zero variables x_{F_i} at most equal to the number of lines [8, Theorem 9.3, page 145]. For instance, in the case of 8 spots, we know that at most 224 floating variables will be used (896 in the 32 spots case), and therefore a simple ceiling of the variables will give a solution with all variables integer and multiple of 32 at less than 2.1% of the float solution (8.3% in the 32 spots case).

In practice, the resolution of the linear program, using the software Cplex CONCERT 8.0, returns an integer solution, which we arbitrarily fix at 1% of the optimal solution of the float problem. Note that solving completely the problem $\mathcal{P}$, using the columns' candidates, cannot be achieved in a reasonable time.

5.9 Algorithm Wrapup

This part sums up the whole behavior of our algorithm. Each part is represented in Figure 14 by a rectangle (resp., an oval) corresponding to a part of the process (resp., an action or a decision). We also show the interaction between the master and the slave explained in Sections 5.7 and 5.6. The algorithm starts in the leftmost rectangle. We first generate valid but nontypified families as described in Section 5.1. Then, the master program directly gives these families to the slave. The slave program operates as described in Section 5.6: the families are typified, the linear program $\mathcal{P}$ is solved, and the slave iterates until reaching optimality. The families involved in the solution are stored for the final integer computation. Afterwards, the master program checks the optimality of the solution given by the slave using the criterion (14). If the optimality is not reached, the pruning technique described in Section 5.7 is performed,

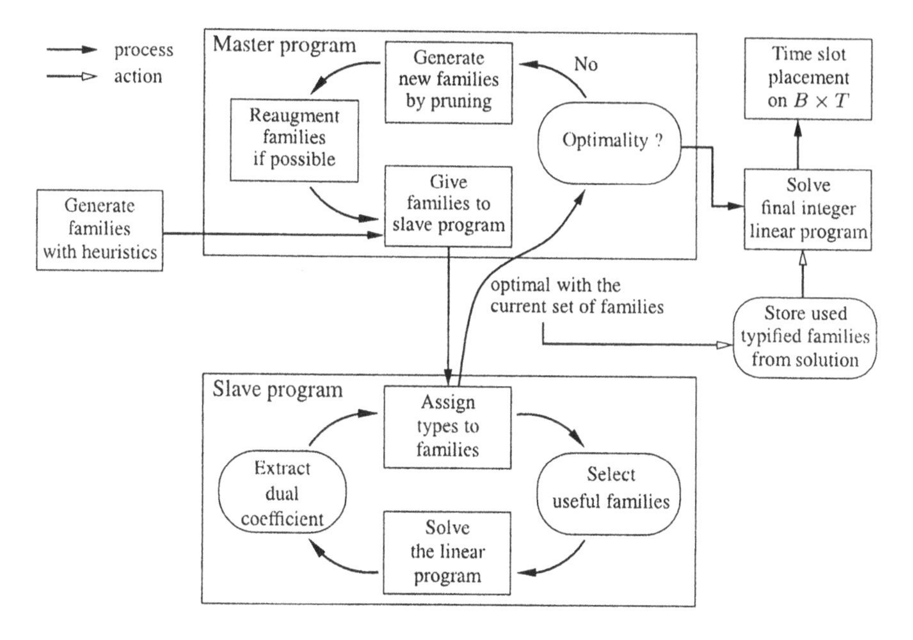

Figure 14: Algorithm overview.

generating new families. The master program then calls the slave again, giving it the new families generated. The master/slave process continues until optimality is reached. Next, the final integer linear program is solved as explained in Section 5.8. Finally, we achieve the placement of the resulting number of typified time slots as described in Algorithm 1 in Section 5.2.

6 Numerical Results

This section provides some numerical results returned by our approach. We have considered two configurations, the first consisting of 8 spots and the other of 32 spots. The zone demand has been generated according to examples previously provided by ALCATEL SPACE INDUSTRIES. The interference (in dB) as well as the gains (also in dB) was drawn from uniform distributions, according to specifications provided by ALCATEL SPACE INDUSTRIES. The global interference was considered to be generated mostly by the spots in the vicinity, as the interference generated by remote spots was reduced by 15% ($\gamma = 0.85$).

Our program outputs a time–frequency plan showing the slots allocated, as can be seen in Figure 15. The time–frequency space therein depicted shows

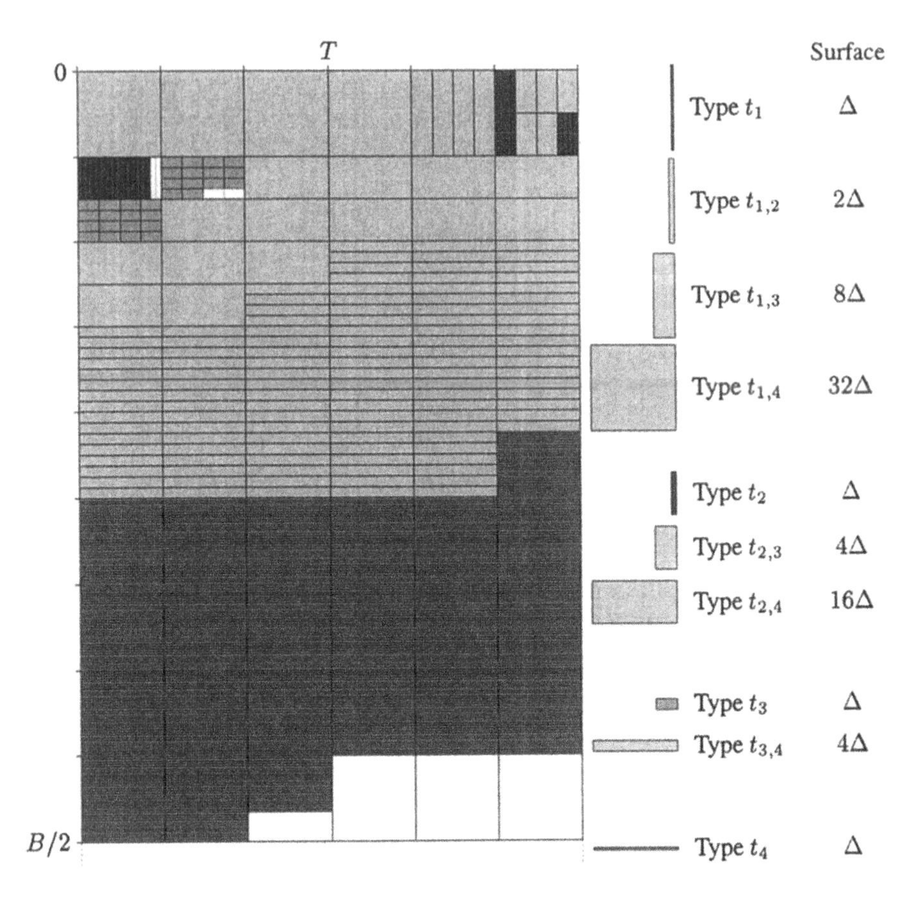

Figure 15: A sample resource allocation.

results in the same way as in Figure 10 and Figure 11 (the diagram at the right showing the satellite point of view). Real data, provided by ALCATEL SPACE INDUSTRIES, were used as input to our program and the results are drawn to scale. Figure 15 illustrates well how combinations of types can be used together. For instance, the surface of a block where both types t_1 and t_4 (denoted type $t_{1,4}$ in Figure 15) could be used is 32Δ, where Δ is the surface of a time slot (recall Section 3.4). Observe that $B = 18t_1^b = 36t_2^b = 144t_3^b = 576t_4^b$ and $T = 6t_4^t = 24t_3^t = 96t_2^t = 192t_1^t$ as indicated in Table 1. The lost space here consists of 2 time slots of type t_2 and 2 others of type t_3 (4 white "holes" in Figure 15).

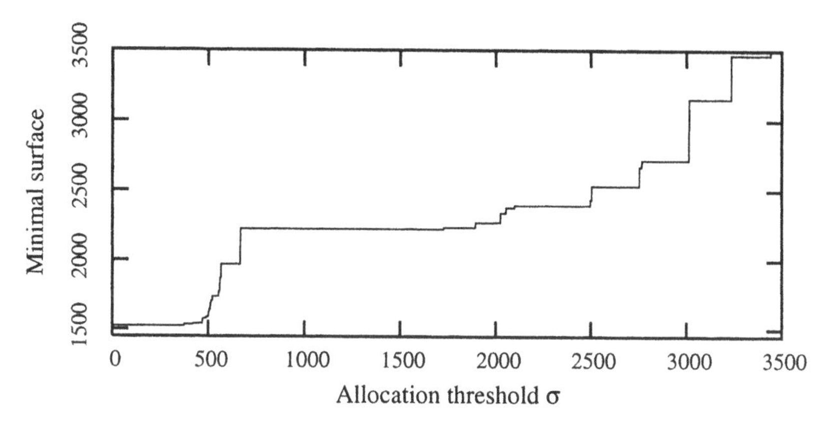

Figure 16: Minimal surface required to satisfy the demand vs. σ.

6.1 Results for 8 Spots

In the case where there are only 8 spots per color, our program succeeds in computing the optimal floating solution in about one minute when running on Pentium III machines. This case is particularly interesting as it enables a precise analysis of the effect of the allocation threshold.

We have computed the minimal surface, in the time–frequency plan, that is needed to satisfy the demand, for several values of the allocation threshold σ. The results are plotted in Figure 16. This figure clearly highlights the fact that the minimal surface increases abruptly around certain values of the threshold. Indeed, at some point, the threshold becomes too high impairing the use of some families that will no longer be valid at the considered threshold. The "loss" of these families degrades the solution, yielding a larger minimal surface. Table 6 reports which families become no longer valid at some threshold values. As a consequence, one is able to highlight the configurations of interference that block the generation of good solutions. This result has obviously a very strong impact on the design of antennas.

6.2 Results for 32 Spots

For a configuration with 32 spots, we recommend a nonoptimal approach using a restricted number of families. We stick to our real-time constraints that consist in obtaining a solution in a few minutes.

Figure 17 depicts the amount of time slots needed to satisfy the demand as a function of the number of valid families used, for several threshold values.

Table 6: Threshold values and families invalidated (X = zone off)

	Spot 0			Spot 1			Spot 2			Spot 3			Spot 4			Spot 5			Spot 6			Spot 7		
σ	0	1	2	0	1	2	0	1	2	0	1	2	0	1	2	0	1	2	0	1	2	0	1	2
373	X					X						X				X								
373	X					X						X					X							
418	X					X						X												
423	X					X										X								
450		X		X							X					X								
450		X		X								X					X							
450		X		X							X										X			
469			X	X								X					X							
472			X	X							X					X								
472			X	X							X										X			
472				X							X					X					X			
472					X						X					X					X			
490		X		X							X													
496				X								X				X					X			
501			X	X								X				X								
505			X	X												X					X			
510				X				X										X						

Observe that when the pool of families used is larger, the required amount of time slots to satisfy the demand gets smaller. It is therefore more efficient to use a larger pool of families. Observe as well that the solution is more efficient when the allocation threshold σ is smaller, regardless of the number of families used. This observation does not come as a surprise. It is obvious that smaller thresholds would allow a larger number of simultaneous transmissions. Every family would therefore include a larger number of zones that could be active, increasing the efficiency of their use.

As written previously, a larger pool of families improves the solution as it lessens the minimal amount of time slots to be allocated. However, this enhancement comes at the cost of an increased solving time, as it can be seen in Figure 18. This figure plots the solving time (over Pentium III machines) as a function of the pool size, for several threshold values. Observe that, for the same number of families used, the solving time increases as the threshold values increase. This is mainly due to the time taken for generating the required amount of valid families. For larger thresholds, much more time is needed to generate valid families, as the number of nonvalid families gets larger. This is why the difference, between solving times for different thresholds, increases as the number of families to generate increases (see Figure 18).

In practice, there is a tradeoff between the solving time and the minimal

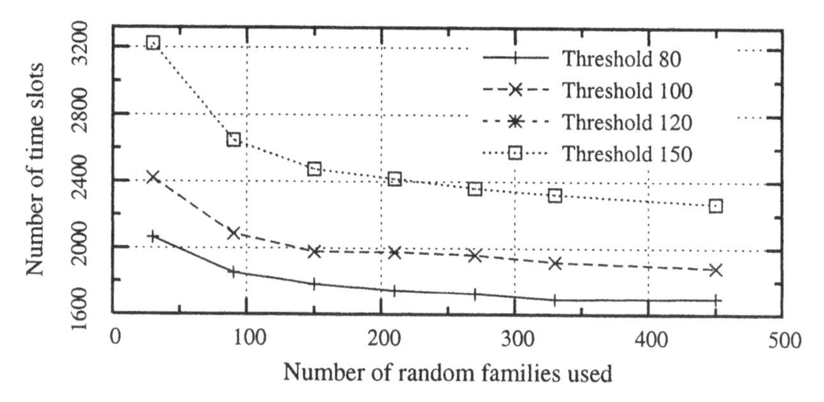

Figure 17: Time slots required to satisfy the demand vs. the number of random families used (32 spots case).

amount of time slots to allocate. For the same number of families used, a small solving time yields a large amount of time slots to satisfy the demand, whereas large solving times yield resource economy. It is then up to the satellite operators to decide for the optimal number of families to use, according to their priorities.

7 Conclusion

In this chapter we have devised a novel resource allocation algorithm for MFT-DMA satellites. We have considered a more accurate model for satellite communications, first by introducing a realistic modeling of the interference that is generated by active terminals. Second, we have considered the fact that terminals have specific transmission capabilities, which are translated into demands of different types of communications. In this context, our model is much more general than the ones presented in Section 2.

We have first introduced the concept of nonconcurrent transmissions with the use of families of spots that could transmit simultaneously at the same frequency. These families are then used to allocate time slots to multiple terminals increasing the efficiency of the algorithm. The total demand is satisfied by judiciously placing the different carriers in the radio channel, and the time slots in the corresponding time frames. A linear program is used to compute the number of typified families to use. A column generation process improves these families and selects the good candidates for the last integer programming.

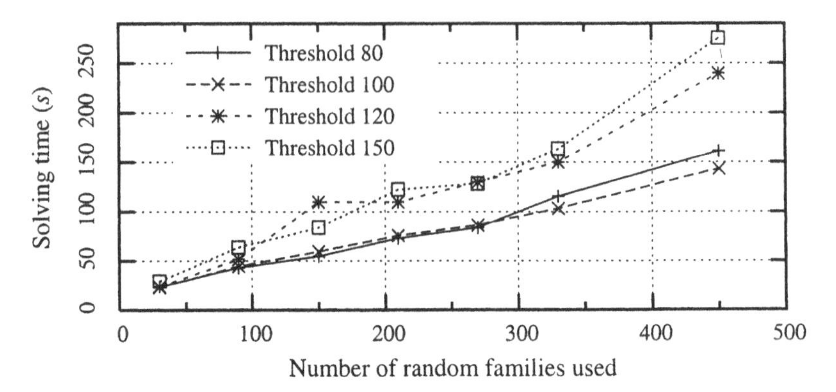

Figure 18: Solving time vs. the number of random families used (32 spots case).

We have shown that with this solution, we can arrange the different carriers in the bandwidth with a less than 1% waste. Our numerical results for a relatively small number of spots have shown that some interference configurations are harmful, in the sense that they impair the use of some families, hence, degrading the efficiency of the solution. For a large number of spots, our results show that a large number of families can improve the efficiency of the solution at the cost of increasing the solving time. Therefore, a tradeoff has to be found according to the priorities of the satellite operator.

Acknowledgment

We would like to thank Benoît Fabre, Cécile Guiraud, and Isabelle Buret, from ALCATEL SPACE INDUSTRIES, for providing many technical explanations. They have contributed to the modeling of the problem, and have shown a high interest in its resolution.

References

[1] E. Altman, J. Galtier, and C. Touati. A survey on TDMA satellite systems and slot allocation, June 2002. `http://www-sop.inria.fr/maestro/personnel/Eitan.Altman/PAPERS/tdmasurve%y.pdf`.

[2] J. Ben-Hur. Technology Summary: Project Nemo. Gaiacomm International Corporation, `http://www.gaiacomminternational.com`, December 2003.

[3] G. Bongiovanni, D. Coppersmith, and C. K. Wong. An optimum time slot assignment algorithm for an SS/TDMA system with variable number of transponders. *IEEE Transactions on Communications*, 29(5):721–726, May 1981.

[4] M. A. Bonuccelli. A fast time slot assignment algorithm for TDM hierarchical switching systems. *IEEE Transactions on Communications*, 37(8):870–874, August 1989.

[5] M. A. Bonuccelli, I. Gopal, and C. K. Wong. Incremental time-slot assignment in SS/TDMA satellite systems. *IEEE Transactions on Communications*, 39(7):1147–1156, July 1991.

[6] S. Chalasani and A. Varma. Efficient time-slot assignment algorithms for SS/TDMA systems with variable-bandwidth beams. *IEEE Transactions on Communications*, 42(2/3/4):1359–1370, Feb/Mar/Apr 1994.

[7] W. Chen, P. Sheu, and J. Yu. Time slot assignment in TDM multicast switching systems. *IEEE Transactions on Communications*, 42(1):149–165, January 1994.

[8] V. Chvatal. *Linear programming*. W. H. Freeman, San Francisco, 1983.

[9] G. B. Dantzig and P. Wolfe. Decomposition principle for linear programs. *Operations Research*, 8:101–111, 1960.

[10] T. ElBatt and A. Ephremides. Frequency reuse impact on the optimum channel partitioning for hybrid wireless systems. In *Proceedings of IMSC '99, Ottawa, Canada*, June 1999.

[11] L. R. Ford and D. R. Fulkerson. *Flows in Networks*. Princeton University Press, Princeton, NJ, 1962.

[12] N. Funabiki and Y. Takefuji. A parallel algorithm for time-slot assignment problems in TDM hierarchical switching systems. *IEEE Transactions on Communications*, 42(10):2890–2898, October 1994.

[13] M. R. Garey and D. S. Johnson. *Computers and Intractability: A Guide to the Theory of NP-Completeness*. W.H. Freeman, San Franciso, 1979.

[14] H. F. Geerdes and H. Karl. The potential of relaying in cellular networks. In *Proceedings of INOC '03, Evry/Paris, France*, pages 237–242, October 2003.

[15] M. X. Goemans and D. P. Williamson. Improved approximation algorithms for maximum cut and satisfiability problems using semidefinite programming. *Journal of the ACM*, 42:1115–1145, 1995.

[16] I. S. Gopal, M. A. Bonuccelli, and C. K. Wong. Scheduling in multibeam satellites with interfering zones. *IEEE Transactions on Communications*, 31(8):941–951, August 1983.

[17] I. S. Gopal, D. Coppersmith, and C. K. Wong. Minimizing packet waiting time in a multibeam satellite system. *IEEE Transactions on Communications*, 30(2):305–316, February 1982.

[18] M. Grötschel, L. Lovász, and A. Schrijver. The ellipsoid method and its consequences in combinatorial optimization. *Combinatorica*, 1:169–197, 1981.

[19] T. Inukai. An efficient SS/TDMA time slot assignment algorithm. *IEEE Transactions on Communications*, 27(10):1449–1455, October 1979.

[20] Y. Ito, Y. Urano, T. Muratani, and M. Yamaguchi. Analysis of a switch matrix for an SS/TDMA system. *Proceedings of the IEEE*, 65(3):411–419, March 1977.

[21] D. J. Kennedy *et al.* TDMA burst scheduling within the INTELSAT system. In *Proceedings of GLOBECOM '82, Miami, Florida*, pages 1263–1267, November 1982.

[22] C. King, P. Trusty, J. Jankowski, R. Duesing, and P. Roach. INTELSAT TDMA/DSI burst time plan development. *International Journal of Satellite Communications*, 3(1–2):35–43, 1985.

[23] J. L. Lewandowski, J. W. S. Liu, and C. L. Liu. SS/TDMA time slot assignment with restricted switching modes. *IEEE Transactions on Communications*, 31(1):149–154, January 1983.

[24] K. Menger. Zur allgemeinen kurventheorie. *Fundamenta Mathematicae*, pages 96–115, 1927.

[25] M. Minoux and C. Brouder. Models and algorithms for optimal traffic assignment in SS/TDMA switching systems. *International Journal of Satellite Communications*, 5(1):33–47, 1987.

[26] T. Mizuike, Y. Ito, D. J. Kennedy, and L. N. Nguyen. Burst scheduling algorithms for SS/TDMA systems. *IEEE Transactions on Communications*, 39(4):533–539, April 1991.

[27] T. Mizuike, Y. Ito, L. N. Nguyen, and E. Maeda. Computer-aided planning of SS/TDMA network operation. *IEEE Journal on Selected Areas in Communications*, 9(1):37–47, January 1991.

[28] R. Ramaswamy and P. Dhar. Comments on "An efficient SS/TDMA time slot assignment algorithm." *IEEE Transactions on Communications*, 32(9):1061–1065, September 1984.

[29] A. K. Sinha. A model for TDMA burst assignment and scheduling. *COMSAT Technical Review*, 6:219–251, November 1976.

[30] Y. K. Tham. Burst assignment for satellite-switched and Earth-station frequency-hopping TDMA networks. *Communications, Speech and Vision, IEE Proceedings I*, 137(4):247–255, August 1990.

[31] Y. K. Tham. On fast algorithms for TDM switching assignments in terrestrial and satellite networks. *IEEE Transactions on Communications*, 43(8):2399–2404, August 1995.

[32] B. Toft. Coloring, stable sets and perfect graphs. In R. L. Graham, M. Grötschel, and L. Lovász, editors, *Handbook of combinatorics*, volume 1, chapter 4, pages 233–288. North Holland, Amsterdam, 1995.

[33] K. L. Yeung. Efficient time slot assignment algorithms for TDM hierarchical and nonhierarchical switching systems. *IEEE Transactions on Communications*, 49(2):351–359, February 2001.

Part II

Combinatorial Optimization in *Optical and Interconnection Networks*

Chapter 13

Optimization Techniques for Survivable Optical Networks

Abdelhamid E. Eshoul
School of Information Technology and Engineering SITE
University of Ottawa, ON, Canada
E-mail: `aeshoul@site.uottawa.ca`

H. T. Mouftah, Fellow, IEEE
School of Information Technology and Engineering SITE
University of Ottawa, ON, Canada
E-mail: `mouftah@site.uottawa.ca`

1 Introduction

In order to exploit the existing fiber's huge bandwidth, Wavelength Division Multiplexing (WDM) introduces the possibility of concurrency among multiple-user transmissions within network architectures. WDM is a technique by which a number of optical signals, each using a unique wavelength, are transmitted in a single optical fiber. In this manner, it is possible to use the huge capacity of the fiber optics efficiently by multiplexing signals from different end-users, each using a different WDM channel, into a single fiber, while at the same time allowing end-user's equipment to operate at their current electronic rates, say 10 Gb/s.

In wavelength-routed WDM networks, for a source-destination (s-d) pair to communicate, a lightpath in the optical layer between the two nodes must be established. A lightpath is a unidirectional connection between two end nodes (source and destination) that may span multiple links and use single or multiple wavelengths. Lightpath establishment, also known as Routing and Wavelength Assignment (RWA), is accomplished by selecting a route between the two end nodes and assigning a suitable wavelength. The aim of the RWA process is to find routes and assign wavelengths for connection requests in a way that minimizes the consumption of network resources, while at the same time ensuring that no two lightpaths are assigned the same wavelength on a shared link. Furthermore, if a network lacks wavelength converters, a lightpath must be assigned the same wavelength on all the links in its path, a constraint known as wavelength continuity constraint.

The traffic applied to wavelength-routed WDM networks is mainly confined to two types: static traffic and dynamic traffic. Numerous research studies have investigated the RWA problem under the two different traffic environments. Under the static traffic environment, all connection requests are known in advance, so the typical objective is to set up all the required lightpaths while at the same time minimizing the number of wavelengths needed. On the other hand, under the dynamic traffic environment, connection requests arrive at and depart from the networks at random times; so the objective of the dynamic RWA algorithm is to minimize the blocking rate of connection requests. Throughout this chapter, we focus on solving the problem of RWA under a static traffic environment.

A node failure or fiber cut in a wavelength-routed WDM network can cause the breakdown of all lightpaths that traverse the failed node or broken link. Due to the huge amount of data that can be lost and the large number of users that can be disrupted as a result of a fiber cut or node failure, network survivability has become a key issue during the RWA process. Network survivability requires the protection of lightpaths against failures by reserving a spare bandwidth during connection setup and restoration during which the spare bandwidth is utilized upon the occurrence of a failure. During connection setup, in addition to setting up a working lightpath (primary lightpath) to carry traffic during the normal operation, a backup lightpath is also set up to carry traffic in case the primary lightpath fails. The working lightpath and the backup lightpath must be link-disjoint in order to protect against fiber cut or node-disjoint in order to protect against node failure.

Based on the rerouting choice, protection schemes can be either link-based, where the traffic is rerouted around the end nodes of the failed link or path-based where a backup lightpath is pre-determined between the source and the destination nodes. Furthermore, protection schemes are also classified based on the possibility of resource sharing as dedicated protection or shared protection. Dedicated protection has fast restoration time at the expense of higher resource redundancy. On the other hand, shared protection significantly reduces resource redundancy on the expense of increased restoration time.

Research on survivable RWA in WDM wavelength-routed mesh networks has come a long way over the past few years, evidenced by the recent publications on the topic [1][2][3]. Researchers in [1] solved the RWA problem for static traffic by formulating it as an integer linear programming problem (ILP). The objective is to minimize the number of wavelengths required to establish a given set of connection requests. This formulation unnecessarily used the variable λ_{sdw} to denote the traffic from any source s to any destination d on wavelength w. As shown in Section 2, only the flow variable that represents the flow in link from request k on wavelength is required to formulate the RWA problem under static traffic.

To protect from single failure, numerous protection and restoration schemes have been developed. In [4], researchers consider solving the RWA problem for shared protection by formulating it as ILP. The idea is to partition the problem into two subproblems: the routing subproblem and the wavelength assignment subproblem. Although partitioning the problem reduces the complexity of the problem by reducing the number of variables required to formulate the problem, it does not always guarantee the optimum solution. For example, the solution obtained by partitioning the problem may require a larger number of wavelengths than the optimum solution to satisfy a given demand matrix. Although partitioning the problem reduces the number of variables required to formulate the problem, it does not always guarantee the optimum selection. For example, the solution obtained by partitioning the problem may require a larger number of wavelengths than does the optimum solution to satisfy a certain demand matrix. The main reason is the reduced search space which restricts lightpaths to pass through a limited subset of links as dictated by the routing tables. As a result, during the optimization process, some of the possible routes are not considered as part of the search space. Therefore, if the excluded routes are part of the optimum solution, the generated solution will not be optimum. Furthermore, the way that the routes are determined greatly affects the complexity of the problem as well as the optimality of the solution. Furthermore, the work in [4] did not show how the alternate routes are determined as they greatly affect the complexity of the problem as well as the optimality of the solution. Also, the formulation did not address the wavelength continuity constraint.

In order to make a network survivable against duct failure, researchers in [5] solve the RWA problem in a WDM mesh network under duct-layer constraints for different path protection schemes. Duct-layer constraint implies that all links that are buried in the same duct under the ground belong to the same Shared Risk Link Group (SRLG). In addition to formulating the RWA problem as ILP, this research proposes a three-stage heuristic to solve the problem. In the first stage, the heuristic computes two duct-disjoint routes for each source-destination pair and in the second stage, it assigns a wavelength to each path. In the final stage, the heuristic performs an iterative optimization by rerouting some of the paths. Similar to [4], researchers in [6] solve the problem of the RWA problem for both dedicated protection and shared protection schemes by partitioning it into two subproblems. However, it is not clear how the wavelength continuity constraint is satisfied. Furthermore, the paper did not show how the routing tables are generated.

Researchers in [7] also formulated the RWA problem as ILP where an integer optimal solution can be obtained in most cases of interest by solving the corresponding relaxed linear programming model using efficient commercial or special-purpose simplex methods with fast running times. This result motivated us to consider solving the RWA jointly without having to partition it into smaller subproblems. Formulating the RWA problem jointly as one problem guarantees an optimal solution (assuming there is a feasible solution). Formulating the RWA for a shared protection scheme is a lot more complicated than it is for a dedicated protection scheme. The complication is mainly due to the resource-sharing constraint. Resource sharing implies that the cost of choosing a link for the protection path of a request depends on the physical location of its working path. Moreover, a wavelength in a link may protect many different lightpaths given that they do not share a common link in their corresponding working paths.

In order to solve the RWA for shared protection, researchers used the two-step approach to generate the two disjoint paths. In this approach, the working path is calculated first using the shortest-path algorithm. Using the updated resource-sharing status calculated based on the location of the working path, the protection path is then derived. However, in some network topologies, although the two disjoint routes are feasible, the two-step approach cannot find them. Furthermore, even for those cases where the two disjoint routes are selected using the two-step approach, they may not be the most optimum pair (in terms of the minimum total bandwidth allocated). As a result, the studies in [3][8][9][10][11] inspect the k-shortest path between each s-d pair and for each case the backup path is derived. Out of the k choices, the least cost of working and shared backup paths is selected. It is worth noting that the modified two-step approach can also be used under dynamic traffic environment.

To improve on the restoration time, researchers in [12][13] propose to divide the working path into a number of segments and find the backup path for each segment separately. Although the idea may reduce the restoration time, it certainly increases the computational complexity of the backup paths and the signaling overhead. Furthermore, it imposes the wavelength continuity constraint between the working path and all the backup paths for the different working segments. In this chapter, we propose an optimization technique that gives the optimum solution to the RWA problem by considering the routing subproblem and wavelength assignment subproblem jointly. In Section 2, we describe how multi commodity flow techniques can be applied to formulate the RWA as an LP problem. In Section 3, we formulate some practical network optimization problems such as RWA in survivable optical networks. In Section 4, we describe the problem complexity and in Section 5, we present a sample of the results obtained by applying the formulation to some network topologies. In Section 6, we comment on how an integer solution is often possible using faster LP solvers and we end the chapter by some concluding remarks in Section 7.

2 Related Theory

Optimization is the art of allocating scarce resources to the best possible effect. In network optimization problems, it is required to optimize a certain objective function (minimize cost or maximize output) subject to some constraints. It requires finding the optimum solutions (for example, the least-cost route) out of all feasible solution that satisfies the constraint equations. In Linear Programming (LP) optimization problems, both the objective function and the constraint equations are linear. All network optimization problem of interest are linear optimization problems. However, most of them have an added constraint that requires all variables to take integer values (individual flow cannot be split). This type of optimization problem is known as integer linear programming problems. So the solution to the ILP may not be the optimum solution because the optimum solution may contain variables with real values (non integer values). Therefore, the solution of an ILP problem requires the search for the most optimum integer combination out of all feasible integer combinations that the variables can take. Consequently, solving an ILP problem can take considerable time depending on the number of variables in the problem. In most network optimization problems, it is required to minimize the consumption of network resources such as bandwidth subject to capacity and demand constraints; for example, minimizing the number of wavelengths per fiber required to establish a given demand matrix in an optical network. As shown in the next section, optimization techniques can also be used to find the shortest path and the k-shortest link-disjoint paths in a network.

To formulate the optimization problem for a network, we use the multi commodity flow approach. A multi commodity network flow problem implies the flow of several different commodities (connection requests) in various links of the network simultaneously. These flows are subject to some constraints such as link capacity, flow conservation, etc. A multi-commodity network flow problem is defined on a directed graph $G(N, E)$, where N and E are the number of nodes and the number of directed edges in the network, respectively (links and edges are used interchangeably throughout the chapter). It is important to clarify that link L_{ij} represents the directed link connecting node i to node j whereas L_{ji} represents the directed link connecting node j to node i, so that L_{ij} is used to transfer flow from node i to node j and L_{ji} is used to transfer flow from node j to node i.

In the context of connection-oriented communications networks, different commodities correspond to different connection requests (demands). For the general circuit-switched networks or optical networks with full wavelength conversion, we only need to distinguish between the flows of different connection requests inside the network. For example, F_{ijk} represents the amount of flow from request k in L_{ij}, whereas F_{ijn} represents the amount of flow from request n in L_{ij}. It requires the use of different symbols to denote the flow of each connection request on each link. However, in the more complicated optical networks with wavelength continuity constraints, we also need to distinguish between the flows of the same connection request in the same link on different wavelengths. For example, F_{ijwk} represents the amount of flow from lightpath k on wavelength w, in L_{ij}, whereas, F_{ijck} represents the amount of flow from lightpath k on wavelength c, in L_{ij}. Although the flow belongs to the same connection request and it flows through the same link, different symbols are used to distinguish between the same flows on different wavelengths to ensure the wavelength continuity constraint. Therefore, every lightpath is considered aa a different flow, and every s-d pair may request as integer multiple number of lightpaths. As a result, we use a different symbol to represent each lightpath on each wavelength in each link.

Each link in the network has an associated cost of transferring one unit of flow through it. The cost can vary from one link to another; however, it is sometimes (especially in an optical network) set to unity. Also given is the maximum capacity on each link in the network. Although the link capacity can vary from one link to another, it is assumed constant (W) throughout the chapter for simplicity . Finally we need to know the demand matrix. A double dimensional array ($NbyN$) is used to represent all the demands in the network where Λ_{ij} denotes the number of connections (number of lightpaths) that are required from source i to destination j in the network.

As this chapter shows, unlike other formulations proposed in the literature, the only unknown variables needed to formulate network optimization problems are the flow variables. Moreover, the problem complexity is a function of the number of variables used in the formulation. As the number of variables increases, the complexity of the problem increases and it gets more difficult to formulate and consequently it takes more time to solve.

The most general form of a network optimization problem takes the following form

Minimize the total flow in the network which is mathematically written as

$$\text{Minimize} \sum_{ij \in E} C_{ij} F_{ij}$$

where F_{ij} is the total flow in L_{ij} and C_{ij} is the cost of transfering one unit of flow along L_{ij}

Subject to the following constraints:

1. Flow conservation constraints which imply the total flow into a network node equal to the total flow out of it.

2. Link capacity constraint which is mathematically written as $F_{ij} \leq W$.

To obtain the optimum solution, first we allow all connection requests (all different commodities) to flow in all the links in the network. This is done by assuming a flow of each connection request in each link which explains the need for a different symbol to represent each connection request in each link. Then we minimize the total flow (cost) that satisfies the specific constraints which the problem imposes such as flow conservation and capacity constraints.

3 Problem Formulation

In this section, we show how practical problems such as RWA in survivable optical networks can be formulated as LP problems. We start with simple problems such as finding the shortest route between any s-d pair and gradually build up to more complicated problems such as formulating the RWA for dedicated protection in wavelength-routed optical networks.

The following notations are used in all formulations
Network $G(N, E)$
where N = number of nodes in the network
E = number of directed links in the network
Λ_{ij} = units of flow sent from source node i to destination node j
W = link capacity (assumed equal in all links)
C_{ij} = cost of sending 1 unit of flow along link L_{ij}

3.1 Shortest K-Link-Disjoint Routes

The following notation is only used for this problem.
F_{ij} = units of flow along link L_{ij}.

In order to find the shortest route or the shortest k-link-disjoint routes between two nodes in a network, it is sufficient to use the single commodity flow technique. Because only one commodity flows in the network, one symbol is enough to represent the units of flow in each link. The following example shows how to formulate an optimization problem to find the shortest route between any two nodes in the network.

Example: for the network shown in Figure 1, show how to formulate an optimization problem to find the shortest route between node 1 and node 3.

In order to find the shortest route between any $s - d$ pair, only one unit of flow is allowed to flow through the network subject to the flow conservation constraint. This unit of flow enters the network from the

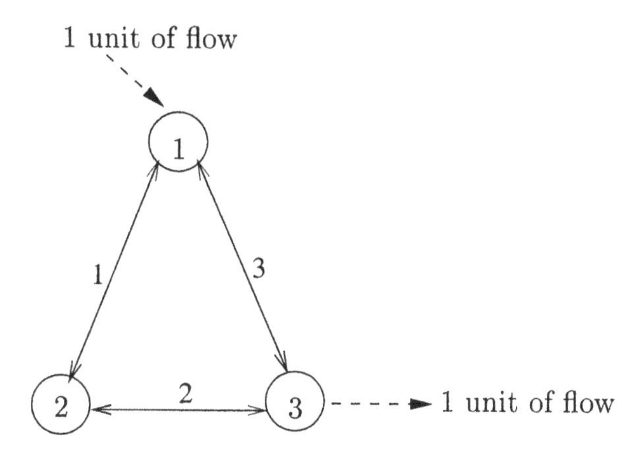

Figure 1: Single commodity flow in a 3-node network.

source node s and leaves the network from the destination node d. The link capacity must be at least one unit so that the unit of flow can take any link. However, for simplicity, we set it equal to unity. The unit of flow is then allowed to flow in all links of the network and the shortest route (least cost route) between node 1 and node 3 is found by selecting the least cost links possible between the required s-d pair. This implies minimizing the total cost required to send one unit of flow from the source node to the destination node subject to the flow conservation constraints. In the solution the variable F_{ij} indicates if the unit of flow passes through L_{ij}. Therefore, $F_{ij} = 1$ if there is a flow along the directed link L_{ij}; 0 Otherwise.

The formulation of the problem takes the following form

1. Objective function
   The general form of the objective function is:

$$\text{Minimize} \sum_{ij \in E} C_{ij} F_{ij}$$

   The objective function for the above example can simply be written as Minimize $F_{12} + 3F_{13} + F_{21} + 2F_{23} + 3F_{31} + 2F_{32}$

2. Constraints:

(a) Flow conservation (supply/demand)

$$\sum_{j \in neighbors\ of\ i} F_{ij} - F_{ji} = \begin{cases} 1, & \text{if } i = \text{source node} \\ -1, & \text{if } j = \text{destination node} \\ 0, & \text{otherwise} \end{cases} \tag{1}$$

For the above example, where node 1 is the source and node 3 is the destination, the flow conservation constraint equations are as follows:

$$\begin{array}{lcll} F_{12} + F_{13} - F_{21} - F_{31} & = & 1 & \text{(source node)} \\ F_{21} + F_{23} - F_{12} - F_{32} & = & 0 & \text{(inter-mediate node)} \\ F_{31} + F_{32} - F_{13} - F_{23} & = & -1 & \text{(destination node)} \end{array}$$

(b) Capacity
Capacity constraint ensures that the total flow in a link is no greater than the link capacity

$$F_{ij} \leq W \quad \forall (ij \in E)$$

The capacity constraint equations for the above example are:

$$F_{12} \leq 1,\ F_{13} \leq 1,\ F_{21} \leq 1,\ F_{23} \leq 1,\ F_{31} \leq 1,\ F_{32} \leq 1$$

(c) Non-negativity $F_{ij} \geq 0\ \forall ij \in E$

$$F_{12} \geq 0,\ F_{13} \geq 0,\ F_{21} \geq 0,\ F_{23} \geq 0,\ F_{31} \geq 0,\ F_{32} \geq 0$$

To find the K-shortest routes between node i and node j, both nodes must have at least K neighbors. The same formulation as for the shortest route can be applied with a slight modification to the flow conservation constraint equations, where the right-hand sides of the source and destination node are changed to K and $-K$, respectively. Furthermore, the capacity constraint for each link must be restricted to one unit to ensure that the k routes are link-disjoint.

3.2 Routing in Circuit-Switched Networks

In circuit-switched networks, we consider a static traffic environment, where a number of connection demands between various s-d pairs are known in advance, and the objective is to establish circuits to satisfy these requests using minimum network resource (minimum bandwidth). To formulate this problem, we use multi commodity flow formulation, where each demand is considered as a different commodity. As explained in Section 2, we assume that there is a different flow representing each demand passing through each link. So it is essential to denote the flow belonging to different demands by different symbols (different variables) to distinguish among them. After the problem is solved, all the flows that are part of the solution are assigned positive values whereas the rest are assigned zero.

The following notations are used
F_{ijk} = units of bandwidth assigned to s-d pair k along link L_{ij}
K = number of s-d pairs
Λ_{ij} = units of bandwidth requested from node i to node j.
W = Link bandwidth (assumed equal in all links)

Figure 2 shows a typical example where Λ_{13} units of bandwidth are requested from node 1 to node 3 and Λ_{21} units of bandwidth are requested from node 2 to node 1.

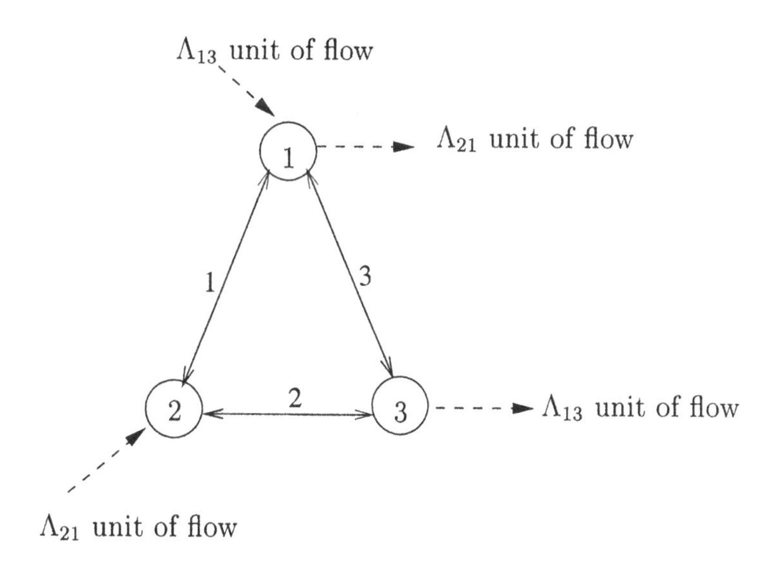

Figure 2: Multi commodity flow in a 3-node network

For the above example, the problem formulation takes the following form

1. Objective function
   The objective function can be written in a general form as

$$\text{Minimize} \sum_{ij \in E} \sum_{k=1}^{K} C_{ij} F_{ijk}$$

   Alternatively, the objective function can be written as
   Minimize $F_{121} + F_{122} + 3F_{131} + 3F_{132} + F_{211} + F_{212} + 2F_{231} + 2F_{232} + 3F_{311} + 3F_{312} + 2F_{321} + 2F_{322}$

2. Constraints

   (a) Flow conservation (supply/demand)

$$\sum_{\forall (j \in neighbors\ of\ i) F_{ijk} - F_{jik}} = \begin{cases} \Lambda_{ij}, & \text{if } i = \text{source node} \\ -\Lambda_{ij}, & \text{if } j = \text{destination node} \\ 0, & \text{otherwise.} \end{cases}$$

   Alternatively, the flow conservation constraint equations can be written as:

$$\begin{aligned}
F_{121} + F_{131} - F_{211} - F_{311} &= \Lambda_{13} && \text{(request 1 source node)} \\
F_{211} + F_{231} - F_{121} - F_{321} &= 0 && \text{(request 1 node 2)} \\
F_{311} + F_{321} - F_{131} - F_{231} &= -\Lambda_{13} && \text{(request 1 destination node)} \\
F_{122} + F_{132} - F_{212} - F_{312} &= -\Lambda_{21} && \text{(request 2 destination node)} \\
F_{212} + F_{232} - F_{122} - F_{132} &= \Lambda_{21} && \text{(request 2 source node)} \\
F_{312} + F_{322} - F_{132} - F_{322} &= 0 && \text{(request 2 node 3)}
\end{aligned}$$

   (b) Capacity

$$\sum_{k=1}^{K} F_{ijk} \leq W \quad \forall (ij \in E)$$

   Alternatively, the capacity constraint equations for the network shown in Figure 2 can be written as

$$\begin{aligned}
F_{121} + F_{122} &\leq W & L_{12} \\
F_{211} + F_{212} &\leq W & L_{21} \\
F_{131} + F_{132} &\leq W & L_{13} \\
F_{311} + F_{312} &\leq W & L_{31} \\
F_{231} + F_{232} &\leq W & L_{23} \\
F_{321} + F_{322} &\leq W & L_{32}
\end{aligned}$$

(c) Non negativity
All variables (flows) cannot be negative; i.e.;
$F_{ijk} \geq 0 \;\; \forall(ij \in E)$ and $\forall(k \in K)$

3.3 RWA Without Protection

The aim of the RWA process in WDM networks is to find routes and assign wavelengths to connection demands (lightpaths) in a way that minimizes the consumption of network resources, while at the same time ensuring that no two lightpaths are assigned the same wavelength on a shared link. Furthermore, under the wavelength continuity constraint, a lightpath must be assigned the same wavelength on all the links in its path. This added constraint requires the distinctions between flows belonging to different wavelengths as well as flows belonging to different requests in a link. As a result, different variables are used to distinguish between the same flow (commodity or lightpath). Use the same network shown in Figure 2, where Λ_{13} lightpaths are requested from node 1 to node 3 and Λ_{21} lightpaths are requested from node 2 to node 1. Assume that each lightpath requires the full capacity of a wavelength and that $W = 2$, $K = 2$. In this formulation:
$F_{ijwk} = 1$ if wavelength w is assigned on L_{ij} for s-d pair k, 0 otherwise
K = number of s-d pairs
Λ_{ij} = number of lightpaths required between node i and node j
W = number of wavelengths available in each link

The problem of RWA under wavelength continuity constraints takes the following form

1. Objective function
The objective function can be written in a general form as:

$$\text{Minimize} \sum_{\forall(ij \in E)} \sum_{w=1}^{W} \sum_{k=1}^{K} C_{ij} F_{ijwk}$$

Alternatively, the objective function can be written as:
$Minimize \;\; F_{1211} + F_{1212} + F_{1221} + F_{1222} + 3F_{1311} + 3F_{1312} + 3F_{1321} + 3F_{1322} + F_{2111} + F_{2112} + F_{2121} + F_{2122} + 2F_{2311} + 2F_{2312} + 2F_{2321} + 2F_{2322} + 3F_{3111} + 3F_{3112} + 3F_{3121} + 3F_{3122} + 2F_{3211} + 2F_{3212} + 2F_{3221} + 2F_{3222}$

2. Constraints

   (a) Flow conservation for s-d nodes

$$\sum_{j \in (neighbors\ of\ i)} \sum_{w=1}^{W} \sum_{k=1}^{K} F_{ijwk} - F_{jiwk} = \begin{cases} \Lambda_{ij} & \text{if } i = \text{s} \\ -\Lambda_{ij} & \text{if } j = \text{d} \end{cases}$$
$$\forall (i \in s - d\ pairs)$$

The flow conservation equations for node 1 are displayed below as an example

(Node 1 source node to request 1)

$$F_{1211} + F_{1221} + F_{1311} + F_{1321} - F_{2111} - F_{2121} - F_{3111} - F_{3121} = \Lambda_{13}$$

(Node 1 destination node to request 2)

$$F_{1212} + F_{1222} + F_{1312} + F_{1322} - F_{2112} - F_{2122} - F_{3112} - F_{3122} = -\Lambda_{21}$$

   (b) Wavelength continuity constraint

All published literature we have encountered did not explain how the wavelength continuity constraint can be ensured. Furthermore, although some of it assumes the wavelength continuity constraint, the formulations do not achieve it. Other formulations come up with some vague equations that are difficult to understand. This motivated us to explain in depth how the wavelength continuity constraint is ensured when the RWA is formulated as an LP problem. The reader may have noticed that the flow conservation constraints for the intermediate nodes are not included in a. Intermediate nodes in a lightpath are part of the route and are neither source nor destination nodes. Therefore, for each lightpath, the net outflow in an intermediate node is zero. However, to ensure the wavelength continuity constraint for a lightpath on a specific wavelength, the net outflow of that lightpath on that specific wavelength in each intermediate node along the route must be zero. This ensures that if a lightpath in a wavelength enters an intermediate node from a link, the same lightpath must exit the node on the same wavelength on a different link. As a result, in each node the number of wavelength continuity constraint equations is equal to the product of the number of wavelengths and the number of s-d pairs for which that node is an intermediate node. In general, the wavelength continuity constraint is expressed as follows

$$\sum_{j \in (neighbor\ of\ i)} F_{ijwk} - F_{jiwk} = 0 \begin{cases} \forall (i \in N) \text{ and } \neq s-d \text{ of } k \\ \forall W \\ \forall K \end{cases}$$

The wavelength continuity constraint equations for the above example is expressed as follows:
For both demands (s-d pairs), there is only one possible intermediate node. For demand 1 (1-3 pair), node 2 is a possible intermediate node and so the wavelength continuity constraint equations are:

$$\begin{aligned} F_{2111} + F_{2311} - F_{1211} - F_{3211} &= 0 \quad \text{..........}(wavelength\ \lambda_1) \\ F_{2121} + F_{2321} + F_{1211} + F_{3221} &= 0 \quad \text{..........}(wavelength\ \lambda_2) \end{aligned}$$

For demand 2 (2-1 pair), node 3 is a possible intermediate node and so the wavelength continuity constraint equations are:

$$\begin{aligned} F_{3112} + F_{3212} - F_{1312} - F_{2312} &= 0 \quad \text{..........}(wavelength\ \lambda_1) \\ F_{3122} + F_{3222} + F_{1322} + F_{2322} &= 0 \quad \text{..........}(wavelength\ \lambda_2) \end{aligned}$$

(c) Capacity
Each wavelength in any directed link can carry a maximum of one lightpath

$$\sum_{k=1}^{K} F_{ijwk} \leq 1 \ \ \forall (ij \in E) \ and \ \forall W$$

The above capacity constraint also ensures that the total number of wavelengths available in each link is W. As an example, the capacity constraint equations for link 1-2 are

$$\begin{aligned} F_{1211} + F_{1212} \leq 1 \\ F_{1221} + F_{1222} \leq 1 \end{aligned}$$

(d) Non negativity
The non negativity constraint is expressed as shown previously.

3.4 RWA in Dedicated Protection scheme

This formulation only addresses survivability against link failure such as a fiber cut. In the dedicated protection formulations, two different lightpaths (two commodities) for each unit of flow must be set up. One flow represents the working path and the other flow represents the backup path. Similarly, the objective is to minimize the total flow in the network for both working and backup paths. However, an added constraint is required to ensure that the working path and the backup path belonging to the same flow do not pass through the same bidirectional link. To easily formulate the problem, each s-d demand is divided into a integer number of lightpaths and each individual lightpath is considered separately. It is also assumed that each lightpath takes the full bandwidth of a wavelength.

In this formulation, the following notations are also used
$F_{ijwn} = 1$ if wavelength w is used on L_{ij} in the working path for the nth lightpath , 0 otherwise
$B_{ijwn} = 1$ if wavelength w is used on L_{ij} in the backup path for lightpath n, 0 otherwise
Λ_{ij} = number of lightpaths requeseted from node i to node j
W = number of wavelengths on each link
R =total number of lightpaths in the network

$$R = \sum_{i=1}^{N}\sum_{j=1}^{N}\Lambda_{ij}$$

Problem formulations

1. Objective function
   The objective function can be written in a general form as

$$\text{Minimize} \sum_{ij\in E}\sum_{w=1}^{W}\sum_{r=1}^{R} C_{ij}(F_{ijwk} + B_{ijwk})$$

2. Constraints

(a) Flow conservation for s-d nodes

i. Working paths

$$\sum_{j \in (neighbor\ of\ i)} \sum_{w=1}^{W} \sum_{r=1}^{R} F_{ijwk} - F_{jiwk} = \begin{cases} 1 & \text{if } i = \text{s} \\ -1 & \text{if } j = \text{d} \end{cases}$$

$\forall(i \in$ s-d pairs)

ii. Backup path

$$\sum_{j \in (neighbor\ of\ i)} \sum_{w=1}^{W} \sum_{r=1}^{R} B_{ijwk} - B_{jiwk} = \begin{cases} 1 & \text{if } i = \text{s} \\ -1 & \text{if } j = \text{d} \end{cases}$$

(W)

(b) Wavelength continuity constraint

i. Working paths
$\sum F_{ijwk} - F_{jiwk} = 0 \quad \forall((i \in N)$ and $i \neq$ s or d of $k)$
$\forall(w)$ and $\forall(k)$

ii. Backup path
$\sum B_{ijwk} - B_{jiwk} = 0 \quad \forall((i \in N)$ and $i \neq$ s or d of $k)$
$\forall(w)$ and $\forall(k)$

(c) Capacity
A wavelength on a link can only be assigned to a working path or a backup path.
$\sum F_{ijwr} + B_{ijwr} \leq 1 \ \ \forall w \in W \ \ and \ \ \forall ij \in E$

(d) Link-disjoint
Working path and backup path of the same lightpath cannot use the same bidirectional link
$\sum F_{ijwr} + F_{jiwr} + B_{ijwr} + B_{jiwr} \leq 1$
$\forall(r \in R)$ and $\forall(bidirectional\ links)$

A more efficient way is to consider the working lightpath and the protection lightpath as one commodity instead of two different commodities.

In this manner, the number of variables required to formulate the dedicated protection problem can be reduced by half. The idea is to consider each lightpath as two units of flow. An added constraint is required to ensure that a maximum of one unit of flow belonging to the same lightpath is allowed to pass through any bidirectional link.

To further reduce the number of variables required to the formulate the RWA problem, a number of researchers partitioned the problem into two smaller subproblems: the routing subproblem and the wavelength assignment sub-problem. Although partitioning the problem reduces the number of variables required to formulate the problem, it does not always guarantee the optimum selection. For example, the solution obtained by partitioning the problem may require a larger number of wavelengths than does the optimum solution to satisfy a certain demand matrix. The main reason is the reduced search space that restricts lightpaths to pass through a limited subset of links as dictated by the routing tables. As a result, during the optimization process, some of the possible routes are not considered as part of the search space. Therefore, if the excluded routes are part of the optimum solution, the generated solution will not be optimum. Furthermore, the way that the routes are determined greatly affects the complexity of the problem as well as the optimality of the solution.

3.5 RWA in Shared Protection Scheme

The LP formulation for the RWA problem with a shared protection scheme is much more complicated than that of the dedicated protection scheme. The difficulties lie in the fact that any working paths under the same risk of a link failure (such as a fiber cut) must not share the same resources in their corresponding backup paths. For example, to formulate this constraint, consider any two working lightpaths out of the many possible different combinations; say lightpath x and y; the following constraint is required:

If $\sum F_{ijwx} + F_{jiwx} + F_{ijwy} + F_{jiwy} \geq 2$

$\forall$(possible combinations of x and y) and $\forall(ij \in E)$

Then

$B_{ijwx} + B_{ijwy} \leq 1$

It can be seen that as the number of links and the number of lightpaths in the network increase, the number of constraint equations increases considerably due to the different combinations of lightpaths that can share a link in their working paths. Furthermore, the cost of the backup path depends on the location of its corresponding working path. For example, if the backup path of a working path is chosen such that it does not consume any new network resources (due to the resource sharing), this backup path has a cost of zero. To overcome these complications, a number of researchers divided the RWA into two subproblems: the routing subproblem and wavelength assignment subproblem. However, even dividing the problem into small sub-problems does not make it easy to formulate and solve. Most researchers solve the problem using heuristic algorithms such as the modified iterative two-step algorithm proposed in [7].

4 Problem Complexity

The problem complexity is a function of the number of variables used in the formulation. As the number of variables increases, the complexity of the problem increases and it gets more difficult and consequently it takes more time to solve. The big O notation of the problem complexity depends on the algorithm used to solve the problem. However, solving the above problems using fast LP solvers makes all of the problems tractable. On the other hand, using an ILP solver is only adequate for small-sized problems due to the exponential complexity of the problem. The number of variables used to formulate the problems in Section 3 is given in Table 1.

From Table 4, it can be seen that in the case of circuit-switched networks, the number of variables used to formulate the problem increases with the number of links in the network and the number of s-d pairs. So the upper bound on the number of variables required in any network topology with the maximum number of s-d pairs is calculated as

Table 1: Number of variables required to formulate a problem

Problem	Number of variables
Shortest k-link-disjoint routes	E
Circuit switched network	$E \times K$
RWA (no protection)	$E \times k \times W$
RWA with dedicated protection	$2 \times E \times W \times R$

Upper bound = $E \times N(N-1)$
On the other hand, for the WDM wavelength-routed networks with wavelength continuity constraint and no protection, the number of variables is greater than that of the circuit-switched networks by a factor of W. The number of variables required to formulate the RWA with a dedicated protection scheme is equal to twice the product of the number of links, the total number of lightpaths, and the number of wavelengths.

5 Results

A computer program (written in Java) has been developed to formulate the various connection-oriented network problems described in Section 3. The program can also be used to solve small-sized problems using the simplex method. However, for larger problems, it is recommended to use more efficient and faster LP and ILP solvers such as CPLEX. In all four cases, the user is only required to provide the input data such as network topology and demand matrix. For problems with a large number of variables, our program can formulate the ILP problem in a form accepted by the CPLEX solver. The program is freely available to researchers at a request from the author. Samples of some of the results are also shown below. To thoroughly test our approach, we used different network topologies. To save space, we only show two of them: the NSF network with 14 nodes and 21 bidirectional links shown in Figure 3, and another network with 6 nodes and 10 bidirectional links shown in Figure 4. In both topologies all links have a constant cost of one unit. The demand matrix used for both cases is randomly selected.

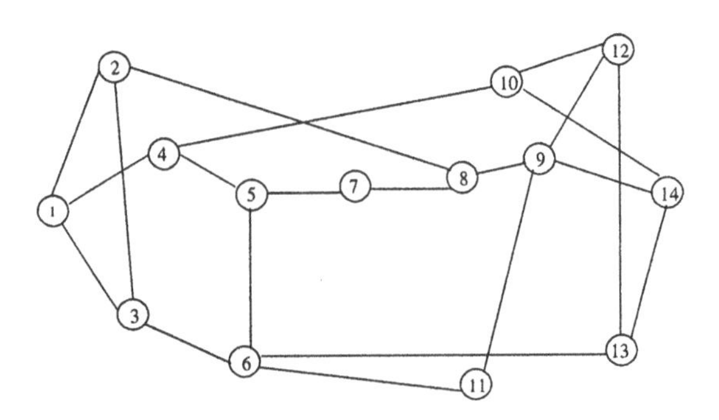

Figure 3: The NSF network with unity cost on each link.

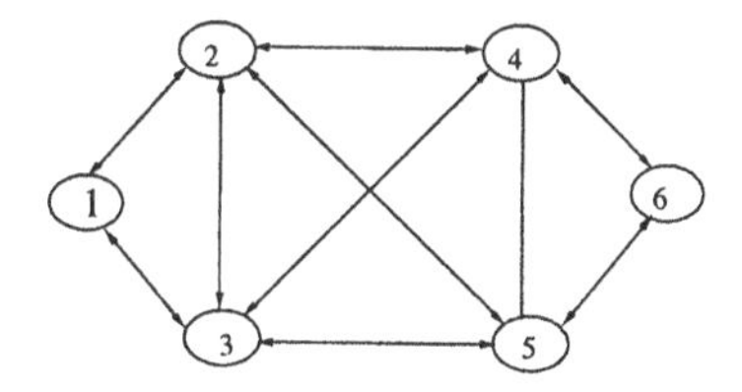

Figure 4: 6-node network with unity cost on each link

Table 2: Sample of the K-shortest link-disjoint routes for Figure 3

s-d	K	routes
1-2	3	1-2 1-3-2 1-4-5-7-8-2
5-14	3	5-4-10-14 5-6-13-14 5-7-8-9-14
9-6	4	9-11-6 9-12-13-6 9-8-2-3-6 9-14-10-4-5-6

Table 3: Sample of the K-shortest link-disjoint routes for Figure 4

s-d	K	routes
1-6	2	1-2-4-6 1-3-5-6
2-3	4	2-3 2-1-3 2-4-3 2-5-3
5-1	2	5-2-1 5-3-1
5-2	4	5-2 5-3-2 5-4-2 5-6-4-3-1-2

Table 2 shows a sample of the K shortest link-disjoint routes from the NSF network shown in Figure 3' and Table 3 shows a sample of the K shortest link-disjoint routes from the network shown in Figure 4. Table 4 shows the demand matrix and the optimum routing solution found by formulating the problem as an LP problem. The problem is to find the minimum bandwidth required to set up the given set of demands in a circuit-switched network for the topology shown in Figure 4, Where s-d denotes the source destination pair, BR denotes the units of bandwidth required by each s-d pair, route denotes the optimum route found by the solution and BA denotes the bandwidth assigned to each route. The optimum solution requires 3 units of bandwidth per link to set up the connection demands.

Table 4: routing in circuit switched network

s-d	BR	Route	BA
1-2	2	1-2	2
1-3	1	1-3	1
1-4	1	1-2-4	1
1-5	1	1-3-4-5	1
1-6	1	1-3-5-6	1
2-1	1	2-1	1
2-3	1	2-3	1
2-4	1	2-4	1
2-5	2	2-5	2
2-6	1	2-4-6	1
3-1	1	3-1'	1
3-2	1	3-2	1
3-4	1	3-4	1
3-5	3	3-5	2
		3-4-5	1
3-6	1	3-2-5-6	1
4-1	2	4-3-1	2
4-2	1	4-2	1
4-3	1	4-3	1
4-5	1	4-5	1
4-6	2	4-6	2
5-1	1	5-2-1	1
5-2	1	5-2	1
5-3	3	5-3	3
5-4	1	5-4	1
5-5	1	5-6	1
6-1	1	6-5-2-1	1
6-2	1	6-4-2	1
6-3	1	6-4-2-3	1
6-4	1	6-4	1
6-5	2	6-5	2

Table 5: Results of routing in circuit switched network

s-d	LR	Route	ω
1-2	2	1-2	2
		1-3-2	3
1-3	1	1-3	1
1-4	1	1-2-4	1
1-5	1	1-3-4-5	2
1-6	1	1-2-5-6	3
2-1	1	2-1	1
2-3	1	2-3	2
2-4	1	2-4	2
2-5	2	2-5	1
		2-4-5	3
2-6	1	2-5-6	2
3-1	1	3-1'	1
3-2	1	3-2	2
3-4	1	3-4	3
		3-5	1
3-5	3	3-5	2
		3-5	3
3-6	1	3-4-6	1
4-1	2	4-2-1	3
		4-3-1	2
4-2	1	4-2	2
4-3	1	4-3	1
4-5	1	4-5	1
4-6	2	4-6	2
		4-6	3
5-1	1	5-2-1	2
5-2	1	5-2	1
		5-3	1
5-3	3	5-3	2
		5-2-3	3
5-4	1	5-4	2
5-5	1	5-6	1
6-1	1	6-5-3-1	3
6-2	1	6-4-2	1
6-3	1	6-4-3	3
6-4	1	6-4	2
6-5	2	6-5	1
		6-5	2

Table 5 shows the demand matrix and the optimum routing solution found by formulating the problem as an LP problem. The problem is to find the minimum number of wavelengths necessary to set up the required lightpaths in a wavelength-routed network with the wavelength continuity constraint for the topology shown in Figure 4. Where s-d denotes the source-destination pair, LR denoted the number of lightpaths required by each s-d pair, route denotes the optimum route obtained by the solution and ω denotes the wavelength assigned to each lightpath. The optimum solution requires 3 wavelengths to set up the lightpaths demands.

6 Comments on the integerity

The reader may have noticed that the integrity constraint has been relaxed throughout the formulations. The reason for this has been that in many cases an integer-feasible optimum solution (when there is one) can still be found using the simpler linear programming solver. This result seems to agree with the result found in [6] even though the cost function used in our formulation is constant. The reason for obtaining an optimum integer solution using the linear programming solver is the fact that optimum solution of the problem happens to have integer values for all the variables. Another reason could be that in our formulation, all the variables used in the RWA problem are binary variables (take either 0 or 1). It is a well-known fact that the techniques used to solve linear programming problems such as the simplex method take advantage of the fact that the optimum feasible solution point must be a feasible corner point. Consequently, the solver checks the value of the objective function only at feasible corner points as it pivots from one feasible corner point to another. It eventually stops when the objective function is optimum at an optimum feasible corner point. Because the number of the feasible corner points in any LP problem is finite, obtaining the optimum solution using the LP solver is much faster than obtaining it using the ILP solver.

However, in some cases, although there is a feasible optimum integer solution, the solver may come up with a non integer optimum solution. The reason for this is that an LP problem can sometimes have more than one optimum solution. In fact, many RWA problems can have more than one integer optimum solution. The question of interest is how can an optimum integer solution (given there is one) always be obtained using an LP solver.

Gerald L. Thompson in [14] proposes the integral simplex algorithm in which initially only pivots on one are carried out until no more such pivots are possible. Pivoting on one ensures that the resulting entries for the simplex tableau are all integers. However, when no more pivoting on one is possible, the algorithm is said to have reached a local optimum point. In some cases, the local optimum is also a global optimum in which case the optimum integer solution is found. On the other hand, if the local optimum is found and the objective function can still be improved, another algorithm called a global integral simplex method is used to find the global integer optimum solution. In this research, we have developed a computer program to find the local optimum solution for the RWA problem we have formulated. We have found out that in many cases the local optimum solution is the optimum integer solution of the problem. We also think that further research is needed to explore this idea further to solve the RWA problem using simpler LP solvers.

7 Summary

In this chapter, we have presented a different approach to solve the problems of RWA in wavelength-routed survivable optical networks. This approach solves the problems of RWA jointly by formulating the problem as a linear programming problem. Although this approach increases the number of variables required to formulate the problem, it ensures obtaining the optimum solution. The study also found that many network optimization problems can be solved using an LP solvers. However, more research is needed to investigate the problem and find out ways to ensure integer solutions using LP solver when an optimum integer solution is feasible. The following problem formulation has been performed:

1. LP formulation to find the shortest k-link-disjoint routes in a network.

2. LP formulation to establish a given set of connection requests (static traffic) using minimum network resources for circuit-switched networks.

3. LP formulation to solve the routing and wavelength assignment problem for wavelength-routed WDM optical networks under wavelength continuity constraints.

4. LP formulation to establish a given set of connection requests (static traffic) using minimum network resources for a dedicated protection scheme in wavelength-routed WDM networks with a wavelength continuity constraint.

Acknowledgment

We are very grateful to Dr. Norman Rice from the department of mathematics and statistics at Queens University for sharing his knowledge and expertise on the two-phase Simplex Method to solve LP problems. We are also very grateful to Marwan Fayed and Tarek Saad from the optical laboratory. for sharing their knowledge and experience in Latex.

References

[1] Hui Zang, Jason P. Jue, and Biswanath Mukherjee. A review of routing and wavelength assignment approaches for wavelength-routed optical WDM networks. *SPIE Optical Networks Magazine*, Vol 1(No. 1), January 2000.

[2] Hui Zang, Canhui Ou, and B. Mukherjee. Path-protection routing and wavelength assignment (RWA) in WDM mesh networks under shared-risk-group constraints. In *Proceedings SPIE (APOC '2001)*, November 2001.

[3] X. Dahai, Q. Chunming, and X. Yizhi. An ultra-fast shared path protection scheme- distributed partial information management, part ii. In *Proceedings IEEE 10th International Conference on Network Protocols (ICNP'2002)*, pages 344–353, November 2002.

[4] L. Sahasrabuddhe, S. Ramamurthy, and Biswanath Mukherjee. Fault management in ip-over-WDM networks: WDM protection versus ip restoration. *IEEE Journal on Selected Areas In Communications*, Vol 20(No 1):21–33, January 2002.

[5] Hui Zang, Ou. Canhui, and B. Mukherjee. Path-protection routing and wavelength assignment RWA in WDM mesh networks under duct-layer constraints. *IEEE/ACM Transaction on Networking*, Vol 11(No 2):248 - 258, 2003.

[6] S. Ramamurthy, Laxman Sahasrabuddhe, and Biswanath Mukherjee. Survivable WDM mesh networks. *Lightwave Technology*, Vol 21(No. 4):870 - 883, April 2003.

[7] A. E. Ozdaglar and D. P. Bertsekas. Routing and wavelength assignment in optical networks. *IEEE/ACM Transaction on Networking*, Vol 11(No 2):259 - 272, April 2003.

[8] Pin-Han Ho, J. Tapolcai, and H. T. Mouftah. Issues on diverse routing for WDM mesh networks with survivability. In *Proceedings IEEE 10th International Conference on Computer Communications and Networks (ICCCN'2001)*, pages 61–66, October 2001.

[9] C. Xin, Y. Ye, S. Dixit, and C. Qiao. A joint lightpath routing approach in survivable optical networks. *Optical Network Magazine*, pages 23–32, May-June 2002.

[10] E. Bouillet, J. F. Labourtee, G. Ellina, R Ramamurthy, and S. Chaudhuri. Stochastic approaches to compute shared mesh restored lightpaths in optical networks architectures. In *Proceedings IEEE (INFOCOM'2002))*, volume 2, pages 801–807, June 2002.

[11] Pin-Han Ho, J. Tapolcai, and H. T. Mouftah. On achieving optimal survivable routing for shared protection in survivable next-generation internet. *IEEE Transactions on Reliability*, Vol 53(No 2):216–225, June 2004.

[12] Pin-Han Ho and H. T. Mouftah. A framework of service guaranteed shared protection for optical networks. *IEEE Communications Magazine*, Vol 40(No 2):97–103, February 2002.

[13] Pin-Han Ho, J. Tapolcai, and H. T. Mouftah. Allocation of protection domains in dynamic WDM mesh networks. In *Proceedings IEEE of the 10th International Conference on Network Protocols (ICNP'2002)*, pages 188–189, November 2002.

[14] G. L. Thompson. An integral simplex algorithm for solving combinational optimization problems. *Computational Optimization and Applications*, Vol 22(No 3):352–367, September 2002.

Chapter 14

WDM Switching Networks: Complexity and Constructions

Hung Q. Ngo
Computer Science and Engineering,
201 Bell Hall
State University of New York at Buffalo,
Amherst, NY 14260, USA.
E-mail: hungngo@cse.Buffalo.EDU

1 Introduction

With the advances of dense wavelength division multiplexing (DWDM) technology [16, 28, 34], the number of wavelengths in a wavelength division multiplexed (WDM) network increases to hundreds or more per fiber, and each wavelength operates at 10 Gbps (OC-192) or higher [12–14]. Although raw bandwidth has increased by more than four orders of magnitude over the last decade or so, capacity of switches has only been up by a factor of ten. Switching speed is the bottleneck at the core of the optical network infrastructure [33]. Consequently, a challenge is to design cost-effective WDM cross-connects (WXC) that can scale in size beyond a hundred inputs and outputs, and at the same time, switch fast (e.g., tens of nanoseconds or less).

The notion of "cost-effectiveness" is difficult to capture. One can analyze and compare WDM switches both qualitatively and quantitatively.

Qualitatively, we need to know if a design is strictly nonblocking (SNB), rearrangeably nonblocking (RNB), and/or wide-sense nonblocking (WSNB) under different request models [20, 27, 29, 30, 37, 38] and different traffic patterns (unicast [20], multicast [40]). Presumably each new design is guided by a particular qualitative feature. For example, one might come up with an RNB design under one request model, which may or may not be SNB under another request model. One might also have an intuitively good design, and hence need to know what qualitative feature the design possesses. This question is challenging in general. We show later that the graph models introduced in this work help, in several ways, answer these types of questions.

Quantitatively, comparing different designs, or asking how close to optimal a new design is, are very important questions. This is a multidimensional problem, as there are many factors effecting the "cost" of a switch. Some factors such as actual cost in dollars are business matters. Other factors include: the numbers of different types of switching components (such as MUX, DEMUX, AWGR, FWC, LWC, SOA, OADM, WSC, WI, etc.), cross-talk, power consumption and attenuation, integratability and scalability, blocking probabilities, and even other non-conventional factors such as the multicast capacity [40].

It should be apparent that we cannot hope to have a cost model that fits all needs. However, one can devise cost models that give good approximated measures on how "complex" a construction is. The notion of complexity should roughly capture as many practical parameters as possible.

In this chapter, we outline an intriguing approach to model switch complexity that not only helps analyze WDM switches quantitatively and qualitatively, but also suggests interesting generalizations of classical switching network theory [2, 24]. Then, we address a few sample problems arising from the graph model.

The following phenomena are samples of what our model suggests:

(a) Designing WXCs in one request model is in a sense the same as that of designing circuit switches. Hence, many old ideas on circuit switching networks can be readily reused.

(b) Being SNB in one request model is equivalent to being SNB in another request model.

(c) There is an inherent tradeoff between a WXC's "depth" (which is proportional to signal attenuation, cross-talk) and its "size" (which approximates the WXC's complexity).

(d) Different designs of WXCs that make use of different optical components can now be viewed in a unified manner. We can tell if two different-looking designs are equivalent, for example.

The framework proposed here gives rise to many interesting mathematical and networking problems, many of which are generalized versions of the well-studied circuit switching problems. We address several of these problems in the later part of the chapter.

Materials from this chapter are drawn from two recent works by the author [17,18]. The rest of the chapter is organized as follows. Section 2 presents basic concepts of WDM switching networks. Section 3 motivates the introduction of complexity models for WDM switching networks. Next, Section 4 rigorously defines the complexity graphs. Section 5 gives several complexity bounds for various types of networks under different request models. Lastly, some explicit network constructions are shown in Section 6.

2 WDM Cross-Connects, Request Models, and Non-blockingness

A general WDM cross-connect consists of f input fibers each of which can carry a set $\Lambda = \{\lambda_1, \dots, \lambda_w\}$ of w wavelengths, and f' output fibers each of which can carry a set $\Lambda' = \{\lambda'_1, \dots, \lambda'_{w'}\}$ of w' wavelengths, where $fw = f'w'$. (See Figure 1.) This setting is referred to as the *heterogeneous* case

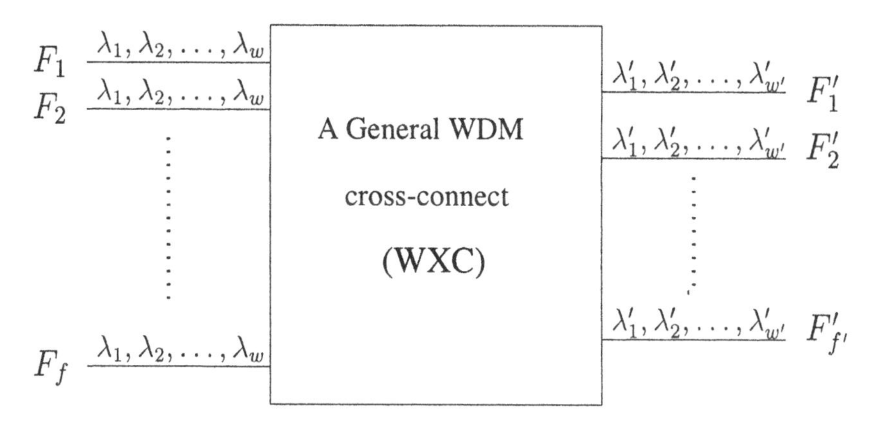

Figure 1: Heterogeneous WDM cross-connect.

[30], which is needed to connect subnetworks from different manufacturers. Henceforth, let $n = fw = f'w'$, unless specified otherwise.

Let $\mathcal{F} = \{F_1, \dots, F_f\}$ and $\mathcal{F}' = \{F'_1, \dots, F'_{f'}\}$ denote the set of input and output fibers, respectively. There are two common types of request models [20, 37]. In the (λ, F, F')*-request model*, a connection request is of the form (λ, F, F'), where $\lambda \in \Lambda, F \in \mathcal{F}$, and $F' \in \mathcal{F}'$. The request asks to establish a connection from wavelength λ in input fiber F to any free wavelength in output fiber F'. In the $(\lambda, F, \lambda', F')$-*request model*, the difference is that the output wavelength λ' in F' is also specified.

In the next sections, we define the concepts of strictly nonblocking, wide-sense nonblocking, and rearrangeably nonblocking for both request models. We are somewhat informal in our definitions; however, the idea should be clear to readers who have been exposed to switching theory [2,9,19].

Henceforth, for any positive integer p, let $[p]$ denote the set $\{1,\dots,p\}$ and S_p denote the set of all permutations on $[p]$. Graph-theoretic terminologies and notations we use here are fairly standard (see [36]). For any set X, let 2^X denote the set of all subsets of X.

2.1 Unicast traffic

Consider a WXC with a few connections already established. For the (λ, F, F')-model, a new request (λ, F, F') is said to be *valid* if and only if λ is a free wavelength in fiber F, and there are at most $w' - 1$ existing connections to F'. Under the $(\lambda, F, \lambda', F')$-model, a new request $(\lambda, F, \lambda', F')$ is *valid* if and only if λ is free in F and λ' is free in F'.

A *request frame* under the (λ, F, F') model is a set of requests such that no two requests are from the same wavelength in the same input fiber, and that there are at most w' requests to any output fiber. A *request frame* under the $(\lambda, F, \lambda', F')$-model is a set of requests such that no two requests are *from* the same input wavelength/fiber pair, nor *to* the same output wavelength/fiber pair.

The following definitions hold for both request models. A request frame is *realizable* by a WXC if all requests in the frame can be routed simultaneously. A WXC is *rearrangeably nonblocking* if and only if any request frame is realizable by the WXC. A WXC is *strictly nonblocking* if and only if a new valid request can always be routed through the WXC without disturbing existing connections. A WXC is *wide-sense nonblocking* if and only if a new valid request can always be routed through the WXC without disturbing existing connections, provided that new requests are routed according to some routing algorithm. When the routing algorithm is known, we also say the WXC is WSNB with respect to the algorithm.

2.2 Multicast Traffic

Under the $(\lambda, F, \lambda', F')$-model, a multicast request is of the form $(\lambda, F, \mathcal{P})$ where $\lambda \in \Lambda$, $F \in \mathcal{F}$, and $\mathcal{P} \subseteq \Lambda' \times \mathcal{F}'$ such that no $F' \in \mathcal{F}'$ appears more than once in $\mathcal{P}$. (That is, if (a, b) and (c, d) are different pairs in $\mathcal{P}$, then $b \neq d$.) This restriction was made because in practical networks it is often not necessary to have a multicast connection going to the same output fiber on two different wavelengths [35,40].

Under the (λ, F, F')-model, a multicast request is of the form $(\lambda, F, \mathcal{S})$, where $\lambda \in \Lambda, F \in \mathcal{F}$, and $\mathcal{S} \subseteq \mathcal{F}'$. Basically, in this case we do not indicate the precise output wavelengths to which the request should be routed. We are only interested in the output fibers $\mathcal{S}$. A multicast tree satisfying this request must have a leaf representing one wavelength from each fiber in $\mathcal{S}$.

For each type of request model, three degrees of non-blockingness can be defined: rearrangeably nonblocking, wide-sense nonblocking, and strictly nonblocking. The basic idea is that a RNB switching network should be able to route a set of compatible requests given in advance. In the WSNB case, requests are nonblocking provided they are routed according to some algorithm. In the SNB case, a new request compatible with any valid network state can always be routed.

One might expect that the complexity of a switching network could be less under the (λ, F, F')-model than under the $(\lambda, F, \lambda', F')$-model, because nonblocking under the former model implies nonblocking under the latter. What is interesting is that this is not always the case, as we show later.

We have been informal in the descriptions above. The reader familiar with switching theory [8, 9] should not have difficulties understanding these concepts. We are more rigorous in our graph definitions to come.

3 Motivations for Having the Graph Models

Main known results on the constructions of (different types of) nonblocking WXCs can be found in [20,21,26,29,30,37–40]. The constructions from these references made use of various different types of optical components, such as arrayed waveguide grating routers (AWGR) and LWCs in [20], SOAs and LWCs in [38], OADMs and FWCs in [37], and wavelength selective cross-connects (WSC) and FWCs in [29, 30]. It is clear that the task of comparing different designs is not easy. Different designs make use of different optical switching components which oftentimes are tradeoffs. For instance, the designs in [38] made use of SOAs and LWCs that have lower wavelength conversion cost than those in [20]. On the other hand, the ones in [20] preferred AWGRs over SOAs because AWGRs consume virtually no power. The newer designs in [21] use the least number of LWCs of a certain kind.

We now propose an approach to uniformly model all designs by graphs, and then discuss switch complexity from the graphs' standpoint.

We classify optical switching components into fibers and other switching components. For any switch design, we apply the following procedure to construct a directed acyclic graph (DAG) from the design: (a) replace each fiber by a set of vertices $\Lambda \cup \Lambda'$, which represents all possible wavelengths that can be carried on the fiber; (b) the edges of the DAG are defined according to the capacity of switching components in the design. The edges connect wavelengths (i.e., vertices) on the inputs of each switching component to the wavelengths on the outputs in accordance with the functionality of the switching component.

We are somewhat brief on this construction. However, the reader can undoubtedly see the basic idea. As an example, Figure 2 shows how to turn an AWGR, an FWC, and a MUX into edges. Figure 3 shows a complete con-

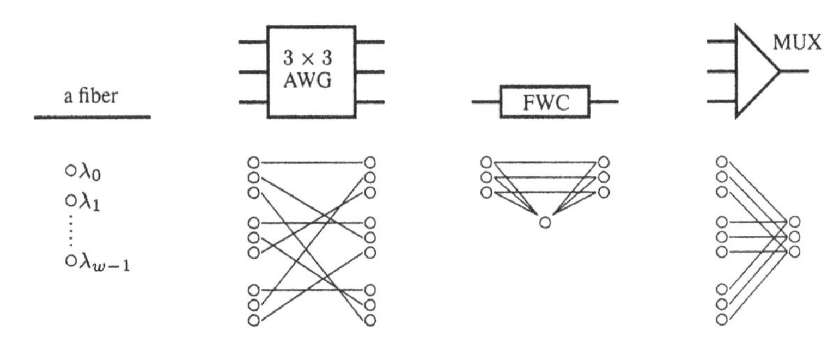

Figure 2: Turning optical components into parts of a graph. A fiber is replaced by a set of vertices representing the wavelengths it can carry. Other components define edges connecting input wavelengths to output wavelengths. For the AWGR, MUX, and FWC, we illustrate with $w = 3$. Edges are directed from left to right.

struction of the DAG from the design on the left. The key point is that *a set of*

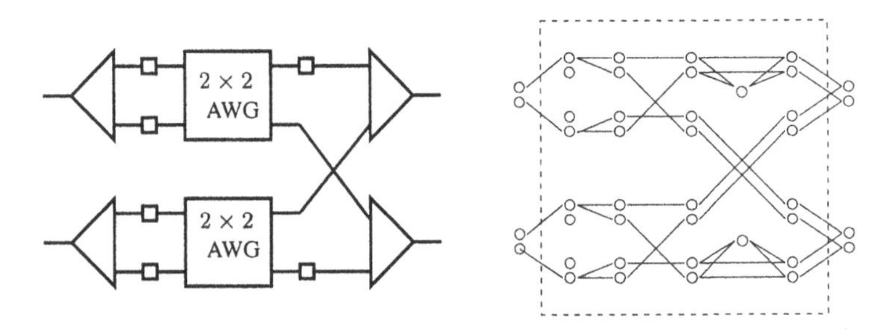

Figure 3: A WDM switch design and its corresponding DAG.

compatible routes from input wavelengths to output wavelengths correspond to a set of vertex disjoint paths from the inputs to the outputs of the DAG.

There are two main parameters of the DAG that capture the notion of "switch complexity" discussed earlier. The number of edges of the DAG, called the *size* of the DAG, is roughly proportional to the total cost of various components in the design. For example, a full-range wavelength converter (FWC) corresponds to $3w$ edges and a wavelength interchanger (WI, [30]) corresponds to w^2 edges; a $w \times w$ AWGR corresponds to w^2 edges, and a $w \times w$ WDM crossbar corresponds to w^4 edges; etc. As WIs and WDM crossbars are more expensive than FWCs and AWGRs, this model makes sense. Other components follow the same trend.

The reader might have noticed that different components contribute different "weights" to the total cost, hence summing up the number of edges may not give the "right" cost. To answer this doubt, we make three points. First, as argued earlier one cannot hope to have a perfect model that fits all needs, and part of the notion of cost is a business matter. Our first aim is at a more theoretical level. Second, this is the first step toward a good cost model. One certainly can envision weighted graphs as the next step. Third, we certainly can and should still use more traditional cost functions such as the direct counts of the number of each component and compare them individually.

The second measure on the DAG is its *depth*, i.e. the length of a longest path from any input to any output. As signals pass through different components of a design, they lose some power. The depth of the DAG hence reflects power loss, or in some cases even the signal delay. Again, different components impose different power loss factors. Hence, other information needs to be taken into account to estimate power loss. However, it is clear that network depth is an important measure.

Last but not least, this DAG model provides a nice bridge between classical switching theory and WDM switching theory. As we show in later sections, this model helps us tremendously in answering qualitative questions about a particular construction. For example, if a wf-input wf-output DAG must have size $\Omega(f^2w^2)$ to be strictly nonblocking, then we know for sure that a construction of cost $o(f^2w^2)$ (reflected by the DAG's size) cannot be strictly nonblocking (for sufficiently large values of fw.)

4 Rigorous Settings

An (n_1, n_2)-*network* is a directed acyclic graph $\mathcal{N} = (V, E; A, B)$, where V is the set of vertices, E is the set of edges, A is a set of n_1 nodes called *inputs*, and B, disjoint from A, is a set of n_2 nodes called *outputs*. The vertices in $V - A \cup B$ are *internal* vertices. The in-degrees of the inputs and the out-degrees of the outputs are 0. The *size* of a network is its number of edges. The *depth* of a network is the maximum length of a path from an input to an output. An n-*network* is an (n, n)-network.

An n-network is meant to represent the DAG from the last section under the $(\lambda, F, \lambda', F')$-request model. (Recall $n = wf = w'f'$.)

Next, we define $[w, f]$-networks that represent the DAG under the (λ, F, F')-request model. In this request model, each pair (λ, F) with $\lambda \in \Lambda$ a wavelength in input fiber $F \in \mathcal{F}$ can still be thought of as an "input" to our graphs as in the previous request model. However, on the output side we do have to indicate the number f of fibers and the number of wavelengths w on each fiber.

A $[w, f]$-network is a wf-network (i.e., an n-network) $\mathcal{N} = (V, E; A, B)$ in which the set B of outputs is further partitioned into f subsets $B_1, \ldots, B_f$ of size w each. We implicitly assume the existence of the partition in a $[w, f]$-network, in order to simplify notations. Sometimes, for the sake of clarity we also write $\mathcal{N}$ as

$$\mathcal{N} = (V, E; A, B; B_1, \ldots, B_f).$$

Each set B_i represents an output fiber in the WDM switch. (There is a slightly subtle point to be noticed here. The inputs are not distinguishable in this request model, whereas we do care from which fiber an output wavelength comes. The parameters w and f in the above definition and henceforth should be thought of as w' and f' in the original discussion of a generic WXC.)

4.1 The $(\lambda, F, \lambda', F')$-Request Model

4.1.1 Unicast Traffic

Given an n-network $\mathcal{N} = (V, E; A, B)$, a pair $D = (a, b)$ in $A \times B$ is called a *request* (or *demand*) for $\mathcal{N}$. A set $\mathcal{D}$ of requests is called a *request frame* iff no two requests share neither an input nor an output. A request $D = (a, b)$ is *compatible* with a request frame $\mathcal{D}$ iff $\mathcal{D} \cup \{D\}$ is also a request frame. A *route* R for a request $D = (a, b)$ is a (directed) path from a to b. We also say R *realizes* D. A *state* of $\mathcal{N}$ is a set $\mathcal{R}$ of vertex disjoint routes. Each state of $\mathcal{N}$ realizes a request frame, one route per request in the frame. A request frame $\mathcal{D}$ is *realizable* iff there is a network state realizing it.

A *rearrangeable n-connector* (or just n-connector for short) is an n-network in which the request frame $\mathcal{D} = \{(a, \pi(a)) \mid a \in A\}$ is realizable, for any one-to-one correspondence $\pi : A \to B$.

A *strictly nonblocking n-connector* is an n-network $\mathcal{N}$ in which given any network state $\mathcal{R}$ realizing a request set $\mathcal{D}$, and given a new request D compatible with $\mathcal{D}$, there exists a route R such that $\mathcal{R} \cup \{R\}$ is a network state realizing $\mathcal{D} \cup \{D\}$.

As requests come and go, a strategy to pick new routes for new requests is called a *routing algorithm*. An n-network $\mathcal{N}$ is called a *wide-sense nonblocking n-connector* with respect to a routing algorithm $\mathbf{A}$ if $\mathbf{A}$ can always pick a new route for a new request compatible with the current network state. We can also replace $\mathbf{A}$ by a class of algorithms $\mathcal{A}$. In general, an n-network $\mathcal{N}$ is WSNB iff it is WSNB with respect to *some* algorithm.

We often consider two classes of functions on each network type: (a) the minimum size of a network, and (b) the minimum size of a network with a given depth. The main theme of research on classical switching networks is to investigate the tradeoff between size and depth [19, 24].

Let $rc(n)$, $wc(n)$, and $sc(n)$ denote the minimum size of an RNB, WSNB, and SNB n-connector, respectively. Let $rc(n,k)$, $wc(n,k)$, and $sc(n,k)$ denote the minimum size of an RNB, WSNB, and SNB n-connector with depth k, respectively. Note that $rc(n) \leq wc(n) \leq sc(n)$, and $rc(n,k) \leq wc(n,k) \leq sc(n,k)$.

These classes of functions are well studied in the context of circuit-switching networks (see, e.g., [19, 24] for nice surveys). The point we are making is that studying WDM switches under the $(\lambda, F, \lambda', F')$-request model is in a sense the same as studying classical switching networks. A lot of results can be readily re-used. For example, using our DAG construction, it is easy to see that many of the constructions in [20, 29, 30, 38] are various forms of the Clos network idea [5], which were discussed in a slightly more general form in Pippenger [22]. In fact, under this request model, the author does not know any result which does not make use of classical circuit designs. The situation under the (λ, F, F')-model is different, though. The RNB design in [20], and the designs presented in this chapter require several new themes. Particularly, this is because the (λ, F, F')-model is not equivalent to the classical switching case, as we show in the next section.

4.1.2 Multicast Traffic

Several preliminary results for the multicast case (under both request models) have been worked out in [17]. We give here a more systematic treatment.

Given a $[w, f]$-network $\mathcal{N} = (V, E; A, B)$, a pair $D = (a, S) \in A \times 2^B$, where $|S \cap B_i| \leq 1, \forall i \in [f]$, is called a *distribution request under the $(\lambda, F, \lambda', F')$-model* for $\mathcal{N}$. When the request model is implicitly understood from context, we do not mention the lengthy "under model ..."; similarly, we do not mention the word "distribution" if it is implicit.

A *distribution request frame* is a set $\mathcal{D}$ of distribution requests where no two requests share neither an input nor an output. A distribution request $D = (a, S)$ is *compatible* with a a distribution request frame $\mathcal{D}$ iff $\mathcal{D} \cup \{D\}$ is also a distribution request frame. A *distribution route* R for a request $D = (a, S)$ is a (directed) tree rooted at a whose leaves are nodes in S. We also say R *realizes* D. A *state* of $\mathcal{N}$ is a set $\mathcal{R}$ of vertex disjoint distribution routes. Each state of $\mathcal{N}$ realizes a unique request frame, with one route per request. A request frame $\mathcal{D}$ is *realizable* iff there is a network state realizing it. A request is *compatible* with a state if it is compatible with the frame realized by the state.

Under the $(\lambda, F, \lambda', F')$-request model, we can define RNB, WSNB, and SNB $[w, f]$-distributors in a similar fashion as the unicast traffic situation. Let $rd(w, f)$, $wd(w, f)$, and $sd(w, f)$ denote the minimum size of an RNB, WSNB, and SNB $[w, f]$-distributor, respectively. Let $rd(w, f, k)$, $wd(w, f, k)$, and $sd(w, f, k)$ denote the minimum size of an RNB, WSNB, and SNB $[w, f]$-distributor with depth k, respectively.

Remark 4.1. In the classical switching literature, distributors are also called *generalized connectors*.

4.2 The (λ, F, F')-Request Model

4.2.1 Unicast Traffic

Given a $[w, f]$-network $\mathcal{N}$, a pair $D = (a, k) \in A \times [f]$ is called a (connection) *request* for $\mathcal{N}$. The number k is called the *output fiber number* of D. A set $\mathcal{D}$ of requests is called a *request frame* iff no two requests share an input, and for any $k \in [f]$, we have $|\{a \mid (a, k) \in \mathcal{D}\}| \leq w$. A request $D = (a, k)$ is *compatible* with a request frame $\mathcal{D}$ iff $\mathcal{D} \cup \{D\}$ is also a request frame.

A *route* R for a request $D = (a, k)$ is a path from a to some vertex b in B_k. We also say R *realizes* D. A *state* of $\mathcal{N}$ is a set $\mathcal{R}$ of vertex disjoint routes. Each state of $\mathcal{N}$ realizes a request frame. A request frame $\mathcal{D}$ is *realizable* iff there is a network state realizing it.

We are interested in (WSNB, SNB, RNB) connectors under this request model. A *(rearrangeable)* $[w, f]$*-connector* is a $[w, f]$-network in which the request frame

$$\mathcal{D} = \{(a, \sigma(a)) \mid a \in A\}$$

is realizable for any mapping $\sigma : A \to [f]$ such that

$$|\{a \mid \sigma(a) = k\}| = w, \ \forall k \in [f].$$

A *strictly nonblocking* $[w, f]$*-connector* is a $[w, f]$-network $\mathcal{N}$ in which given any network state $\mathcal{R}$ realizing a request set $\mathcal{D}$, and given a new request D

compatible with $\mathcal{D}$, there exists a route R such that $\mathcal{R} \cup \{R\}$ realizes $\mathcal{D} \cup \{D\}$. As requests come and go, a strategy to pick new routes for new requests is called a *routing algorithm*. An $[w, f]$-network $\mathcal{N}$ is called a *wide-sense nonblocking* $[w, f]$*-connector* with respect to a routing algorithm **A** if **A** can always pick a new route for a new request compatible with the current network state. We can also replace **A** by a class of algorithms $\mathcal{A}$. In general, an $[w, f]$-network $\mathcal{N}$ is WSNB iff it is WSNB with respect to *some* algorithm.

The different $[w, f]$-networks are generalized versions of the corresponding n-networks.

Proposition 4.2. *A network is an SNB, WSNB, RNB* $[1, f]$*-connector if and only if it is an SNB, WSNB, RNB* f*-connector, respectively.*

Let $\overline{rc}(w, f)$, $\overline{wc}(w, f)$, and $\overline{sc}(w, f)$ denote the minimum size of an RNB, WSNB, and SNB $[w, f]$-connector, respectively. Similarly, for a fixed depth k, we define $\overline{rc}(w, f, k)$, $\overline{wc}(w, f, k)$, and $\overline{sc}(w, f, k)$. These functions have not been studied before. Some trivial bounds can be summarized as follows:

Proposition 4.3. *Let* $n = wf$*, then*

(i) $\overline{rc}(w, f) \leq \overline{wc}(w, f) \leq \overline{sc}(w, f)$

(ii) $\overline{rc}(w, f, k) \leq \overline{wc}(w, f, k) \leq \overline{sc}(w, f, k)$

(iii) An RNB, WSNB, SNB n*-connector is also an RNB, WSNB, SNB* $[w, f]$*-connector, respectively. Consequently,* $\overline{rc}(\cdot) \leq rc(\cdot)$, $\overline{wc}(\cdot) \leq wc(\cdot)$, *and* $\overline{sc}(\cdot) \leq sc(\cdot)$*, where the dots on the left-hand sides can be replaced by* (w, f) *or* (w, f, k)*, and the dots on the right-hand sides by* (n) *or* (n, k)*, correspondingly.*

4.2.2 Multicast Traffic

Given a $[w, f]$-network $\mathcal{N}$, a *distribution request under the* (λ, F, F')*-model* is a pair $D = (a, T) \in A \times 2^{[f]}$. A *distribution route* R for $D = (a, T)$ is a tree rooted at a with exactly $|T|$ leaves, one in each $B_j, j \in T$. A *distribution request frame* is a set $\mathcal{D}$ of requests such that no two inputs appear twice in $\mathcal{D}$, and that for each $j \in [f]$,

$$|\{T : (a, T) \in \mathcal{D} \text{ and } j \in T\}| \leq w. \tag{1}$$

In words, no output fiber is involved in more than w requests. The rest of the concepts are defined similarly to the ones under the other request model: RNB, SNB, WSNB $[w, f]$-distributors, and corresponding complexity functions $\overline{rd}$, $\overline{wd}$, and $\overline{sd}$. The following observations are straightforward from definitions.

Proposition 4.4. *An SNB, WSNB, RNB $[w, f]$-distributor is also an SNB, WSNB, RNB $\overline{[w, f]}$-distributor, respectively. Consequently, $\overline{\mathbf{x}d}(\cdot) \leq \mathbf{x}d(\cdot)$, where $\mathbf{x}$ stands for r, w, or s, and the argument represented by the dot is either (w, f) or (w, f, k).*

Proposition 4.5. *We have*

$$rd(\cdot) \leq wd(\cdot) \leq sd(\cdot),$$

$$\overline{rd}(\cdot) \leq \overline{wd}(\cdot) \leq \overline{sd}(\cdot),$$

where the $\cdot$ is either (w, f) or (w, f, k).

5 Complexity Bounds

5.1 Strictly Nonblocking $[w, f]$-Connectors

We study SNB $[w, f]$-connectors in this section. For $f = 1$, it is easy to see the following.

Proposition 5.1. *Let w and k be positive integers; then $\overline{sc}(w, 1, k) = w+k-1$.*

Proof. In a depth-k SNB $[w, 1]$-connector, there must be a path of length k and $w - 1$ other vertex disjoint paths from the inputs to the outputs. Hence, $w + k - 1$ edges are necessary. Conversely, the network consists of a path of length k from some input to some output, and a matching of size $w - 1$ between the rest of the inputs and the rest of the outputs is clearly an SNB $[w, 1]$-connector. □

Intuitively, an optimal SNB $[w, f]$-connector might have (strictly) smaller size than an optimal SNB wf-connector, because the $(\lambda, F, \lambda', F')$-request model is more restrictive than the (λ, F, F')-request model. However, the following theorem shows a somewhat surprising result that we can do no better than a wf-connector when $f \geq 2$.

Theorem 5.2. *Let $n = wf$, where n, w, f are positive integers, and $f \geq 2$. An n-network $\mathcal{N} = (V, E; A, B)$ is a strictly nonblocking n-connector iff it is a strictly nonblocking $[w, f]$-connector.*

Proof. Clearly an SNB n-connector is also an SNB $[w, f]$-connector, no matter how the fiber partitioning is done. For the converse, let $\mathcal{N}$ be an SNB $[w, f]$-connector. Let $B = B_1 \cup \cdots \cup B_f$ be the partition of B. (Recall, by definition, that $|B_i| = w, \forall i \in \{1, \ldots, f\}$ and that $|A| = wf$.)

Consider a state $\mathcal{R}$ of this network. We show that if a is a free input and b is a free output, then there exists a route R from a to b such that $\mathcal{R} \cup \{R\}$ is a network state.

Let X be the set of free inputs and Y the set of free outputs. Note that $a \in X, b \in Y$, and $|X| = |Y|$.

Suppose $b \in B_k$, for some $k \in \{1, \ldots, f\}$. Without loss of generality, we assume that there are no free output in any B_j for $j \neq k$. This can be accomplished by creating as many requests of the form (x, j) as possible, where $x \in X - \{a\}$ and $j \neq k$, until there is no more free outputs at the B_j with $j \neq k$. Then, let $\mathcal{R}$ be the new network state (which satisfies all new requests and also contains the old network state). An (a, b)-route compatible with $\mathcal{R}$ is certainly compatible with the old network state.

We can now assume $Y \subseteq B_k$. Create $|X|$ requests of the form (x, k), one for each $x \in X$. Because $\mathcal{N}$ is a $[w, f]$-connector, there is a route R_x for each x in X satisfying the following. (i) R_x starts from x and ends at some vertex in B_k, and (ii) $\mathcal{R} \cup \{R_x \mid x \in X\}$ is a network state.

If R_a is an (a, b)-route, then we are done. Moreover, if $|X| = |Y| = 1$, then R_a must be an (a, b)-route. Consequently, we can assume the following:

- $|X| = |Y| \geq 2$.
- R_a goes from a to some vertex $y \in B_k - \{b\}$.
- There is some $x \in X - \{a\}$ such that R_x ends at b.
- There is some vertex $a' \notin X$ and a route $R' = (a', v_1, \ldots, v_p, b') \in \mathcal{R}$ that goes from a' to a vertex $b' \in B_j$, where $j \neq k$. (The route $R' \in \mathcal{R}$ exists because we assumed that the vertices in $B_j, j \neq k$, are all busy.)

To this end, let

$$\mathcal{R}' = \mathcal{R} \cup \{R_t \mid t \in X\} - \{R_a, R_x, R'\}.$$

We show that there exists an (a, b)-route R for which $\mathcal{R}' \cup \{R', R\}$ is a state. The route R is then the route we are looking for, because $\mathcal{R} \subseteq \mathcal{R}' \cup \{R'\}$.

We first claim that there exists an (x, y)-route R_{xy} compatible with $\mathcal{R}'$. Consider the state $\mathcal{R}' \cup \{R_a\}$. The request (a', k) is valid (i.e., compatible with the request frame realized by $\mathcal{R}' \cup \{R_a\}$), and b is the only free output in B_k; hence, there is an (a', b)-route $R_{a'b}$ such that $\mathcal{R}' \cup \{R_a, R_{a'b}\}$ is a state. Now, in the state $\mathcal{R}' \cup \{R_{a'b}\}$ the request (x, k) is valid, and y is the only free output in B_k. Hence, there is an (x, y)-route R_{xy} such that $\mathcal{R}' \cup \{R_{a'b}, R_{xy}\}$ is a state. Consequently, there is an (x, y)-route R_{xy} compatible with $\mathcal{R}'$ as claimed.

Now, consider two cases as follows.
Case 1: Among all the (x, y)-routes that are compatible with $\mathcal{R}'$ there is some R_{xy} which is vertex also disjoint from R'. Then, in the state $\mathcal{R}' \cup \{R', R_{xy}\}$ the request (a, k) is valid, and b is the only free output in B_k. Hence, there is an (a, b)-path compatible with $\mathcal{R}' \cup \{R'\}$ as desired.
Case 2: Every (x, y)-route compatible with $\mathcal{R}'$ intersects R' at some point. Let R_{xy} be such an (x, y)-route whose last intersection vertex on $(v_1, \dots, v_p)$ has the largest index, say v_j, for some $j \in \{1, \dots, p\}$. Then, R_{xy} is composed of two parts: the part from x to v_j and the part from v_j to y.

Let $R_{a'y}$ be an (a', y)-path consisting of the part $(a', v_1, \dots, v_j)$ of R' and the (v_j, y)-part of R_{xy}. Then, certainly $R_{a'y}$ is compatible with $\mathcal{R}'$. In the state $\mathcal{R}' \cup \{R_{a'y}\}$ the request (a, k) is valid, and b is the only free output in B_k. Hence, there is an (a, b)-path R_{ab} compatible with $\mathcal{R}' \cup \{R_{a'y}\}$.

If R_{ab} does not intersect R', then we are done. Otherwise, R_{ab} must intersect R' at some $v_{j'}$ for which $j' > j$. Similarly to the previous paragraph, we can form an (a', b)-path $R_{a'b}$ compatible with $\mathcal{R}'$ consisting of $(a', v_1, \dots, v_{j'})$ and the part of R_{ab} from $v_{j'}$ to b. Now, consider the state $\mathcal{R}' \cup \{R_{a'b}\}$ in which y is the only free vertex in B_k. The request (x, k) is valid; hence there is some (x, y)-path compatible with $\mathcal{R}' \cup \{R_{a'b}\}$. This (x, y)-path must then intersect R' (because we are in case 2) at some vertex after $v_{j'}$ (for compatibility with $R_{a'b}$), contradicting our choice of R_{xy} earlier. □

Corollary 5.3. *The following hold for $f \geq 2$.*

(i) $\overline{sc}(w, f, 1) = w^2 f^2$.

(ii) $\Omega\left((wf)^{1+1/(k-1)}\right) = \overline{sc}(w, f, k) = O((wf)^{1+1/\lfloor (k+1)/2 \rfloor})$.

(iii) $\overline{sc}(w, f) = \Theta(wf \lg(wf))$.

Proof. Let $n = wf$; then $\overline{sc}(w, f, k) = sc(n, k)$ by Theorem 5.2. The first equality is obvious. The fact that $sc(n, k) = O(n^{1+1/\lfloor (k+1)/2 \rfloor})$ can be seen from the constructions by Clos (1953, [5]), Cantor (1971, [4]), and Pippenger (1978, [22]). The lower bound $\Omega\left((wf)^{1+1/(k-1)}\right)$ was shown by Friedman (1988, [7]). That $sc(n) = \Theta(n \lg n)$ can be seen from Bassalygo and Pinsker (1973, [1]) and Shannon (1950, [32]). The reader is referred to Pippenger (1990, [24]) and Ngo (2001, [19]) for more details on these functions. □

5.2 Rearrangeable $[w, f]$-Connectors

In this section, we first devise lower bounds for the optimal size of RNB $[w, f]$-connectors and connectors of a fixed depth. The upper bounds follow from explicit constructions presented in Section 6.

We use an argument of Pippenger [23] to show the following theorem.

Theorem 5.4. *Every rearrangeable $[w, f]$-connector must have size at least*

$$\frac{45}{7} wf \log_6 f + O(f) - O(f \lg w).$$

In particular, $\overline{rc}(w, f) = \Omega(wf \lg f)$.

Proof. Let $G = (V, E; A, B)$ be a $[w, f]$-connector.

If a vertex $v \in V$ has only one out-edge (v, w) or only one in-edge (w, v), then we can shrink v and w into one vertex without changing the rearrangeability of the network and without increasing the network size. Thus, we may assume each $v \in V$ has in-degree at least 2 and out-degree at least 2.

Now, if some vertex v has exactly two in-edges (u_1, v), (u_2, v) and exactly two out-edges (v, w_1) and (v, w_2), then we could delete v and add edges (u_1, w_1), (u_1, w_2), (u_2, w_1), and (u_2, w_2) without changing the rearrangeability of the network while keeping the network size the same. Consequently, we may also assume each vertex in $V - A \cup B$ has total (in and out) degree at least 5.

Let π be any one-to-one correspondence between A and B. Let $G' = (V', E')$ be the graph obtained from G by identifying v with $\pi(v)$ for every $v \in A$, and adding a loop to every vertex $v \in V - A \cup B$. Let $d(v)$ be the total degree of vertex v in G. Then

$$\begin{aligned}
|E'| &= \frac{1}{2}\left(\sum_{v \in A \cup B} d(v) + \sum_{v \notin A \cup B} d(v)\right) + |V - A \cup B| \\
&\leq \frac{1}{2}\left(\sum_{v \in A \cup B} d(v) + \sum_{v \notin A \cup B} d(v)\right) + \frac{1}{5} \sum_{v \notin A \cup B} d(v) \\
&\leq \frac{7}{10}\left(\sum_{v \in A \cup B} d(v) + \sum_{v \notin A \cup B} d(v)\right) \\
&= \frac{7}{5}|E|. \qquad (2)
\end{aligned}$$

A *full request frame* for $\mathcal{N}$ is a request frame of size wf. The total number of different full request frames for $\mathcal{N}$ is the multinomial coefficient

$$\binom{wf}{\underbrace{w, \ldots, w}_{f \text{ times}}} = \frac{(wf)!}{(w!)^f} \geq \frac{\sqrt{2\pi wf}(wf/e)^{wf}}{e^{f/12w}(2\pi w)^{f/2}(w/e)^{wf}}$$

$$= (2\pi w)^{1-f/2} e^{-f/12w} f^{wf+1/2}, \quad (3)$$

where the inequality follows from a good version of Stirling's approximation [31]:

$$\sqrt{2\pi n}\left(\frac{n}{e}\right)^n \leq n! \leq e^{\frac{1}{12n}}\sqrt{2\pi n}\left(\frac{n}{e}\right)^n. \tag{4}$$

A *cycle decomposition* of G' is a 1-regular spanning subgraph of G'. (Note that a 1-regular directed graph is a union of cycles, not a matching. Also, loops are considered to be cycles.) Let C be the set of cycle decompositions of G; then each network state realizing a full request frame induces a different cycle decomposition, because $\mathcal{N}$ is rearrangeable, Hence,

$$|C| \geq (2\pi w)^{1-f/2} e^{-f/12w} f^{wf+1/2}. \tag{5}$$

Let A be the adjacency matrix of G', and per A denote the permanent of A. Then,

$$|C| = \text{per } A. \tag{6}$$

Moreover, let r_v be the sum of row v of A; then

$$\sum_{v\in V'} r_v = |E'|. \tag{7}$$

The famous Minc's conjecture [15] and its proof by Brègman (1973, [3]) give

$$\text{per } A \leq \prod_{v\in V'} (r_v!)^{1/r_v}. \tag{8}$$

It is easy to see that the function $\log_6(x!)/x^2$ over the positive integers gets its maximum at $x = 3$. Identities (6), (7), and (8) give

$$\begin{aligned} |E'| = \sum_{v\in V'} r_v &\geq 3^2 \sum_{v\in V'} \frac{\log_6(r_v!)}{r_v} \\ &= 9\sum_{v\in V'} \frac{\log_6(r_v!)}{r_v} \geq 9\log_6(\text{per } A) = 9\log_6|C|. \end{aligned} \tag{9}$$

Lastly, (2), (5), and (9) yield

$$\begin{aligned} |E| \geq \frac{45}{7}\log_6|C| &\geq \frac{45}{7}\log_6\left((2\pi w)^{1-f/2} e^{-f/12w} f^{wf+1/2}\right) \\ &= \frac{45}{7} wf \log_6 f + O(f) - O(f \lg w). \end{aligned}$$

□

Note that the original theorem by Pippenger [23] is now a direct consequence of this theorem.

Corollary 5.5. *We have* $rc(n) \geq \frac{45}{7} n \log_6 n + O(n)$. *In particular,* $rc(n) = \Omega(n \lg n)$.

Proof. When $w = 1, n = f$, an RNB $[w, f]$-connector is an n-connector. □

The bound $\Omega(wf \lg f)$ implies that for $w \leq f$, $[w, f]$-connectors must have size at least $\Omega(wf \lg(wf))$, which is asymptotically no better than a wf-connector. This confirms our intuition that for small values of w, $[w, f]$-connectors are almost the same as wf-connectors.

Fortunately, in WDM networks it is often the case that $w \geq f$, i.e. the number of wavelengths per fiber (in the hundreds) is often much larger than the number of fibers (in the tens). The next section shows that we can construct $[w, f]$-connectors that are asymptotically less expansive than all known constructions of wf-connectors.

We next give lower bounds for fixed depth $[w, f]$-connectors.

Theorem 5.6. *The optimal size of a depth-*1 $[w, f]$*-connector is* $wf(wf - w + 1)$, *namely,*

$$\overline{rc}(w, f, 1) = wf(wf - w + 1).$$

Proof. In the next section, we construct depth-1 $[w, f]$-connectors of size $wf(wf - w + 1)$, which proves the upper bound

$$\overline{rc}(w, f, 1) \leq wf(wf - w + 1).$$

For the lower bound, let $\mathcal{N} = (A \cup B, E; A, B)$ be a depth-1 $[w, f]$-connector. Then, $\mathcal{N}$ is a (directed) bipartite graph where $|A| = wf$, $|B| = wf$, and B has a partition into $B_1 \cup \cdots \cup B_f$, such that $|B_i| = w, \forall i$. The network $\mathcal{N}$ is a $[w, f]$-connector iff for every partition of A into $A_1, \ldots, A_f$ with $|A_i| = w, \forall i$, there exist f complete matchings from each A_i to each B_i.

It follows that each vertex $b \in B$ must have a neighbor in every w-subset of A. Consequently, each vertex $b \in B$ must be of degree at least $|A| - w + 1$. Hence, the number of edges of $\mathcal{N}$ is at least

$$|B|(|A| - w + 1) = wf(wf - w + 1).$$

□

For $k \geq 2$, we can use an argument by by Pippenger and Yao [25] on the so-called *n-shifters* to find a lower bound for depth-k $[w, f]$-connectors.

Let $T_k(f)$ be a directed rooted tree with f leaves and depth at most k where all edges are directed to the direction of the leaves. Let $P_1, \ldots, P_f$ be the f paths from the root to the leaves of $T_k(f)$. Define

$$\Delta(T_k(f)) := \sum_{j=1}^{f} \sum_{v \in P_j} \text{out-degree}(v). \tag{10}$$

The following lemma is from [25].

Lemma 5.7. $\Delta(T_k(f)) \geq kf^{1+1/k}$.

Theorem 5.8. *Let $k \geq 2$ be an integer; a depth-k $[w, f]$-connector must have size at least $kwf^{1+1/k}$. Specifically, $\overline{rc}(w, f, k) = \Omega(kwf^{1+1/k})$.*

Proof. Let $\mathcal{N} = (V, E; A, B)$ be a $[w, f]$-connector of depth k, where $A = \{a_1, \ldots, a_n\}$. For each q in $\{1, \ldots, f\}$, define a function

$$\phi_q(i) = (i + q - 1 \pmod{f}) + 1,\ 1 \leq i \leq n.$$

Also define the following request frame, for each q,

$$\mathcal{D}_q := \{(a_i, \{\phi_q(i)\}) \mid 1 \leq i \leq n\}.$$

Because $\mathcal{N}$ is a $[w, f]$-connector, for each $q = 1, \ldots, f$ there are n vertex disjoint paths P_{iq}, $i = 1, \ldots, n$, such that P_{iq} joins a_i to some vertex in $B_{\phi_q(i)}$.

To this end, for $1 \leq i \leq n$, $1 \leq q \leq f$, and $e \in E$, let

$$\mu(i, q, e) := \begin{cases} 1 & \text{if } e \text{ is an arc emitted from a node on } P_{iq} \\ 0 & \text{otherwise.} \end{cases}$$

Fix an i, assemble all f paths P_{iq} into a tree T_i (keeping only the initial common segments of the paths); then T_i is a tree with f leaves and depth at most k.

For each vertex $v \in V$, let out-degree$_{T_i}(v)$ denote the out-degree of v in T_i. It is easy to see the following.

$$\sum_{e \in E} \mu(i, q, e) \geq \sum_{v \in P_{iq}} \text{out-degree}_{T_i}(v). \tag{11}$$

Basically, the left-hand side also counts some arcs not in T_i (but starts on P_{iq}).

Summing (11) over $i = 1, \ldots, n$ and $q = 1, \ldots, f$, we get

$$\begin{aligned} \sum_{i=1}^{n}\sum_{q=1}^{f}\sum_{e\in E}\mu(i,q,e) &\geq \sum_{i=1}^{n}\sum_{q=1}^{f}\sum_{v\in P_{iq}} \text{out-degree}_{T_i}(v) \\ &= \sum_{i=1}^{n}\Delta(T_i) \\ &\geq nkf^{1+1/k}. \end{aligned} \tag{12}$$

The last inequality comes from Lemma (5.7).

On the other hand, because the paths P_{iq} for a fixed q are vertex disjoint, we have

$$\sum_{i=1}^{n}\mu(i,q,e) \leq 1.$$

Consequently,

$$\sum_{i=1}^{n}\sum_{q=1}^{f}\sum_{e\in E}\mu(i,q,e) = \sum_{q=1}^{f}\sum_{e\in E}\sum_{i=1}^{n}\mu(i,q,e) \leq f|E|. \tag{13}$$

Together, (12) and (13) lead to $|E| \geq kwf^{1+1/k}$ as desired. □

We can also get the result $\overline{rc}(w, f) = \Omega(wf \lg f)$ from the previous theorem (with a worse constant than $45/7$):

Corollary 5.9. *For $k \geq 2$, $\overline{rc}(w, f) \geq ewf \ln f$, where e is the base of the natural log.*

Proof. The function $kwf^{1+1/k}$ is minimized at $k = \ln(f)$. □

5.3 Strictly Nonblocking Distributors

The following theorem essentially shows that being SNB in the more relaxed (λ, F, F')-request model gives us no advantage as far as network cost is concerned. The theorem is the distributor analogue of Theorem 5.2

Theorem 5.10. . *Let w, f be positive integers where $f \geq 2$. Then, a $[w, f]$-network is an SNB $[w, f]$-distributor if and only if it is an SNB $\overline{[w, f]}$-distributor.*

Proof. That an SNB $[w, f]$-distributor is also an SNB $\overline{[w, f]}$-distributor is obvious. We now show the converse.

Let $\mathcal{N} = (V, E; A, B; B_1, \ldots, B_f)$ be an SNB $\overline{[w, f]}$-distributor. Let $\mathcal{R}$ be a state of $\mathcal{N}$; namely, $\mathcal{R}$ is a set of vertex disjoint trees whose roots are inputs, and whose leaves are outputs of $\mathcal{N}$. Let $D = (a, S)$ be a distribution request frame compatible with $\mathcal{R}$, where $S \subseteq B$ and $|S \cap B_i| \leq 1, \forall i \in [f]$. We show that there is a tree R rooted at a with leaves S, and R is vertex disjoint from trees in $\mathcal{R}$. For each $s \in S$, let $B_{j(s)}$ denote the output band of which s is a member. Notice that $B_{j(s)} \neq B_{j(s')}$, for members $s \neq s'$ of S.

The main idea is that we show there is a state $\mathcal{S}$ of $\mathcal{N}$ such that $\mathcal{R} \subseteq \mathcal{S}$, a is free in $\mathcal{S}$, and that each s in S is the only free output in $B_{j(s)}$. Suppose such a state $\mathcal{S}$ can be constructed. Consider the request (a, T), where $T = \{j(s) \mid s \in S\}$. This request is compatible with $\mathcal{S}$ under the (λ, F, F')-model. Because $\mathcal{N}$ is an SNB $\overline{[w, f]}$-distributor, there is a tree R realizing (a, T). This is the tree we are looking for, as the leaves of the tree have to be precisely those in S.

To show the existence of such a state $\mathcal{S}$, let us consider two cases as follows.
Case 1: *There is some route in $\mathcal{R}$ with more than one leaf.* Let X (Y) be the number of free inputs (outputs) in $\mathcal{R}$. Then, $a \in X$ and $|X| > |Y|$, because the total numbers of inputs and outputs are the same. Now, let k be such that B_k has some free output in $\mathcal{R}$. Let x be a member of $X - \{a\}$. The request (x, k) is compatible with $\mathcal{R}$; hence, there is a route R_1 from x to some output in B_k for which $\mathcal{R} \cup \{R_1\}$ is a state. Repeat this process $|Y|$ times; we have a state $\mathcal{R}' = \mathcal{R} \cup \{R_1, \ldots, R_{|Y|}\}$ in which there are no more free outputs, yet a is still free. Now, remove from $\mathcal{R}'$ all routes whose endpoints are those in S; we get the desired state $\mathcal{S}$.
Case 2: *All routes in $\mathcal{R}$ are one-to-one routes.* This is a much trickier case, as $|X| = |Y|$ and a has to be involved in the "filling up" process. Fortunately, we can make use of the proof of Theorem 5.2, because all routes in $\mathcal{R}$ are one-to-one, and because unicast requests are also multicast requests under any request model.

Similar to the proof of Theorem 5.2, we can show that if x is a free input and y is a free output, then there will be an (x, y)-route R_{xy} such that $\mathcal{R} \cup \{R_{xy}\}$ is a state. Now, pick arbitrarily a vertex $s_a \in S$ and add route R_{as_a} to state $\mathcal{R}$. Then, create arbitrary one-to-one requests to fill up the outputs with new routes. We obtain a very large state $\mathcal{R}'$ such that $\mathcal{R} \subset \mathcal{R}'$. Remove from $\mathcal{R}'$ all routes ending at vertices in S and we get the state $\mathcal{S}$ as desired. □

Corollary 5.11. *Given positive integers w, f, and k, we have $sd(w, f) = \overline{sd}(w, f)$ and $sd(w, f, k) = \overline{sd}(w, f, k)$.*

Corollary 5.12. $sd(w, f, 1) = \overline{sd}(w, f, 1) = (wf)^2$.

Proof. It is easy to see that $sd(w, f, 1) = (wf)^2$. □

5.4 Rearrangeable $[w, f]$-Distributors

The results in this section are the distributor analogues of Theorem 5.8. Let $A = \{a_0, \dots, a_{n-1}\}$ and $B = \{b_0, \dots, b_{n-1}\}$. An *$n$-shifter* is an n-network $G = (V, E; A, B)$ such that for each $k \in \{0, \dots, n-1\}$, there are n vertex disjoint paths joining a_i to $b_{(i+k) \bmod n}$, for $i = 0, \dots, n$.

The following lemma was shown by Pippenger and Yao [25].

Lemma 5.13. *An n-shifter of depth k has at least $kn^{1+1/k}$ edges.*

Theorem 5.14. *For $k \geq 2$, a depth-k $[w, f]$-distributor must have size at least $k(wf)^{1+1/k}$. Specifically*

$$rd(w, f, k) \geq k(wf)^{1+1/k}. \tag{14}$$

Proof. Let $\mathcal{N} = (V, E; A, B)$ be a depth-k $[w, f]$-distributor. Let $n = wf$. Arbitrarily assign labels to the inputs in A and outputs in B so that $A = \{a_0, \dots, a_{n-1}\}$ and $B = \{b_0, \dots, b_{n-1}\}$. For each $q = 0, \dots, n-1$, consider the following set

$$\mathcal{D}_q = \left\{(a_i, \{b_{(i+q) \bmod n}\}) \mid i \in \{0, \dots, n-1\}\right\}.$$

Clearly $\mathcal{D}_q$ is a distribution request frame. Hence, there exist n vertex disjoint paths joining a_i to $b_{(i+q) \bmod n}$. Consequently, $\mathcal{N}$ is an n-shifter of depth k. Our result now follows from Lemma 5.13. □

Corollary 5.15. *For $k \geq 2$, $rd(w, f) \geq ewf(\ln f + \ln w)$, where e is the base of the natural log.*

Proof. The function $g(k) = k(wf)^{1+1/k}$, with $k \geq 1$, is minimized at $k = \ln(wf)$. □

Because a $\overline{[w, f]}$-distributor is also a $[w, f]$-connector, we obtain the following result as a consequence of Theorem 5.8.

Theorem 5.16. *For $k \geq 2$, a depth-k $\overline{[w, f]}$-distributor must have size at least $kwf^{1+1/k}$, and a $\overline{[w, f]}$-distributor must have size at least $ewf \ln f$. Specifically*

$$\overline{rd}(w, f, k) \geq kwf^{1+1/k} \tag{15}$$

$$\overline{rd}(w, f) \geq ewf \ln f. \tag{16}$$

6 Explicit Constructions

For any network $\mathcal{N}$, let $A(\mathcal{N})$ and $B(\mathcal{N})$ denote the set of inputs and outputs of $\mathcal{N}$, respectively. For any $[w, f]$-network $\mathcal{N}$, we always use $B_1(\mathcal{N}), \ldots, B_f(\mathcal{N})$ to denote the partition of $B(\mathcal{N})$.

6.1 Basic Networks

Definition 6.1 (Crossbars). Let $\mathcal{B}(x, y) = (A \cup B; E)$ denote the complete $x \times y$ directed bipartite graph; i.e., $|A| = x, |B| = y$, and $E = A \times B$. The (x, y)-network $\mathcal{B}(x, y)$ is called an *(x, y)-crossbar*. When $x = y$, we use the shorter notation $\mathcal{B}(x)$, and call it the *x-crossbar*.

Definition 6.2 (Matchings). For any positive integer m, let $\mathcal{M}(m) = (A \cup B; E)$ denote a perfect matching of size m from A into B. (Therefore, $|A| = |B| = m$.)

Definition 6.3 (Concentrators). An *(n, m)-concentrator* is an (n, m)-network where $n \geq m$, such that for any subset S of m inputs there exists a set of m vertex disjoint paths connecting S to the outputs.

6.2 Union-Networks and Optimal Depth-1 Connectors

Definition 6.4 (The $\triangleleft$-Union). Let $\mathcal{N}_1, \ldots, \mathcal{N}_f$ be (wf, w)-networks, with input sets $A_1, \ldots, A_f$, and output sets $B_1, \ldots, B_f$, respectively. For each $i = 1, \ldots, f-1$, let $\phi_i : A_i \to A_{i+1}$ be some one-to-one mapping. A *left union* or *$\triangleleft$-union* of $\mathcal{N}_1, \ldots, \mathcal{N}_f$ is a $[w, f]$-network $\mathcal{N}$ constructed by identifying each vertex $a \in A_1$ with all vertices $\phi_1(a), \phi_2 \circ \phi_1(a), \ldots, \phi_{f-1} \circ \cdots \circ \phi_1(a)$ (to become an input of $\mathcal{N}$), and let $B_1, \ldots, B_f$ be, naturally, the partition of the outputs of $\mathcal{N}$ (see Figure 4). We denote $\mathcal{N}$ as $\mathcal{N} = \triangleleft(\mathcal{N}_1, \ldots, \mathcal{N}_f)$.

Definition 6.5 (The $\triangleright$-Union). Let $\mathcal{N}_1, \ldots, \mathcal{N}_k$ be any k (m, n)-networks. An (mk, n)-network

$$\mathcal{N} = \triangleright(\mathcal{N}_1, \ldots, \mathcal{N}_k)$$

constructed by identifying outputs of the $\mathcal{N}_i$ in some one-to-one manner is called a *right union* (or *$\triangleright$-union*) of the $\mathcal{N}_i$. The picture is virtually symmetrical to the left union picture.

The next theorem summarizes a few important properties of the union constructions.

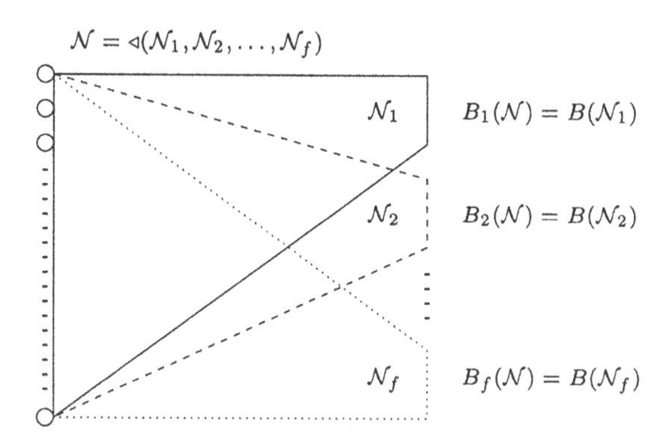

Figure 4: The left union $\mathcal{N}$ of f (wf, w)-networks is a $[w, f]$-network.

Theorem 6.6 (Optimal Depth-1 Construction). *Let w, f be positive integers; then the following hold*

(i) Suppose $\mathcal{N}_1, \dots, \mathcal{N}_f$ are (wf, f)-concentrators; then $\triangleleft(\mathcal{N}_1, \dots, \mathcal{N}_f)$ is a $[w, f]$-connector.

(ii) The network $\mathcal{C}_1(w, f) = \triangleright(\mathcal{B}(wf - w, w), \mathcal{M}(w))$, is a depth-1 (wf, w)-concentrator of size $w(wf - w + 1)$.

(iii) Let $\mathcal{S}_1(w, f)$ be a left union of f copies of $\mathcal{C}_1(w, f)$. Then, $\mathcal{S}_1(w, f)$ is a depth-1 $[w, f]$-connector of size $wf(wf - w + 1)$, which is optimal!

Proof. Parts (i) and (ii) follow from definitions. Assertion (iii) follows from (i) and (ii). Note that (iii) also completes the proof of Theorem 5.6. □

6.3 Product Networks, Constructions of Depth-2 [w,f]-Connectors

Definition 6.7 (The $\times\times$-Product). Let $\mathcal{N}_1$ be an m-network, and $\mathcal{N}_2$ be a $[w, f]$-network; define the *ordered product* (for lack of better term) $\mathcal{N} = \mathcal{N}_1 \times \times \mathcal{N}_2$ as follows. We "connect" wf copies of $\mathcal{N}_1$, denoted by $\mathcal{N}_1^{(1)}, \dots, \mathcal{N}_1^{(wf)}$, to m copies $\mathcal{N}_2^{(1)}, \dots, \mathcal{N}_2^{(m)}$ of $\mathcal{N}_2$. For each $i \in \{1, \dots, wf\}$ and $j \in \{1, \dots, m\}$, we identify the jth output of $\mathcal{N}_1^{(i)}$ with the ith input of $\mathcal{N}_2^{(j)}$. The output partition for $\mathcal{N}$ is defined by

$$B_k(\mathcal{N}) = \bigcup_{j=1}^{m} B_k(\mathcal{N}_2^{(j)}).$$

Naturally, $A(\mathcal{N}) = \cup_{i=1}^{wf} A(\mathcal{N}_1^{(i)})$. Figure 5 illustrates the construction.

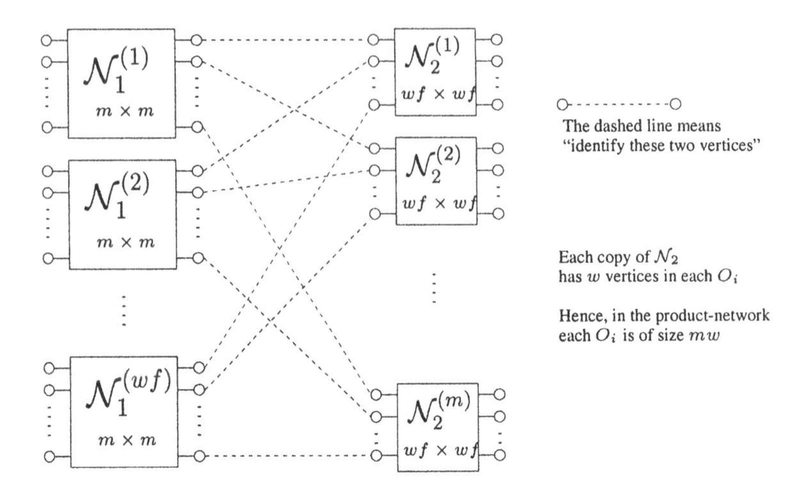

Figure 5: Product of two networks: $\mathcal{N}_1$ is an m-network and $\mathcal{N}_2$ is a $[w, f]$-network.

The following proposition summarizes a few trivial properties of the product network.

Proposition 6.8. *Let $\mathcal{N}_1$ be an m-network of size s_1 and depth d_1, and $\mathcal{N}_2$ a $[w, f]$-network of size s_2 and depth d_2. Then, the network $\mathcal{N} = \mathcal{N}_1 \times \times \mathcal{N}_2$ is an $[mw, f]$-network of size*

$$s = wfs_1 + ms_2,$$

and depth

$$d = d_1 + d_2.$$

Before proving a crucial property of this construction, we need a simple yet important lemma.

Lemma 6.9. *Let $G = (X \cup Y; E)$ be a bipartite multigraph where the degree of each vertex $x \in X$ is m and the degree of each vertex $y \in Y$ is mw. Then, there is an edge coloring for G with exactly m colors such that vertices in X are incident to different colors, and vertices in Y are incident to exactly w edges of each color.*

Proof. Split each vertex $y \in Y$ into w copies $y^{(1)}, \ldots, y^{(w)}$ such that each copy has degree m. The resulting graph is an m-regular bipartite graph, which can be m-edge-colored, by König's line coloring theorem [11]. This induces a coloring of G as desired. □

The following lemma is the point of the ordered-product construction.

Lemma 6.10. *If $\mathcal{N}_2$ is a rearrangeable $[w, f]$-connector and $\mathcal{N}_1$ is a rearrangeable m-connector, then $\mathcal{N} = \mathcal{N}_1 \times \times \mathcal{N}_2$ is a rearrangeable $[mw, f]$-connector.*

Proof. Consider a request frame $\mathcal{D}$ for $\mathcal{N}$. We use $(a^{(i)}, k)$ to denote a request $(a, k) \in \mathcal{D}$ if $a \in A(\mathcal{N}_1^{(i)})$. This is to signify the fact that the request was from the ath input of $\mathcal{N}_1^{(i)}$ to B_k. By definition of a request frame, $|\{(a^{(i)}, k) \mid (a^{(i)}, k) \in \mathcal{D}\}| = m$ for a fixed i. We find vertex disjoint routes realizing requests in $\mathcal{D}$.

Construct a bipartite graph $G = (X \cup Y; E)$ where $X = \{\mathcal{N}_1^{(1)}, \ldots, \mathcal{N}_1^{(wf)}\}$ is the set of all copies of $\mathcal{N}_1$, and $Y = \{B_1, \ldots, B_f\}$. There is (a copy of) an edge of G between $\mathcal{N}_1^{(i)}$ and B_k for each request $(a^{(i)}, k)$ Clearly G is a bipartite graph satisfying the conditions of Lemma 6.9.

As each edge of G represents a request $D \in \mathcal{D}$, Lemma 6.9 implies that there is an m-coloring of all the requests such that, for a fixed i, requests of the form $(a^{(i)}, k)$ get different colors. Moreover, for a fixed k, requests of the form $(a^{(i)}, k)$ can be partitioned into m classes, where each class consists of exactly w requests of the same color.

Let $C = \{1, \ldots, m\}$ be the set of colors. Let $c(a, k)$ denote the color of request $(a, k) \in \mathcal{D}$. Without loss of generality, we number the m outputs of $\mathcal{N}_1^{(i)}$ with numbers from 1 to m, i.e., $B(\mathcal{N}_1^{(i)}) = C$, for all $i \in \{1, \ldots, wf\}$.

Fix an $i \in \{1, \ldots, wf\}$. As the m requests coming out of $\mathcal{N}_1^{(i)}$ have different colors, the correspondence $a^{(i)} \leftrightarrow c(a^{(i)}, k)$, where $(a^{(i)}, k) \in \mathcal{D}$, is a one-to-one correspondence between the inputs and the outputs of $\mathcal{N}_1^{(i)}$. Hence, for the m requests $(a^{(i)}, k)$, there exist m vertex disjoint routes $R_1(a^{(i)}, k)$ connecting input $a^{(i)}$ to the output numbered $c(a^{(i)}, k)$ of $\mathcal{N}_1^{(i)}$.

Fix a $j \in \{1, \ldots, m\}$. The ith input of $\mathcal{N}_2^{(j)}$ is the jth output of $\mathcal{N}_1^{(i)}$, which is the end point of some route $R_1(a^{(i)}, k)$ for which $c(a^{(i)}, k) = j$. Let $k(i, j)$ be the number k such that the request $(a^{(i)}, k) \in \mathcal{D}$ has color $c(a^{(i)}, k) = j$. Then, for the fixed j and any $k \in \{1, \ldots, f\}$, Lemma 6.9 ensures that there are exactly w of the $k(i, j)$ with value k, namely, $|\{i : k(i, j) = k\}| = w$. Thus,

$$\mathcal{D}' = \{(i, k(i, j)) \mid 1 \leq i \leq wf\}$$

is a valid request frame for the rearrangeable $[w, f]$-connector $\mathcal{N}_2^{(j)}$. Consequently, we can find vertex disjoint routes $R_2(i, k(i, j))$ connecting input i to some output in $B_{k(i,j)}$ of $\mathcal{N}_2^{(j)}$.

The concatenation of $R_1(a^{(i)}, k)$ and $R_2(i, k)$ completes a route realizing request $(a^{(i)}, k)$. These routes are vertex disjoint as desired. □

We now illustrate the use of the Lemma 6.10 by a simple construction of depth-2 $[w, f]$-connectors.

Theorem 6.11 (Depth-2 constructions). *Let w, f be positive integers, then*

(i) When $w \leq f - 1$, we can construct depth-2 $[w, f]$-connectors of size $wf(w + f)$.

(ii) In the other case when $w \geq f$, we can construct depth-2 $[w, f]$-connectors of size $2wf\sqrt{w(f-1)}$.

Proof. We ignore the issue of integrality for the sake of a clean presentation.

Write $w = mx$. By Theorem 6.6, $\mathcal{S}_1(x, f)$ is an $[x, f]$-connector of depth 1 and size $xf(xf - x + 1)$. By Proposition 6.8 and Lemma 6.10, the network $\mathcal{B}(m) \times \times \mathcal{S}_1(x, f)$ is a $[w, f]$-connector of depth-2 and size

$$s(x) = xfm^2 + mxf(xf - x + 1) = wf(w/x + (f - 1)x + 1).$$

Minimizing $s(x)$ as a function of x, with $1 \leq x \leq w$, we get the desired results. We pick $x = 1$ in case (i) and $x = \sqrt{w/(f-1)}$ in case (ii). □

6.4 Recursive Constructions

Toward the constructions of $[w, f]$-connectors, we need a few more definitions and properties.

Definition 6.12. Let $w_0, w_1, \ldots, w_k$ and f be positive integers and G be any $[w_0, f]$-network. Let $\mathcal{N}(w_k, \ldots, w_1; G)$ denote the recursively constructed network defined as follows.

$$\begin{aligned} \mathcal{N}(\cdot; G) &= G \\ \mathcal{N}(w_k, \ldots, w_1; G) &= \mathcal{B}(w_k) \times \times \mathcal{N}(w_{k-1}, \ldots, w_1; G). \end{aligned}$$

Lemma 6.13. *Given positive integers $w_0, w_1, \ldots, w_k$ and f, let $w = \prod_{i=0}^{k} w_i$, and G be any $[w_0, f]$-connector of size $s(G)$ and depth $d(G)$. Then, the network $\mathcal{N} = \mathcal{N}(w_k, \ldots, w_1; G)$ is a $[w, f]$-connector of size*

$$s(\mathcal{N}) = \begin{cases} s(G) & k = 0 \\ w_0 \ldots w_k f \cdot (w_1 + \cdots + w_k) + w_1 \cdots w_k \cdot s(G) & k \geq 1 \end{cases}$$

and depth $d(\mathcal{N}) = (k + d(G))$.

Proof. This follows from Proposition 6.8 and Lemma 6.10. □

The following proposition is trivially true.

Proposition 6.14. *For any positive integers w and f, the following hold.*

(i) Let $\mathcal{N}$ be any wf-connector. The $[w, f]$-network $\mathcal{N}'$ obtained by partitioning the outputs of $\mathcal{N}$ arbitrarily into f subsets of size w is a $[w, f]$-connector.

(ii) $\mathcal{B}(f)$ is an f-connector and also a $[1, f]$-connector.

Basically, Proposition 6.14 implies that one can use good $w_0 f$-networks to serve as the network G in Lemma 6.13. A general $[w, f]$-network can then be constructed by decomposing $w = w_0 \cdots w_k$ with the right set of divisors $w_0, \ldots, w_k$. As $[w, f]$-networks of depth 2 have been constructed, we attempt to construct good networks of general depth and networks of a fixed depth at least 3.

Pippenger (1978, [22]) has constructed a rearrangeable n-network, which we call $\mathcal{P}(n)$, of size

$$6n \log_3 n + O(n).$$

Pippenger also constructed rearrangeable n-networks of depth $2i + 1$, $i \geq 1$, and size

$$2(i+1)n \left(\frac{n}{2}\right)^{1/(i+1)} + O(n).$$

An n-connector of depth $2i + 2$ can be constructed by concatenating an n-matching with a depth-$(2i + 1)$ n-connector. Hence, we can construct an n-connector of depth $j \geq 3$ and size

$$2\lceil j/2 \rceil n \left(\frac{n}{2}\right)^{1/\lceil j/2 \rceil} + O(n).$$

We denote this network $\mathcal{P}_j(n)$. For $j = 2$, Feldman and Friedman and Pippenger (1988, [6]) gave a construction of size $O(n^{5/3})$. Hwang and Richards (1993, [10]) also obtained the same result. Abusing notation, we also use $\mathcal{P}_2(n)$ to denote an n-connector of depth-2 and size $O(n^{5/3})$.

In the following results, we ignore the issue of integrality for the sake of clarity. We first address the general depth case.

Theorem 6.15. *We can construct rearrangeable $[w, f]$-connectors of size*

$$wf \ln w + \frac{6}{\ln 3} wf \ln f + O(f).$$

Proof. Let $w = xw_1 \cdots w_k$. By Lemma 6.13 the network

$$\mathcal{N} = \mathcal{N}(w_k, \ldots, w_1; \mathcal{P}(xf))$$

is a $[w, f]$-connector of size

$$\begin{aligned} s(\mathcal{N}) &= wf(w_1 + \cdots + w_k) + 6fx \log_3(fx) + O(fx) \\ &\geq wf \cdot k \cdot (\frac{w}{x})^{1/k} + 6fx \log_3(fx) + O(fx). \end{aligned}$$

The right hand side is minimized at $x = 1$ and $k = \ln w$. Equality can be obtained by setting $w_i = w^{1/k}, \forall i$. □

We now consider the fixed depth case. The networks $\mathcal{P}_j(n)$ are to be used. The following three theorems apply Lemma 6.13 with $G = \mathcal{P}_1, \mathcal{P}_2$, or $\mathcal{P}_j$ with $j \geq 3$. Depending on the relative values among f, w and k, one theorem may be better than the others.

Theorem 6.16. *Let w, f, and $k \geq 3$ be positive integers.*

(i) If $w < (f-1)^{k-1}$, then we can construct a $[w, f]$-connector of depth k and size

$$(k-1)fw^{1+1/(k-1)} + wf^2 = O(kwf^2). \quad (17)$$

(ii) If $w \geq (f-1)^{k-1}$, then we can construct a $[w, f]$-connector of depth k and size

$$k \cdot [w(f-1)]^{1+1/k} + wf = O\left(k(wf)^{1+1/k}\right). \quad (18)$$

Proof. Write $w = xw_1 \cdots w_{k-1}$. Then, Lemma 6.13 implies that

$$\mathcal{N}(w_{k-1}, \ldots, w_1; \mathcal{S}_1(x, f))$$

is a $[w, f]$-network of depth k and size

$$\begin{aligned} s = wf(w_1 + \cdots + w_{k-1}) + w_1 \cdots w_{k-1} \cdot xf(xf - x + 1) \\ \geq wf(k-1)(w/x)^{1/(k-1)} + wf(x(f-1)+1). \quad (19) \end{aligned}$$

Minimizing the right hand side with respect to x and we get the desired results.

In case (i), equality can be obtained when $w_i = w^{1/(k-1)}, \forall i$, and $x = 1$. In case (ii), equality can be obtained when $w_i = w^{1/(k-1)}, \forall i$, and $x = \left(\frac{w}{(f-1)^{k-1}}\right)^{1/k}$. □

Theorem 6.17. *Let w, f, and $k \geq 3$ be positive integers. Then, there are constants c_1, c_2, and c_3 such that*

(i) If $w < c_1 f^{2(k-2)/3}$, then we can construct a $[w, f]$-connector of depth k and size

$$(k-2)f \cdot w^{1+1/(k-2)} + c_2 w f^{5/3} = O\left(kwf^{5/3}\right). \qquad (20)$$

(ii) If $w \geq c_1 f^{2(k-2)/3}$, then we can construct a $[w, f]$-connector of depth k and size

$$\begin{aligned}(k-2) \cdot (wf)^{1+1/(k-3/2)} &+ c_3 w^{1+1/(k-3/2)} f^{(6k-7)/(6k-9)} \\ &= O\left(k(wf)^{1+1/(k-3/2)}\right). \qquad (21)\end{aligned}$$

Proof. Recall that $\mathcal{P}_2(n) \approx cn^{5/3}$ for some constant c. Write $w = xw_1 \cdots w_{k-2}$. Then, Lemma 6.13 implies that

$$\mathcal{N}(w_{k-2}, \ldots, w_1; \mathcal{P}_2(xf))$$

is a $[w, f]$-network of depth k and size

$$\begin{aligned} s &= wf(w_1 + \cdots + w_{k-2}) + w_1 \cdots w_{k-2} \cdot (xf)^{5/3} \\ &\geq wf\left((k-2)(w/x)^{1/(k-2)} + c(xf)^{2/3}\right). \qquad (22)\end{aligned}$$

Minimizing the right hand side with respect to x and we get the desired results.
In case (i), equality can be obtained when $w_i = w^{1/(k-2)}, \forall i$, and $x = 1$. In case (ii), equality can be obtained when $w_i = w^{1/(k-2)}, \forall i$, and

$$x = \frac{w^{3/(2k-3)}}{c' f^{(2k-4)/(2k-3)}}$$

for some constant c'. □

The following result can be improved by finer analysis. We give a somewhat "cleaner" version.

Theorem 6.18. *Let w, f, and $k \geq 4$ be positive integers.*

(i) If $w < f$, then we can construct a $[w, f]$-connector of depth k and size

$$O\left(kw^{1+1/(k+1)} f^{1+2/(k+1)}\right). \qquad (23)$$

(ii) If $w \geq f$, then we can construct a $[w, f]$-connector of depth k and size

$$O\left(k(wf)^{1+3/(2(k+1))}\right). \tag{24}$$

Proof. For both cases, we use the network

$$\mathcal{N}\left(w_{k-2i-1}, \ldots, w_1; \mathcal{P}_{2i+1}(xf)\right),$$

where $w = xw_1 \cdots w_{k-2i-1}$, and $i \leq \frac{k}{2} - 1$. Case (ii) is obtained by setting $x = \sqrt{w/f}$, $i = (k-2)/3$, and $w_j = (wf)^{\frac{1}{2(k-2i-1)}}$. Case (i) is obtained by setting $x = 1$, $i = \frac{k-1-\lg w/\lg f}{2+\lg w/\lg f}$, and $w_j = w^{1/(k-2i-1)}$. □

Acknowledgments

Dr. Hung Q. Ngo is supported in part by NSF CAREER Award CCF-0347565.

References

[1] BASSALYGO, L. A., AND PINSKER, M. S. The complexity of an optimal non-blocking commutation scheme without reorganization. *Problemy Peredači Informacii 9*, 1 (1973), 84–87.

[2] BENEŠ, V. E. *Mathematical Theory of Connecting Networks and Telephone Traffic*. Academic Press, New York, 1965. Mathematics in Science and Engineering, Vol. 17.

[3] BRÈGMAN, L. M. Certain properties of nonnegative matrices and their permanents. *Dokl. Akad. Nauk SSSR 211* (1973), 27–30.

[4] CANTOR, D. G. On non-blocking switching networks. *Networks 1* (1971/72), 367–377.

[5] CLOS, C. A study of non-blocking switching networks. *Bell System Tech. J. 32* (1953), 406–424.

[6] FELDMAN, P., FRIEDMAN, J., AND PIPPENGER, N. Wide-sense nonblocking networks. *SIAM J. Discrete Math. 1*, 2 (1988), 158–173.

[7] FRIEDMAN, J. A lower bound on strictly nonblocking networks. *Combinatorica 8*, 2 (1988), 185–188.

[8] HINTON, H. *An Introduction to Photonic Switching Fabrics*. Plenum, New York, 1993.

[9] HWANG, F. K. *The Mathematical Theory of Nonblocking Switching Networks*. World Scientific, River Edge, NJ, 1998.

[10] HWANG, F. K., AND RICHARDS, G. W. A two-stage network with dual partial concentrators. *Networks 23*, 1 (1993), 53–58.

[11] KÖNIG, D. Über graphen und ihre anwendung auf determinantentheorie und mengenlehre. *Math. Ann. 77* (1916), 453–465.

[12] Lucent Technologies Press Release. Lucent Technologies unveils untra-high-capacity optical system; Time Warner Telecom first to announce it will deploy the system, 2001. http://www.lucent.com/press/0101/010117.nsa.html.

[13] Lucent Technologies Press Release. Lucent Technologies engineer and scientists set new fiber optic transmission record, 2002. http://www.lucent.com/press/0302/020322.bla.html.

[14] Lucent Technologies Website. What is dense wave division multiplexing (DWDM), 2002. http://www.bell-labs.com/technology/lightwave/dwdm.html.

[15] MINC, H. Upper bounds for permanents of (0, 1)-matrices. *Bull. Amer. Math. Soc. 69* (1963), 789–791.

[16] MUKHERJEE, B. *Optical Communication Networks.* McGraw-Hill, New York, 1997.

[17] NGO, H. Q. Multiwavelength distribution networks. In *Proceedings of the 2004 Workshop on High Performance Switching and Routing (HPSR 2004, Phoenix, Arizona)* (2004), IEEE, pp. 186–190.

[18] NGO, H. Q. WDM switching networks, rearrangeable and nonblocking $[w, f]$-connectors. SIAM J. Comput., to appear.

[19] NGO, H. Q. AND DU, D.-Z. Notes on the complexity of switching networks. In *Advances in Switching Networks*, D.-Z. Du and H. Q. Ngo, Eds. Kluwer Academic, Hingham, MA, 2001, pp. 307–367.

[20] NGO, H. Q., PAN, D., AND QIAO, C. Nonblocking WDM switches based on arrayed waveguide grating and limited wavelength conversion. In *Proceedings of the 23rd Conference of the IEEE Communications Society (INFOCOM'2004, HongKong)* (2004), IEEE.

[21] NGO, H. Q., PAN, D., AND YANG, Y. Optical switching networks with minimum number of limited range wavelength converters. In *Proceedings of the 24rd Conference of the IEEE Communications Society (INFOCOM'2005, Miami, U.S.A.)* (2005), IEEE.

[22] PIPPENGER, N. On rearrangeable and nonblocking switching networks. *J. Comput. System Sci. 17*, 2 (1978), 145–162.

[23] PIPPENGER, N. A new lower bound for the number of switches in rearrangeable networks. *SIAM J. Algebraic Discrete Methods 1*, 2 (1980), 164–167.

[24] PIPPENGER, N. Communication networks. In *Handbook of Theoretical Computer Science, Vol. A.* Elsevier, Amsterdam, 1990, pp. 805–833.

[25] PIPPENGER, N. AND YAO, A. C. C. Rearrangeable networks with limited depth. *SIAM J. Algebraic Discrete Methods 3*, 4 (1982), 411–417.

[26] QIN, X. AND YANG, Y. A cost-effective construction for WDM multicast switching networks. In *Proceedings of the 2002 IEEE International Conference on Communications (ICC 2002)* (New York, 1979), vol. 5, IEEE, pp. 2902–2906.

[27] RAMAMIRTHAM, J. AND TURNER, J. S. Design of wavelength converting switches for optical burst switching. In *Proceedings of the 21st Annual Joint Conference of the IEEE Computer and Communications Societies (INFOCOM)* (2002), vol. 2, IEEE, pp. 1162–1171.

[28] RAMASWAMI, R. AND SIVARAJAN, K. *Optical Networks: A Practical Perspective (Second Edition).* Morgan Kaufmann, San Francisco, 2001.

[29] RASALA, A. AND WILFONG, G. Strictly non-blocking WDM cross-connects. In *Proceedings of the Eleventh Annual ACM-SIAM Symposium on Discrete Algorithms (SODA'2000, San Francisco, CA)* (New York, 2000), ACM, pp. 606–615.

[30] RASALA, A. AND WILFONG, G. Strictly non-blocking WDM cross-connects for heterogeneous networks. In *Proceedings of the Thirty-Second Annual ACM Symposium on Theory of Computing (STOC'2000, Portland, OR)* (New York, 2000), ACM, pp. 513–524.

[31] ROBBINS, H. A remark on Stirling's formula. *Amer. Math. Monthly 62* (1955), 26–29.

[32] SHANNON, C. E. Memory requirements in a telephone exchange. *Bell System Tech. J. 29* (1950), 343–349.

[33] SINGHAL, A. AND JAIN, R. Terabit switching: A survey of techniques and current products. *Computer communications 25*, 8 (2002), 547–556.

[34] STERN, T. E. AND BALA, K. *Multiwavelength Optical Networks: A Layered Approach.* Prentice Hall, Upper Saddle River, NJ, 1999.

[35] WANG, Y. AND YANG, Y. Multicasting in a class of multicast-capable WDM networks. *J. Lightwave Technol. 13*, 2 (Feb 2002), 128–141.

[36] WEST, D. B. *Introduction to graph theory.* Prentice Hall, Upper Saddle River, NJ, 1996.

[37] WILFONG, G., MIKKELSEN, B., DOERR, C., AND ZIRNGIBL, M. WDM cross-connect architectures with reduced complexity. *J. of Lightwave Technology 17*, 10 (Oct 1999), 1732–1741.

[38] YANG, Y. AND WANG, J. Designing WDM optical interconnects with full connectivity by using limited wavelength conversion. In *Proceedings of the 18th IEEE International Parallel and Distributed Processing Symposium (IPDPS'04)* (2004), IEEE.

[39] YANG, Y. AND WANG, J. WDM optical switching networks using sparse crossbars. In *Proceedings of the 23th Annual Joint Conference of the IEEE Computer and Communications Societies (INFOCOM'2004, Hong Kong)* (2004), IEEE.

[40] YANG, Y., WANG, J., AND QIAO, C. Nonblocking WDM multicast switching networks. *IEEE Trans. Para. and Dist. Sys. 11*, 12 (Dec 2000), 1274–1287.

Chapter 15

Topological Properties of Interconnection Networks

Ivan Stojmenovic
SITE
University of Ottawa, Ottawa, Ont K1N6N5, Canada
E-mail: ivan@site.uottawa.ca

1 Introduction

Computers are widely used in everyday life and business. Currently almost all of them have one processor that is able to perform instructions one after the other, in a sequence. These computers are usually referred to as sequential computers. For about fifty years computer manufacturers were able to design computers that were every year approximately twice as fast or with twice the memory than in the previous year. This trend still continues, and the demand for computing grows even faster. There soon will be a future time when sequential computers reach their physical limits in speed (they cannot perform faster than the speed of electron transmission). Computers based on a different computing philosophy have already emerged. One such alternative is parallel computers, based on a very simple idea of using several processors that cooperate to solve a problem instead of using only one. Starting from the general idea, various specific models of parallel computation were considered. Each particular model assumes either shared or local memory, simple or powerful processors, synchronous or asynchronous operation of processors, and defines the ways in which processors can communicate.

In the shared-memory model processors use a common memory. A dynamic network built out of crossbar switching devices is used to connect processors to a system of memory modules. A comprehensive account of dynamic networks can be found in [25]. Local memory models consist of processors, each having its own memory (of relatively small size), which are interconnected by a fixed scheme. Such a static network is commonly called an interconnection network. An interconnection network consists of a set of processors, each with a local memory, and a set of bidirectional links that serve for the exchange of data between processors. A convenient representation of an interconnection network is by an undirected (in some cases directed) graph $G = (V, E)$ where each processor A_i is a vertex in V, and two vertices A_i and A_j are connected by an edge $(A_i, A_j) \in E$ if and only if there is a direct (bidirectional for undirected and unidirectional for directed graphs) communication link between processors A_i and A_j. Such processors are called neighbors. The interconnection graph of a network is often referred to as its topology. We use the terms 'interconnection network' and 'graph' interchangeably; similarly we use 'node', 'vertex' and 'processors' with the same meaning, and, finally, the terms 'edge', and 'link' are synonyms.

Usually all processors in a network are identical, and each is assumed to have input and output ability. Sometimes a master processor is recognized as a leader and more powerful than others, often as the only one with i/o functions. Processors may execute the same or different programs, the network may be regular or irregular, and the number of neighbors of each processor may be a constant or a function of the size of the network. The time complexity of any algorithm has two components: computation time which covers local computation by every processor, and the communication time which is the time needed for the exchange of data between processors. The state of the art in technology is that a message exchange takes considerably more time than a computing step inside a processor. Thus there is a higher demand for reducing communication than computation. The situation may change with the development of fast communication technology.

Each processor is assumed to know its own coordinates within the network of n processors and to have a certain number of registers of size $O(\log n)$; in unit time, every processor performs some arithmetic or Boolean operation, or communicates with one of its neighbors using a local link. Processors in a network may operate synchronously or asynchronously. Synchronous computation is also referred to as the single instruction multiple data (SIMD) mode. It means that at each time unit, the same instruction is broadcast to all processors, which execute it and wait for the next instruction. We refer here to such a computing mode as parallel computation. Asynchronous computation corresponds to the multiple instruction multiple data (MIMD) mode, where each processors has its own program and runs it on its own data; the processors occasionally communicate with each other to exchange data. This mode is referred here as the distributed computation.

We outline some of the most widely used networks. For each of them we discuss some of its basic topological properties. A topology is evaluated in terms of the following main parameters: degree (called also the fanout), diameter, symmetry, bisection width (see section on lower bounds), and wire length (see section on layout). Although all characteristics are important, it seems that the network cost, defined as the product of the degree and diameter (measured with respect to the number of nodes) is the most important parameter of a network. Further important measures of 'goodness' of a network are existence of optimal data communication techniques such as routing and broadcasting, node disjoint paths, embeddability, and recursive decomposition (or scalability). The distance between two processors is the smallest number of communication links that a message has to traverse in order to be routed between the two processors. It corresponds to the number of edges in the shortest path between two processors (where each edge has unit weight). The diameter of a network is the farthest distance between any two processors. The degree of a network is the maximal number of neighbors of a processor.

In the coming subsections we describe some interconnection network models. Except meshes with multiple broadcasting (MMB), all mentioned networks are direct networks, which are networks that allow only direct communication between neighboring processors. Some networks, such as MMBs, are enhanced by adding buses that allow quick broadcasting of a datum from one processor to a group of other processors. There are other models with buses that are not studied here. In indirect networks, nodes in the underlying graph are distinguished as either processors or switches or network controllers, etc. Switches are simple elements that can perform a comparison and forward the message in one or another direction. They are considerably less powerful than processors. Indirect networks, in general, have high bisection width, low degree, low diameter, and long wires, and are in practice considerably less popular than direct networks. They are not studied in this chapter. Lately, one more kind of network was introduced, called reconfigurable networks.

They allow processors to dynamically change the links to their neighbors during the execution times, forming various buses for fast broadcasting of data. They are also not studied here.

Two extreme models of interconnection networks are fully connected graph (with every node connected to every other node) and independent processors (no edge in the graph exists; in this model there exists a central processor that distributes the job to others and collects results from them, therefore in reality it is a graph in which one node is connected to all other nodes with no additional edges). The implementation of the fully connected graph has physical limits with the current technology.

2 Network Topologies

A simple way to arrange n processors in a network is a one-dimensional linear array. Complete binary trees can be considered as interconnection networks. In the sequel we describe other, both traditional and recent, types of interconnection networks. A number of networks are described briefly in exercises. A survey article [29] gives more details on some networks and also summarizes routing and broadcasting algorithms on a number of networks.

2.1 Mesh-Connected Computer

Processors in a mesh-connected computer occupy vertices of an $m \times m$ grid, where $n = m^2$. For easy reference, a processor in row i and column j is denoted $p(i,j)$, $1 \leq i,j \leq m$. Processor $p(i,j)$ is connected via bidirectional links to its four neighbors $p(i+1,j)$, $p(i-1,j)$, $p(i,j+1)$, and $p(i,j-1)$, whichever of them exists; in other words, each node is connected with its four neighbors immediately above, below, to the left and right of it. Figure 1 shows a mesh-connected computer (MCC) with 36 processors. A MCC has $n = m^2$ nodes, with degree 4 and diameter $2(m-1) = 2(\sqrt{n}-1) = O(\sqrt{n})$. The number of edges is $2(n-\sqrt{n})$.

In many applications it is important to define an ordering among processors, i.e. an indexing from 1 to n (for example, in order to call some data 'sorted'). There are two common indexing schemes. In row major order, processor i is in row j and column k such that $i = (j-1)\sqrt{n}+k$. In snakelike row-major order, processor i is placed in row j and column k of MCC such that $i = (j-1)\sqrt{n}+k$ when j is odd, and $i = (j-1)\sqrt{n}+\sqrt{n}-k+1$ when j is even.

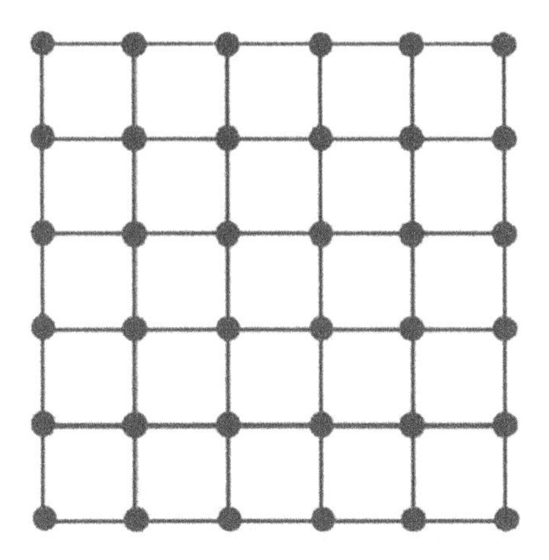

Figure 1: Mesh-connected computer.

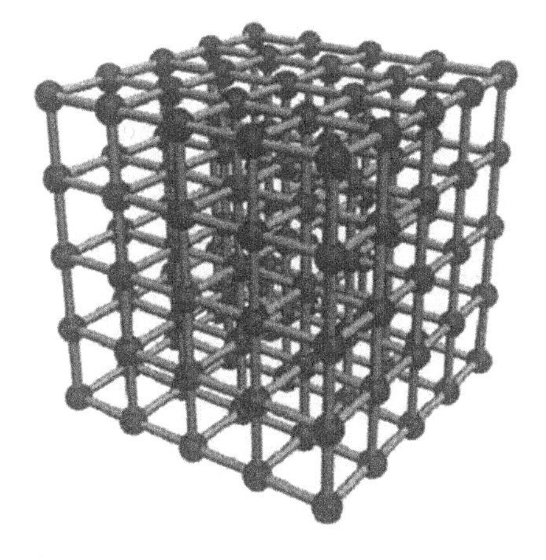

Figure 2: 3-dimensional 5-sided array.

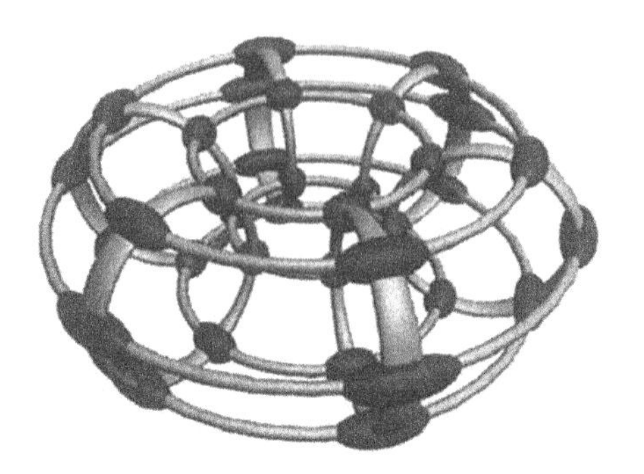

Figure 3: 6 x 6 torus.

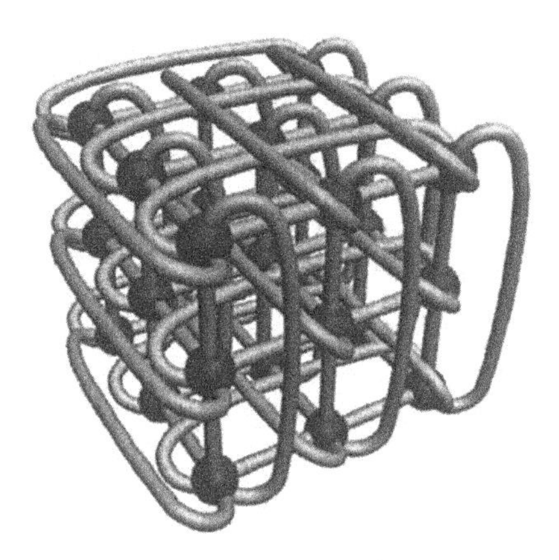

Figure 4: 3-ary 3-cube.

The MCC architecture (and its variants presented below) can be nicely embedded in the plane making it useful for VLSI implementation, and is commercially available. The definition of two-dimensional MCC can be easily generalized to higher-dimensional arrays. A k-dimensional r-sided array has r^k nodes that correspond to vectors $(c_1, c_2, \ldots, c_k)$ where $0 \leq c_j \leq r-1$ for $1 \leq j \leq k$. Two nodes are linked by an edge if they differ in precisely one coordinate and if the absolute value of the difference in that coordinate is 1. Figure 2 shows a 3-dimensional 5-sided array. These networks may also have rectangular shapes rather than square one.

MCCs are asymmetric networks. They can be made symmetric by adding additional links. This can be done in several ways. The most popular enhanced mesh network is the torus network [21], obtained from a mesh-connected computer when the topmost and bottommost processor in each column (and similarly the rightmost and leftmost processor in each row) are connected. The torus network is vertex and edge symmetric. A k-ary n-cube is a generalization of a torus network. Each node has an address of the form $(a_1, a_2, \ldots, a_n)$, where $0 \leq a_i \leq k-1$. Each such node has $2n$ neighbors, that is, two neighbors along each dimension. The neighbors along the i-th dimension share the $n-1$ coordinates, and the i-th coordinates are $a_i + 1 \pmod k$ and $a_i - 1 \pmod k$, respectively. That is, the neighbors of node $(a_1, a_2, \ldots, a_n)$ are $(a_1, \ldots, a_{i-1}, a_i + 1 \bmod k, a_{i+1}, \ldots, a_n)$ and $(a_1, \ldots, a_{i-1}, a_i - 1 \bmod k, a_{i+1}, \ldots, a_n)$. Therefore this is the n-D mesh model with added wraparound edges. Note that a 2-ary n-cube is an n-dimensional hypercube. Figure 3 shows a 6 x 6 torus, while Figure 4 shows a 3-ary 3-cube.

The (r, k)-MC (multiplicative circulant) graph is defined in [27] as follows. The vertices of the graph are addressed $0, 1, 2, \ldots, n-1$. Two vertices x and y are connected by an edge if and only if $|x - y| = r^i \bmod n$ for some integer i, $0 \leq i \leq k-1$. Each node m of an (r, k)-MC network is connected by an edge to the following nodes: $m+1, m-1, m+r, m-r, m+r^2, m-r^2, \ldots, m+r^{k-1}, m-r^{k-1}$. These nodes are all different, because any two neighbors differ by at most $2r^{k-1} \leq r^k$, with equality valid only when $r = 2$. Therefore each node is connected to $2k$ neighbors for $r > 2$ and to $2k-1$ neighbors for $r = 2$. This defines the degree of the (r, k)-MC graph. All (r, k)-MC graphs are vertex symmetric. In general, they are not edge symmetric.

It follows easily that an (r, k)-MC network has a Hamiltonian cycle, consisting of nodes $0, 1, 2, 3, \ldots, n-1$, because the nodes with consecutive addresses are neighbors. One can build an $(r, k+1)$-MC network from r copies of an (r, k)-MC network by assigning copy numbers from 0 to $r-1$ to them, changing the address of node y in copy x to $ry + x$, and adding edges between nodes (with their new addresses) which are at distance ± 1. The latter is equivalent to adding all edges in a Hamiltonian cycle $0, 1, 2, \ldots, r^{k+1} - 1$.

The diameter of the (r, k)-MC network is as follows: for $r = 2$ it is $\lceil k/2 \rceil$ [3], for r odd it is $k\lfloor r/2 \rfloor$ [33], and for $r > 2$ it is $kr/2 - \lfloor k/2 \rfloor$ [27].

Figure 8 shows a (5,2)-MC graph. In general, a k-dimensional r-sided array is a subgraph of an (r, k)-MC graph [6], and the latter can be considered as a kind of 'twisted' torus.

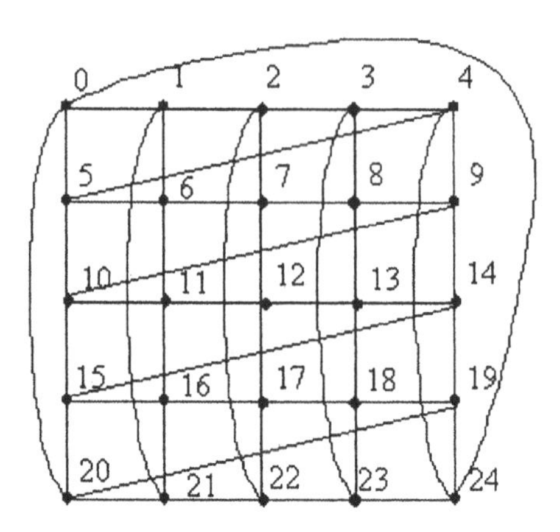

Figure 5: (5,2)-multiplicative circulant network.

2.2 Meshes with Multiple Broadcasting

Interconnection networks are sometimes enhanced by adding additional features; in particular, by adding buses that can serve for broadcasting a datum from one processor in unit time to any other processor that is linked to the bus. Several variations of this idea exist and are investigated in the literature: meshes with one bus connecting all processors, meshes with horizontal buses in each row, meshes with horizontal and vertical buses, and reconfigurable meshes (in this model each processor is connected or disconnected along a separate link with its neighbors, and the decision can be made by a program; in other words, programs are able to create various buses at each step, in addition to permanent links between neighboring processors). We consider here one of these variants of MCCs.

A mesh with multiple broadcasting (MMB, for short) consists of a mesh connected computer enhanced by the addition of row and column buses. It is proposed in [20]. More formally, a MMB of size $n^{\beta} \times n^{1-\beta}$ (without loss of generality, we may assume $0 \leq \beta \leq 0.5$) consists of n identical processors positioned on a rectangular array overlaid with a bus system. In every n^{β} row of the mesh the processors are connected to a horizontal bus; similarly, in every $n^{1-\beta}$ column the processors are connected to a vertical bus (see Figure 6).

For practical reasons, only one processor is allowed to broadcast on a given bus at any one time. By contrast, all the processors on the bus can simultaneously read the value being broadcast. Experiments with

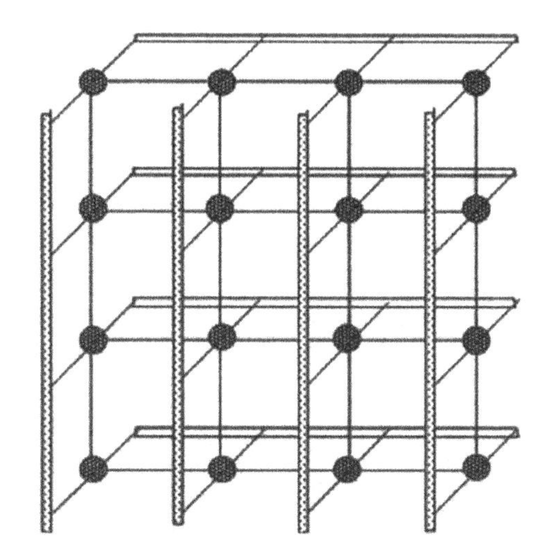

Figure 6: Mesh with multiple broadcasting.

existing multiprocessor arrays indicate that the communication along buses take $O(1)$ time.

Due to the bus characteristics, the MMB operates in SIMD mode. In unit time, every processor performs some arithmetic or Boolean operation, communicates with one of its neighbors using a local link, broadcasts a value on a bus, or reads a value from a specific bus.

2.3 Hexagonal and Honeycomb Meshes

Hexagonal [11] and honeycomb [28] meshes are obtained by using tessellations with regular triangles and hexagons, respectively (note that a hexagonal mesh is obtained from triangular tessellation; the original network names from literature are preserved here). Both networks have similar characteristics as MCCs. Figure 7 shows these two networks. Both of them can be expanded to the corresponding tori. It appears that [11] proposed a cumbersome addressing scheme, which has led to a page-long complicated routing algorithm, and similarly complex broadcasting scheme. We describe here the coordinate system that was proposed for a honeycomb network in [28], and for a hexagonal one in [16]. In this scheme, three axes, x, y, and z, parallel to three edge directions, and at a mutual angle of 120 between any two of them are introduced, as indicated in Figure 8. Let i, j, and k be three unit vectors in these axes. These three vectors are, obviously, not independent. More precisely, they

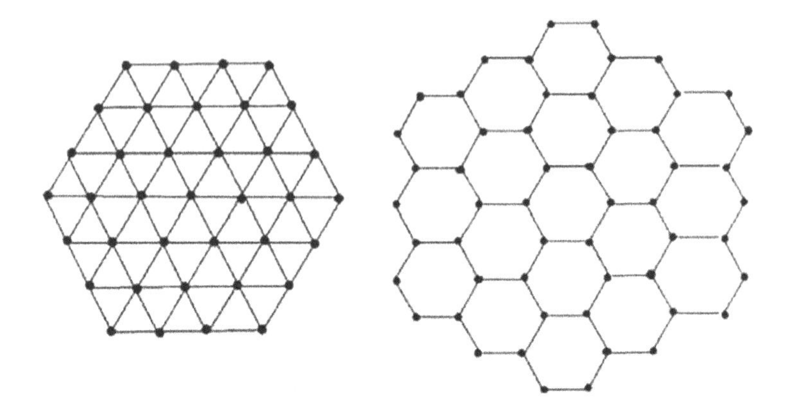

Figure 7: Hexagonal and honeycomb meshes.

are related by $i + j + k = 0$. However, this redundancy greatly simplifies addressing, the distance formula, and the routing algorithm. Using three such axes seems to be the only common property of addressing schemes for honeycomb and hexagonal networks, because details are different.

Given a hexagonal network, choose any node as the origin and assign $(0,0,0)$ as its address. For any other node A on the network, if there is a path from the origin to node A, and the path has altogether a units of vector i , b units of vector j, and c units of vector k, then the address for node A is $(a,b,c) = ai + bj + ck$. An address (a,b,c) for node A is of the shortest-path form if there is a path from the origin to node A, consisting of a units of vector i, b units of vector j, and c units of vector k, and the path has the shortest length. A node address (a,b,c) is of the shortest path form if and only if the following conditions are satisfied [16]: (1) At least one component is zero (that is, $abc = 0$); (2) Any two components cannot have the same sign (that is, $ab \leq 0$, $ac \leq 0$, and $bc \leq 0$).

The node address of the shortest-path form is unique. If $D - S = (a,b,c)$ is the shortest path from S to D, then the length of the shortest path between S and D is $|D - S| = |a| + |b| + |c|$. If $D - S = ai + bj + ck$ then $|D - S| = \min\ (|a-c| + |b-c|, |a-b| + |b-c|, |a-b| + |a-c|)$.

A node address (a,b,c) is of the zero-positive form if and only if the following conditions are satisfied [16]: (1) At least one component is zero (that is, $abc = 0$); (2) All components are nonnegative (that is, $a \geq 0$, $b \geq 0$, and $c \geq 0$). The node address of the zero-positive form is unique. If $D - S = (a,b,c)$ which is of the zero-positive form, then $|D - S| =$

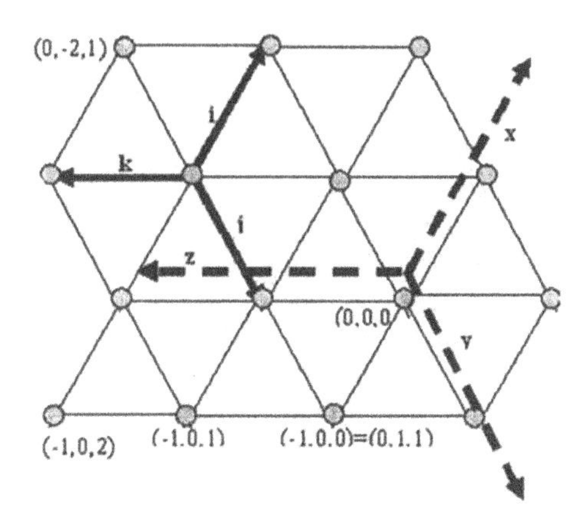

Figure 8: Addressing scheme for hexagonal networks.

max $(|a|, |b|, |c|)$.

Nodes of a honeycomb of size t can be coded by integer triples (u, v, w) such that $-t+1 \leq u, v, w \leq t$, and $1 \leq u+v+w \leq 2$. Two nodes $(u', v', w'$ and (u'', v'', w'') are connected by an edge iff $|u'u''| + |v'v''| + |w'w''| = 1$ [28]. The distance between nodes (u', v', w') and (u'', v'', w'') of a honeycomb is $|u'u''|+|v'v''|+|w'w''|$, and the diameter is $4t-1$ (nodes $(t, -t+1, 0)$ and $(-t+1, t, 1)$ are at distance $|2t-1|+|2t-1|+1 = 4t-1$). Thus, a honeycomb mesh network with n nodes has degree three and diameter $\approx 1.63\sqrt{n} - 1$. Honeycomb meshes have 25% smaller degree and 18.5% smaller diameter than the mesh-connected computer with the same number of nodes.

Higher-dimensional honeycombs and hexagons are based on the following observation. It has been shown in [9] that there exist $k+1$ vectors $X_1, X_2, \ldots, X_k, X_{k+1}$ in a k-dimensional space such that $X_1 + X_2 + \ldots + X_k + X_{k+1} = 0$ and the dot product $X_i X_j = -1/k$ for any pair for distinct coordinates i and j (in other words, these $k+1$ vectors are fully symmetrical in space). For instance, in three dimensions, these vectors are those perpendicular to the tetrahedron faces, and their exact values are listed in [9]. Higher-dimensional hexagonal networks can be constructed as follows [17]. Starting from the origin, unit size vectors will be added and will lead to new nodes along $k+1$ symmetrically positioned vectors (in both positive and negative orientations along these vectors), so that each node has at most $2k+2$ neighbors. Two nodes are neighbors in such a graph if and only if they differ by one of such $2k+2$ unit size

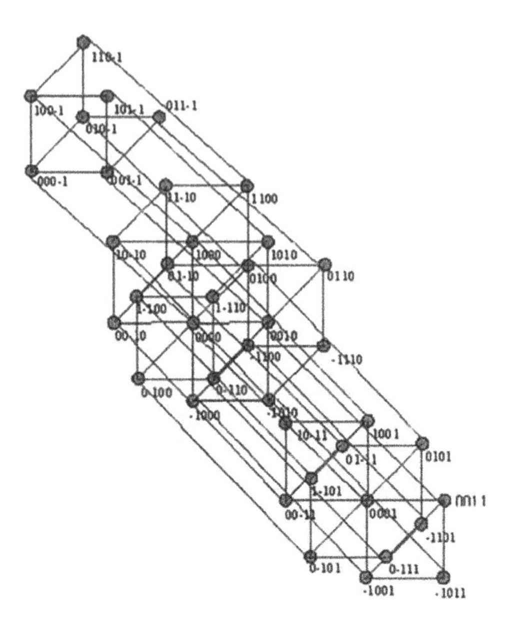

Figure 9: A three-dimensional hexagonal network of size 1.

vectors (in the vector sense). Although such a construction is intuitively clear, the addressing of nodes in such networks is not obvious. Choose any node as the origin and assign $(0, 0, \ldots, 0)$ ($k+1$ zeros) as its address. For any other node A on the network, if there is a path from the origin to node A, and the path has altogether $|a_i|$ units of vector $sign(a_i)X_i$ (that is, X_i for $a_i > 0$ and $-X_i$ otherwise), $1 \leq i \leq k+1$, then an address for node A is $(a_1, \ldots, a_{k+1}) = a_1X_1 + \ldots + A_{k+1}X_{k+1}$. An address $(a_1, \ldots, a_{k+1})$ for node A is of the shortest-path form if there is a path from the origin to node A, consisting of $|a_i|$ units of either vector X_i, (for $a_i > 0$) or vector $-X_i$ (for $a_i < 0$), $1 \leq i \leq k+1$, and the path has the shortest possible length. The distance between two nodes A and B is $|a_1| + \ldots + |a_{k+1}|$, where $B - A = (a_1, \ldots, a_{k+1})$ is in the shortest-path form. The selection of a unique address for each node is discussed in [17]. Figure 9 shows a three-dimensional hexagonal network of size 1.

Nodes of a k-D honeycomb are $(k+l)$-tuples $(u_1, u_2, \ldots, u_{k+1})$ such that $u_1 + u_2 + \ldots + u_{k+1} = 1$ (black node) or $= 2$ (white node). Two nodes $(u'_1, u'_2, \ldots, u'_{k+1})$ and $(u''_1, u''_2, \ldots, u''_{k+1})$ are connected if and only if $|u'_1 - u''_1| + |u'_2 - u''_2| + \ldots + |u'_{k+1} - u''_{k+1}| = 1$. This means that a k-D honeycomb is a bipartite graph. The distance between two such nodes

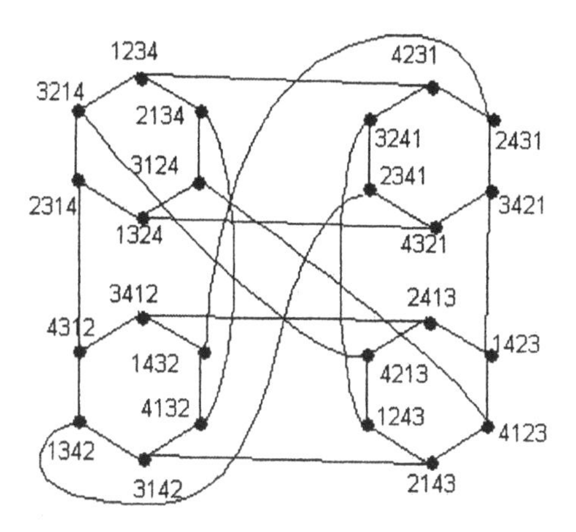

Figure 10: Star.

is $|u'_1 - u''_1| + |u'_2 - u''_2| + \ldots + |u'_{k+1} - u''_{k+1}|$. The formulas for diameter of k-D honeycomb are given in [17].

2.4 Star, Pancake, Arrangement, and Rotator Graphs

In this section we describe some networks that are based on permutations and permutations of combinations. Each processor of both star and pancake networks corresponds to a distinct permutation of k symbols; for simplicity, the symbols are $1, 2, \ldots, k$. Therefore the number of processors is $n = k!$. It is easier to label processors by the corresponding permutation. To number processors as integers from 1 to $k!$, one has to rank the permutations in a certain way (say, lexicographic or a minimum change order).

In the star network, denoted by S_k, a processor u is connected to a processor u if and only if the corresponding permutation of u can be obtained from that of v by exchanging the first symbol with the ith symbol, for some i, $2 \leq i \leq k$. For example, processors 3421 and 2431 are neighbors because they are obtained by exchanging the first and third symbols. Figure 10 shows S_4. The star network is introduced in [1].

In the pancake network, denoted by P_k, a processor v is connected to a processor u if and only if the label of u can be obtained from that of v by flipping the first i symbols, for some i, $2 \leq i \leq k$. For example, 3142 and 2413 are neighbors because they are obtained from each other by flipping all four symbols. Figure 11 shows P_4.

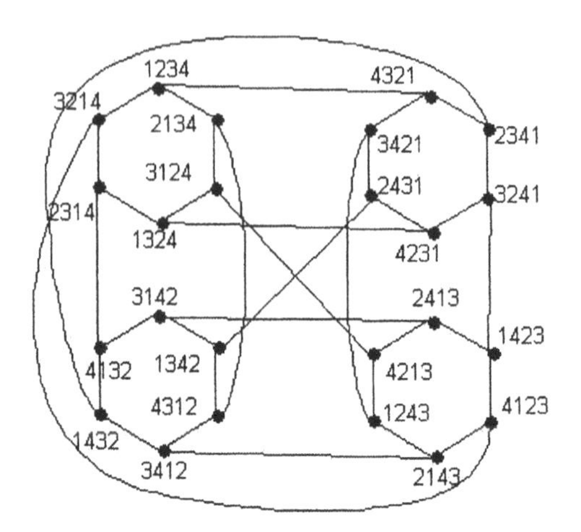

Figure 11: Pancake.

A graph is vertex (edge) symmetric if the graph looks the same from each of its vertices (edges, respectively). Star and pancakes are vertex and edge symmetric. It allows for all the processors to be identical, and minimizes the congestion when messages are routed in the graph.

Each vertex of the star or pancake has exactly $k-1$ adjacent edges, and thus the total number of edges is $k!(k-1)/2$. It is well known that $\log k! = \Theta\ (k \log k)$. Therefore $k = o(\log k!)$, and the degree of the star and pancake, being $k-1$, is sublogarithmic in the number of processors.

Consider now the diameter of the star graph. Because stars are vertex and edge symmetric, the furthest distance from a node to any other node is the same as the furthest distance from the identity permutation to any other permutation. It is proved in [1] that the diameter of the star graph with $n = k!$ vertices is $\lfloor 3(k-1)/2 \rfloor$. The proof is as follows. First, we present a greedy algorithm that will transform an arbitrary permutation to the identity one. Each step in the transformation will correspond to moving from one permutation to another one which is its neighbor in the star network. The whole process corresponds to the routing of a message from a given node to the node corresponding to the identity permutation. In each step of the transform, if 1 is the first element then exchange it with any position not occupied by the correct symbol (if any); else exchange the first element $x \neq 1$ in the permutation with the element at the xth position. The 'else' case moves x to its position in the identity permutation. Consider, for example, the permutation 86324517. The sequence of transformations is: 86324517 $\rightarrow$ 76324518 $\rightarrow$ 16324578 $\rightarrow$

61324578 → 51324678 → 41325678 → 21345678 → 12345678. Note that the first elements in these permutations follow the cyclic structure of initial permutation 86324517 = (871)(6542). We now calculate the number of steps $d(p)$ needed to transfer permutation p to the identity permutation. According to the algorithm, elements that are invariant (already in their correct position) are never exchanged. Let m be the total number of noninvariant (displaced) elements and c be the number of cycles of length ≥ 2. The 'else' case is performed m times if $p_1 = 1$ and $m - 1$ times if $p_1 > 1$. Each 'then' case is followed by placing, one by one, all elements of a cycle in their correct positions. Therefore the 'then' case is performed c times if $p_1 = 1$ and $c - 1$ times otherwise (in this case the first cycle is resolved without the preliminary 'then' case). Therefore $d(p) = c + m$ if $p_1 = 1$ and $d(p) = c + m - 2$ otherwise. When $p_1 = 1$, $m \leq k - 1$ and $c \leq (k-1)/2$ and $d(p) = m + c \leq 3(k-1)/2$. Otherwise $m \leq k$ and $c \leq k/2$ and $d(p) = m + c - 2 \leq (3k-4)/2$. We now need to show that there are two permutations that are at distance $\lfloor 3(k-1)/2 \rfloor$. For k odd, consider permutation $1325476 \ldots k(k-1)$, and for k even consider $214365 \ldots k(k-1)$. Every pair $t(t-1)$ for $t > 2$ requires at least three steps to be transformed to the correct position $(t-1)t$ (pair 21 needs only 1): exchange 1st and tth position; exchange 1st and $(t-1)$th position; exchange 1st and tth position. A simple count separately for odd and even cases verifies that the maximal distance is reached in both cases.

Let us look now at the structure of star and pancake networks. If the symbol in the last position is held fixed at value i, then there are $(k-1)!$ permutations and they with all the other permissible interchanges will constitute a $(k-1)$-star graph $S_{k-1}(i)$. Thus the vertices of S_k can be partitioned into k groups, each containing $(k-1)!$ vertices, based on the symbol in the last position. Each group is isomorphic to S_{k-1}. These groups will be interconnected by edges corresponding to interchanging the symbol in the first position with the symbol in the last position. Figure 10 shows S_4 as 4 interconnected copies of S_3. Analogous decomposition properties are valid for pancakes.

It should be noted that in S_k, the position at which we fix the symbols to get instances of S_{k-1} could be at any i, $2 \leq i \leq k$ (and not just the last). Thus, in general, we can define S to be an S_{k-1} such that all the vertices in it have the same symbol j at position i, $2 \leq i \leq k, 1 \leq j \leq k$. Therefore $S_{k-1}(j) = S$. It follows that there are $k-1$ ways of decomposing S_k into k copies of S_{k-1}: $S_{k-1}^i(j), 2 \leq i \leq k, 1 \leq j \leq k$ [23].

Two neighboring nodes of a star differ by a transposition of corresponding permutations. Thus each edge of a star graph joins two vertices that have an even and an odd corresponding permutation. Therefore S_k is bipartite graph [1], where sets of odd and even permutations split the graph according to the definition of bipartite graphs. By contrast, the pancake graph P_k is not bipartite. It contains a cycle of odd length in P_4; one such cycle (of length 7) is as follows [2]: 1234, 2134, 4312, 1342, 2431, 3421, 4321.

A new topology called the arrangement graph has been defined in [14] as a class of generalized star graphs. The arrangement graphs $A(k,m)$ are defined as graphs whose vertex set is the set $P(k,m)$ of all (k,m)-permutations (permutations of k out of m elements $1,2,\ldots,m$). Two vertices are adjacent if they differ in exactly one position. $A(k-1,k)$ is obviously the same as k-star (the equivalence is obtained when the first element in each permutation in the star nodes is dropped). The number of nodes in $A(k,m)$ is $m!/(m-k)!$, and each node has $k(m-k)$ neighbors. $A(k,m)$ is vertex and edge symmetric. Figure 12 shows $A(2,4)$. The diameter of the arrangement graph $A(k,m)$ is $\lfloor 3k/2 \rfloor$ [14].

The k-rotator graph [12] is a directed graph with $k!$ nodes that correspond to permutations of k symbols. The in-degree and out-degree of each node is $k-1$. An edge exists from node $(a_1,a_2,\ldots,a_k)$ to the following nodes: $(a_2,a_3,\ldots,a_i,a_1,a_{i+1},\ldots,a_k)$ for $1<i\leq k$. Thus there is an edge from one node to the other if the later corresponding permutation is obtained by inserting the first element of the permutation corresponding to the former node to any position, shifting to the left the elements that precede the position. For example, node 563214 is linked to nodes 653214, 635214, 632514, 632154, and 632145. Figure 13 shows a 3-rotator graph.

The generalized (k,m)-rotator graph $(m\leq k)$ has $k!/(k-m)!$ nodes that correspond to all permutations of m out of k elements $1,2,\ldots,k$. A node $(a_1,\ldots,a_m)$ is connected via outgoing links to nodes $(a_2,a_3,\ldots,a_i,a_1,a_{i+1},\ldots,a_m)$ for $1<i\leq m$ and to nodes $(a_2,a_3,a_4,\ldots,a_m,a_j)$ where $m<j\leq k$; i.e. the symbol a_j was not present in the original selection of m symbols. (The (k,k)-rotator graph is the same as the k-rotator graph. According to the definition, each node of the (k,m)-rotator graph is connected to $k-1$ edges.

The diameter of the k-rotator graph is $k-1$ whereas the diameter of the (k,m)-rotator graph is m for $m<k$. The proof is as follows [12]. Because the graphs are symmetric, it suffices to consider the distance from any node $(a_1,\ldots,a_m)$ to the origin $123\ldots m$. The problem corresponds to sorting the symbols in the permutations using the operations

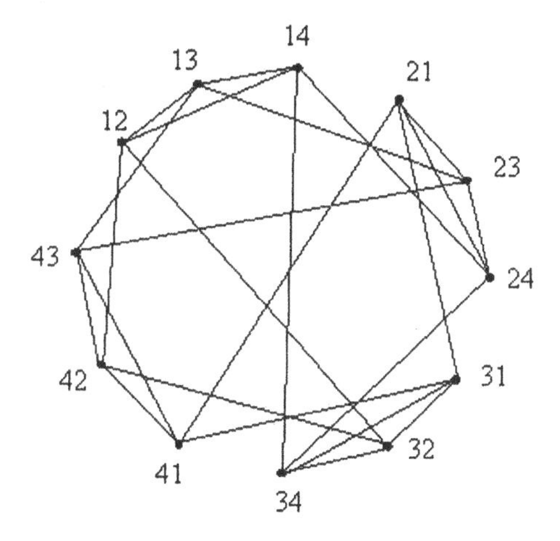

Figure 12: Arrangement graph.

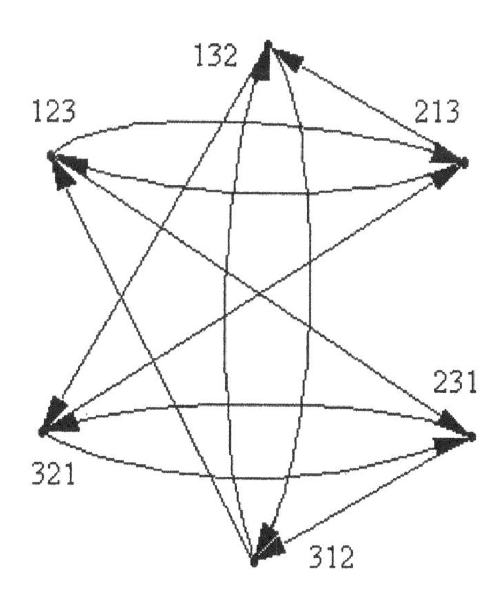

Figure 13: Rotator graph.

corresponding to existing edges, which here are rotations. The shortest path can be found using the insertion sort technique. Consider the complete permutation $(a_1, \ldots, a_m, a_{m+1}, \ldots, a_k)$ where the remaining elements are added in increasing order $(a_{m+1} < \ldots < a_k)$. Let i be chosen such that $a_i > a_{i+1} < a_{i+2} < \ldots < a_k$; i.e. the last $k-i$ elements in the permutation are sorted in increasing order (obviously $i \leq m$). The rotation moves a_1 to a new position while keeping the relative order of $a_2, a_3, \ldots, a_k$ unchanged. In order to place a_i after a_{i+1}, one has to perform i (left) rotations, placing the element at the first position, which allows it to be placed after a_{i+1} (more precisely, to its correct position). Therefore the shortest path has the length $\geq i$. We now show that i such rotations actually suffice to sort the permutations. First we place the a_1 in its correct position in the sorted list $a_{i+1} < a_{i+2} < \ldots < a_k$, increasing the length of the monotone increasing sequence at the end of the permutation by one. We repeat this operation i times, each time placing the first element in its correct position in the sorted sequence. At the end the sequence will be sorted. Thus the distance from node $(a_1, \ldots, a_m)$ to origin $123 \ldots m$ is i where i is the greatest index of a symbol that is greater than its right neighbor. The greatest value of i is $k-1$ for the k-rotator graph and m for the (k,m)-rotator graph, $m < k$.

Consider an example of the routing in a $(9,6)$-rotator graph, from node 583492 to node 123456. Complete the source representation to 583492167 and route as follows: 583492167 $\rightarrow$ 834921567 $\rightarrow$ 349215678 $\rightarrow$ 492135678 $\rightarrow$ 921345678 $\rightarrow$ 213456789 $\rightarrow$ 123456789.

It is possible to determine the number of nodes at a given distance from a given node. The last $i-1$ symbols in the sorted order can be selected by choosing i symbols out of k symbols, putting any of $i-1$ symbols (except the smallest) in the first position (in $i-1$ ways), putting the rest of symbols in sorted order, and putting nonchosen $k-i$ items in any order. Therefore the number of nodes at distance $k-i+1$ from an origin in a k-rotator graph is $C(i,k)(i-1)(k-i)! = k!(i-1)/i!$ [12].

3 Hypercubic Networks

3.1 Hypercube

A hypercube computer consists of $n = 2^d$ processors (or nodes, numbered 0 to $n-1$), linked together in a d-dimensional binary cube network. Each node u $(0 \leq u < n)$ has associated local memory and is assigned its d-bit binary representation $u_{d-1} \ldots u_1 u_0$ (obviously $u = 2^{d-1}u_{d-1} + \ldots + 2u_1 +$

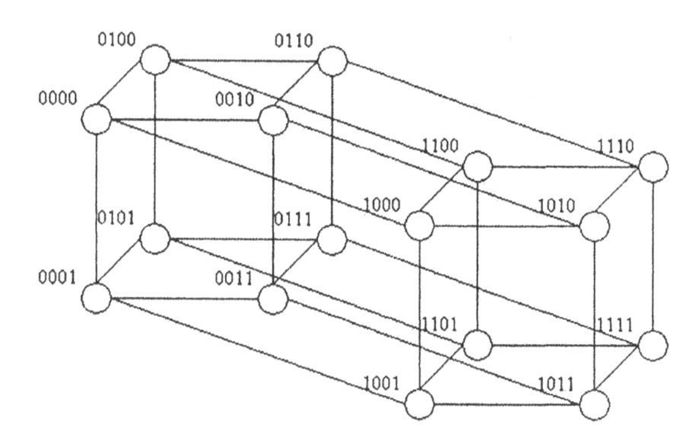

Figure 14: 4-dimensional hypercube

u_0 where the jth bit of the address of a node u is denoted u_j). We define the relative address, $a \oplus b$, of two nodes a and b as the bitwise exclusive-OR of their binary representations, and define the Hamming distance $\rho(a, b)$ between two nodes a and b as the number of 1s in $a \oplus b$. For example, $01101 \oplus 11001 = 10100$, $\rho(01101, 11001) = 2$. Two nodes a and b in a hypercube share a (communication) link if and only if $\rho(a, b) = 1$ (such nodes are called neighbors). Nodes communicate by passing messages on these links. The notation $\mathcal{R}(u, j)$ is used to denote the node obtained from u by flipping the jth bit of u. Flipping is a complement function; i.e., $\overline{0} = 1$, $\overline{1} = 0$. For example, $\mathcal{R}(01001, 3) = 00001$. The only possible neighbors of a node u are exactly $\mathcal{R}(u, 0), \ldots, \mathcal{R}(u, d-1)$. A link has link number j if, for some u, it connects two nodes u and $\mathcal{R}(u, j)$.

Therefore, the vertices of a d-dimensional hypercube network correspond to all binary strings of length d (equivalently, to subsets of a d-element set in binary representation, or $(d, 2)$-variations out of 0,1) and whose edges correspond to the neighboring processors. The d neighbors of processor i are those processors j such that the binary representation of the numbers i and j differs in exactly one bit. Figure 14 shows a hypercube with $n = 2^4 = 16$ processors.

Hypercubes have emerged as one of the most effective and popular architectures for parallel machines. They have several highly desirable properties, including edge and vertex symmetric layout (the hypercube has the same shape from any of its vertices or edges). The network allows very elegant and simple parallel solutions to a number of problems. We study this network in more detail because of its popularity.

The diameter of a d-dimensional hypercube is d and is achieved by any pair of nodes with complement binary string representations. For example, 1001 and 0110 are at distance 4 and a part between them is given as follows: $1001 \to 0001 \to 0101 \to 0111 \to 0110$.

The number of nodes at distance i from any node of a hypercube is $C(i,d)$. These are nodes that differ from a given node in exactly i out of d bits in their address. Because each node has d=log n neighbors, the degree of hypercube is log n. An important characteristic of a d-dimensional hypercube is that it is constructed recursively from two $(d-1)$-dimensional ones. More precisely, consider two identical $(d-1)$-cubes whose vertices are numbered from 0 to $2^{d-1}-1$. By joining every vertex of the first $(d-1)$-dimensional hypercube to the vertex of the second one having the same number, one obtains a d-dimensional hypercube. The nodes of the first cube receive a 0 at the beginning of their binary address and the nodes of the second one receive a 1, which effectively increases their decimal address for 2^{d-1}. The subdivision of a d-dimensional hypercube into $(d-1)$-dimensional ones can be made along any of d dimensions. For given $i < d$, nodes having ith bit 1 belong to one $(d-1)$-dimensional hypercube, and nodes having ith bit 0 belong to the other.

Hypercubes have nonconstant degree, and the number of processors must be a power of two. Modifications of a hypercube network are therefore investigated to improve these characteristics. Some of these modifications are described in subsequent subsections.

3.2 De Bruijn Network

The de Bruijn graph is defined by de Bruijn [13] as follows. The vertices of a de Bruijn graph $D(m,k)$ are all (k,m)-variations, i.e., all variations $d_1d_2\dots d_k$ such that $d_i \in 0,1,\dots,m-1$ for $1 \le i \le k$. Thus the number of nodes is $n=m^k$. Two variations (nodes) are connected by an edge if and only if they are obtained from each other by a shift from one position to the left or right and appending an arbitrary element to the position that becomes empty by shifting. Therefore the shift is not cyclic. More precisely, the neighbors of node $d_1d_2\dots d_k$ are nodes $pd_1d_2\dots d_{k-1}$ and $d_2d_3\dots d_kp$ for each $p, 0 \le p \le m-1$. Therefore each node has at most $2m$ neighbors. Some nodes have less than $2m$ neighbors because some of the $2m$ variations may coincide. A binary de Bruijn graph $D(2,k)$ is shown in Figure 15. For instance, node 110 is connected to the following nodes: 011 (shift to the right and append 0 at the beginning), 111 (right

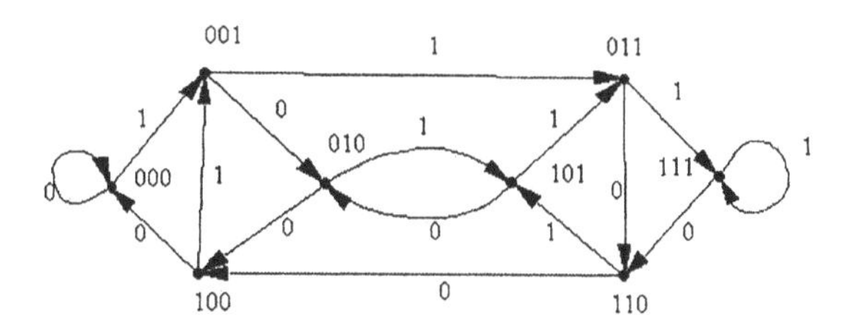

Figure 15: Binary de Bruijn graph.

shift, append 1), 100 (left shift, append 0) and 101 (left shift, append 1). The de Bruijn graph is normally considered as a directed graph (with some loops), as in Figure 16, which facilitates the control of information flow and the design of some algorithms. An edge leads from node u to node v if v is obtained from u by a left shift and appending a new element. An edge receives link number p if it joins a node $d_1 d_2 \dots d_k$ to node $d_2 d_3 \dots d_k p$. Fig. 15 contains also link numbers of edges.

The diameter of $D(m, k)$ is $k = \log_m n$ because any variation can be obtained easily from any other by a sequence of k shifts (say, to the left) and appending the appropriate element of the other variation each time. On the other hand, the distance between nodes $00 \dots 0$ and $11 \dots 1$ is k because in one step at most one 0 can be turned to 1; thus k bit changes are necessary to transform one node to the other. The binary de Bruijn graph has twice more (directed) edges than the nodes. It also has a very interesting recursive structure. Nodes of $D(2, k+1)$ correspond to edges of $D(2, k)$, and two nodes of $D(2, k+1)$ are connected by a directed edge if the endpoint of the first coincides with the beginning of the second edge.

3.3 Fibonacci Cubes

Fibonacci cubes are introduced in [18] and are a type of incomplete hypercube. The description given here is more elegant and is taken from [30]. They are subgraphs of hypercube graphs induced by nodes that have no two consecutive 1s in their binary representation. More precisely, a Fibonacci cube Γ_k of dimension k is an undirected graph of f_k nodes, each labeled by a $k-2$ bit binary number such that no two 1s occur consecutively (where f_k is the kth Fibonacci number). Two nodes are connected iff their labels differ in exactly one bit position. It is a kind of incomplete hypercube consisting of f_k out of 2^{k-2} nodes (the selected nodes corresponding to Fibonacci codes). Figure 16 shows

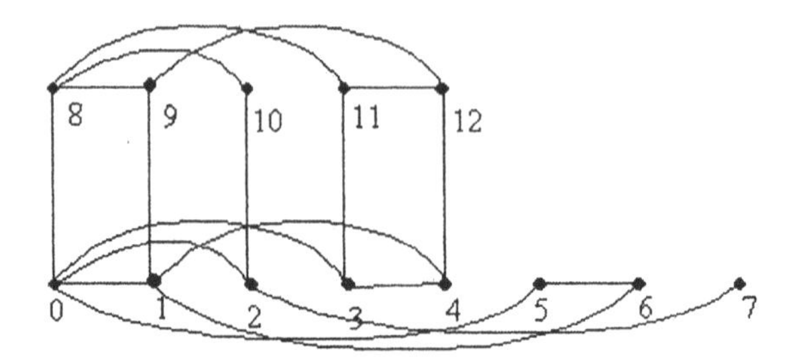

Figure 16: Fibonacci cube of size 7.

Fibonacci cube of size 7.

The diameter of a Fibonacci cube of order k is $k-2$; for example, nodes $0101\ldots0(1)$ and $1010\ldots1(0)$ are at distance $k-2$ and both are Fibonacci codes of some integers. The Fibonacci cube of size k can be decomposed into two Fibonacci cubes of sizes $k-1$ and $k-2$, respectively, with f_{k-2} edges between them. These two smaller cubes are distinguished by their first bit b_{k-1} being 0 and 1, respectively. Let C_k denote the set of order k Fibonacci codes; e.g., $C_4 = 00, 01, 10$. Then clearly $C_k = (0C_{k-1}) + (10C_{k-2})$, where '+' is the concatenation operation. This corresponds to decomposing Γ_k into Γ_{k-1} and Γ_{k-2}.

The number of edges E_k of Γ_k satisfies the following recurrence relation: $E_k = E_{k-1}+E_{k-2}+f_{k-2}$. The solution is $E_k = \frac{1}{5}(2(k-1)f_k - kf_{k-1})$ [18], which can be proven by induction. We now prove that the degree of each node of Γ_k is between $k-2$ and $\lfloor(k-3)/2\rfloor$. Node $00\ldots0$ has degree $k-2$ and node $001001\ldots$ has degree $\lfloor(k-3)/2\rfloor$. To prove that no node has degree less than $\lfloor(k-3)/2\rfloor$, it suffices to prove that among any three consecutive bits in a code, at least one of them corresponds to a neighbor. This is easy to show because combinations 111, 011, and 110 are not permitted as part of a code; in combination 000 the middle bit 0 can be turned to a 1, while for each of the remaining combinations 101, 100, 010, 001 any of the 1s can be turned to a 0 which will lead to the Fibonacci code of an integer. Thus at least one third of bits link each node to a neighbor, which suffices to claim the lower bound of the degree.

4 Paths, Cycles, and Fault Tolerance

In this section we study the existence of cycles, Hamiltonian cycles and paths, and other related questions for interconnection networks. Hamiltonian cycles or paths correspond to generating all nodes of a network such that successive nodes are neighbors. In the case of hypercubes, stars, pancakes, arrangement graphs, de Bruijn graphs etc., they correspond to generating all subsets, permutations, permutations of combinations, and variations with certain restrictions on the possible next element of each element.

Consider the following data communication problem in a network. One node, the source, has to send data to every other node in the network; however, the data to be sent are different for each destination. Let a network consist of T nodes. Because the source sends $T-1$ different data, and can disseminate one datum at a time, there is a lower bound of $\Omega(T)$ for the problem, even in an all-port communication model. Suppose now that the network has a Hamiltonian path or cycle. The source sends, one by one, data along the cycle (or each datum toward its destination, in the case of the Hamiltonian path). The data are moved forward in pipelined fashion. At each step, each node receives a new datum, checks whether it is the destination for the datum, receives the message if so and forwards it to the next node in the cycle (path) otherwise. Clearly there is no congestion in the data movement, and each node receives its message in $O(T)$ time, which is therefore optimal. Thus Hamiltonian paths and cycles in the network suffice to solve the important data communication problems.

Moreover, $O(T)$ time suffices to solve a more general problem. Suppose that each node wishes to send a different (personalized) message to every other node in the network (all-to-all broadcast). All nodes can send their message along the Hamiltonian cycle in one direction. Messages rotate along the cycle and are picked up in the process by the correct destinations. The special case of the problem is multinode broadcasting, where each node sends the same message to all other nodes.

The routing algorithm gives a path between any two nodes a and b of an interconnection network. One important question is whether there are different paths between a and b. The existence of such paths might be useful for speeding up transfers of a large amount of data between two nodes. It also provides a way of selecting alternative routes in case a given node or edge in a path fails. The different paths must not cross each other; i.e., they must not share any node except a and b. Such paths are called node disjoint paths or parallel paths. The question now is how many parallel paths are there between any two nodes a and b? The number of parallel paths cannot exceed the degree of the network because paths must start from a and lead to different nodes.

An important property of interconnection networks is their fault tolerance. It indicates the maximum number of vertices that can be removed and still have the graph connected. Thus fault tolerance is one less than the connectivity, where the connectivity of a graph is the minimum number of vertices that need to be removed to disconnect the graph. Clearly, any graph can be disconnected by removing all the vertices adjacent to a given vertex. Thus, its connectivity is less than the minimal degree of a node. A graph is maximally fault tolerant if the fault tolerance is one less than this minimal degree.

The fault diameter of a graph with fault tolerance f is the maximum diameter of any graph obtained from G by deleting at most f vertices. Clearly, if the fault diameter of a graph is very close to its normal diameter then the performance of the graph, in terms of the communication delay, does not drastically worsen in the presence of a tolerable number of faults.

4.1 Hypercubes

We can define the parity of a node to be even if the number of ones in its binary label is even and odd otherwise. It is clear that neighboring nodes have different parity. It follows from this that there are no cycles in a hypercube of odd length [24] (if $a_1, a_2, \ldots, a_t, a_1$ is a cycle then a_t has opposite parity to a_1, i.e. t is even).

Gray code orders allow the creation of cycles in a hypercube of any even length l, $4 \leq l \leq 2^d$. Let $m = (l-2)/2$ and denote $G_{d-1}(m)$ the first m elements of $(d-1)$-dimensional hypercube in Gray code order. Then $\{0G_{d-1}(m), 1G_{d-1}(m)^R\}$ is the desired cycle, where $G_{d-1}(m)^R$ is the reverse sequence of $G_{d-1}(m)$ [24].

We give a construction [24] of $\rho(a,b)$ parallel paths between nodes a and b, each of length $\rho(a,b)$. To simplify the description, without loss of generality we assume that the first $i = \rho(a,b)$ bits of a and b are different, and denote $a = a_1a_2\ldots a_ia_{i+1}\ldots a_d$, $b = \overline{a_1a_2}\ldots\overline{a_i}a_{i+1}\ldots a_d$. The jth path $(1 \leq j \leq i)$ is constructed as follows:

$$\begin{aligned} a = {} & a_1a_2 ... a_{j-1}a_j ... a_d \rightarrow a_1a_2 ... a_{j-1}\overline{a_j}a_{j+1} ... a_d \\ & \rightarrow a_1a_2 ... a_{j-1}\overline{a_ja_{j+1}}a_{j+2} ... a_d \rightarrow ... \rightarrow a_1a_2 ... a_{j-1}\overline{a_ja_{j+1}} ... \overline{a_i}a_{i+1} ... a_d \\ & \rightarrow \overline{a_1}a_2 ... a_{j-1} ... \overline{a_ja_{j+1}} ... \overline{a_i}a_{i+1} ... a_d \rightarrow \overline{a_1a_2} ... a_{j-1}\overline{a_ja_{j+1}} ... \overline{a_i}a_{i+1} ... a_d \\ & \rightarrow ... \rightarrow \overline{a_1a_2} ... \overline{a_{j-1}a_ja_{j+1}} ... \overline{a_i}a_{i+1} ... a_d. \end{aligned}$$

In other words, one might start correcting jth bit, then the $(j+1)$th bit, and so forth until the ith bit is reached, after which we correct, in turn, bits $1, 2, \ldots, j-1$. It is obvious that any two nodes on any two paths are different.

The maximal number of parallel paths is d. The $d - \rho(a,b)$ paths of length $\rho(a,b) + 2$ can be constructed as follows [24]. Let these paths be marked as $j = i+1, i+2, \ldots, d$, where we assume that these $n-i$ bits are exactly those where the addresses of a and b coincide. For given j, send the message from node a to node $\mathcal{R}(a,j)$, then route the message from node $\mathcal{R}(a,j)$ to node $\mathcal{R}(b,j)$, and finally send the message from node $\mathcal{R}(b,j)$ to node b. Routing can be performed by the standard or any other algorithm. Note that $\mathcal{R}(a,j)$ and $\mathcal{R}(b,j)$ differ in exactly $\rho(a,b)$ bits. Thus there exist d parallel paths between any two nodes a and b, each with length at most $\rho(a,b) + 2$.

The existence of d parallel paths between any two nodes of a d-dimensional hypercube proves that the network is maximally fault tolerant. The lengths of these paths are ($\leq d+2$) and prove that the fault diameter of the network is $\leq d+2$.

We find the number of t-cubes (subcubes of dimension t) in a d-dimensional hypercube. Choose $d-t$ dimensions among d dimensions and fix them. All nodes having these dimensions fixed (and the same among themselves) form a t-cube. $d-t$ dimensions can be chosen in $C(d-t,d)$ ways, and for each such way, the dimensions can be fixed in 2^{d-t} ways.Thus the number of t-cubes is $C(d-t,d)2^{d-t}$ [7].

Consider now the question of how many faults are necessary to make every t-cube faulty (i.e., to contain one or more faulty nodes). Let us denote the number with $f(d,t)$. Every node of hypercube belongs to $C(d-t,d)$ t-cubes. Thus it follows that $f(d,t) \geq 2^{d-t}$ [7], which is a significant fault tolerance. Let $s(x)$ denote the sum of digits of node x. If S is the sum of digits of $d-t$ fixed coordinates then $s(x)$ is between S and $S+t$. If the sums are considered modulo $t+1$, the obtained sums are $0, 1, \ldots, t+1$. Each of these $t+1$ different values is achieved on the t-cube. Partition all nodes of the hypercube according to these sums (modulo $t+1$). The smallest of these sets has $\leq \frac{2^d}{t+1}$ nodes. If all nodes in the smallest set are removed then no t cube exists. Thus $f(d,t) \leq \frac{2^d}{t+1}$ [15].

4.2 Stars and Pancakes

There exists a recursive Hamiltonian cycle for P_k [34] that can be described as follows. Let f_i denote flipping the first i symbols in a permutation. Each permutation can be obtained by a sequence of flippings from the identity permutation $I = 12\ldots k$. Each edge in the pancake corresponds to a flipping. If we define a sequence p_k of flippings recursively as follows: $p_2 = f_2, p_k = (p_{k-1}f_k)^{k-1}p_{k-1}$, then the sequence $p_n f_n$ corresponds to a Hamiltonian cycle in P_k. For example, for $n = 3, p_3 = f_2f_3f_2f_3f_2$, and the sequence $p_3f_3 = f_2f_3f_2f_3f_2f_3$ corresponds to Hamiltonian cycle 123, 213, 312, 132, 231, 321, 123 in P_3. There is no simple recursive Hamiltonian cycle for S_k.

We find the number of t-stars (or t-pancakes) in S_k (P_k, respectively). Choose $k - t$ positions among $k - 1$ positions $2, 3, \ldots, k$ and fix them. All nodes having these positions fixed (and the same among themselves) form a t-star (or t-pancake). $k-t$ positions can be chosen in $C(k-t, k-1)$ ways, and for each such way, the positions can be fixed in $\frac{k!}{t!}$ ways. Thus the number of t-stars in S_k is $C(k - t, k - 1)\frac{k!}{t!}$ [1].

Consider now the question of how many faults are necessary to make every t-star faulty (i.e. to contain one or more faulty nodes). Let us denote the number with $f(k,t)$. Any single faulty node appears in exactly $C(k - t, k - 1)$ t-stars. Thus it follows that $f(k,t) \geq \frac{k!}{t!}$ [1], which is a significant fault tolerance.

4.3 De Bruijn Graphs

An (m, k) de Bruijn sequence $B_{m,n}$ is a sequence $a_1a_2\ldots a_L$ with each $a_i \in S = 0, 1, \ldots, m - 1$ such that every word w of length k from S is realized as $a_ia_{i+1}\ldots a_{i+k-1} (0 \leq i \leq L)$ for exactly one i, where each subscript is to be interpreted in modulo L. For example, for $m = 2$ and $k = 3$ the sequence 00010111 contains all binary sequences of length 3 with 110 and 100 being obtained by wraparound from the end of the sequence to the beginning.

There exists a one-to-one correspondence between de Bruijn sequences of length m^k and Hamiltonian cycles in the directed de Bruijn graph $D(m, k)$. For example, the sequence 00010111 corresponds to the Hamiltonian cycle of the binary de Bruijn graph given by $000 \to 001 \to 010 \to 101 \to 011 \to 111 \to 110 \to 100 \to 000$.

A Hamiltonian cycle of a (directed) binary de Bruijn graph $D(2, k)$ can be constructed using their recursive structure. Each node of $D(2, k)$ corresponds to an edge of $D(2, k - 1)$. Thus a Hamiltonian cycle of $D(2, k)$ corresponds to an Eulerian tour of $D(2, k - 1)$. The existence of an Eulerian tour follows from the following observation. Because the in-degree of every node equals its out-degree, one can follow an arbitrary path until we return to the initial vertex. If some edges are not visited, it is possible to augment them at nodes that are incident to unused edges.

The fault tolerance of $D(m, k)$ is $m - 1$. It is one of the drawbacks, especially for binary de Bruijn graphs. $DM(m, k)$ has $C(e, m)D(e, k)$ graphs in it [26], because all the nodes that are represented by only e digits out of d digits form a $D(e, k)$, and there are $C(e, m)$ such selections. Every node i in $D(m, k)$ is in $C(e - j, m - j)D(e, k)$ graphs, where j is the number of different digits in the m-ary digit representation of i [26]. It follows from the observation that the remaining $e - j$ digits can be selected in $C(e - j, m - j)$ ways. Therefore a failed node eliminates at most $C(e - j, m - j) \leq C(e - 1, m - 1)D(e, k)$ graphs. This means that the failure of $F < \frac{C(e,m)}{C(e-1,m-1)} = \frac{m}{e}$ nodes leaves at least one fault-free $D(e, k)$ [26].

5 Embeddings

In this section, we are concerned with the problem of mapping network A into network B. A is called the guest graph and B is the host graph. The dilation is the maximum path length of an edge in the guest graph when embedded in the host graph. The expansion is the ratio between the number of nodes in the host and guest graphs. Embeddings with dilation 1 are an assignment of nodes so that the proximity property is preserved. Embeddings of rings and linear array of processors, with dilation 1, are equivalent to finding paths and cycles in the network of given length.

One of the most attractive properties of hypercubes is that meshes of arbitrary dimensions can be embedded into them [24]. Consider a 2-dimensional MCC of size p X q. Let d' and d" be chosen such that $2^{d'} \geq p, 2^{d"} \geq q$. Label rows and columns of mesh separately in Gray code order, starting from 0. Labels used form sequences $G_{d'}(p)$ and $G_{d"}(q)$. Let each node of mesh be labeled with a row label followed by a column label. The mesh is embedded into a hypercube of size $2^{d'+d"}$, because clearly neighboring nodes of mesh receive labels that differ in exactly one bit. Figure 17 gives an illustration. A d-dimensional mesh-connected computer can be embedded into a hypercube with dilation 1 [24].

Fibonacci cube Γ_k is directly embedded into a k-dimensional hypercube (as discussed in Section 3), with dilation 1 and expansion $\frac{2^{k-2}}{f_k}$. We show that the opposite is also true, more precisely, that k-dimensional hypercube B_k can be embedded into Fibonacci cube Γ_{2k+1}. Insert a

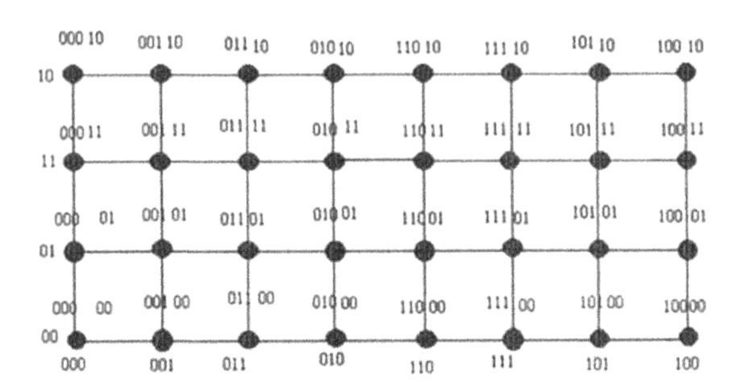

Figure 17: Gray code for a 8 x 4 MCC.

0 between any two bits in the address of a node of a hypercube; i.e., $d_{k-1}d_{k-2}\dots d_1d_0$ becomes $d_{k-1}0d_{k-2}0\dots 0d_10d_0$. All obtained nodes correspond to Fibonacci codes of some integers, and all of them belong to Γ_{2k+1}. The proof is direct while the proof in [18] uses induction without giving explicit correspondence. The dilation is 1 and the expansion is $\frac{f_{2k+3}}{2^k}$.

Embedding of complete binary trees into hypercubes is possible but is not possible with both expansion and dilation 1. It is easy to embed complete binary tree of $2k-1$ nodes into $(k+1)$-dimensional hypercube (expansion 2 and dilation 1). They can be embedded into k-dimensional hypercube with dilation 2 [5]. Binary de Bruijn networks allow an easy embedding of complete binary trees, as follows [13]. Let the root of the tree be $00\dots01$. The left and right children of a node $u = u_1u_2...u_k$ are $u_2u_3\dots u_k0$ and $u_2u_3\dots u_k1$, respectively. Nodes $1u_2\dots u_k$ have no children. Both dilation and expansions are 1 (more precisely, node $00\dots0$ is the only one not used in the embedding).

Reference [19] contains an excellent survey of various embeddings, in particular meshes into meshes and hypercubes, tree, and pyramid embeddings.

6 Lower Bounds

In this section we derive some lower bounds for some problems and some models of parallel computation. Lower bounds depend on the problems that are attempted to be solved. Obviously problems such as reporting a datum already in a processor have no lower bound other than $\Omega(1)$ (except if the datum is in one processor and the output is in another). A similar bound applies to problems the result of which depends only on data that are located in neighboring processors in a bounded degree network. For example, given a black/white image, one pixel per processor on a mesh-connected computer, determining how many neighboring pixels are black can be done in constant time by reading data from all neighboring processors in four steps and counting the black pixels in them.

A straightforward lower bound is obtained when the input or output size is divided by the number of processors that have input or output ports. For example, the multiplication of two $n \times n$ matrices cannot be done faster than $O(\frac{n^2}{n}) = O(n)$ time on an MCC with n processors, assuming that all processors have an input port. Sometimes the restriction is made that only boundary processors (those that have less than four neighbors) have an input port. In such a case the lower bound is $\Omega(\frac{n^2}{\sqrt{n}}) = \Omega(n^{3/2})$ for the matrix multiplication problem.

A standard lower bound argument in parallel computing is that a problem that requires time T to be solved sequentially cannot be solved in time faster than $O(\frac{T}{p})$ on an interconnection network with p processors. An example is sorting some data (one per processor) on the star network. Sorting $k!$ numbers requires sequentially $\Omega((k!) \log(k!))$ time. Thus sorting cannot be done faster than $\Omega(\log(k!)) = \Omega(k \log k)$ on a star with $k!$ processors.

In particular, we study lower bounds for two classes of problems, which we call nontrivial problems and extensive data movement problems. The extensive data movement problems are problems that may require moving many data from one part of a network to another. A nontrivial problem is a problem in which an output of a selected processor depends on data from any other processor; therefore the selected processor needs to receive, directly or indirectly, a datum from every other processor.

A typical example of a nontrivial problem is semigroup computation. The problem is to compute $a_1 H a_2 H \ldots H a_n$ where processor i keeps datum a_i and H is any associative binary operation. Examples of semigroup operations are minimum, maximum, sum, product, logical OR, logical AND, etc. The final result depends on any a_i. For instance, each of a_i, once all other data are fixed, can change the result of minimum, maximum, or product. The observation is trickier for product, logical OR or AND because for many inputs some a_i cannot alter the result (e.g., if there is any 1 among data, 0 or 1 in a_i does not affect the result in logical OR). However, the lower bound is valid even in such cases,

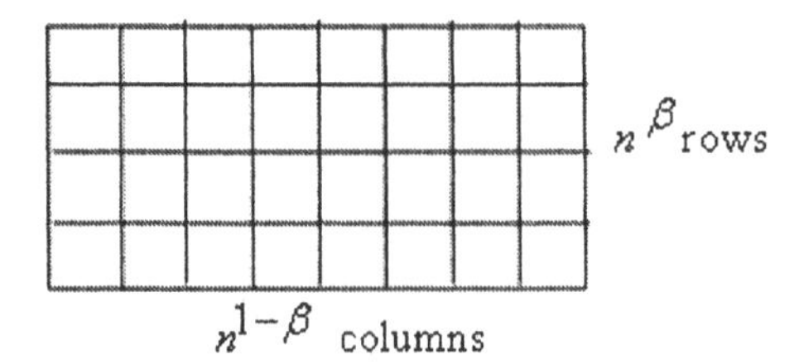

Figure 18: Rectangular MMB.

because in all cases it is possible to create such inputs that the result of semigroup computation indeed depends on a_i, and, because lower bound arguments should apply for every possible input, a selected processor may not produce output without 'consulting' a_i in processor i.

For any nontrivial problem, a lower bound for solving it is the diameter of the network, because a datum cannot be routed from one processor to the other in a lesser number of time units. Therefore, any nontrivial problem takes $O(n)(O(\sqrt{n}))$ time to be solved on a linear array (mesh-connected computer, respectively) with n processors. In many cases the lower bound is matched by an optimal algorithm; in many others the diameter does not give the real lower bound.

The bisection width B of a network is the minimum number of links that, when cut, separate the network into two parts with an equal number of processors. It measures how much wiring density is required by a network. It also affects the computation speed for extensive data movement problems. Such problems may, in particular instances, require moving most of the data from one half of the network to the other. A good tradeoff balance between the wiring density and speed of computation is desired.

6.1 Meshes with Multiple Broadcasting

A lower bound for nontrivial problems on MMBs is found in [4]. We present here an alternative proof of the same result, using combinatorial arguments. The lower bound is for a rectangular MMB, with n^{β} rows and $n^{1-\beta}$ columns, as indicated in Figure 18. We assume ($0 \leq \beta \leq 0.5$), i.e., no more rows than columns. A square mesh is obtained for $\beta = 0.5$, and for $\beta = 0$ one gets a linear array of processors equipped with a bus.

We prove the following two lemmas on the lower bound for MMB. Both lemmas are proved by finding the minimal number of squares (rect-

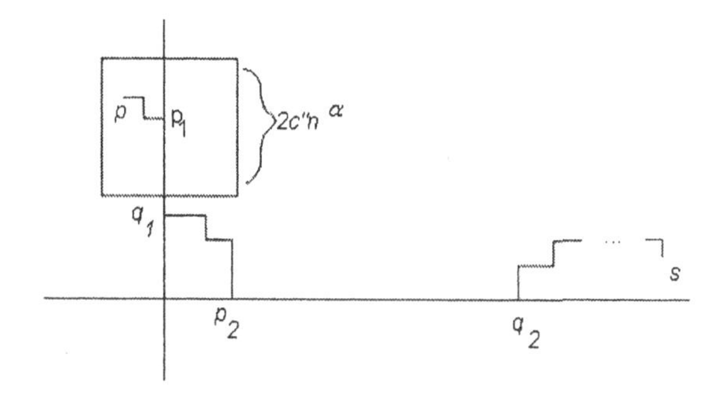

Figure 19: Message path from p to s.

angles, respectively) of given size that are necessary to cover the 'area' of MMB, which is equal to the number of processors n. The size of sides and the 'area' of squares (or rectangles) are both measured in terms of the number of processors along the side or inside the square (rectangle, respectively).

Lemma 1. Any algorithm solving a nontrivial problem on a MMB of size $n^{\beta} \times n^{1-\beta} (0 \leq \beta \leq 0.5)$ takes $\Omega(n^{\beta/3})$ time.

Proof. Suppose that an algorithm used at most $O(n^{\alpha})$ operations. It means that there exist two constants c' and c'' such that each bus is used at most $c'n^{\alpha}$ times for broadcasting, and that neighboring processors are communicated in at most $c''n^{\alpha}$ steps by each processor. Each of the horizontal and vertical buses has been used at most $c'n^{\alpha}$ times, i.e., by at most $c'n^{\alpha}$ processors in the corresponding row or column. Let us call 'bus users' these processors that used a bus for broadcasting data. Therefore the number of bus users is at most $c'(n^{\beta}+n^{1-\beta})n^{\alpha} \leq 2c'n^{1-\beta}$.

Consider a processor p. The message path from p to selected processor s on MMB has the following form. First, the message is routed from p to a bus user p_1; then p_1 broadcasts it to a processor q_1 along the bus, which routes it to a bus user p_2, which then routes it along the bus to a processor q_2, etc. until the message reaches the selected processor s (see Figure 19). To simplify our derivation, we charge only the path from p to p_1 and the bus usage on p_1 (i.e., we do not count the path from q_1 to s).

In time n^{α}, each bus user is able to 'collect' data from processors that are at a distance no more than $c''n^{\alpha}$ from it. Clearly processors that are at a greater distance need more than the maximum number of communication steps to 'reach' the bus user. All processors that are able to send their data to a given bus user in specified time steps are located inside a square, centered on the bus user, whose sides are horizontal and vertical and are of size $2c''n^\alpha$ (see Figure 19 for bus user p_1). The total number of such processors is less than $2c''n^\alpha \cdot 2c''n^\alpha$, per bus user. All together, the number of processors that are able to reach a bus user is less than $2c'n^{1-\beta}n^\alpha 2c''n^\alpha \cdot 2c''n^\alpha = 8c''^2 n^{1-\beta}n^{3\alpha}$. In order that a selected processor has access to data in every other processor, each of n processors must be able to reach a bus user. Therefore $8c'c''^2n^{1-\beta}n^{3\alpha} \geq n$; i.e., $1-\beta+3\alpha \geq 1$. Thus the lower bound for α is $\alpha \geq \frac{\beta}{3}$.

Lemma 2. Any algorithm solving a nontrivial problem on an MMB of size $n^\beta \times n^{1-\beta}$ $(0 \leq \beta \leq 0.5)$ takes $\Omega(n^{0.5-\beta})$ time.

Proof. Define α, c' and c'' as in the previous lemma. To derive the lower bound, we now decide not to charge the use of vertical buses. The maximum number of horizontal bus users is $c'n^\alpha n^\beta$. In time n^α, each horizontal bus user is able to 'collect' data from processors that are at a distance no more than $c''n^\alpha$ from the vertical column containing the user (see Figure 20, where processor p uses processor p_2 as the associated horizontal bus user). This follows from the observation that the use of a vertical bus does not change the distance to the vertical column containing a horizontal bus user. All processors that are able to communicate their data to a given horizontal bus user in limited time steps, using only local links between neighboring processors and broadcasting via vertical buses, are located inside a strip of width $2c''n^\alpha$. The total number (for all horizontal bus users) of such processors is $2c''n^\alpha \cdot n^\beta \cdot c'n^\alpha n^\beta$ (the width times the height times the number of horizontal bus users). This 'area' should cover the MMB of 'area' n, and thus $2c'c''n^{2\alpha}n^{2\beta} \geq n$, which implies $\alpha \geq 0.5-\beta$.

Theorem 1. The lower bound for any nontrivial problem on an $n^\beta \times n^{1-\beta}$ $(0 \leq \beta \leq 0.5)$ MMB is $\Omega(n^{0.5-\beta})$ for $0 \leq \beta \leq \frac{3}{8}$ and $\Omega(n^{\beta/3})$ for $\frac{3}{8} \leq \beta \leq \frac{1}{2}$.

Proof. Follows from Lemmas 1 and 2. Figure 21 illustrates the two lower bounds.

Consider a few special cases of the lower bound result. Square MMBs have $\beta = 0.5$ and lower bound $\Omega(n^{1/6})$. A linear array with a bus has $\beta = 0$ and lower bound $\Omega(n^{1/4})$. The lower bound has its minimum for $\beta = \frac{3}{8}$ and is $\Omega(n^{1/8})$. Therefore, as observed in [4], square meshes are not always optimal.

Sorting is a typical example of an extensive data movement problem. The data may happen to be originally given in almost reverse order, in which case each of them should be moved to another part of the network.

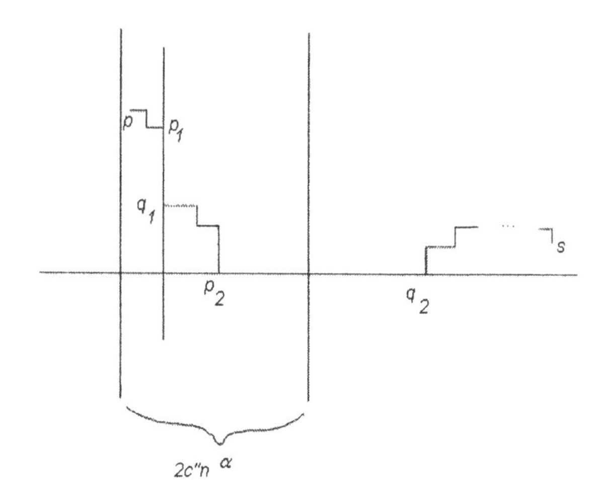

Figure 20: Lower bound on MMB.

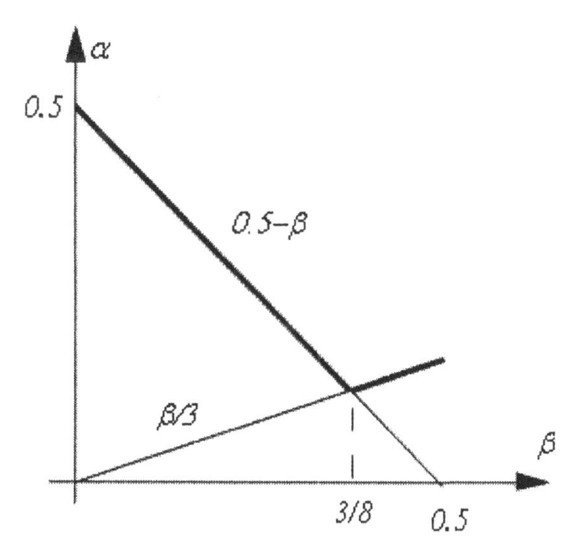

Figure 21: Two lower bounds for MMB.

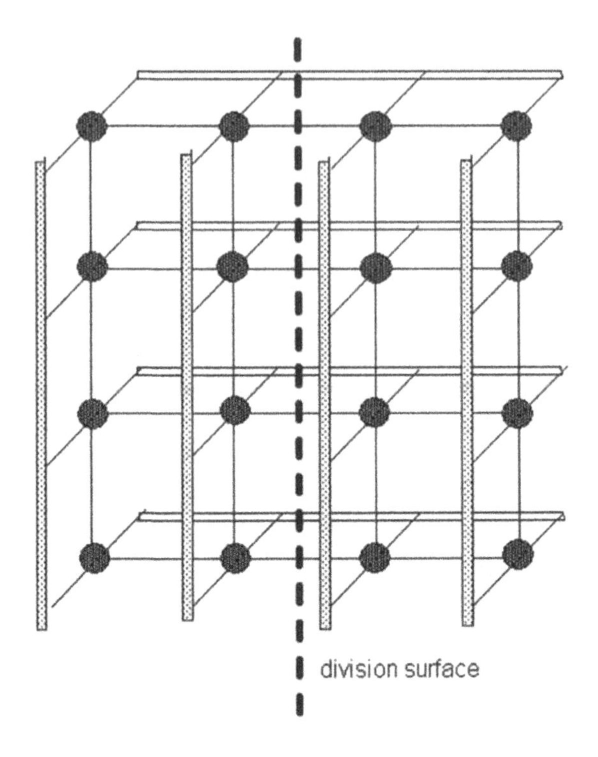

Figure 22: Transferring data over division surface of a MMB.

Consider an MCC or MMB with n processors, and divide it into two parts, each with $\frac{n}{2}$ processors. The division line or surface may be chosen such that it intersects $O(\sqrt{n})$ communication links between neighbors and (for MMBs) buses (see Figure 22). If the data are such that the position of every element in the sorted list is in the opposite part of the network, each of the data has to be transferred from one part to the other. In a unit time, at most $O(\sqrt{n})$ data can be transferred. The presence of buses does not speed the transmission, because at most one datum at a time can be on the bus. Therefore, n data need $O(\frac{n}{\sqrt{n}}) = O(\sqrt{n})$ time to be transferred from one part of the network to the other, and the lower bound $\Omega\sqrt{n}$ applies. In particular, it means that the lower bound for sorting is $\Omega(\sqrt{n})$ on both MCC and MMB with n processors. In other words, the addition of buses does not speed up (asymptotically) the time required to sort an array. This lower bound for sorting is claimed (without proof) in [20].

7 Layout

A digital computer can be viewed as a large collection of interconnected logical gates. These gates come in packages called chips, which are tiny pieces of semiconductor material used to fabricate logical gates and the wires connecting them. VLSI (very large scale integration) is e-technology that enables packing millions of logical gates on a single $1cm^2$ chip. In a physical VLSI implementation of interconnection networks, the area required to place all processors and the wire lengths to interconnect them are particularly important. The length of the wires determines the speed at which the network can operate and the amount of power dissipated driving the wires.

The processors can be placed either on a straight line, or on a plane (board), or they may be configured in a space arrangement. In a planar embedding, one important question is whether two links can cross each other. In the Thompson grid model [32] for planar layout, processing elements can be located only at the intersections of vertical and horizontal tracks, i.e., at the grid points. Tracks are spaced at unit distance. Processor interconnections run along the tracks but no two links share the same track. Links may cross at grid points.

If the physical layout of a complete binary tree network with n nodes is the same as its usual drawing (see Figure 23) then it can be shown that the area requirement is $\mathrm{O}(n \log n)$ and the maximal wire length is $O(n)$ (the shortest wire is assumed to have unit length). A better implementation can be achieved as in Figure 24. The area is $O(n)$ [22] and the wire length is $O(\sqrt{n})$ which is nearly optimal [8].

Hypercubes, butterflies, and cube-connected cycles require a layout area almost quadratic in the number of nodes and wire length that grows almost linearly with the number of nodes.

There exists another layout model for networks, especially for those with higher degree. It is the model of a book with processors located along the book spine and pages attached to it. Each page is a half-plane that contains some edge connections among processors such that no two edges cross each other. The goal in the design is to minimize the total number of pages for a given graph.

We describe an efficient hypercube layout [10] in the page model. The processors are ordered along the edge of the book in the Gray code order. A two-page layout for a three-dimensional hypercube is straightforward. For each remaining dimension a new page suffices. Two copies of a lower-dimensional hypercube are placed along the same book spine

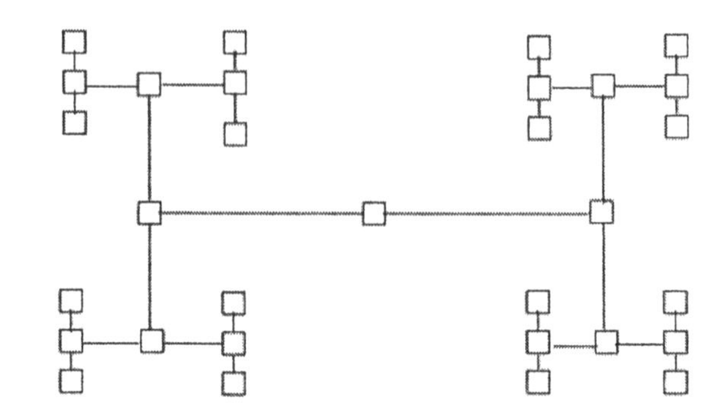

Figure 23: Physical layout of a binary tree.

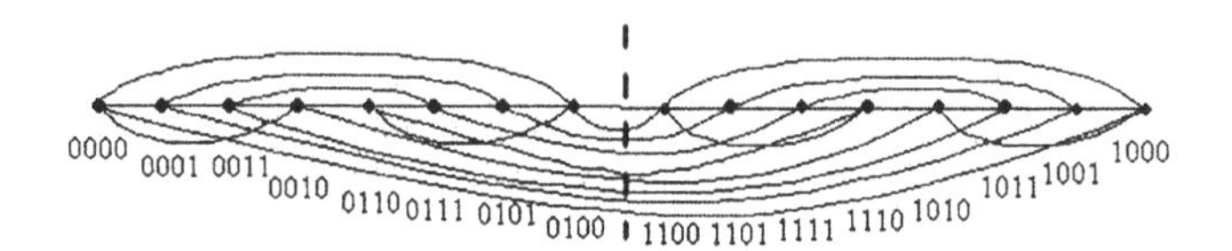

Figure 24: Three-page layout of a four-dimensional hypercube.

next to each other and use the same pages for internal connections. All interconnections between the two copies may fit on the same new page. For example, Figure 24 shows a three-page layout for 4-dimensional hypercube. All processors are placed on a horizontal line. One page is above the line, one is below the line, and the third page consists of links that cross the dashed vertical line, which splits two three-dimensional hypercubes.

References

[1] Akers S.B., Harel D., and Krishnamurthy B., The star graph: An attractive alternative to the n-cube, Proc. Int. Conf. Parallel Processing, 1987, 393-400.

[2] Akl S.G., Qiu K., and Stojmenovic I., Fundamental algorithms for the star and pancake interconnection networks with applications to computational geometry, Networks, Vol. 23, 1993, 215-225.

[3] Arno S. and Wheeler F.S., Signed digit representations of minimal Hamming weight, IEEE Trans. on Computers, 42, 8, 1993, 1007-1010.

[4] Bar-Noy A., and Peleg D., Square meshes are not always optimal, IEEE Trans. on Computers, 40, 2, 1991, 196-204.

[5] Bhatt S.N. and Ipsen C.F., How to embed trees into hypercubes, YALEU/DCS/RR-443, 1985.

[6] Bruck J., Cypher R., and Ho C.-T., Fault-tolerant meshes and hypercubes with minimal numbers of spares, IEEE Trans. Computers, 42, 9, 1993, 1089-1104.

[7] Becker B. and Simon H.U., How robust is the n-cube?, Proc. Symp. Foundation of Comp. Sci., 1986, 283-291.

[8] Brent R.P., and Kung H.T., On the area of binary tree layouts, Information Processing Letters 11, 1, 1980, 44046.

[9] Carle J., Myoupo J.F. and Stojmenovic I., Higher dimensional honeycomb networks, Journal of Interconnection Networks, 2, 4, December 2001, 391-420.

[10] Chung F.R.K., Leighton F.T., and Rosenberg A.L., Embedding graphs in books, a layout problem with applications to VLSI design, SIAM J. Algebr. Discr. Meth., 8, 1987, 33-58.

[11] Chen M.S., Shin K.G., and Kandlur D.D., Addressing, routing, and broadcasting in hexagonal mesh multiprocessors, IEEE Trans. Computers, 39, 1, 1990, 10-18.

[12] Corbett P.F., Rotator graphs: An efficient topology for point-to-point multiprocessor networks, IEEE Trans. Parallel and Distributed Systems, 3, 5, 1992, 622-626.

[13] de Bruijn N.G., A combinatorial problem, Koninklijke Netherlands: Academie Van Wetenschappen, Proc. 49, part 20, 1946, 758-764.

[14] Day K. and Tripathi A., Arrangement graphs: A class of generalized star graphs, Information Processing Letters 42, 1992, 235-241.

[15] Graham N., Harary F., and Livingston M., Subcube fault tolerance in hypercubes, Information and Computation 102, 1993, 280-314.

[16] Garcia Nocetti F., Stojmenovic I. and Zhang J., Addressing and routing in hexagonal networks with applications for location update and connection rerouting in cellular networks, IEEE Trans. Parallel and Distributed Systems, Vol. 13, No. 9, Sept. 2002, 963-971.

[17] Garcia F., Solano J., Stojmenovic I. and Stojmenovic M., Higher-dimensional hexagonal networks, Journal of Parallel and Distributed Computing, Vol. 63, Issue 11, November 2003, 1164-1172.

[18] Hsu W.J., Fibonacci cubes - a new interconnection topology, IEEE Trans. on Parallel and Distributed Systems, 4, 1, 1993, 3-12.

[19] Johnsson S.L., Communication in network architectures, in: VLSI and Parallel Computation, Morgan-Kaufmann, 1990.

[20] Kumar V.P. and Raghavendra C.S., Array processor with multiple broadcasting, J. Parallel and Distributed Computing, 2, 1987, 173-190.

[21] Martin A.J., The torus, an exercise in constructing a processing surface, Proc. 2nd Caltech Conf. VLSI, 1981, 527-537.

[22] Mead C.A. and Conway L., Introduction to VLSI Systems, Addison-Wesley, 1980.

[23] Menn A. and Somani A.K., An efficient sorting algorithm for the star graph interconnection network, Proc. Int. Conf. Parallel Processing, 1990, 1-8.

[24] Saad Y. and Schultz M.H., Topological properties of hypercubes, IEEE Trans. Computers, 37, 7, 1988, 867-872.

[25] Siegel H.J., Interconnection Networks for Large Scale Parallel Processing, Heath and Co., 1985.

[26] Samatham M.R. and Pradhan D.K., The de Bruijn multiprocessor network: A versatile parallel processing and sorting network for VLSI, IEEE Trans. on Computers 38, 4, 1989, 567-581.

[27] Stojmenovic I., Multiplicative circulant networks: Topological properties and communication algorithms, Discrete Applied Mathematics, 77, 1997, 281-305.

[28] Stojmenovic I., Honeycomb networks: Topological properties and communication algorithms, IEEE Trans. Parallel and Distributed Systems, Vol. 8, No. 10, October 1997, 1036-1042.

[29] Stojmenovic I., Direct interconnection networks, in: Parallel and Distributed Computing Handbook (A.Y. Zomaya, ed.), McGraw-Hill, 1996, 537-567.

[30] Stojmenovic I., Optimal deadlock-free routing and broadcasting on Fibonacci cube networks, Utilitas Mathematica, Vol. 53, May 1998, 159-166.

[31] Stojmenovic I., Routing and broadcasting on incomplete and Gray code incomplete hypercubes, Parallel Algorithms and Applications, Vol. 1, No. 3, 1993, 167-177.

[32] Thompson C.D., Area-time complexity for VLSI, Proc. 11th ACM Symp. on Theory of Computing, 1979, 81-88.

[33] Wong C.K. and Coppersmith D., A combinatorial problem related to multimodule memory organizations, J. ACM, 21, 3, 1974, 392-402.

[34] Zaks S., A new algorithm for generation of permutations, BIT, 24, 1984, 196-204.

Chapter 16

Some Bounded Degree Communication Networks and Optimal Leader Election

Pradip K. Srimani
Department of Computer Science
Clemson University, Clemson, South Carolina 29634
E-mail: `srimani@cs.clemson.edu`

Shahram Latifi
Department of Electrical Engineering
University of Nevada, Las Vegas, NV 89154
E-mail: `latifi@ee.unlv.edu`

1 Introduction

For the past many years, much work has been done on a class of graphs called Cayley graphs. These families of graphs have been proven to be very useful and cost effective for designing interconnection networks for distributed systems; they are also very interesting from a pure graph theory point of view in terms of their rich topological properties. Cayley graphs are based on permutation groups and include a large number of families of graphs, such as star graphs [2, 3, 4], hypercubes [5], pancake graphs [2, 6] and others [7, 8, 9]. These graphs are symmetric (edges are bidirectional), regular, and seem to share many of the desirable properties such as low diameter, low degree, high fault tolerance, etc. with the well-known hypercubes (which are also Cayley graphs). An excellent survey of these Cayley graphs (along with an extensive bibliography) can be found in [1]. All Cayley graphs are regular i.e., each node has the same degree;

also, for many Caley graphs as reported in [1], the degree of the nodes increases with the size of the graph (the number of nodes) either logarithmically or sub-logarithmically.

From a VLSI design point of view on the other hand, bounded or constant degree networks are more suitable for area efficient layout and there are also important applications because the computing nodes in the interconnection network can have only a fixed number of I/O ports [10, 11]. There are graphs in the literature that have constant node degrees, such as the cube-connected cycles [12] where the degree of any node is 3 irrespective of the size of the graph. These cube-connected cycle graphs can be viewed as Cayley graphs [13]. Another example would be the graphs introduced in [14]. There also exist graph topologies in the literature that have *almost* constant node degrees such as the De Bruijn graphs [15] or the Möbius graphs [16]. Constant degree network graphs have been of considerable practical importance since De Bruijn graphs were used for designing a 8096 node multiprocessor at JPL for the Galileo project [17]. But, neither De Bruijn graphs nor Möbius graphs are regular; none of them can be viewed as Cayley graphs. Also, these graphs have a low vertex connectivity of only 2 (i.e., fault tolerance is minimal in the sense that the graphs cannot tolerate more than one faulty node) although most of the nodes in those graphs have degree larger than 2 (most of the nodes in De Bruijn graphs have degree 4 and most of the nodes in Möbius graphs have degree 3). Recently, some bounded degree Cayley graphs have been proposed in the literature. Our first purpose in the present chapter is to provide a brief survey of some of those bounded degree Cayley networks with special emphasis on their routing, diameter and fault tolerance. We also briefly explore possibilities of combining the good fault tolerance of unbounded degree Cayley networks with the easy VLSI implementation of bounded degree networks to generate tunable topologies that can satisfy both the needs.

In the second part of the chapter, we consider the problem of anonymous leader election in bounded degree Cayley networks. Leader election in a network is one of the most important problems in the area of distributed algorithm design. Consider any network of N nodes; a *leader* node is defined to be any node of the network unambiguously identified by some characteristics (unique from all other nodes). A leader election process is defined to be a uniform algorithm (code) executed at each node of the network; at the end of the algorithm execution, exactly one node is elected the *leader* and all other nodes are in the nonleader state. Gallager et. al. [18] have developed a leader election algorithm for arbitrary networks whose message complexity is $\mathcal{O}(N \log N + E)$ where N is the number of nodes and E is the number of edges in the network. Santoro in [19] has studied how the knowledge of topology of the network and the orientation of the links in the network affect the message complexity of leader election algorithms in networks. Subsequently, many authors have studied leader election algorithms for various kinds of networks with or without link orientations. Existence of orientation of the links has been shown not to improve the message complexity of leader election for either rings or torii [20],although that helps for cliques; the lower bound on message complexity of leader election in oriented cliques is $\mathcal{O}(N)$ [21] and that for unoriented cliques is $\Omega(N \log N)$ [22]. Authors in [21] have considered complete networks with a sense of direction. Recently, Tel [23, 24] has developed leader election for oriented hypercubes whose message complexity is linear in the number of nodes. Our purpose in the last section of this chapter is to develop a strategy for a leader election algorithm for bounded degree networks. We use Hyper butterfly networks to illustrate our approach.

2 Hypercubes and Star Graphs

An interconnection network (graph), G, is called a Cayley graph iff its vertices are the elements of a group $\mathcal{G}$. The edges are determined by a set $\Gamma \subseteq \mathcal{G}$; namely, the edges are (v, vg) for $v \in \mathcal{G}$, and $g \in \Gamma$. The edge (v, vg) may be labeled with g. If Γ generates $\mathcal{G}$, then the digraph is strongly connected. If Γ is closed under inverse the digraph is symmetric and may be considered undirected. A more complete theoretical treatment of Cayley graphs can be found in [25, 26]. Let's consider a Cayley graph, $G = (V, E)$. We use the set of generators, Ω, to define V and E. The vertices are defined as

$$V = \{x | x \in V \text{ or } x = y \cdot g \text{ where } y \in V, g \in \Omega\}$$

The edge set is given by

$$E = \{(x, y)_g | x, y \in V \text{ and } g \in \Omega \text{ such that } y = x \cdot g\}$$

2.1 Hypercubes

The Cayley graph H_n, the hypercube (Figure 1), is commonly used as an interconnection network. We define a hypercube as the graph $G_{H_n} = (V_{H_n}, E_{H_n})$. The generator for this graph is the set of transpositions defined by

$$\Omega_{H_n} = \{(2j-1\ 2j) | 1 \leq j \leq n\}$$

Observation 1 (Diameter) *This graph has a diameter of n, because there are only n disjoint transpositions that can be performed on any given label. That is, the most that any given label can be incorrect is n transpositions.*

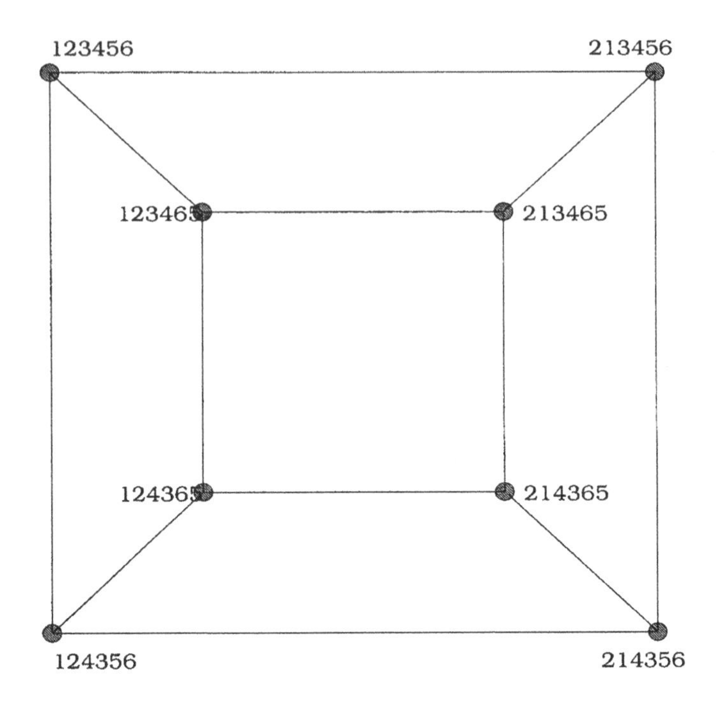

Figure 1: Hypercube, H_3

Observation 2 (Node Degree) *The node degree is n, because there are n generators. Application of each generator leads to a distinct node, because the members of Ω_{H_n} are mutually disjoint. From this, we can notice that H_n is not a bounded degree graph, because the node degree increases with n.*

2.2 Star Graphs

A star graph S_n, of order n, is defined to be a symmetric graph $G = (V, E)$ where V is the set of $n!$ vertices, each representing a distinct permutation of n elements and E is the set of symmetric edges such that two permutations (nodes) are connected by an edge iff one can be reached from the other by interchanging its first symbol with any other symbol. For example, in S_3, the node representing permutation ABC has edges to two other permutations (nodes) bac and cba. Throughout our discussion we denote the nodes by permutations of the English alphabets. For example, the identity permutation is denoted by $\mathbf{I} = (abcd...z)$ (z is the last symbol, not necessarily the 26th).

Properties of $\mathbf{S_n}$

- These star graphs are members of the family of Cayley group graphs. For a star graph S_n of dimension n, there are $n-1$ generators, $g_2, g_3, \cdots, g_n$, where g_i swaps the first symbol with the ith symbol of any permutation. Each generator is its own inverse i.e., the star graph is symmetric. Also, the star graph S_n is a $(n-1)$-regular graph with $n!$ nodes and $n!(n-1)/2$ edges.

- Because star graphs are vertex symmetric [2], we can always view the distance between any two arbitrary nodes as the distance between the source node and the identity permutation by suitably renaming the symbols representing the permutations.

- Diameter of S_n, $\Delta_n = \lfloor 3(n-1)/2 \rfloor$.

3 Bounded Degree Cayley Graphs

3.1 Cube-Connected Cycles CC_n

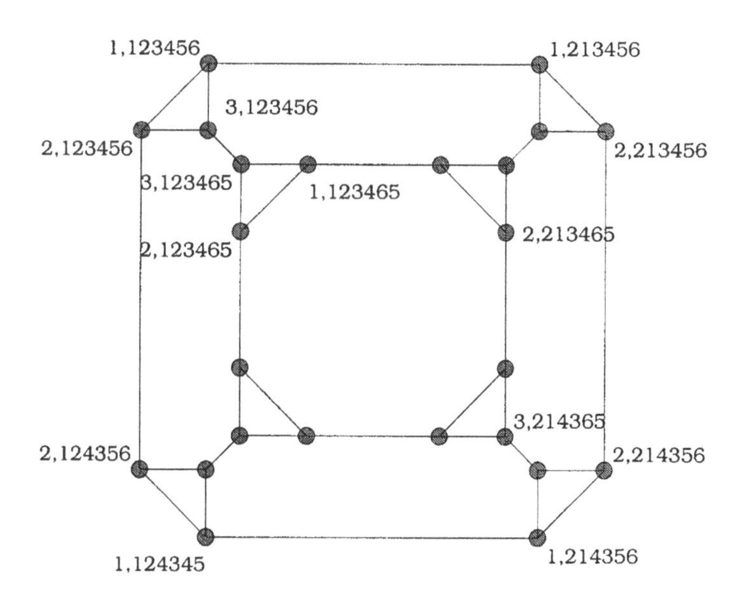

Figure 2: Cube-connected cycles, CC_3

Cube-connected Cycles (CC_n) [25, 26] are a fairly standard fixed degree topology. The basic idea is to replace each node of a hypercube, H_n, with a cycle of length n. To accomplish this, define the generator set

$$\Omega_{CC_n} = \Omega_{H_n} \times \Omega_{C_n}$$

where the generator for the cycle is defined as

$$\Omega_n = \{1, -1\}$$

By doing this, we have defined the set of vertices as

$$V = \{(i, p) | 0 \leq i < n \text{ and } p \in V_{H_n}\}$$

To define the edges, we restrict the use of generators in the following way

$$\begin{aligned} E_{CC_n} = \quad \{((i,p),(j,q))| \quad & 0 \leq i, j < n \text{ and } p, q \in V_{H_n} \text{ and} \\ & (i = j \text{ and } q = p \cdot (2i + 1\ 2i + 2)) \text{ or} \\ & (p = q \text{ and } i = (j+1) \bmod n \text{ or} \\ & \quad i = (j-1) \bmod n)\} \end{aligned}$$

Properties of CC_n

- [Node Degree] Cube-connected cycles have a node degree of 3 ($n \geq 3$). This is obvious because there are 3 distinct edge rules, two for moving around the cycle in opposite directions and one for moving to the next composite node in the hypercube.

- [Diameter] CC_n has a diameter of $\frac{5n-2}{2}$. Consider the optimal routing through CC_n, given in Figure 3. We notice that the inner loop can only be executed n times, because S_i can differ from D_i by at most n transpositions. We can also note that, in this case we must move around the cycles a total of $n - 1$ moves. This is, in fact, the most that we need to move around the cycle, because we are always taking the very next differing transposition. At this point, we have $S_p = D_p$, so we only need to correct for Si and Di. This can be accomplished in, at most, $\frac{n}{2}$ moves, because we can move either direction around the current cycle, which has length n. This yields the D, therefore we can add up the maximal moves and arrive at a diameter of $\frac{5n-2}{2}$.

- Given a source address $S = (S_i, S_p)$ and a destination address $D = (D_i, D_p)$, repeat until $S_p = D_p$:
  - Pick the next transposition, j, from the $S_i^t h$ position in S_p that differs from D_p.
  - Move around the current cycle from S_i to j.
  - Apply the generator to correct S_{p_j}.
- Move around the current cycle either clockwise or counterclockwise, whichever is shortest, so that $S_i = D_i$.

Figure 3: Optimal routing algorithm for CC_n

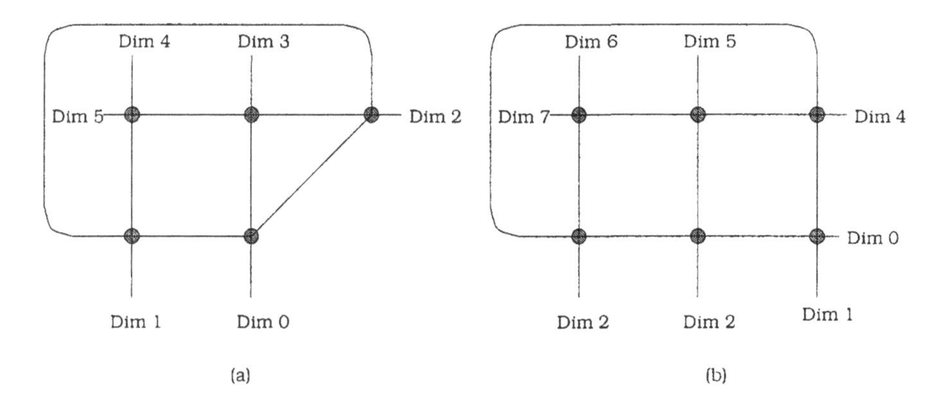

Figure 4: The composite node for Cube-connected Möbius ladders of size (a) 5 (external degree $= 6$) and (b) 6 (external degree $= 8$)

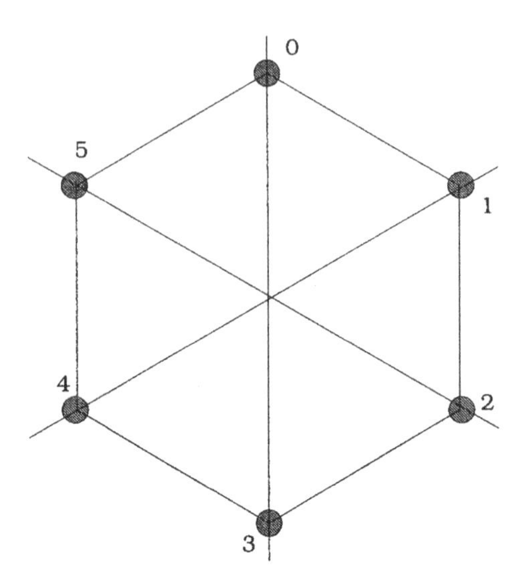

Figure 5: The Cayley form composite node for Cube Connected Möbius Ladders of size 6 (external degree= 6)

3.2 Cube-Connected Möbius Ladders $\mathrm{CCML_n}$

A variation on Cube-connected cycles, Cube-connected Möbius ladders has been explored in [27]. Again, we start out with a hypercube, H_n, but replace the vertices with a Möbius Ladder (see Figures 4 and 5). Pritchard and Nicole initially define the architecture as in Figure 4 but point out that this is not a Cayley graph, because it isn't homogeneous. This can be observed in that some nodes have one connection to the hypercube and some have two. This graph can be defined as a Cayley graph, though, by connecting the nodes with the highest external degree (e.g., Dim 7 and Dim 0 in Figure 4(b)), yielding Figure 5. The generator for this is defined as

$$\Omega_{CCML_n} = \Omega_{ML_n} \times \Omega_{H_n}$$

where the generator for the Möbius ladder composite nodes is defined by

$$\Omega_{ML_n} = \left\{1, -1, \frac{n}{2}, -\frac{n}{2}\right\}$$

- Given a source address $S = (S_i, S_p)$ and a destination address $D = (D_i, D_p)$, repeat until $S_p = D_p$:
  - Pick the next transposition, j, from the $S_i^t h$ position in S_p that differs from D_p.
  - Move around the current cycle from S_i to j. If j is between $\frac{1}{4}$ and $\frac{3}{4}$ of the way around the cycle from i, use the cross edge.
  - Apply the generator to correct S_{p_j}.
- Move around the current cycle either clockwise or counterclockwise, whichever is shortest, so that $S_i = D_i$. If j is between $\frac{1}{4}$ and $\frac{3}{4}$ of the way around the cycle from i, use the cross edge.

Figure 6: Optimal routing algorithm for $CCML_n$

Similarly to Cube-connected cycles, we only consider the edges where

$$E_{CCML_n} = \{((i,p),(j,q))| \quad 0 \leq i,j < n \text{ and } p,q \in V_{H_n} \text{ and } (i = j \text{ and } q = p \cdot (2i+1\ 2i+2)) \text{ or } (p = q \text{ and } i = (j+1) \bmod n \ \text{ or } \ i = (j-1) \bmod n)\}$$

These are essentially cube-connected cycles with the opposite nodes in the cycle connected by an edge. This restricts $n = 2k$, where $k \in \mathcal{Z}^+$.

Properties of $CCML_n$

- [Node Degree] $CCML_n$ has a node degree of 4.
- [Diameter] $CCML_n$ has a diameter of $\frac{9n-4}{4}$. We proceed similarly as before. Arriving at $S_p = D_p$ requires at most $2n - 1$ moves, as it did for CC_n. Arriving at D_i only requires $\frac{n}{4}$, though. To see this, notice that, if we are j nodes away from the target node of the cycle, and $j \leq \frac{n}{4}$ or $j \geq \frac{3n}{4}$ we are only $\frac{n}{4}$ from the destination. If, however, $\frac{n}{4} < j < \frac{3n}{4}$, we can move on the cross edge (distance of 1) and then proceed to the target node (maximum distance of $\frac{n}{4} - 1$). This yields a maximum possible shortest path from S to D of $\frac{9n-4}{4}$.

Because $CCML_n$ is only slightly better from the point of view of diameter and node degree than CC_n, a stronger motivation is necessary for using this graph. In [27], Pritchard and Nicole present a deadlock-free routing algorithm for this topology to justify the use of Cube-connected Möbius ladders.

3.3 deBruijn Graphs

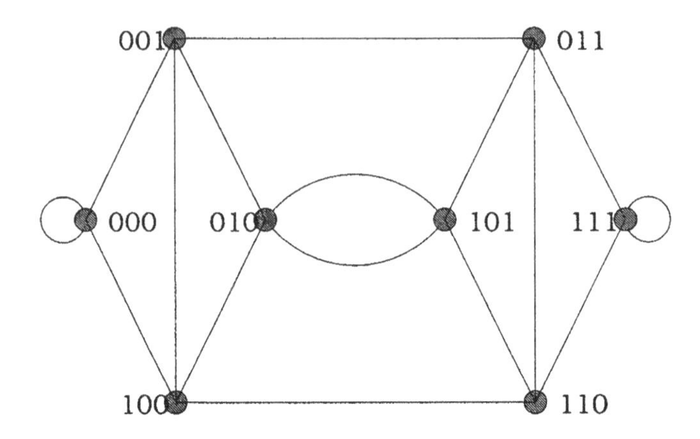

Figure 7: The deBruijn graph, $D3$

A (binary) deBruijn graph, $D_n = (V_{D_n}, E_{D_n})$ can be defined as

$$V_{D_n} = \{x_n ... x_1 | x_i \in \{0,1\}\}$$

$$E_{D_n,1} = \{((\alpha x \beta), (x \beta \alpha)) | x \in V_{D_{n-2}}, \alpha, \beta \in \{0,1\}\}$$
$$E_{D_n,2} = \{((\alpha x \beta), (x \beta \bar{\alpha})) | x \in V_{D_{n-2}}, \alpha, \beta \in \{0,1\}\}$$
$$E_{D_n,3} = \{((\alpha x \beta), (\beta \alpha x)) | x \in V_{D_{n-2}}, \alpha, \beta \in \{0,1\}\}$$
$$E_{D_n,4} = \{((\alpha x \beta), (\bar{\beta} \alpha x)) | x \in V_{D_{n-2}}, \alpha, \beta \in \{0,1\}\}$$
$$E_{D_n} = E_{D_n,1} \cup E_{D_n,2} \cup E_{D_n,3} \cup E_{D_n,4}$$

Observation 3 (Node Degree) *This gives the nodes maximum degree 4, though two nodes, $\vec{0}$ and $\vec{1}$ have two self-loops and there are two parallel edges between* $1010\ldots$ *and* $0101\ldots$, *yielding degrees of 2 and 3, respectively.*

3.4 Shuffle-Exchange Permutation $\mathrm{SEP_n}$ Graph

An n-dimensional SEP graph is an undirected regular graph with $N = n!$ vertices, each vertex corresponding to a distinct permutation of the set $\{1, 2, \cdots, n\}$. Two vertices are directly connected if and only if the label (permutation) of one is obtained from the label of the other by one of the following operations: (1) swapping the first two digits (the leftmost digit is ranked as first); (2) shifting cyclically to left (or right) by one digit. Formally, consider a set of n distinct symbols (we use Arabic numerals as symbols; for example, when $n = 4$, the symbols are $1, 2, 3, 4$). Thus, for n distinct symbols, there are exactly $n!$ different permutations of the symbols or the number of vertices in SEP_n is $n!$; using $a_1 a_2 \cdots a_n$ as the symbolic representation of an arbitrary vertex, the edges of SEP_n are defined by the following three generators in the graph:

$$g_L(a_1 a_2 \cdots a_n) = a_2 a_3 \cdots a_n a_1$$
$$g_R(a_1 a_2 \cdots a_n) = a_n a_1 a_2 \cdots a_{n-1}$$
$$g_{12}(a_1 a_2 \cdots a_n) = a_2 a_1 a_3 \cdots a_n$$

The SEP is an undirected graph as swapping the first two digit is a symmetric operation; and if a vertex u is obtained from another vertex v by shifting to the left, vertex v can be obtained from u by shifting to the right; thus, if u is adjacent to v, so is v to u, (ii) The generator g_{12} is its own inverse, i.e., $g_{12}^{-1} = g_{12}$ and $g_L^{-1} = g_R$ ($g_R^{-1} = g_L$). Figure 8 shows the proposed Cayley graphs SEP_3 and SEP_4 of dimensions 3 and 4, respectively. SEP_n, $n \geq 2$, is a Cayley graph. Note that the two generators $g_{(1,2)}$ and g_L could produce all the elements of the group S_n and so could the two generators $g_{(1,2)}$ and g_R. The third generator is included with two objectives: to make the generator set closed under inversion (a requirement for the graph to be a Cayley graph) and to make the graph undirected (bidirectional).

Properties of SEP_n

- For any n, $n \geq 2$, the graph SEP_n: (1) is a symmetric (undirected) regular graph of degree 3; (2) has $n!$ vertices; and (3) has $3n!/2$ edges.
- The graph $SEP_n$is 3-connected for any given n, $n \geq 3$.

3.5 Star-Connected Cycle SCC_n Graph

The star-connected cycle graph SCC_n is defined for $n \geq 2$, and is obtained from the star graph S_n, by replacing each node by an $(n-1)$-vertex cycle. To define it, a variation of the notation is convenient, which we adopt here. S_n is considered to act on $Z_{n_i} \cup x$ where Z_{n_i} is the cyclic group of order $n-1$. The generators for the graph S_n are $(i\infty)$ for $i \in Z_{n-1}$.Thus, $\{l, 2, \cdots, n\}$ is replaced by $\{\infty, 0, \ldots, n-2)$. One reason the former convention is frequently used is that S_{n-1} is canonically embedded as a subset of S_n as the permutations fixing n. For SCC_n, the latter notation has many advantages.

The nodes of SCC_n, are the pairs $< p, i >$ where $p \in S_n$, and $i \in Z_{n-1}$. There are edges from $< p, i >$ to $< p, i + -1 >$ (called intracycle edges), and from $< p, i >$ to $< po(i \inf), i)$ (called intercycle edges). The subgraph

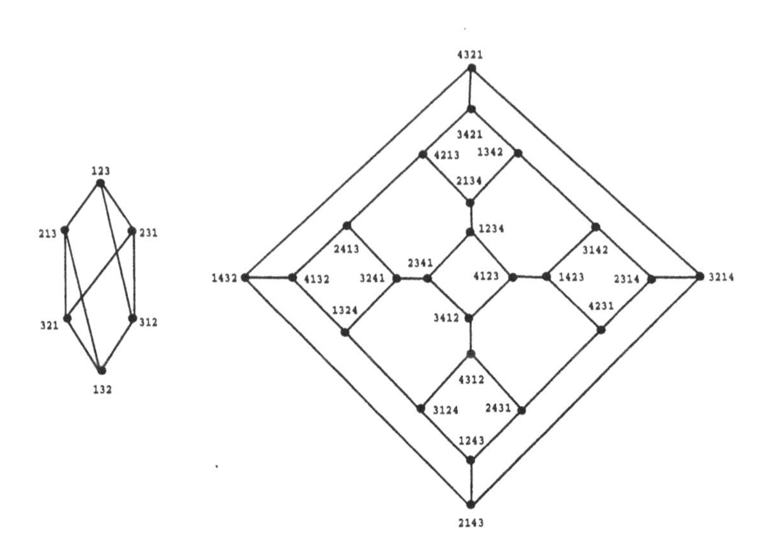

Figure 8: SEP_n graphs for $n = 3$ and $n = 4$

induced on $\{(p, i)\}$ for fixed p is a cycle (called the p-cycle). The relation of belonging to the same p-cycle yields S_n, as a quotient of SCC_n. In particular, a path in SCC_n, induces a path in S_n, namely visiting the p of the (p, i).

SCC_n is analogous to the cube connected cycles HCC_n where each node of the hypercube H_n, is replaced by an n-vertex cycle. It is well known that HCC_n is a Cayley graph [13]; an analogous argument shows that SCC_n, is also a Cayley graph [28].

3.6 Butterfly Networks $\mathbf{B_n}$

A wrapped butterfly network, denoted by B_n, is defined [29] as follows: a vertex is represented as $(z_{n-1} \cdots z_0, \ell)$, where $z_{n-1} \cdots z_0$ is an n-bit binary number and ℓ is an integer, $0 \leq \ell \leq n-1$.

The edges of B_n are defined by a set of four generators. Consider an arbitrary node $(z_{n-1} \cdots z_0, \ell)$ in B_n. We define $\alpha(\ell) = \ell + 1(\text{mod} n)$ and $\beta(\ell) = \ell - 1(\text{mod} n)$. The four edges node $(z_{n-1} \cdots z_0, \ell)$ can be derived by the following four generators,

$$g(z_{n-1} \cdots z_0, \ell) = z_{n-1} \cdots z_0, \alpha(\ell)$$
$$g^{-1}(z_{n-1} \cdots z_0, \ell) = z_{n-1} \cdots z_0, \beta(\ell)$$

$$f(z_{n-1} \cdots z_0, \ell) = z_{n-1} \cdots z_{\ell+1} \bar{z}_\ell z_{\ell-1} \cdots z_0, \alpha(\ell)$$
$$f^{-1}(z_{n-1} \cdots z_0, \ell) = z_{n-1} \cdots z_{\beta(\ell)+1} \bar{z}_{\beta(\ell)} z_{\beta(\ell)-1} \cdots z_0, \beta(\ell)$$

Remark 1 *We refer to the first part in the butterfly label, i.e., $z_{m-1} \cdots z_0$ as the* complementation index*; and the second part, i.e., ℓ, as the* permutation index.

It is interesting to note that these networks are, indeed, Cayley graphs of degree 4. Consider an alternative definition of the same butterfly networks B_n, as given in [30]. B_n is defined as a graph on $n \times 2^n$ vertices for any integer n, $n \geq 3$; each vertex is represented by a circular permutation of n symbols in lexicographic order where each symbol may be present in either uncomplemented or complemented form. Let t_k, $1 \leq k \leq n$ denote the kth symbol in the set of n symbols (we use the English alphabet as symbols; thus for $n = 4$, $t_1 = a$, $t_2 = b$, $t_3 = c$ and $t_4 = d$). We use t_k^* to denote either t_k or $\bar{t}_k$. Thus, for n distinct symbols, there are exactly n different cyclic permutations of the symbols in lexicographic order and because each symbol can be present in either complemented or uncomplemented form, the vertex set of G_n (i.e. the underlying group Γ) has a cardinality of $n.2^n$ (for example, for $n = 3$, the number of vertices in G_3 is 24; abc, cab, $\bar{c}ab$ are valid nodes whereas acb or bac are not). Let I denote the *identity permutation* $t_1 t_2 \cdots t_n$. Because each node is some cyclic permutation of the n symbols in lexicographic order, then if $a_1 a_2 \cdots a_n$ denotes the label of an arbitrary node and $a_1 = t_k^*$ for some integer k, then for all i, $2 \leq i \leq n$, we have $a_i = t^*_{(k+i) \bmod n+1}$. The edges of G_n are defined by the following four generators in the graph.

$$g(a_1 a_2 \cdots a_n) = a_2 a_3 \cdots a_n a_1$$
$$f(a_1 a_2 \cdots a_n) = a_2 a_3 \cdots a_n \bar{a}_1$$
$$g^{-1}(a_1 a_2 \cdots a_n) = a_n a_1 \cdots a_{n-1}$$
$$f^{-1}(a_1 a_2 \cdots a_n) = \bar{a}_n a_1 \cdots a_{n-1}$$

Example: For $n = 3$, the vertex set of B_3 (the underlying group Γ) is given by $\{abc,\ \bar{a}bc,\ a\bar{b}c,\ ab\bar{c},\ cab,\ \bar{c}ab,\ c\bar{a}b,\ ca\bar{b},\ bca,\ \bar{b}ca,\ b\bar{c}a,\ bc\bar{a},\ \bar{a}\bar{b}c,\ a\bar{b}\bar{c},\ \bar{a}b\bar{c},\ \bar{c}\bar{a}b,\ c\bar{a}\bar{b},\ \bar{c}a\bar{b},\ \bar{b}\bar{c}a,\ b\bar{c}\bar{a},\ \bar{b}c\bar{a},\ \bar{a}\bar{b}\bar{c},\ \bar{b}\bar{c}\bar{a},\ \bar{c}\bar{a}\bar{b}\}$ and the generator set Ω is given by $\Omega = \{bc\bar{a},\ \bar{c}ab,\ cab,\ bca\}$. Note that because we are considering groups of permutation of distinct symbols that may be complemented or uncomplemented, the generators themselves are also permutations of complemented or uncomplemented symbols.

Properties of B_n

- The set of four generators, $\Omega = \{f, g, f^{-1}, g^{-1}\}$ closed under inverse; in particular g is the inverse of g^{-1} and f is the inverse of f^{-1}; thus the edges in G_n are bidirectional.

- For an arbitrary n, $n > 2$, for any arbitrary node v of the graph G_n, $\delta(v) \neq v$ where $\delta \in \{g, f, g^{-1}, f^{-1}\}$.

- For any n, $n \geq 3$, the graph G_n: (1) is a symmetric (undirected) regular graph of degree 4; (2) has $n.2^n$ vertices; and (3) has $n.2^{n+1}$ edges.

- [Simple Routing] G_n is a Cayley graph, therefore it is vertex symmetric [2]; i.e., we can always view the distance between any two arbitrary nodes as the distance between the source node and the identity permutation by suitably renaming the symbols representing the permutations. The algorithm S_R (Figure 9) computes a path from an arbitrary source node $a_1a_2 \cdots a_n$ in G_n to the identity node I. For an arbitrary node $a_1a_2 \cdots a_n$ in G_n, the algorithm S_R generates a path of length $\leq \lfloor \frac{3n}{2} \rfloor$. Although the algorithm does generate a path of length $\leq \lfloor \frac{3n}{2} \rfloor$ from an arbitrary node to the identity node, it does not generate the optimal (shortest) path in most cases.

- [Diameter] The diameter of the graph G_n is given by $\mathcal{D}(G_n) = \lfloor \frac{3n}{2} \rfloor$.

- B_n, $n \geq 3$, has a Hamiltonian cycle.

- [Fault Tolerance] The node fault tolerance of an undirected graph is measured by the vertex connectivity of the graph. A graph G is said to have a vertex connectivity ξ if the graph G remains connected when an arbitrary set of less than ξ nodes is faulty. Obviously, the vertex connectivity of a graph G cannot exceed the minimum degree of a node in G. The graph B_n is 4-connected for any given n, $n \geq 3$ and is thus, *maximally fault tolerant.*

- See [30] for the related proofs and an optimal routing algorithm.

B_n has a diameter logarithmic in the number of nodes, is regular and dense, and is maximally fault tolerant. B_n is an attractive alternative to the well known binary De Bruijn graphs for designing interconnection networks especially when the network is needed to be maximally fault tolerant. It is to be noted that although simple routing is very efficient for De Bruijn graphs [15], optimal routing in those graphs is complicated [31]; simple routing in the proposed graphs is very efficient, and optimal routing, although more complicated than simple routing, is relatively simpler than that for De Bruijn graphs.

S1:	Compute k, $1 \leq k \leq n$, such that $a_k = t_n^*$.
S2:	**If** $k > \lfloor \frac{n}{2} \rfloor$ **then** go along successive g^{-1} edges $(n-k)$ times **else** go along successive g edges k times.
S3:	**for** $i = 1$ **to** n **do** go to the node $t_i^* t_{i+1}^* \cdots t_n^* t_1 t_2 \cdots t_i$ by either the g or the f edge;

Figure 9: Simple routing algorithm S_R in B_n

3.7 Trivalent Cayley Graphs Q_n

Trivalent Cayley graph Q_n is defined as a symmetric (undirected) graph on $N = n \times 2^n$ vertices for any integer n, $n \geq 2$; each vertex corresponds to a circular permutation of n symbols in lexicographic order where each symbol may be present in either uncomplemented or complemented form. Let t_k, $1 \leq k \leq n$ denote the kth symbol in the set of n symbols (we use the English alphabets as symbols; thus for $n = 4$, $t_1 = a$, $t_2 = b$, $t_3 = c$ and $t_4 = d$). We use t_k^* to denote either t_k or $\bar{t}_k$. Thus, for n distinct symbols, there are exactly n different cyclic permutations of the symbols in lexicographic order (disregarding the complements) and because each symbol can be present in either complemented or uncomplemented form, the vertex set of G_n has a cardinality of $n.2^n$. For example, for $n = 3$, the number of vertices in G_3 is 24; abc, cab, $\bar{c}ab$ are valid nodes whereas acb or $\bar{b}ac$ are not (because neither acb nor bac is a cyclic permutation of the three symbols in lexicographic order). Let I denote the identity permutation $t_1 t_2 \cdots t_n$. Since each node is some cyclic permutation of the n symbols in lexicographic order, then if $a_1 a_2 \cdots a_n$ denotes the label of an arbitrary node and $a_1 = t_k^*$ for some integer k, then for all i, $2 \leq i \leq n$, we have $a_i = t^*_{(k+i) \bmod n+1}$. The edges of G_n are defined by the following three generators in the graph:

$$f(a_1 a_2 \cdots a_n) = a_2 a_3 \cdots a_n \bar{a}_1$$
$$f^{-1}(a_1 a_2 \cdots a_n) = \bar{a}_n a_1 \cdots a_{n-1}$$
$$g(a_1 a_2 \cdots a_n) = a_1 a_2 a_3 \cdots \bar{a}_n$$

Properties of Q_n

- The three generators are closed under inversion; in particular g is its own inverse ($g = g^{-1}$) and f is the inverse of f^{-1}; thus the edges in Q_n are bidirectional. For an arbitrary n, $n \geq 2$, for any node v of the graph Q_n,

$\delta(v) \neq v$ where $\delta \in \{g, f, f^{-1}\}$. Figure 10 shows the proposed trivalent Cayley graphs Q_2 and Q_3 of dimensions 2 and 3.

- For any n, $n \geq 2$, the graph Q_n: (1) is a symmetric (undirected) regular graph of degree 3; (2) has $n \times 2^n$ vertices; and (3) has $3n \times 2^{n-1}$ edges.
- Because Q_n is a Cayley graph, it is vertex symmetric [2]. The algorithm Simple_Route, shown in Figure 11 computes two different paths from an arbitrary source node $a_1 a_2 \cdots a_n$ in Q_n to the identity node I. For an arbitrary node $a_1 a_2 \cdots a_n$ in G_n, the algorithm Simple_Route generates at least one path of length $\leq 2n$.
- The upper bound on the diameter of Q_n is given by $\mathcal{D}(Q_n) \leq 2n$ and the graph Q_n is 3-connected for any given n, $n \geq 2$ and hence is maximally fault tolerant.

The graph Q_n has a diameter logarithmic in the number of nodes, is regular and dense, and is maximally fault tolerant. We have investigated different algebraic properties of the graph and proposed a simple routing algorithm. Q_n seems to be an attractive alternative (Q_n is regular and has better vertex connectivity) to the well known binary De Bruijn graphs or Mobius graphs (neither is a Cayley graph) for designing interconnection networks especially when the degree of the nodes needs to be constant independent of the size of the network.

4 Hyper-deBruijn and Hyper-butterfly Graphs

4.1 Hyper-De Bruijn Graphs

Ganesan and Pradhan [32] propose the hyper-deBruijn graph, $HD_{n,m}$. Unlike the previous interconnection networks, this network topology isn't a composite node graph. Instead, two network topologies, the hypercube and deBruijn graphs, are woven together, with each node taking part in both types of subnetworks. This is illustrated in Figures 12 and 13 which show $HD_{2,2}$ as a set of deBruijn graphs connected by hypercubes and as a set of hypercubes connected by deBruijn graphs.

Formally we define the hyper-deBruijn graph, $HD_{n,m}$ by the generator set

$$\Omega_{HD_{n,m}} = \Omega_{D_n} \times \Omega_{H_m}$$

The hyper-deBruijn graph has maximum node degree $m + 4$. Also notice that the Hyper-deBruijn graph is defined by two parameters, n and m. These two parameters tune the degree of the hypercube and deBruijn subgraphs. Because the degree of the deBruijn component is always a maximum of 4 (with four nodes of degree two and two of degree three), the value of n does not directly affect the node degree. However, adjusting m alters the degree of the H_m subgraphs, setting the node degree to $m + 4$. Although these graphs provide a desired level of fault tolerance and other features such as logarithmic diameter, optimal routing algorithms, scalability and partitionability and ability to emulate most of existing architectures [33], hyper-deBruijn graphs have two shortcomings: (1) they are not regular (nor they are Cayley graphs; hence optimal routing algorithms and VLSI implementations are significantly complicated; (2) the fault tolerance is lower than the degree of the vast majority of nodes in the graph (due to the existence of few nodes of smaller degrees). The Hyper Butterfly networks seem to overcome these shortcomings.

4.2 Hyper-butterfly Graph $HB_{(m,n)}$

Consider two undirected graphs $G = (V_G, E_G)$ and $H = (V_H, E_H)$; the product graph $G \times H$ has node set $V_G \times V_H$. Let u and v be any two nodes in G, and let x and y be any two nodes in H; then, $(\langle u, x\rangle, \langle v, y\rangle)$ is an edge of $G \times H$ iff either (1) (u, v) is an edge of G and $x = y$, or (2) (x, y) is an edge of H and $u = v$.

Definition 1 *A* **hyper-butterfly** *graph $HB_{(m,n)}$, of order (dimension) $(m+n)$ is defined as the product graph of a hypercube H_m of dimension m and a butterfly B_n of dimension n.*

In $HB_{(m,n)}$, each node is assigned a label $(x_{m-1} \cdots x_0, z_{n-1} \cdots z_0, \ell)$ where each x_i and z_j are binary bit, $0 \leq i \leq m-1$ and $0 \leq j \leq n-1$. ℓ is an integer, $0 \leq \ell \leq n-1$. $x_{m-1} \cdots x_0$ is the *hypercube-part-label* and $(z_{n-1} \cdots z_0, \ell)$ is the *butterfly-part-label*. The edges of the $HB_{(m,n)}$ graph are defined by the following $m + 4$ generators.

$$h_i(x_{m-1} \cdots x_0, z_{n-1} \cdots z_0, \ell) = (x_{m-1} \cdots x_{i+1}\bar{x}_i x_{i-1} \cdots x_0, z_{n-1} \cdots z_0, \ell) \quad \forall i,\ 0 \leq i \leq m-1$$

$$g(x_{m-1} \cdots x_0, z_{n-1} \cdots z_0, \ell) = (x_{m-1} \cdots x_0, z_{n-1} \cdots z_0, \alpha(\ell)) \quad \alpha(\ell) = \ell + 1 (\mathrm{mod}\, n)$$

$$g^{-1}(x_{m-1} \cdots x_0, z_{n-1} \cdots z_0, \ell) = (x_{m-1} \cdots x_0, z_{n-1} \cdots z_0, \beta(\ell)) \quad \beta(\ell) = \ell - 1 (\mathrm{mod}\, n)$$

$$f(x_{m-1} \cdots x_0, z_{n-1} \cdots z_0, \ell) = (x_{m-1} \cdots x_0, z_{n-1} \cdots z_{\alpha(\ell)+1}\bar{z}_{\alpha(\ell)} z_{\alpha(\ell)-1} \cdots z_0, \alpha(\ell))$$

$$f^{-1}(x_{m-1}\cdots x_0, z_{n-1}\cdots z_0, \ell) = (x_{m-1}\cdots x_0, z_{n-1}\cdots z_{\ell+1}\bar{z}_\ell z_{\ell-1}\cdots z_0, \beta(\ell))$$

Remark 2

- *The set of $m+4$ generators of the graph $HB_{(m,n)}$, $\Omega = \{h_i, 0 \le i < m, f, g, f^{-1}, g^{-1}\}$ is closed under inversion; in particular h_i for all i is its own inverse, g is the inverse of g^{-1} and f is the inverse of f^{-1}; thus the edges in $HB_{(m,n)}$ are bidirectional.*
- *For an arbitrary n, $n > 2$, for any arbitrary node v of the graph $HB_{(m,n)}$, $\delta(v) \neq v$ where $\delta \in \Omega = \{h_i, 0 \le i < m, f, g, f^{-1}, g^{-1}\}$; also, for any two $\delta_1, \delta_2 \in \Omega$, $\delta_1(v) \neq \delta_2(v)$.*
- *Hyper-butterfly graph $HB_{(m,n)}$ is a Cayley graph of degree $m+4$.*

Definition 2

- *The m edges generated by the generators h_i are called* hypercube edges *and the 4 edges generated by either of the generators g, f, g^{-1}, f^{-1} are called* butterfly edges.
- *Any arbitrary node $v = (h, b) \in HB_{(m,n)}$ has m* hypercube neighbors *$\{(h^{(i)}, b),\ 1 \le i \le m\}$ (reached from v by the m hypercube edges) and has* 4 butterfly neighbors *$\{(h, b^{(j)}),\ 1 \le j \le 4\}$ (reached from v by the 4 butterfly edges).*

Remark 3 *Along any hypercube edge, only the hypercube-part-label of a node changes, and along any butterfly edges, only the butterfly-part-label changes.*

Remark 4 *The labeling of a hyper-butterfly graph is not unique. There exist many possible different label assignments with the same graph using the traditional labeling scheme. We arbitrarily choose one such traditional labeling and refer to it as canonical labeling and refer to the nodes using its canonical label.*

4.2.1 Topological Properties of Hyper-butterfly Graphs

In this section, we summarize some topological properties that are important to the leader election algorithm.

Lemma 4.1 *In a butterfly graph B_n, the nodes with the same complementation index form a ring of size n.*

Proof : Given an arbitrary complementation index, i.e., $z_0 \cdots z_n$, there are n nodes in B_n that have the same complementation index. The labels of these n nodes are $(z_0 \cdots z_n, \ell)$, $0 \leq \ell \leq n-1$. We refer to these nodes using $(z_0 \cdots z_n, *)$. They are connected by generators g and g^{-1}, and it is easy to see that these n nodes form a ring. □

Definition 3

- *We use $H_m^{(*,z,\ell)}$ to denote an m-dimensional hypercube subgraph of $HB_{(m,n)}$ where each node has the same butterfly-part-label (z, ℓ).*
- *We use $B_n^{(h,*,*)}$ to denote an n-dimensional butterfly subgraph of $HB_{(m,n)}$ where each node has the same hypercube-part-label h.*
- *We use $R_n^{(h,z,*)}$ to denote a ring of n nodes where each node has the same hypercube-part-label h and same complementation index $z = z_0 \cdots z_n$.*
- *We use $HR_{m,n}^{(*,z,*)}$ to denote the set of nodes that have the same complementation index z. This set of nodes is actually the product of $H_m^{(*,z,\ell)}$ and $R_n^{(h,z,*)}$, with the same z value.*

5 Leader Election in Hyper-butterfly Graphs

We start with the following observation.

Remark 5 *If each node knows its canonical label, this election process is trivial. Consider the node having the smallest label, i.e., $(0 \cdots 0, 0 \cdots 0, 0)$ in $HB_{(m,n)}$; we can say that the node with this label will automatically become the leader and all other nodes are nonleaders.*

As in [24], we assume that the nodes in the graph do not know their canonical labels. We still refer to the nodes by some canonical labels for convenience, but these labels or names have no topological significance [24]. We also assume in this chapter that the network is oriented in the sense that each node can differentiate the links incident to it by different generators. (in contrast, a node in an unoriented graph distinguishes its adjacent links by different but uninterpreted names). We define the *direction* of a link as the index of the generator that generates the link. So, the link that is associated with generator h_i has direction i, $0 \leq i \leq n-1$. And the link that is associated with generator g, g^{-1}, f, f^{-1} has direction g, g^{-1}, f, f^{-1}, respectively.

The whole election algorithm in the hyper-butterfly graph consists of three major steps. At different steps, the graph is divided into different regions, and the leader for each region is elected. At first step, the hyper-butterfly graph $H_{(m,n)}$ is divided into $n \times 2^n$ hypercubes. Within each hypercube, the nodes run a formerly proposed election algorithm for hypercubes. After this step, each hypercube will have a leader node. In the second step, the nodes with the same complementation index are considered in one region. For a certain complementation index z, the region is actually a ring of hypercubes, i.e., $HR^{(*,z,*)}_{(m,n)}$. The hypercube leaders elected in the first step will compete with each other and elect one leader in $HR^{(*,z,*)}_{(m,n)}$ for each different z value. The third step is the final step, where the leaders elected in the second step compete with each other and elect one final leader for $HB_{(m,n)}$. In the following sections, we discuss the details of the election algorithm at each step.

Remark 6 *It should be noted that in an oriented hyper-butterfly graph, each node can identify the region with the knowledge of direction of edges, i.e. one node can send a message to some other node in the same region by using a sequence of certain directions. The node does not have to know its canonical label in order to identify the region.*

5.1 Election algorithm in $H^{(*,z,\ell)}_m$

In $HB_{(m,n)}$, there are $n \times 2^n$ different butterfly labels, which we denote as $(z, \ell), 0 \le z < 2^n, 0 \le \ell < n$; and the nodes with the same butterfly-part label form a hypercube of dimension m, which we denote as $H^{(*,z,\ell)}_m$.

In this step, each node first runs a leader election algorithm for hypercube [24]. The algorithm uses only *hypercube edges*, i.e., edges with direction 0 to $n - 1$. At the end of this procedure, there will be one leader elected in $H^{(*,z,\ell)}_m$ for each different (z, ℓ) value pair. The details of the algorithm can be found in [24].

After the leader is set at $H^{(*,z,\ell)}_m$. The leader will broadcast its id to all nodes in $H^{(*,z,\ell)}_m$. Each node receiving this broadcast message will save the leader's id into a variable HL (abbreviation for hypercube leader).

Lemma 5.1 *This step requires less than* $7.24 \times n \times 2^{m+n}$ *messages.*

Proof : The leader election algorithm proposed in [24] takes less than $6.24 \times N$ message for a hypercube with N nodes. Because $H^{(*,z,\ell)}_m$ has 2^m nodes and there are $n \times 2^n$ such hypercubes in $HB_{(m,n)}$, the total messages required for the election procedure are less than $6.24 \times n \times 2^{m+n}$. After the election, the leader of $H_m^{(*,z,\ell)}$ also broadcasts its id to all nodes in $H_m^{(*,z,\ell)}$, this takes 2^m for each hypercube. And there are $n \times 2^n$ hypercubes, so the broadcast procedure takes $n \times 2^{m+n}$ messages. The total number of messages needed in this step is the sum of both election and broadcast procedure, and it is less than $7.24 \times n^{m+n}$. □

5.2 Election in $\mathrm{H}R_{(m,n)}^{(*,z,*)}$

After the first step, there will be a leader in each hypercube $H_m^{(*,z,\ell)}$, $(0 \leq z < 2^n,\ 0 \leq \ell < n)$. We use $(h(z,\ell), z, \ell)$ to denote the label of the leader in $H_m^{(*,z,\ell)}$, where $h(z,\ell)$ specifies the hypercube part label of the leader.

Remark 7 *$h(z,\ell)$ does not denote a particular function to derive the hypercube part label from z and ℓ. $h(z,\ell)$ only indicates that the hypercube part label of the leader varies for different hypercubes. Because the leader in $H_m^{(*,z,\ell)}$ is determinate for any (z,ℓ) value pair, the hypercube part label of leader is also determinative and solely depends on the value of z and ℓ.*

In the first step, the leader of each hypercube also broadcasts its id within the hypercube when it becomes the leader. After the broadcast procedure, every node in $H_m^{(*,z,\ell)}$ will be informed with the id of leader, i.e. $(h(z,\ell), z, \ell)$, and has variable HL set to it.

The objective in the second step is to elect leaders in larger regions. In this step, the nodes in the hyper-butterfly graph $HB_{(m,n)}$ is considered to be grouped into 2^n new regions: $HR_{(m,n)}^{(*,z,*)}$, $0 \leq z < 2^n$, each consists of n hypercubes with the same complementation index, $H_m^{(*,z,\ell)}$, $0 \leq \ell < n$. Each hypercube has one leader elected from the first step, and there are in total $n \times 2^n$ such hypercube leader, with n in each new region $HR_{(m,n)}^{(*,z,*)}$. In the second step, each hypercube leader $(h(z,\ell), z, \ell)$ invokes procedure $HRElect$ to compete with other $n-1$ hypercube leaders in the same region $HR_{(m,n)}^{(*,z,*)}$. Only one of them becomes the new leader. After every hypercube leader finishes procedure $HRElect$, there will be only 2^n leaders left, with one in each $HR_{(m,n)}^{(*,z,*)}$, $0 \leq z < 2^n$. The *** code of procedure $HRElect$ is listed below.

Procedure $HRElect(h(z,\ell), z, \ell)$

Initial Conditions:

1. Node $(h(z,\ell), z, \ell)$ is the leader of $H_m^{(*,z,\ell)}$.

2. All nodes in $HB_{(m,n)}$ have variable HL set to the label of the hypercube leader.

Invocation of the Procedure:

Node $(h(z,\ell),z,\ell)$ sends message $RingTest((h(z,\ell),z,\ell),0,True)$ along direction g.

Upon receiving message $RingTest(id,i,b)$ from direction g^{-1}:

```
//id is the label of the node which invokes procedure HRElect.
//i is an integer from 0 to n − 1 and b is a boolean
//with the value of either True or False.
if (i < n − 1)
    {
    if (b == False||HL > id)
        send message RingTest(id, i + 1, False) through direction g.
    else
        send message RingTest(id, i + 1, True) through direction g.
    }
else
    {
    // This means the message gets back to the leader node (h(z,ℓ), z, ℓ)
    if (b == True)
        // This node passed all tests in the ring, becomes the new leader.
        Current node (h(z,ℓ), z, ℓ) becomes the leader of HR_(m,n)^(*,z,*).
    else
        // Failed the test, becomes nonleader.
        Current node becomes nonleader in HR_(m,n)^(*,z,*).
    }
```

As we can see, the procedure can be invoked from any hypercube leader $(h(z,\ell),z,\ell)$. It consists of sending message $RingTest(id,i,b)$ carrying three parameters. The first parameter is the id of the hypercube leader that invokes the procedure. The second parameter i is an integer that counts the number of nodes message $RingTest(id,i,b)$ has passed except the origin node. b is a Boolean to indicate if id is large enough to be the leader of the part of $HR_{(m,n)}^{(*,z,*)}$ that the message has passed so far.

Because every node increments i and relays the message through direction g, message $RingTest(id,i,b)$ will go through a ring of n nodes and get back to the origin node $(h(z,\ell), z, \ell)$ after that. At each intermediate node, the variable HL is compared to id that comes from the message. If HL is larger than id, then b is set to $False$ to indicate that the node invoking the procedure is not large enough to be the leader of $HR_{(m,n)}^{(*,z,*)}$. When message $RingTest(id, i, b)$ gets back to the origin node that invokes the procedure, the node checks the value of b and becomes the leader of $HR_{(m,n)}^{(*,z,*)}$ or a nonleader accordingly.

Example 1 *Figure 14 shows the structure of* $HB_{(2,3)}^{(*,0,*)}$, *which is a part of* $HB_{(2,3)}$. *All nodes in the figure have the same* complementation index *which is* 0, *and there are three hypercubes of dimension* 2, *each with a distinct* permutation index. *We assume that each of hypercubes already has a leader elected from the first step (The leader is shown with filled circles). For convenience, we use the canonical label as the id of leader nodes when they compete with each other. In real applications, it uses some distinct id, but not its canonical label because it is not known to the node.*

After the first step, all nodes have variable HL *set to be the id of the leader of the hypercube that the node is in. For example, the nodes in* $(*,0,1)$ *will have* HL *set as* $(1,0,1)$, *which is the leader of* $(*,0,1)$.

In the second step, all three hypercube leaders, $(0,0,0)$, $(1,0,1)$, *and* $(3,0,1)$, *will execute procedure* $HBElect()$. *We first look at node* $(0,0,0)$. *It sends a message* $RingTest((0,0,0),0,True)$ *to node* $(0,0,1)$. *In message* $RingTest((0,0,0),0,True)$, *the first parameter* $(0,0,0)$ *is the id of the hypercube leader that invokes this procedure and is the id to be tested throughout the ring. After node* $(0,0,1)$ *gets the message, it compares* HL *with the id included in the message. Because the value of* HL *at* $(0,0,1)$, *which is* $(1,0,1)$ *is larger than the id included in the message, which is* $(0,0,0)$, *the test failed at node* $(0,0,1)$. *It then sends message* $RingTest((0,0,0),1,False)$ *to node* $(0,0,2)$ *and the message will finally be routed back to node* $(0,0,0)$. *After getting the feedback message, node* $(0,0,0)$ *checks the Boolean parameter in the message. Because the boolean value is* $False$ *which indicates the id is not the largest, the node will become a nonleader.*

Another hypercube leader, $(1,0,1)$, *will also invoke procedure* $HBElect()$. *A similar process goes through and the node will become a nonleader because its id is not the largest in all three hypercube leaders. Among all three hypercube leaders elected in the first step, only node* $(3,0,2)$ *will receive message* $RingTest()$ *with the Boolean parameter set to be* $True$. *So only this node will become the leader in* $HB_{(2,3)}^{(*,0,*)}$.

Lemma 5.2 *For any arbitrary value of* z, $0 \leq z < 2^n$, *after all* n *hypercube leaders in* $HR_{(m,n)}^{(*,z,*)}$ *execute procedure* $HBElect$ *and get back the message*

RingTest, only one of them will become the leader of $HR_{(m,n)}^{(*,z,*)}$ *and all others will become nonleader.*

Proof : For any arbitrary value z, there are n hypercube leaders in $HR_{(m,n)}^{(*,z,*)}$. They are $(h(z,\ell),z,\ell)$, $0 \leq \ell < n$. Each of them invokes procedure $HRElect$ and will determine which becomes a leader or nonleader depending on the value of b returned from message $RingTest$.

Consider an arbitrary leader $(h(z,\ell),z,\ell)$ among them. This node starts procedure $HRElect$ by sending message $RingTest((h(z,\ell),z,\ell),0,True)$ through direction g. Because every node gets the message also relays it through direction g, the message will traverse every node in $R_n^{(h(z,\ell),z,*)}$ which includes n nodes: $(h(z,\ell),z,j)$, $0 \leq j < n$. And because node $(h(z,\ell),z,j)$ has variable HL set to the id of leader of hypercube $H_m^{(*,z,j)}$, i.e. $(h(z,j),z,j)$. The id $(h(z,\ell),z,\ell)$ is compared with the id of leaders of other hypercubes $H_m^{(*,z,j)}$, i.e. $(h(z,j),z,j)$, $0 \leq j < n$. If some node becomes the leader after executing procedure $HRElect$, it is assured that its id is larger than all other leaders in $H_m^{(*,z,j)}$, $0 \leq j < n$. Therefore, from n hypercube leaders $(h(z,\ell),z,\ell)$, $0 \leq \ell < n$, only one of them can claim to be the leader of $HR_{(m,n)}^{(*,z,*)}$ after executing procedure $HRElect$; all other $n-1$ nodes will become nonleaders. □

Remark 8

- *After every hypercube leaders (elected from first step) complete procedure HRElect, there will be* 2^n *nodes remaining as leader, with each from* $HR_{(m,n)}^{(*,z,*)}$, $0 \leq z < 2^n$.

- *After a node becomes the leader in* $HR_{(m,n)}^{(*,z,*)}$, *it broadcasts its id to all nodes in* $HR_{(m,n)}^{(*,z,*)}$. *And the node that receives the broadcast message will save the leader's id into variable* HRL.

Lemma 5.3 *There will be* $n \times 2^{m+n} + n^2 \times 2^n$ *number of messages generated in the second step, including procedure HRElect and the broadcast process afterwards.*

Proof : There are $n \times 2^n$ hypercube leaders elected from the first step. Each leader executing procedure $HRElect$ generates n messages. And there will be 2^n leaders elected afterwards. Each leader in $HR_{(m,n)}^{(*,z,*)}$, $0 \leq z < 2^n$ will broadcast its id, which takes $n \times 2^m$ message. So the total number of messages needed in this step is $n \times 2^{m+n} + n^2 \times 2^n$. □

5.3 Leader Election in $\mathbf{HB_{(m,n)}}$

After the second step, there will be one leader left in each $HR^{(*,z,*)}_{(m,n)}$, $0 \leq z < 2^n$. We use $(h(z), z, \ell(z))$ to denote the label of the leader in $HR^{(*,z,*)}_{(m,n)}$.

Remark 9 *Similar to the second step, $h(z)$ or $\ell(z)$ does not specify the function to derive the hypercube part label or permutation index of the leader node. They only indicates the dependency relationship between those labels with z.*

In the third step, which is the final step, the objective is to elect one leader for the entire hyper-butterfly graph $HB_{(m,n)}$. Because we already have 2^n leaders in each $HR^{(*,z,*)}_{(m,n)}$, we use similar approach to that of the second step to elect one node from the those leaders to become the final leader.

In this step, each of the 2^n leaders from the second step sends $TreeTest$ message through a tree structure in the butterfly graph. We ensure that the message will get to the nodes with different complementation index, so that the id of each leader will be tested to see if it is larger than the ids of all other leaders. As in procedure $HRElect$, only the node that passes all tests will become the leader which is the leader of the entire hyper-butterfly graph $HB_{(m,n)}$. The following code lists the details of procedure $HBElect$ which is invoked by every leader of $HR^{(*,z,*)}_{(m,n)}$.

Procedure $\boldsymbol{HBElect(h(z), z, \ell(z))}$

Initial Conditions:

1. Node $(h(z), z, \ell(z))$ is the leader of $HR^{(*,z,*)}_{(m,n)}$.

2. All nodes in $HB_{(m,n)}$ have variable HRL set to the id of the leader of $HR^{(*,z,*)}_{(m,n)}$ in which it resides.

Invocation of the Procedure:

Node $(h(z), z, \ell(z))$ sends message $TreeTest((h(z), z, \ell(z)), 0, True)$ along directions g and f.

Upon receiving message $TreeTest(id, i, b)$ from direction dir:

//id is the label of the node that invokes procedure $HBElect$.

//i is an integer from 0 to $n-1$ and b is a Boolean with the value of either $True$ or $False$.

//dir is the direction through which the node receives the message.

It can be either g^{-1} or f^{-1}.

```
    if (i == n)
        {
        //We have reach the leaf of tree. Compare HRL with id at here,
and send the result back.
        if (HRL > id)
            send message TreeReply(id, i − 1, False)
            through direction dir^{-1}.
        else
            send message TreeReply(id, i − 1, True)
            through direction dir^{-1}.
        }
    else
        {
        // We first pass the test to lower level down the tree and
wait for answers.
        //if we get positive answer from both children, then return positive.
        send message TreeTest(id, i + 1, b) through direction g and f.
        wait message TreeReply from direction g^{-1} and f^{-1}.
            if the message TreeReply(id, i, b) from both
g^{-1} and f^{-1} have b value as True
    then send message TreeReply(id, i − 1, True) through dir^{-1}.
    otherwise send message TreeReply(id, i − 1, False) through dir^{-1}.
        }
```

As we can see from the above procedure, there are two types of messages used in the procedure. The first type of message is $TreeTest$ which travels down a binary tree because each node distributes the message through direction g and f. The parameter id and b have the same meaning as in the second step, while i indicates the current level of the tree. The other type of message is $TreeReply$ which goes through the reversal path of $TreeTest$. Only the leaf nodes will compare the value of HRL with id that comes from the message. The intermediate nodes only act to transmit message $TreeTest$ to both children and collect $TreeReply$ from them.

Example 2 *Figure 15 shows the structure of* $B_3^{(*,0,0)}$, *which is a part of* $HB_{(2,3)}$. *We use this figure to show the execution of procedure* $HBElect()$ *from node* $(0,0,0)$. *After node* $(0,0,0)$ *invokes procedure* $HBElect()$, *message* $TreeTest()$ *travels along a complete binary tree rooted from node* $(0,0,0)$.

The message will reach 8 leaf nodes of the tree. These 8 leaf nodes all have a different complementation index *and cover all possible complementation indices that $HB_{(2,3)}$ can have.*

After these leaf nodes get message $TreeTest()$, they compare variable HBL with the id included in the message and send $TreeReply()$ message back to their parents with the boolean parameter in the messages properly set. $TreeReply()$ messages will travel along the reversal tree structure and finally get back to root node $(0,0,0)$. The root node becomes a leader or nonleader according to the Boolean values returned in $TreeReply()$ messages from its two children.

Lemma 5.4 *Consider the execution of procedure $HBElect$ from any node $(h(z), z, \ell(z))$; there are 2^i nodes that receive both message $TreeTest(id, i, b)$ and $TreeReply(id, i, b)$. These 2^i nodes have the same permutation index and hypercube label, but a different complementation index.*

Proof: We prove by induction. And because $HB_{(m,n)}$ is vertex symmetric, we can assume that the node invoke procedure $HBElect$ is the identity node. i.e. $(h(z), z, \ell(z)) = (0,0,0)$. We only prove for message $TreeTest(id, i, b)$. The proof for message $TreeReply(id, i, b)$ can be similarly established. Obviously, the conclusion holds when $i = 0$ because there is only one node.

When $i = 0$, the 2^i nodes that receive message $TreeTest(id, i, b)$, have a permutation index of 0 and hypercube part label 0 and a complementation index from 0 to $2^i - 1$. These nodes will send message $TreeTest(id, i + 1, b)$ through directions g and f. Now consider the nodes that receive message $TreeTest(id, i + 1, b)$. From the definition of g and f generators, we know that these nodes will all have permutation index $i + 1$. It is also easy to see that these $2 \times 2^i = 2^{i+1}$ nodes have different complementation index from 0 to $2^{i+1} - 1$. And because we only use butterfly edges, the hypercube part label won't change for these 2^{i+1} nodes. Hence, the proof is established. □

Lemma 5.5 *After every leader from the second step completely executes procedure $HBElect$, only one node will become leader of $HB_{(m,n)}$. All other leaders will become non-leader.*

Proof: n procedure $HBElect$, there are 2^n nodes receive message $TreeTest$ (id, n, b). From Lemma 5.4, we know that these 2^n nodes all have different complementation labels. Since these node will compare HRL with id from the message, the id will be compared with ids of all other leaders in $HR^{(*,z,*)}_{(m,n)}$, $0 \leq z < n$. Therefore, in order for a node to become leader after procedure

$HBElect$, its id must be larger than all other leaders in $HR_{(m,n)}^{(*,z,*)}$, $0 \leq z < n$. We know that there's only one such node, and only this node can become the leader of $HB_{(m,n)}$. All others will become non-leader.

□

Lemma 5.6 *There are 2^{2n+2} number of messages generated in the third step.*

Proof : There are 2^n leaders elected from the second step. Each of them executes procedure $HBElect$ which sends the $TreeTest$ message to 2^{n+1} nodes and $TreeReply$ message to 2^{n+1} nodes. So the total number of messages used in this step is $2^n \times 2 \times 2^{n+1} = 2^{2n+2}$. □

Theorem 5.7 *The total number of messages needed for a leader election algorithm in $HB_{(m,n)}$ is $8.24 \times n \times 2^{m+n} + (n^2 + 2^{n+2}) \times 2^n$.*

Proof : The result can be easily derived by adding the number of messages used in three steps. □

6 Acknowledgment

The work of Srimani was partially supported by a NSF grant # ANI-0073409.

References

[1] S. Lakshmivarahan, J. S. Jwo, and S. K. Dhall. Symmetry in interconnection networks based on Cayley graphs of permutation groups: a survey. *Parallel Computing*, 19:361–407, 1993.

[2] S. B. Akers and B. Krishnamurthy. A group-theoretic model for symmetric interconnection networks. *IEEE Transactions on Computers*, 38(4):555–566, April 1989.

[3] S. B. Akers and B. Krishnamurthy. The star graph: an attractive alternative to n-cube. In *Proceedings of International Conference on Parallel Processing (ICPP-87)*, pages 393–400, St. Charles, Illinois, August 1987.

[4] K. Qiu, H. Meijer, and S. G. Akl. Decomposing a star graph into disjoint cycles. *Information Processing Letters*, 39(3):125–129, 1991.

[5] L. Bhuyan and D. P. Agrawal. Generalized hypercube and hyperbus structure for a computer netwrk. *IEEE Transactions on Computers*, 33(3):323–333, March 1984.

[6] K. Qiu, H. Meijer, and S. G. Akl. On the cycle structure of star graphs. Technical Report 92-341, Department of Computer Science, Queen's University, Ontario, Canada, November 1992.

[7] B. W. Arden and K. W. Tang. Representation and routing of Cayley graphs. *IEEE Transactions on Communications*, 39:1533–1537, December 1991.

[8] I. D. Scherson. Orthogonal graphs for the construction of interconnection networks. *IEEE Transactions on Parallel and Distributed Systems*, 2(1):3–19, 1991.

[9] K. Day and A. Tripathi. Arrangement graphs: a class of generalized star graphs. *Information Processing Letters*, 42:235–241, July 1992.

[10] M. R. Samatham and D. K. Pradhan. The De Bruijn multiprocessor network: a versatile parallel processing and sorting network for VLSI. *IEEE Transactions on Computers*, 38(4):567–581, April 1989.

[11] C. Chen, D. P. Agrawal, and J. R. Burke. dBCube: a new class of hierarchical multiprocessor interconnection networks with area efficient layout. *IEEE Transactions on Parallel and Distributed Systems*, 4(12):1332–1344, December 1993.

[12] F. Preparata and J. Vuillemin. The cube-connected cycles: a versatile network for parallel computation. *Communications of ACM*, 24(5):30–39, May 1981.

[13] G. E. Carlsson, J. E. Cruthirds, H. B. Sexton, and C. G. Wright. Interconnection networks based on a generalization of cube-connected cycles. *IEEE Transactions on Computers*, C-34(8):769–772, 1985.

[14] D. J. Pritchard and D. A. Nicole. Cube connected Möbius ladders: an inherently deadlock free fixed degree network. *IEEE Transactions on Parallel and Distributed Systems*, 4(1):111–117, January 1993.

[15] D. K. Pradhan. On a class of fault-tolerant multiprocessor network architectures. In *3rd International Conference on Distributed Computing Systems*, pages 302–311, Miami, FL, October 1982. IEEE.

[16] W. E. Leland and M. H. Solomon. Dense trivalent graphs for processor interconnection. *IEEE Transactions on Computers*, 31(3):219–222, March 1982.

[17] D. K. Pradhan. Fault tolerant VLSI architectures based on de Bruijn graphs (Galileo in the mid nineties). *DIMACS Series in Discrete Mathematics*, 5, 1991.

[18] R. G. Gallagar, P. A. Humblet, and P. M. Spira. A distributed algorithm for minimum weight spanning trees. *ACM Transactions on Programming Languages and Systems*, 5:67–77, 1983.

[19] N. Santoro. Sense of direction, topological awareness and communication. *ACM SIGACT News*, 16:50–56, 1984.

[20] H. L. Bodlaender. New lower bound techniques for distributed leader finding and other problems on rings of processors. *Theoretical Computer Science*, 81:237–256, 1991.

[21] M. C. Loui, T. A. Matsuhita, and D. B. West. Election in a complete network with a sense of direction. *Information Processing Letters*, 22:185–187, 1986.

[22] E. Korach, S. Moran, and S. Zaks. Tight upper and lower bounds for some distributed algorithms for a complete network of processors. In *Proceedings of Symposium on Principles of Distributed Computing*, pages 199–207, 1984.

[23] G. Tel. Linear election in oriented hypercubes. Technical Report RUU-CS-93-39, Computer Science, Utrecht University, 1993.

[24] G. Tel. Linear election in hypercubes. *Parallel Processing Letters*, 5:357–366, 1995.

[25] S. B. Akers and B. Krishnamurthy. A group theoretic model for symmetric interconnection networks. In *Proceedings of the 1986 International Conference on Parallel Processing*, pages 216–223. IEEE, August 1986. U Ma. and Tektronix.

[26] S. Lakshmivarahan, J.-S. Jwo, and S. K. Dhall. Symmetry in interconnection networks based on Cayley graphs of permutation groups: A survey. *Parallel Computing*, 19:361–407, 1993.

[27] D. J. Pritchard and D. A. Nicole. Cube connected mobius ladders: An inherently deadlock-free fixed degree network. *IEEE Transactions on Parallel and Distributed Systems*, 4(1):111–117, January 1993.

[28] S. Latifi, M. M. Azevedo, and N. Bagherzadeh. The star connected cycles: a fixed degree network for parallel processing. In *Proceedings of the International Conference on Parallel Processing*, volume 1, pages 91–95, 1993.

[29] F. T. Leighton. *Introductions to Parallel Algorithms and Architectures: Arrays, Trees and Hypercubes*. Morgan Kaufman, 1992.

[30] P. Vadapalli and P. K. Srimani. A new family of Cayley graph interconnection networks of constant degree four. *IEEE Transactions on Parallel and Distributed Systems*, 7(1), January 1996.

[31] S. Guha and A Sen. On fault tolerant distributor communication architecture. *IEEE Transactions on Computers*, C-35(3):281–283, March 1986.

[32] Elango Ganesan and Dhiraj K. Pradhan. The hyper-deBruijn networks: Scalable versatile architecture. *IEEE Transactions on Parallel and Distributed Systems*, 4(9):962–978, September 1993.

[33] S. Ohring and S. K. Das. Dynamic embeddings of trees and quasi-grids into hyper DeBruijn networks. In *Proceedings of the 7th International Parallel Processing Symposium*, pages 519–523, 1993.

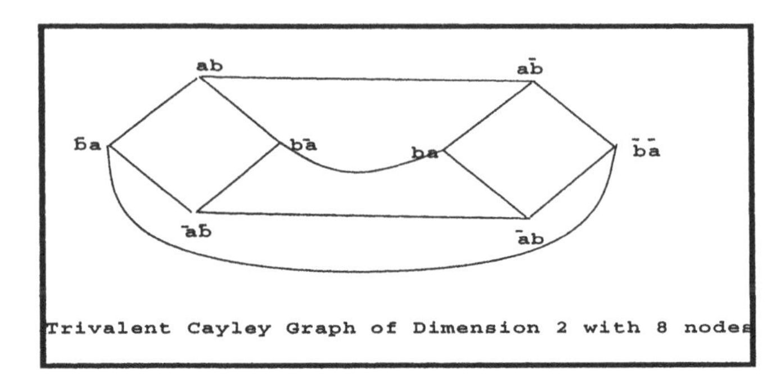

Trivalent Cayley Graph of Dimension 2 with 8 nodes

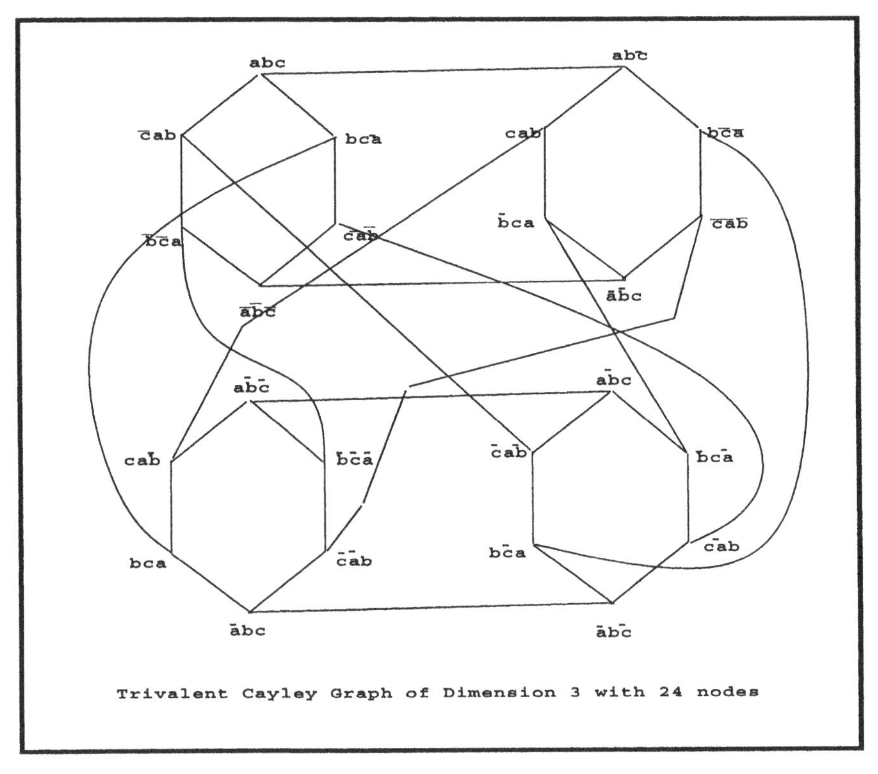

Trivalent Cayley Graph of Dimension 3 with 24 nodes

Figure 10: Q_n Graphs for $n = 2$ and $n = 3$

Algorithm Simple_Route

S1: For a given node $s = (a_1 a_2 \cdots a_n)$, find k, $1 \leq k \leq n$, such that $a_k = t_1^*$.

S2: **Path (a)** Go along successive f edges $(k-1)$ times to reach the node
$u = a_k a_{k-1} \cdots a_n \bar{a}_1 \bar{a}_2 \cdots \bar{a}_{k-1}$.
Path (b) Go along successive f^{-1} edges $(n-k+1)$ times to reach the node $v = \bar{a}_k \bar{a}_{k-1} \cdots \bar{a}_n a_1 a_2 \cdots a_{k-1}$.

S3: **For** each Path (a) and (b) **do**
for $i = 1$ **to** n **do**
Go to the node $t_{i+1}^* t_{i+2}^* \cdots t_n^* t_1 t_2 \cdots t_i$ by either the f edge or the f edge followed by the g edge;

Figure 11: Algorithm Simple_Route for Q_n

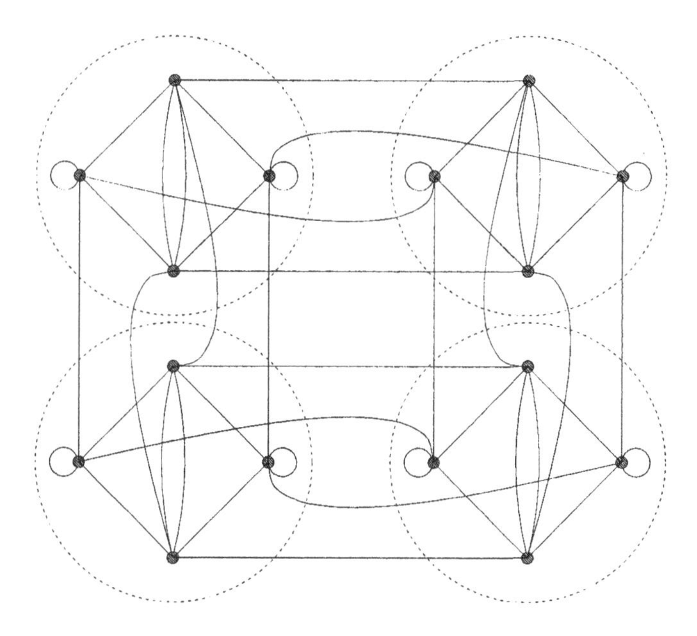

Figure 12: The Hyper-deBruijn graph, $HD_{2,2}$, showing the deBruijn subgraphs.

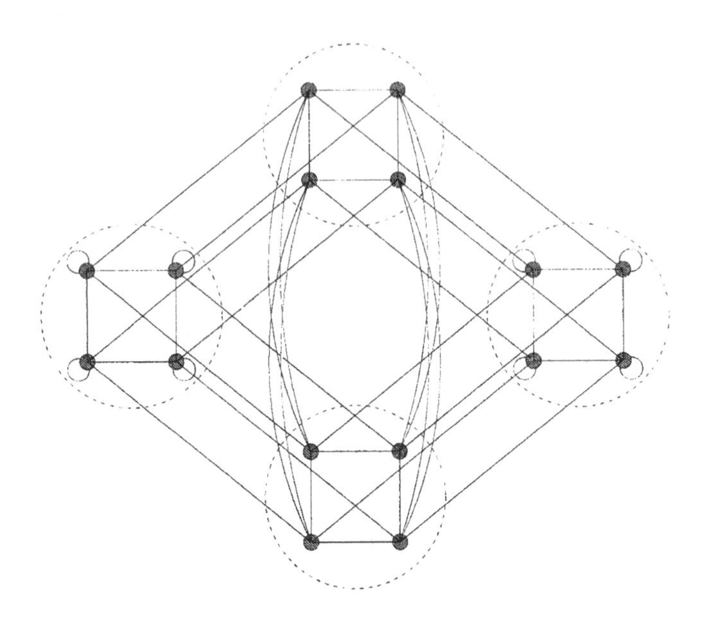

Figure 13: The Hyper-deBruijn graph, $HD_{2,2}$, showing the Hypercube subgraphs.

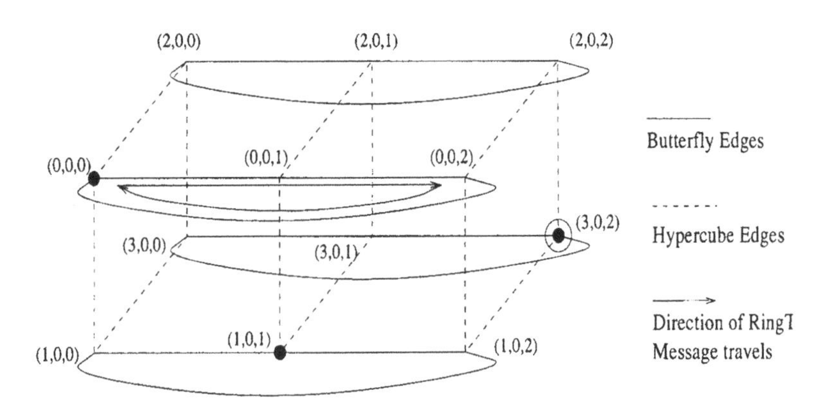

Figure 14: Example of leader election in $HR^{(*,0,*)}_{(2,3)}$(subgraph of $HB_{(2,3)}$)

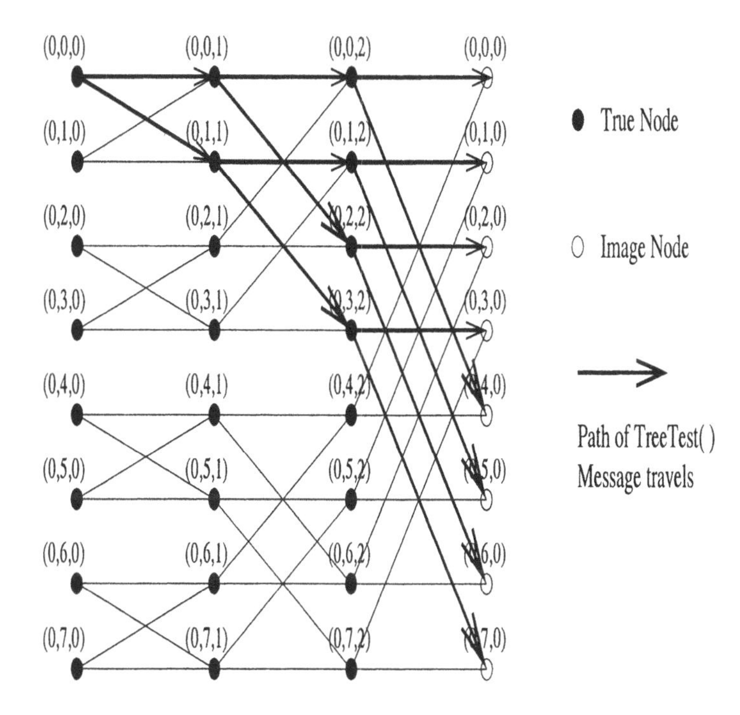

Figure 15: Example of execution of $HBElect()$ in $B_3^{(0,*,*)}$(subgraph of $HB_{(2,3)}$)

Part III

Combinatorial Optimization in *Other Network Applications*

Chapter 17

Routing Optimization in Communication Networks

Dirceu Cavendish
NEC Labs America
10080 North Wolfe Road, SW3-350
Cupertino, CA 95014
E-mail: dirceu@sv.nec-labs.com

Mario Gerla
Computer Science Department
University of California, Los Angeles 90095-1596
E-mail: gerla@cs.ucla.edu

1 Introduction

Routing has always been an important area in communication networks. Regardless of the type of network, a bad routing strategy may prove disastrous for the network overall performance. Viable paths may not be found, tortuous paths may quickly exhaust network resources, etc.

A routing strategy involves the definition of a set of one or more paths over which communication between end-devices takes place over a network. Given the multiplicity of paths in any realistic network, the problem of finding paths is clearly of a combinatorial nature. Therefore, the selection of paths is always guided by some optimization criteria: minimizing network resource usage, minimizing network blocking, etc.

This chapter addresses various combinatorial strategies commonly used to solve routing problems across different types of communication networks. Rather than focusing on the network technologies, which are simply too many to address in a single chapter, we focus on the combinatorial approaches used to solve routing problems, using specific network technologies for illustrative purposes.

We identify the following types of routing problems. Depending on the number of communicating devices, we have unicast and multicast routing problems. Unicast routing problems consist in finding a path between a pair of devices only. In multicast routing, we seek paths between three or more communicating devices. We can further differentiate between one-to-many and many-to-many multicasting: i.e., single or multiple sources, respectively.

Regarding routing objectives, we have unconstrained and constrained routing problems. In the unconstrained routing problems, an optimization criterion is the only strategy observed for finding paths. In constrained routing problems, besides an optimization objective, constraints on the path sought are imposed for feasibility.

Finally, there are online and offline routing problems. Online routing considers a set of end-devices that need to establish communications among themselves one at a time. This is typical of many communication networks where "connection requests" arrive along the time in a dynamic way. In offline routing, a set of "connection requests" is assumed to be known in advance, before routing takes place. This is typical of networks with "circuit provisioning", where communication pipes are provisioned to last a long time; hence the communication system can be assumed quasi-static.

2 Dynamic Programming

Dynamic programming [1] is a well-known technique for solving optimization problems of most diverse types. A problem is represented by a set of equations on real numbers, where the left side of each equation contains the unknown quantities and the right side contains expressions typically with a minimum operator and monotone functions. The equations are called functional equations, and are usually solved via an iterative method. The approach can be seen as a divide and conquer strategy, where solutions for large instances of a problem depend on solutions of "subproblems", and it has proven to be powerful in solving a great number of combinatorial optimization problems. In routing optimization problems, perhaps the most used dynamic programming algorithms are the Dijkstra and Bellman–Ford algorithms. In what follows, we analyze these two algorithms in solving routing problems in packet networks.

2.1 Bellman–Ford Algorithm

Dijkstra and Bellman–Ford shortest-path algorithms have been used to solve shortest-path routing problems. Shortest-path routing strategy makes sense when there is no prior knowledge about traffic patterns in the network, because each path consumes a minimum amount of network resources if routed through a shortest path. It is worth noting that the Bellman–Ford algorithm has a worst case complexity of $O(N^3)$, which is higher than the Dijkstra algorithm complexity of $O(N^2)$ [2]. However, Dijkstra does not classify the paths it generates on the basis of hop count, as the BF algorithm does. This feature comes in handy in many routing problems, in particular when dealing with multiple constrained routing problems.

Recently, circuit and packet networks have motivated the search for shortest paths that satisfy a given elegibility criteria, such as Quality of Service (QoS) in a packet network. The problem of finding shortest paths subject to multiple metrics is known to be NP-complete [13]. In what follows, we use a Bellman–Ford type of algorithm to find shortest paths with QoS bounds as constraints [5].

2.1.1 Routing with Multiple Bounds

We model a network as a directed graph $G(V, E)$, where V is the set of nodes ($|V| = N$) and E the set of directed *edges*. Every edge $(i, j) \in V$ is associated with a set $\mathcal{M}$ of non negative values, referred to as edge metrics. Thus an edge (i, j) has cost $m(i, j)$ with respect to metric $m \in \mathcal{M}$. Examples of edge metrics are: link cost (c), bandwidth (b), delay (d), delay jitter (j), and loss probability (l). A path $Pa(s, t)$ is a sequence of vertices $v_s, \ldots, v_i, v_{i+1}, \ldots, v_t$ such that $\forall s \leq i \leq t$, edge $(v_i, v_{i+1}) \in E$. Upon each metric $m \in \mathcal{M}$, we further define the cost $c_m(s, t)$ of a path $Pa(s, t)$ as a generic function f of the path's edge metrics, or:

$$c_m Pa(s,t) = f(m(v_s, v_{s+1}), \ldots, m(v_i, v_{i+1}), \ldots, m(v_{t-1}, v_t)) \qquad (1)$$

Examples of path cost functions are:

$$c_{bw}Pa(s,d) = \min_{(i,j)\in Pa(s,d)} b(i,j) \qquad < bandwidth > \tag{2}$$

$$c_{bf}Pa(s,d) = \min_{(i,j)\in Pa(s,d)} bf(i,j) \qquad < buffer > \tag{3}$$

$$c_{h}Pa(s,d) = \#\, edges\ of\ Pa(s,d) \qquad < hop\ count > \tag{4}$$

$$c_{d}Pa(s,d) = \sum_{(i,j)\in Pa(s,d)} d(i,j) \qquad < delay > \tag{5}$$

$$c_{j}Pa(s,d) = \sum_{(i,j)\in Pa(s,d)} j(i,j) \qquad < jitter > \tag{6}$$

$$c_{l}Pa(s,d) = 1 - \prod_{(i,j)\in Pa(s,d)} [1 - l(i,j)] \qquad < packet\ loss > \tag{7}$$

where $b(i,j)$, $bf(i,j)$, $d(i,j)$, $j(i,j)$, $l(i,j)$ are the bandwidth, buffer, delay, delay jitter, and packet loss probability of link (i,j), respectively.[1]

The original Bellman–Ford (BF) algorithm was designed to compute shortest paths between a given vertex s and multiple destinations. It did not include multiple metrics. Moreover, path cost function was always considered to be of additive nature, as in the path delay cost. We wish to extend BF algorithms to handle as many metrics of the types previously defined as possible.

Without loss of generality, we assume vertex 1 as the vertex from which we wish to compute distances to all other network vertices. As usual in the BF algorithm, we define D_i^h as the minimum distance with respect to some metric d between vertex 1 and vertex i with at most h number of hops. The Bellman–Ford equation [9] is:

$$D_i^{h+1} = \min_{j\in N(i)} [d(i,j) + D_j^h], \qquad \forall i \neq 1 \tag{8}$$

where $N(i)$ is the set of neighbors of vertex i. Starting with initial conditions $D_i^0 = \infty$ and $D_1^h = 0$, the BF algorithm iterates Eq. (8) until a predefined number of hops $H_{\max}$ has been reached, $H_{\max} \leq N$ (if the shortest possible path is sought, regardless of number of hops, $H_{\max} = N$).

In a regular BF algorithm, it is well known that:

[1]We assume that a QoS application requires a minimum amount of buffering at each router along its path

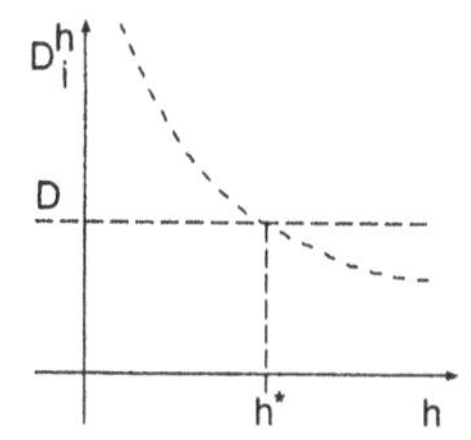

Figure 1: Non increasing cost function

Theorem 2.1 *[2] D_i^h is non increasing with h, regardless of the cost metric d used.*

Essentially, the non increasing property of the distance D^h is provided by the *min* operator of Eq. (8). Notice also that the search direction defined by the *min* operator is the steepest descent of D^h. Figure 1 illustrates D_i^h dependence on h.

The first routing problem of interest is:

Problem 2.1 *Find a minimum-hop route $Pm(s,t)$ with a delay d cost upper bounded by $c_d Pm(s,t) \leq D$*

It is not difficult to see that we can use the BF algorithm to solve Problem 2.1. More specifically, if h^* is defined to be the shortest number of hops in which $D_i^h \leq D$ [2], then:

Theorem 2.2 *A regular BF algorithm outputs the minimum-hop path with $c_d Pa(s,t) \leq D$ at the first time $D_i^h \leq D$.*

Proof of Theorem 2.2 We run the BF algorithm until the first time D_i^h falls below D, say at h^*. The number of hops always increases as BF progresses, and by assumption h^* is the point in which D_i^h falls below D for the first time, therefore the theorem is proven (Figure 1).

Figure 2 illustrates successive iterations of the Bellman–Ford algorithm. It is worth noting that the Bellman–Ford algorithm has a worst case complexity of $O(N^3)$ [2], which is higher than the Dijkstra algorithm complexity of $O(N^2)$. However, the regular Dijkstra does not classify the paths it generates on the basis of hop count, as the BF algorithm does. This feature comes in handy when dealing with multiple constrained routing problems.

[2] We use $*$ to denote optimal paths

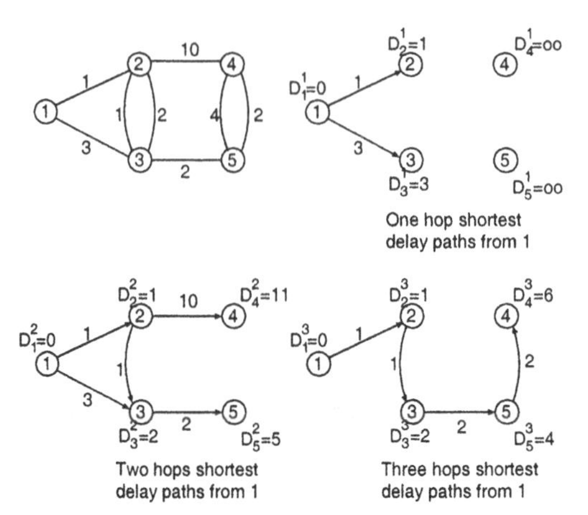

Figure 2: Successive iterations of the Bellman–Ford algorithm.

We now introduce a second distance M in the BF algorithm, in the following way.

$$M_i^{h+1} = f[m(i,k), M_k^h], \qquad k \; s.t. \; \min_{j \in N(i)} [d(i,j) + D_j^h] \tag{9}$$

i.e., M_i^h is the path distance with regard to metric m of the path chosen in order to minimize distance D_i^h. For example, metric m could be any of the metrics defined in Eqs. (2) through (7). Notice that a priori we cannot claim any property of M_i^h, as stated in Theorem 2.1 for D_i^h, unless we assume some correlation between metrics m and h. If such a correlation exists, two cases are of interest: M_i^h is non increasing with h; M_i^h is non decreasing with h. Figure 3 illustrates such cases.

For both cases, we further assume that M_i^h has no discontinuities; namely, any two paths with the same number of hops have the same distance M. So vertical jumps on Figure 3 are forbidden. Given that the path cost functions dealt with are of the form depicted in Figure 3, one can use the same strategy as before to compute minimum hop paths with bounds on multiple constraints. More specifically, we define the following routing problems.

Problem 2.2 *Find a minimum-hop route $Pm(s,t)$ with d cost upper bound $c_d Pm(s,t) \leq D$ and m cost upper bound $c_m Pm(s,t) \leq M$.*

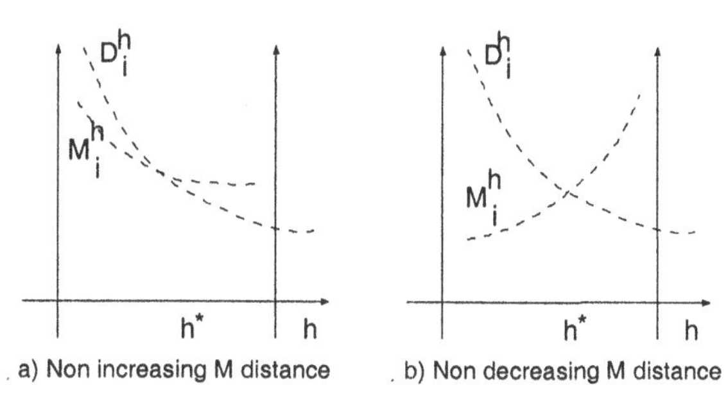

Figure 3: Multiple cost functions.

Problem 2.3 *Find a minimum-hop route $Pm(s,t)$ with d cost upper bound $c_dPm(s,t) \leq D$ and m cost lower bound $c_mPm(s,t) \geq M$.*

It is not difficult to see that:

Theorem 2.3 *A regular BF algorithm outputs the minimum-hop path with $c_dPa(s,t) \leq D$ and $c_mPm(s,t) \geq M$ if at the first time the condition $(D_i^h \leq D)$ we have $(M_i^h \geq M)$ for a non increasing M^h distance. If it so happens that $(M_i^h \leq M)$, the problem has no solution.*

Proof of Theorem 2.3 It follows from the non increasing property of distance M^h and Theorem 2.2.

Theorem 2.4 *A regular BF algorithm outputs the minimum-hop path with $c_dPa(s,t) \leq D$ and $c_mPm(s,t) \leq M$ if at the first time the condition $(D_i^h \leq D)$ we have $(M_i^h \leq M)$ for a non decreasing M^h distance. If it so happens that $(M_i^h \geq M)$, the problem has no solution.*

Proof of Theorem 2.4 It follows from the non decreasing property of distance M^h and Theorem 2.2.

Figure 4 illustrates two metric constrained routing problems and their solutions.

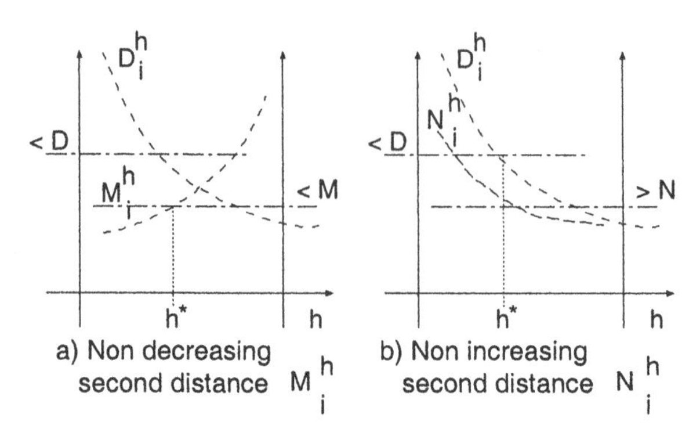

Figure 4: Two constrained BF solvable routing problems.

Notice that if we are seeking an upper bound for a non increasing distance N_i^h, or a lower bound for a non decreasing distance M_i^h, the doubly constrained minimum-hop problem cannot be solved by the Bellman–Ford algorithm. Also, the monotonic non decreasing / non increasing properties of the distances M_i^h and N_i^h are key to the claim that the problem has no solution if the two bounds are not met. For instance, let's assume two distances D_i^h and M_i^h, based on generic metrics d and m, whose dependency with the number of hops is displayed in Figure 5.

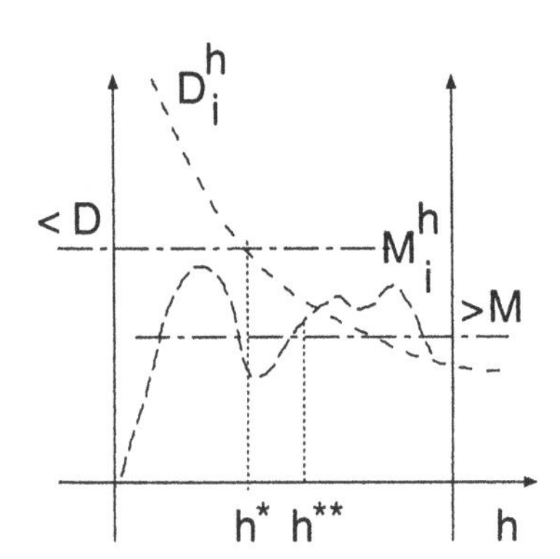

Figure 5: When BF algorithm fails.

The path represented by h^* satisfies the upper bound D, although it violates the lower bound M. However, according to the figure, it is possible to continue lowering distance D^h to find a path with $h^{**} > h^*$ satisfying the lower bound M. Notice that we cannot claim that the path determined by h^{**} is the shortest one (in number of hops) sought satisfying both bounds. This is so because, once the upper bound D is satisfied, a search on the steepest descent direction on distance D_i^h might no longer be the best search strategy towards finding a minimum-hop path respecting the bound M. Conversely, although a search on the steepest descent direction on distance M_i^h will lead to a shortest path on h satisfying the lower bound M, this search strategy is not guaranteed to yield a monotonic decrease on the distance D_i^h (see the definition of distances D_i^h and M_i^h).

Now for the main theorem. For a multiple constrained routing problem, define distance D_i^h as in Eq. (8) wrt a given metric $d(i,j)$. Then:

Theorem 2.5 *Partition the multiple constrained minimum-hop routing problem into subproblems by pairwising metric d with each other metric. The original multiple constrained problem can be solved by a multiple metric BF algorithm if and only if each subproblem is of the form defined by either Theorem 2.3 or 2.4.*

Proof of Theorem 2.5 It follows from Theorems 2.3 and 2.4 and the fact that multiple metrics interact only through the distance D_i^h which defines the search direction (see Eq. (9)).

One more observation is due. For a generic metric, the verification whether a given distance has non increasing or non decreasing property might take exponential time (see [15] for further discussions on this issue). However, this property is called upon only when the multiple metric BF algorithm fails to find a path, to claim that the problem has no solution. If a path satisfying the multiple constraints is found, Theorem 2.1 implies that this path is indeed the shortest hop path sought.

In [5], a simulation study shows how a Bellman–Ford algorithm with delay and loss probability bounds performs in comparison against a Dijkstra algorithm with regard to network blocking.

2.2 Dynamic Programming in Complex Routing Problems

In some instances, we are interested in finding paths that minimize a given metric, and, in case several paths are present, we select the path with minimum cost under a different metric. In this type of problem, multiple metrics are involved, and a given hierarchy of metrics is defined, where we optimize on a metric at higher rank before optimizing on metrics of lower rank. Notice that in this case there are no bounds to be satisfied, so an optimal path always exists in a connected network. Dynamic programming is very useful in dealing with complex routing problems of such nature, as we exemplify with a routing problem in optical networks.

2.2.1 Routing and Wavelength Assignment in Optical Networks

Routing and Wavelength Assignment (RWA) is the procedure by which routing paths are determined and wavelengths are assigned to connections to be provisioned by a Wavelength Division Multiplexing (WDM) Optical Transport Network (OTN). The RWA should be done so as to conserve OTN resources as much as possible, resulting in a lower blocking probability when additional connections are requested. Designing the system so as to directly minimize blocking probability is quite a difficult problem.

In what follows, our primary objective is to minimize the number of wavelength conversions used. This criterion is motivated by the fact that wavelength conversion requires special hardware, involving optical-electrical-optical (OEO) conversions, and hence it consumes expensive network resources. Note that the use of this criterion may result in a connection request to be routed on a long path using one single wavelength along the path, rather than on a short path using different wavelengths on different links.

A WDM mesh network of N nodes is modeled as a directed graph $\mathcal{G}(\mathcal{V},\mathcal{E})$, where $|\mathcal{V}| = N$, and $|\mathcal{E}| = J$. Nodes are labeled by n, where $0 \leq n \leq N-1$. A link (n, m) connects node n to node m. The presence of link (n, m) in $\mathcal{E}$ means that communication can take place from node n to node m. We represent graph $\mathcal{G}$ by means of adjacency lists. So, for all $j \in \mathcal{V}$, we define $\mathcal{A}(j)$ to be a subset of $\mathcal{V}$, containing the nodes adjacent to node j; i.e., $n \in \mathcal{A}(j)$ if and only if $(n, j) \in \mathcal{E}$. A wavelength set $\Lambda = \{0, 1, \cdots, K-1\}$ of size $|\Lambda| = K$ represents the number of wavelengths (colors) available in the WDM network. For each node j of the mesh network, we specify a number κ_j, which denotes the maximum number of wavelengths that can be converted at node j. An OEO switch fabric at node j has κ_j input/output ports and tunable lasers, and therefore, κ_j is typically a power of 2. This implies that any wavelength in the set Λ can be converted into any other, provided that the number of such conversions is limited to κ_j. Note that this corresponds to full wavelength conversion capability only when $\kappa_j = K$ for all j. In view of the cost of the OEO switch, we expect κ_j to be much smaller than K.

A request set $\mathcal{R} = \{(s_0, d_0), \cdots, (s_{|\mathcal{R}|-1}, d_{|\mathcal{R}|-1})\}$ is defined as a set of pairs s_i and d_i, of size $|\mathcal{R}|$, representing the source and destination nodes of the connections to be provisioned onto the optical network. We require $s_i \neq d_i$. From an element (s_i, d_i) (also called a connection) of the request set, we must construct a *lightpath* $\mathcal{L}_i$, which is defined as a sequence of vertices $\mathcal{L}_i = (s_i, I_i(1), I_i(2), \cdots, I_i(\ell_i), d_i)$, where $\{I_i(j); j = 1, \cdots, \ell_i\}$ represent intermediate nodes in the path from s_i to d_i for request i. Note that it is possible for ℓ_i to be zero. The size of a path $\mathcal{L}_i$ is defined to be the number of vertices minus one, or the number of links contained in the path or $\ell_i + 1$. This definition of lightpath implies that any two adjacent nodes in a lightpath must be a valid link in $\mathcal{E}$.

The coloring of lightpath $\mathcal{L}$ is defined as the assignment of wavelength labels to all links of $\mathcal{L}$. A valid coloring must fulfill the following conditions. (a) Every link belonging to each lightpath must be colored. (b) If two consecutive links in a lightpath are colored with different colors, say colors r and s, the node j attached to these two links must be capable of performing this wavelength conversion. This means that the value of κ_j must be greater than or equal to the number of lightpaths that require wavelength conversion at node j. (c) No two distinct lightpaths $\mathcal{L}_i$ and $\mathcal{L}_j$, with a common link and in the same direction, can use the same wavelength at this common link. The problem now is to route and color each and every element of the request set so as to minimize the number of wavelength conversions, subject to satisfying the three constraints stated above.

The global optimization problem (when all requests are considered together as in the offline version of the problem) is NP-hard (see [6, 22] for rings) and therefore, we do not directly consider the global problem in this chapter. We also do not attempt to formulate the global problem as an integer linear programming problem due to its large computation time (see comparison of exact ILP solutions and heuristics for rings in [7]). Instead, we examine a class of algorithms with polynomial complexity, which are sequential. Quite apart from the computational difficulty of finding a globally optimal solution, we state that the sequential optimization problem is more natural because in a real setting, connection requests are expected to come in one at a time.

We describe an algorithm to minimize the number of wavelength conversions used. If there are multiple ways in which this can be achieved, the algorithm further minimizes the length of the lightpath.

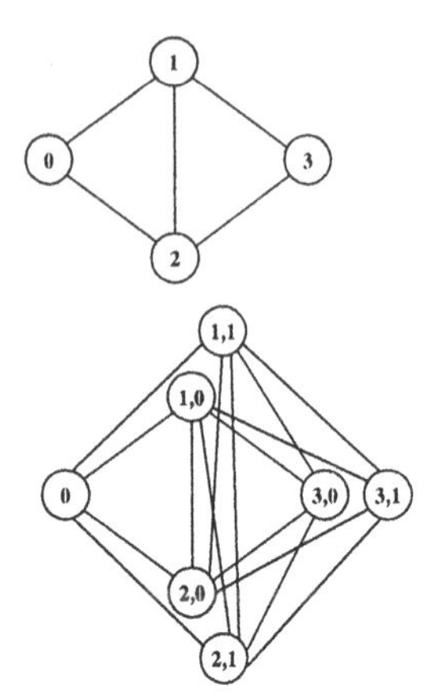

Figure 6: An example of a renumbered and new network with $N = 4$ and $K = 2$.

We start by renumbering the nodes of the network so that node 0 is the source and node $N - 1$ is the destination node of a lightpath. In the discussion to follow, we refer to this network as the *renumbered* network. Then we replace the renumbered network of N nodes by a *new* network of $(N-1)K+1$ vertices. Note that to avoid confusion between the renumbered and new networks, we use the words *node* and *link* in the context of the renumbered network and correspondingly, we use the words *vertex* and *edge* in the context of the new network hereafter. We define the set $\mathcal{V}^* = \{1, \cdots, N-1\}$, i.e., $\mathcal{V}^* = \mathcal{V} - \{0\}$. Node 0 of the renumbered network is left unchanged in the new network; i.e., it is called vertex 0. However, every node j ($j \neq 0$) of the renumbered network is replaced by K vertices of the form $[j, k]$ for $k \in \Lambda$. Every link of the form (n, j) in the renumbered network is replaced by K^2 edges from vertex $[n, k]$ to vertex $[j, l]$ for $n, j \in \mathcal{V}^*$ and $k, l \in \Lambda$ in the new network. A link of the form $(0, j)$ in the renumbered network is replaced by K edges from vertex 0 to vertex $[j, k]$ for $j \in \mathcal{V}^*$ and $k \in \Lambda$ in the new network. Thus the new network has a total of $(J - |\mathcal{A}(0)|)K^2 + |\mathcal{A}(0)|K$ edges. This number is always upper bounded by JK^2. Figure 6 shows an example of a renumbered and new network, where $N = 4$ and $K = 2$. The renumbered network is shown on the top part of the figure and the corresponding new network is shown in the bottom part. Suppose we want to examine a path in the renumbered network from node 0 to node 1 and then to node 3, using wavelength 1 on the link (0,1) and wavelength

0 on link (1,3). This same path in the new network is from vertex 0 to vertex [1,1] and then to vertex [3,0]. Notice that by increasing the number of nodes and links in the network, we have been able to include in the vertex label all information about the wavelengths used in the path.

Consider the edge from vertex $[n, k]$ to vertex $[j, l]$ for $n, j \in \mathcal{V}^*$ and $k, l \in \Lambda$ in the new network. The meaning of this edge is that if a path contains this link, then the path represents the use of wavelength l on the link (n, j) of the renumbered network and the use of wavelength k in entering node n. Referring to our example in Figure 6, the edge from $[1, 0]$ to $[3, 1]$ in the new network represents the fact that wavelength 0 is used in entering node 1 and wavelength 1 is used in going from node 1 to 3 in the renumbered network.We now place a "cost" on each edge of the new network. The cost of the edge from vertex $[n, k]$ to vertex $[j, l]$ in the new network is denoted by $d(n, k; j, l)$, when $n, j \in \mathcal{V}^*$ and $k, l \in \Lambda$. For example, if we are solving the problem of minimizing the number of wavelength conversions for the network of Figure 6, the value of $d(1, 1; 3, 0)$ can be set to 1 to reflect the fact that one wavelength conversion has been used to switch from wavelength 1 to wavelength 0 at node 1. Similarly, for the same network, $d(2, 0; 3, 0) = 0$ to reflect the fact that no wavelength conversion is used. So, if the values of $d(n, k; j, l)$ are set in this manner, we can solve the problem of minimizing the number of wavelength conversions by simply solving a shortest-path problem in the new network. There is, however, one difference. In the renumbered network, we are interested in finding an optimum path from node 0 to node 3, which corresponds to finding the shortest path from vertex 0 to any element of the set of vertices $\{[3, 0], [3, 1]\}$. In a general setting, this is equivalent to solving a shortest-path problem from vertex 0 to the set of vertices of the form $[N - 1, k]$ for $k \in \Lambda$. To summarize, by increasing the number of nodes and links, we have done away with the need for separately keeping track of the wavelengths and converted the problem to the well-known shortest-path problem. Accordingly, we apply a dynamic programming approach to solve the routing and wavelength assignment problem, as follows. A set $\mathcal{M}$ is used to keep track of the vertices closest to vertex 0. Starting with the empty set, in each iteration of step 4 below, a new vertex is added to the set $\mathcal{M}$. Furthermore, $f(j, k)$ represents the current estimate of the smallest cost of the shortest path from vertex 0 to vertex $[j, k]$ of the new network, when we are restricted to the use of the vertices in the set $\mathcal{M}$ as intermediate nodes. The variables $\Gamma(j, k)$ and $\Pi(j, k)$ keep track of the optimum wavelength used and the node number respectively in the step immediately prior to entering the vertex $[j, k]$ of the new network. Going back again to the example of Figure 6, if the path from vertex 0 to vertex [1,0] and then to vertex [3,1] happens to be an optimal one, then $\Pi(3,1) = 1$ and $\Gamma(3,1) = 0$.

Algorithm Start

Step 1: First order the request set in some order. One way of ordering it would be to find the shortest paths for each element of the set and order them by the length of the shortest path (increasing or decreasing). If ties remain, then order them further by the index of the source node. If ties still remain, then order them further by the index of the destination node. At this time, if ties remain, it implies that there are multiple requests with the same source and destination nodes, which are indistinguishable. For all $n, m \in \mathcal{V}$ and $k \in \Lambda$ define $\alpha_{nm}(k) = 1$, if wavelength k is available for use on that link, and 0 otherwise. Usually, $\alpha_{n,m}(k) = 1$ for $k = 0, \cdots, K-1$ and $(n, m) \in \mathcal{E}$, unless for some reason, certain wavelengths are not available on some links (e.g., out of service). Also, if $(n, m) \notin \mathcal{E}$, then $\alpha_{n,m}(k) = 0$. Let $\beta_k = 1$ for $k = 0, \cdots, K-1$. Let H denote the total number of hops in all paths selected so far and initialize H to 0. Let ϵ denote a small quantity satisfying $\epsilon < (KN)^{-1}$. For all $j \in \mathcal{V}$, let $\mathcal{A}(j)$ be a subset of $\mathcal{V}$ containing nodes adjacent to node j; i.e., $n \in \mathcal{A}(j)$ if and only if $(n, j) \in \mathcal{E}$. Let $\rho_j = \kappa_j$ for all $j \in \mathcal{V}$. Select the first request of the ordered request set.

Step 2: Renumber the nodes of the network so that the source node of the request under consideration is numbered 0 and the destination node is numbered $N-1$. Because this renumbering is a permutation of the indices of the original network, permute the quantities $\alpha_{nj}(k)$, ρ_j, and $\mathcal{A}(j)$ in a similar manner, for all $j, n \in \mathcal{V}$. Let the set $\mathcal{V}^* = \{1, \cdots, N-1\}$, i.e., $\mathcal{V}^* = \mathcal{V} - \{0\}$. Let $\mathcal{W}$ denote the set of all tuples of the form (n, k) where $n \in \mathcal{V}^*$ and $k \in \Lambda$. Then, $\mathcal{M}$ denotes a subset of $\mathcal{W}$, which we initialize with the empty set. For all $j \in \mathcal{V}^*$ and $k \in \Lambda$, let $\Gamma(j, k) = 0$ and $\Pi(j, k) = 0$.

Step 3: For all $n \in \mathcal{V}^*$, $j \in \mathcal{A}(n)$, and $k, l \in \Lambda$, let

$$d(n, k; j, l) = \begin{cases} \epsilon \text{ if } k = l \text{ and } \alpha_{nj}(l) = 1 \\ 1 + \epsilon \text{ if } k \neq l, \alpha_{nj}(l) = 1 \text{ and } \rho_n \geq 1 \\ \infty \text{ otherwise} \end{cases} \tag{10}$$

Remark: As described earlier, $d(n, k; j, l)$ is the cost between vertices $[n, k]$ and $[j, l]$ in the new network. Notice that this link cost is set to a small value of ϵ for every link in the path to keep track of the number of hops for the lightpath. Thus a lightpath with m hops has a cost of $m\epsilon$, if there are no wavelength conversions. This is done so that a path using the smallest number of hops is favored when there is a tie between multiple paths with the same number of wavelength conversions. Also, note that the cost is $1+\epsilon$ when a wavelength conversion is needed (the 1 accounts for the number of wavelength conversions and ϵ is needed to keep track of the number of hops). Finally, note that according to the definition, this cost is infinity in the following cases: (a) wavelength l is not available between nodes n and j or $\alpha_{nj}(l) = 0$ (this could happen if nodes n and j are not adjacent or if wavelength l has been used previously by some other request) and (b) nodes n and j are adjacent, but a wavelength conversion is needed and no wavelength conversion capability is available (ρ_n has dropped to 0). For all $j \in \mathcal{V}^*$ and $k \in \Lambda$, let

$$f(j,k) = \begin{cases} \epsilon \text{ if } \alpha_{0j}(k) = 1 \\ \infty \text{ otherwise} \end{cases} \tag{11}$$

Remark: As explained earlier, $f(j,k)$ represents the current estimate of the cost of the shortest path from vertex 0 to vertex $[j,k]$ in the new network. In this step, only the immediate neighbors of the source node with available wavelengths have a cost of ϵ, and all other nodes have a cost of infinity. Again, the small cost of ϵ is included in the cost to account for the fact that the path from vertex 0 to vertex $[j,k]$ (where node j is a neighbor of node 0) is exactly one hop long. This is used for breaking ties. Note also that there is no cost associated with wavelength conversion in this initializing step.

Step 4: Let

$$g = \min_{(j,k)\in\mathcal{W}-\mathcal{M}} f(j,k) \tag{12}$$

Let (μ,ν) be the value of (j,k) that minimizes (12). Resolve ties by first taking the lowest index of the j variable and if ties still remain, then the lowest index of the k variable. Let $\mathcal{M} \leftarrow \mathcal{M} \cup \{(\mu,\nu)\}$. If $\mu = N-1$ then go to step 5; otherwise continue. For all $j \in \mathcal{A}(\mu)$ and $k \in \Lambda$:

$$\text{if } (f(j,k) > f(\mu,\nu) + d(\mu,\nu;j,k)) \tag{13}$$

$$\text{then } \{f(j,k) = f(\mu,\nu) + d(\mu,\nu;j,k) \tag{14}$$

$$\Pi(j,k) = \mu \tag{15}$$

$$\Gamma(j,k) = \nu\} \tag{16}$$

Repeat step 4.

Step 5: Let $m = 1$, $\xi_m = \mu$, and $\phi_m = \nu$. If $f(\xi_m, \phi_m) = \infty$, stop. There is no path available for this request. Select the next request (if any left) from the request set and go to step 2. Otherwise, continue.

$$\begin{aligned} \text{while } (\xi_m &\neq 0) \\ \{m &\leftarrow m+1 \\ \phi_m &= \Gamma(\xi_{m-1}, \phi_{m-1}) \\ \xi_m &= \Pi(\xi_{m-1}, \phi_{m-1})\} \end{aligned}$$

Remark: This step computes m, which is the number of nodes in the optimal path (including the source and destination nodes). This step also reconstructs the sequence of nodes and wavelengths used in the optimal path.

Step 6: Let $\alpha_{\xi_{j+1},\xi_j}(\phi_j) = 0$ for $j = 1, \cdots, m-1$. Let $\beta_{\phi_j} = 0$ for $j = 1, \cdots, m-1$. Let $\rho_{\xi_j} \leftarrow \rho_{\xi_j} - \Delta(\phi_{j-1}, \phi_j)$ for $j = 2, \cdots, m-1$, where $\Delta(x, y) = 1$ if $x \neq y$ and 0 otherwise. Let $H = H + m - 1$. If requests are left, then pick a new request from the request set and go to step 2. If no requests are left, then stop.

End of Algorithm

For the request under consideration, the optimal lightpath consists of m nodes and equals $(\xi_m, \xi_{m-1}, \cdots, \xi_1)$, where $\xi_m = 0$ and $\xi_1 = N-1$. The wavelength used on the link (ξ_{i+1}, ξ_i) is ϕ_i for $i = 1, \cdots, m-1$. The number of wavelength conversions needed by the request under consideration is given by $\sum_{i=2}^{m-1} \Delta(\phi_{i-1}, \phi_i)$. The variable β_k is 1 if wavelength k has been used by any request provisioned so far and it is 0 otherwise. Hence the number of wavelengths used by all requests processed so far is given by $K - \sum_{k \in \Lambda} \beta_k$. The hop count for the request under consideration is given by $m - 1$. The hop count of all requests processed so far is given by H. The value of ρ_j for $j = 0, \cdots, N-1$ denotes the wavelength conversion capability still left at node j. This means that no wavelength conversion capability is left at node j if ρ_j has dropped to a value of 0. The value of $\sum_{k \in \Lambda} \alpha_{nm}(k)$ represents the total number of wavelengths left for use on the link (n, m) for all $n, m \in \mathcal{V}$, after taking into account the wavelengths used by all requests processed so far. In [8], we analyze the performance of the algorithm above, as compared against similar formulations using other routing objective functions besides the minimization of wavelength conversion.

3 Linear Programming

Linear programming has been used to solve optimization problems of various domains. In this case, the problem calls for the optimization of a linear function over a discrete set of solution vectors. Solution constraints are expressed as linear equalities and inequalities.

Linear programming finds usage in many routing problems. In what follows, we exemplify its usage with a wavelength routing problems applied to wavelength division multiplexing rings [7].

3.1 Routing in WDM Rings

A WDM ring of N nodes is modeled as a graph $G(V, E)$, where $|V| = N$, and $|E| = N$. Vertices are labeled $0 \leq n \leq N-1$ in the clockwise direction of the ring. An edge $(n, \overline{n+1})$ connects vertex n with its clockwise neighbor $\overline{n+1}$, where the overbar notation $\overline{x}$ denotes $x \bmod N$. Edges $(n, \overline{n+1})$ and $(\overline{n+1}, n)$ represent the same fiber, connecting nodes n and $\overline{n+1}$. The fiber represented by edges $(n, \overline{n+1})$ and $(\overline{n+1}, n)$ is referred to as link n.

A request set $R = \{(s_0, d_0), \cdots, (s_{|R|-1}, d_{|R|-1})\}$ is defined as a vector of pairs s_i and d_i, of size $|R| = S$, representing the source and destination nodes of the lightpaths to be provisioned. We require $s_i \neq d_i$. From an element of the request set (s_i, d_i) (also called a connection), we must construct a lightpath L, which could be either the sequence of vertices $L = (s_i, \overline{s_i+1}, \cdots, \overline{d_i-1}, d_i)$ or the sequence $L = (s_i, \overline{s_i-1}, \cdots, \overline{d_i+1}, d_i)$, in which no vertex appears more than once in a path L. The size of a path L is defined to be the number of vertices minus one, or the number of edges contained in the path. Moreover, we say that path L spans vertex(edge) $n(e)$ if it contains vertex(edge) $n(e)$. We refer to a lightpath request and its corresponding path simply by an arc, which can be either clockwise or counterclockwise.

We define the load of a ring, with respect to a collection of arcs, as follows. We count the number of clockwise arcs that span each edge of the ring. The maximum of this count over all edges gives us the ring load in the clockwise direction. We do the same in the counterclockwise direction. The maximum over these two load values is the overall load of the ring.

A wavelength set $\Lambda = \{0, 1, \cdots, K-1\}$ of size $|\Lambda| = K$ represents the wavelengths (colors) available in the WDM ring system. For each node n of the ring, a square matrix Q^n of size $K \times K$ is associated, which describes the wavelength conversion capability of node n. The (i, j) element $q^n_{i,j}$ of Q^n is one if node n is capable of performing the $i \rightarrow j$ wavelength conversion, and zero otherwise. Thus, a node n that has full color conversion capability has all elements of the matrix Q^n equal to 1. Note that we are allowing different nodes of the network to have different color conversion capabilities.

The coloring of arc A is defined as the assignment of color labels to all edges spanned by arc A. The coloring of a request set involves the coloring of all arcs resulting from a given routing of each element of the request set. A valid coloring must fulfill the following conditions. (a) Every edge belonging to each arc must be colored. (b) If two consecutive edges in an arc are colored with different colors, say colors i and j, the node m shared by these two edges must be capable of performing the corresponding wavelength conversion ($i \rightarrow j$). This implies that $q^m_{i,j} = 1$. (c) No two arcs A and B, with common edges and in the same direction, share a common color at a shared edge.

Given all the constraints for a valid routing and color assignment, the problem is to route and color each and every arc of the request set so as to minimize the number of wavelengths used. A wavelength is considered to have been used if it is used either in the clockwise or in the counterclockwise direction. More specifically, let μ_c represent the number of colors used in the coloring of request set elements that are routed in a clockwise manner, and μ_{cc} the number of colors used in the coloring of request set elements routed in a counterclockwise manner. Then, the total number of colors used in the coloring of the request set is $\mu = \max(\mu_c, \mu_{cc})$. We seek to minimize μ. Notice that although only single-fiber bidirectional rings are considered, multiple fibers can be easily incorporated into the formulation by representing a given wavelength over parallel fibers as distinct wavelengths. In what follows, we provide an integer linear programming formulation of the problem.

Let N denote the number of nodes and links of the ring and the nodes are numbered from 0 to $N - 1$. We define $\overline{j} = j - \lfloor j/N \rfloor N$; i.e., $\overline{j}$ is the remainder when j is divided by N. The link l for $0 \leq l \leq N - 1$, connects node l to node $\overline{l+1}$. There are K colors available, numbered from 0 to $K - 1$. Let s_i and d_i denote the source and destination, respectively, of connection i. Let R denote the request set, where $R = \{(s_0, d_0), \cdots, (s_{|R|-1}, d_{|R|-1})\}$ and let the cardinality of R be denoted by $|R|$. Let us further define the following: $a_{i,j} = 1$ if link j is in the clockwise path of connection i or 0 otherwise; $x_{i,j,k} = 1$ if color k is used on link j for the clockwise path of connection i or 0 otherwise; $y_{i,j,k} = 1$

if color k is used on link j for the counterclockwise path of connection i or 0 otherwise; and $w_i = 1$ if connection i is routed through the clockwise path or 0 otherwise.

For each node j, define L_j permutation matrices (each of size $K \times K$) P_l^j, for $l = 0, \cdots, L_j - 1$. These matrices all satisfy three properties: $e' P_l^j = e'$, $P_l^j e = e$, and $P_l^j \leq Q^j$, where the matrix Q^j defines node j wavelength conversion capabilities, e is a vector of ones, and e' its transpose. In addition, the elements of each matrix P_l^j can be 0 or 1 only. Here, L_j is the total number of distinct matrices satisfying the properties above. For example, if all elements of Q^j equal 1 (meaning that node j has full wavelength conversion capability), then $L_j = K!$. Note that Q^j defines the color conversion capability of node j. At any time, node j must be configured to perform a specific set of color changes only. For instance, even though node j may be capable of making many different types of color changes, it may be actually configured to pass on each color without any change. Each matrix P_l^j represents a specific configuration of the color changes allowed in node j. Thus, the (r, s) element of P_l^j is $P_l^j(r, s)$, which is 1 if the l configuration of node j allows the color r to be converted into color s and 0 otherwise. For example, if the l configuration of node j does not allow *any* color changes, then P_l^j is the identity matrix. We now further define the following: $u_{j,l} = 1$ if node j uses permutation matrix P_l^j in the clockwise direction for color conversion or 0 otherwise; $v_{j,l} = 1$ if node j uses permutation matrix P_l^j in the counterclockwise direction for conversion or 0 otherwise; and $z_k = 1$ if color k is used in the clockwise or counterclockwise direction by any connection or 0 if color k is not used by any connection. Then the routing and wavelength assignment problem is to

$$\text{Minimize} \sum_{k=0}^{K-1} z_k \tag{17}$$

where z_k is given by

$$z_k = \begin{cases} 1 & \text{if} \sum_{i=0}^{|R|-1} \sum_{j=0}^{N-1} x_{i,j,k} + \\ & \quad \sum_{i=0}^{|R|-1} \sum_{j=0}^{N-1} y_{i,j,k} \geq 1 \\ 0 & \text{otherwise} \end{cases} \tag{18}$$

from the definition of z_k. Although Eq. (18) is not linear, it is easily verified that (18) is equivalent to the following linear constraints.

$$z_k \leq \sum_{i=0}^{|R|-1} \sum_{j=0}^{N-1} (x_{i,j,k} + y_{i,j,k}) \text{ and} \tag{19}$$

$$\sum_{i=0}^{|R|-1} \sum_{j=0}^{N-1} (x_{i,j,k} + y_{i,j,k}) \leq M z_k \tag{20}$$

for $0 \leq k \leq K-1$, where M is any number greater than or equal to $2|R|N$. Each node j is configured to conform to only one of the L_j permutation matrices. Thus, we have

$$\sum_{l=0}^{L_j-1} u_{j,l} = 1 \quad \text{and} \quad \sum_{l=0}^{L_j-1} v_{j,l} = 1 \tag{21}$$

for $0 \leq j \leq N-1$. If $a_{i,j} = 1$ and $w_i = 1$, then $x_{i,j,k}$ must be one for *exactly* one color k, because exactly one color must be used for a connection/link pair. Also, if $a_{i,j}$ or $w_i = 0$, then $x_{i,j,k}$ must be zero for all k. This is

$$a_{i,j} w_i = \sum_{k=0}^{K-1} x_{i,j,k} \tag{22}$$

for $0 \leq i < |R|, 0 \leq j < N$. Similarly, in the counterclockwise direction, we have

$$(1 - a_{i,j})(1 - w_i) = \sum_{k=0}^{K-1} y_{i,j,k} \tag{23}$$

for $0 \leq i < |R|, 0 \leq j < N$. If links j and $\overline{j+1}$ are used by connection i in the clockwise direction, then their color change at node j is determined by the use of one of the L_j permutation matrices. That is equivalent to

$$x_{i,j} \sum_{l=0}^{L_j-1} P_l^j u_{j,l} x'_{i,\overline{j+1}} = a_{i,j} a_{i,\overline{j+1}} w_i \tag{24}$$

for $0 \leq i < |R|, 0 \leq j < N$, where $x_{i,j} = [x_{i,j,0}, x_{i,j,1}, \cdots, x_{i,j,K-1}]$. Note that the right-hand side of Eq. (24) is 1 if links j and $\overline{j+1}$ are both used by connection i in the clockwise direction, and it is 0 otherwise. Note also that the right-hand side of Eq. (24) is not linear. We now show how to convert this into a set of linear constraints. Let us define

$$s_{i,j,r,s} = \sum_{l=0}^{L_i-1} P_l^j(r,s) u_{j,l} x_{i,j,r} x_{i,\overline{j+1},s} \tag{25}$$

for $0 \leq i < |R|, 0 \leq j < N, 0 \leq r, s < K$. We define the set $A_l^j = \{(r,s) | \sum_{l=0}^{L_j-1} P_l^j(r,s) = 1\}$. Then, it is easy to verify that Equations (24) and (25) are equivalent to

$$\begin{aligned} s_{i,j,r,s} &\leq \sum_{l=0}^{L_i-1} P_l^j(r,s) u_{j,l} && (26) \\ s_{i,j,r,s} &\leq x_{i,j,r} && (27) \\ s_{i,j,r,s} &\leq x_{i,\overline{j+1},s} && (28) \\ \text{and } 2 + s_{i,j,r,s} &\geq \sum_{l=0}^{L_j-1} P_l^j(r,s) u_{j,l} + x_{i,j,r} \\ &\quad + x_{i,\overline{j+1},s} && (29) \end{aligned}$$

for $0 \leq i < |R|, 0 \leq j < N$ and $(r,s) \in A_l^j$. We now rewrite Eq. (24) as

$$\sum_{(r,s) \in A_l^j} s_{i,j,r,s} = a_{i,j} a_{i,\overline{j+1}} w_i \tag{30}$$

for $0 \leq i < |R|$ and $0 \leq j < N$. This means that constraints related to $s_{i,j,r,s}$ need to be present in Equations (26) – (29) only when $(r,s) \in A_l^j$. This fact reduces the number of constraints significantly.

There is a corresponding set of constraints in the counterclockwise direction. If we define

$$t_{i,j,r,s} = \sum_{l=0}^{L_i-1} P_l^j(r,s) v_{j,l} y_{i,j,r} y_{i,\overline{j-1},s} \tag{31}$$

for $0 \leq i < |R|, 0 \leq j < N$ and $0 \leq r, s < K$, then

$$\begin{aligned} t_{i,j,r,s} &\leq \sum_{l=0}^{L_i-1} P_l^j(r,s) v_{j,l} && (32) \\ t_{i,j,r,s} &\leq y_{i,j,r} && (33) \\ t_{i,j,r,s} &\leq y_{i,\overline{j-1},s} && (34) \\ \text{and } 2 + t_{i,j,r,s} &\geq \sum_{l=0}^{L_j-1} P_l^j(r,s) v_{j,l} + y_{i,j,r} \\ &\quad + y_{i,\overline{j-1},s} && (35) \end{aligned}$$

for $0 \leq i < |R|, 0 \leq j < N, (r,s) \in A_l^j$ and

$$\sum_{(r,s)\in A_l^j} t_{i,j,r,s} = (1 - a_{i,j})(1 - a_{i,\overline{j-1}})(1 - w_i) \tag{36}$$

for $0 \leq i < |R|$ and $0 \leq j < N$. Again, it is possible to verify that Equation (31) is equivalent to Equations (32) - (35). Finally, we have the following integer constraints

$$z_k, x_{i,j,k}, y_{i,j,k}, u_{j,l}, v_{j,l}, w_i, s_{i,j,r,s} \text{ and } t_{i,j,r,s} = 0 \text{ or } 1 \tag{37}$$

for $0 \leq i < |R|, 0 \leq l < L_j, 0 \leq j < N$ and $(r,s) \in A_l^j$. The ILP is now fully formulated by the objective function (17) and the constraints (19) - (23), (26) - (30) and (32) - (37). If we wish to solve the wavelength assignment problem only, variables $w_i, 0 \leq i \leq |R| - 1$ can be set to 0 or 1, depending on the routing path of each connection, which is a special case of the ILP described above.

In [7], we evaluate various routing and wavelength assignment heuristics against optimum solutions generated by the ILP formulation just presented. The disadvantage of the ILP approach is that it generally leads to a large number of variables on any realistic network. Therefore, ILP solutions are only practical for small networks; hence the need to design efficient heuristics.

4 Steiner Tree Construction

Steiner trees are typically used to solve multicast routing problems [10], for efficiency of network resources. Multicast traffic is defined as traffic originated by one or more sources and destined for multiple destinations. Hence, the design of suitable multicast trees can be modeled as the construction of spanning/Steiner trees. Depending on the application, these trees might be constrained in several ways, such as maximum diameter, capacity, etc. In what follows, we illustrate how multicast trees can be constructed in a packet network scenario where a certain amount of bandwidth capacity must be provisioned between a source and each destination of a multicast group. We refer to such a problem as the bandwidth constrained multicast routing problem.

In general, a Steiner tree is used to connect end nodes of a generic network. A Steiner tree is a tree that spans a subset of the nodes of a graph. The construction of a minimum cost Steiner tree, even unconstrained, is known to have exponential complexity. In what follows, we show how to build capacitated (bandwidth) Steiner trees of low cost in polynomial time.

4.1 Multicast Routing in Packet Networks

We model a packet network as a directed graph $G(V, E)$ where V is the set of nodes and E the set of directed *edges*. We refer to the *link* (i, j) as the pair of edges (i, j) and (j, i). We associate two values with every edge $(i, j) \in V$: a positive weight $w(i, j)$ and a nonnegative (bandwidth) capacity $b(i, j)$. Let $S \subseteq V$, $D \subseteq V$ be source and destination subsets of V, such that $\forall i \in S$ we associate $B_i \geq 0$ as the bandwidth needed by source i.

Define a Multipoint Connectivity Structure (MCS) $\sigma(V', E')$ as a connected subgraph of $G(V, E)$ containing at least the nodes $S \cup D$ and having at least one path from each node $s \in S$ to every node $d \in D$. The bandwidth and weights associated with an edge in σ are those associated with the original edges of G. Let $S(i, j) \subseteq S$ be the subset of sources contributing flow to edge (i, j). An edge (i, j) is said to be underloaded if $b(i, j) \geq \sum_{p \in S(i,j)} B_p$. Otherwise, the edge is said to be overloaded. A link is underloaded if both edges comprising it are underloaded. An MCS is called *feasible* if all its links are underloaded. The *weight* or *cost* of an MCS is the sum of the weights of its edges. We are interested in finding feasible low cost MCSs.

4.1.1 The Bidirectional Connection

The simplest multipoint-to-multipoint connection that one may conceive is a bidirectional unicast connection. In this case, $S = D = \{s, d\}$. If we allow the paths $p(s, d)$ and $p(d, s)$ to be distinct, the problem of finding a minimum-cost feasible connectivity structure can be reduced to computing two single-source shortest-path problems with bandwidth constraint [21]. In our case of interest, we require that the MCS be a single bidirectional path connecting (s, d). To proceed, we need the following simple definitions.

Definition 4.1 *The length $d^c(s, d)$ of a path $p(s, d)$ under cost function c is defined as*

$$d_p^c(s,d) = \begin{cases} \sum_{(i,j)\in p(s,d)} c(i,j) & \\ \quad\quad \textit{if}\ \forall (i,j) \in p(s,d), & b(i,j) \geq B_s \ \textit{and}\ b(j,i) \geq B_d \\ \infty \quad \textit{otherwise} & \end{cases}$$

i.e., the path length is the sum of its link lengths only if its edge components are all underloaded, or otherwise ∞.

Definition 4.2 *The shortest path between (s,d) with respect to the cost function c is:*

$$\delta^c(s,d) = \min_{p\in P} d^c(s,d)$$

where P is the set of all paths $p(s,d)$.

We propose a Dijkstra type of algorithm to solve the single bidirectional path min-cost problem. The algorithm is called BiDirect Dijkstra (BD-Dijkstra), and is similar to the original Dijkstra algorithm. The difference is that only links that have sufficient bandwidth in both directions are considered in a relaxation step. BD-Dijkstra's pseudo-code is presented below. Its proof of correctness and complexity analysis can be found in [3].

BD-DIJKSTRA G(V,E)

```
BD-Initialize(G,s);
S ← {}, Q ← V[G];
while Q ≠ 0 do
        u= BD-Extract-Min(Q, d);
        S ← S ∪ {u};
        For each vertex v ∈ Adj[u] do
              BD-Relax(B_s, B_d, u, v, b(u,v), b(v,u), c(u,v));
```

BD-Initialize(G,s)

```
For each vertex v ∈ V[G]
     Do d[v] ← ∞; π[v] = NIL ;
d[s] ← 0;
```

BD-Extract-Min(Q,d)

```
d_min ← ∞; u ← NAN;
For i ∈ Q do
     If (d[i] < d_min)
        { d_min ← d[i]; u ← i;}
RETURN u;
```

```
BD-Relax(B_s, B_d, u, v, b(u,v), b(v,u), c(u,v))

If ((B_s < b(u,v))and(B_d < b(v,u)))
   {w(u,v) = c(u,v);}
else
   {w(u,v) = ∞;}
If (d[v] > d[u] + w(u,v))
   d[v] ← d[u] + w(u,v);
   π[v] ← u;
```

4.1.2 Multicast Tree Problems

We now focus our attention on larger S and D sets. We are interested in a Multipoint Connectivity Structure (MCS), called multicast tree, defined as

Definition 4.3 *A multicast tree (mtree) $MT(E', V')$ is an acyclic MCS $\sigma(E', V'), E' \subseteq E, V' \subseteq V$ providing connectivity to every $m \in S \cup D$.*

One can easily see that an mtree is a Direct Acyclic Graph (DAG). Mtrees inherit the same feasibility definition as for any MCS. Regarding the construction of feasible mtrees, we devise two problems.

Problem 4.1 *Construct a feasible mtree.*

Problem 4.2 *Construct a feasible mtree of minimum cost.*

Generic minimum cost tree problems are known as Steiner tree problems, and are known to be NP-complete. Problems of such nature with additional constraints are called constrained Steiner tree problems. Our approach, therefore, is to provide polynomial time algorithms, for the sake of scalability, with *worst case performance* guarantees. We have established theoretical hard results about the mtree problems defined above from which worst case guarantees can be drawn. These are results regarding both feasibility and cost, which we now state, starting with the former.

Theorem 4.1 *The problem of finding a feasible multicast tree is NP-complete.*

Proof of Theorem 4.1 By reduction from the PARTITION problem. Given a generic instance X of the PARTITION problem, we construct an instance G of the (bandwidth, Steiner tree) problem in polynomial

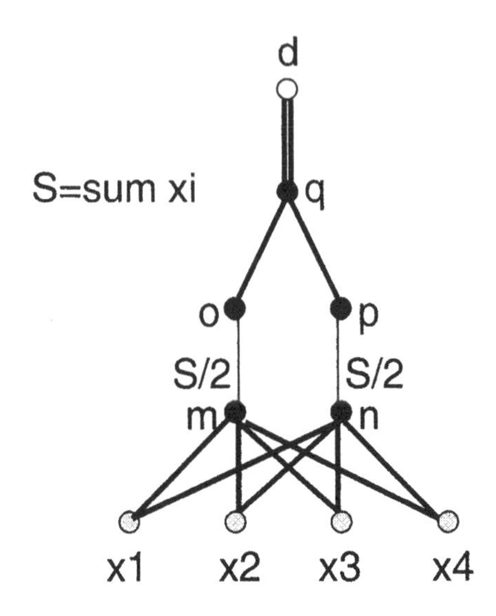

Figure 7: Reduction from PARTITION to feasible multicast problem.

time, as follows. Let the graph G have a vertex x_i for each corresponding x_i element of X. Let there be six extra vertices: m, m, o, p, q, d. For each x_i element, we attach an edge of infinite bandwidth between itself and vertex m, and another between itself and vertex n. We also attach infinity bandwidth edges between vertices o and q, p and q, and between q and d. Moreover, we attach an edge of bandwidth equal to $(\sum_{x_i \in X} s(x_i))/2$ between vertices m and o, and between vertices n and p. All other possible edges are assumed to have zero bandwidth. Finally, let $S = X$ be the source set, where $B_{x_i} = s(x_i)$, for each $x_i \in X$, and $D = d$ the destination set. The arrangement for a particular instance of the PARTITION problem is illustrated in Figure 7. It is easy to see that any feasible mtree will define a solution for the PARTITION problem.

Regarding cost, we use a recent hard result about the minimum set cover problem, which is:

Theorem 4.2 *[12]Unless $NP \subseteq DTIME(n^{\log \log n})$, the MIN SET COVER problem, with a universe of size k, cannot be approximated to better than a $\ln k$ factor.*

In a similar approach to [19], we use an approximation-preserving reduction from the MIN SET COVER problem to Problem 4.2, to prove the following theorem.

Theorem 4.3 *Unless $NP \subseteq DTIME(n^{\log\log n})$, given an instance of the bandwidth constrained minimum cost Steiner tree problem with $k = \mid S \cup D \mid$ sites, there is no polynomial time approximation algorithm that outputs a feasible Steiner tree with cost at most R times that of the minimum cost Steiner tree, for $R < \ln k$.*

Proof of Theorem 4.3 Let (T, X) be a generic instance of the MIN SET COVER problem, where $T = \{t_1, t_2, \cdots, t_n\}$, $X = \{X_1, X_2, \cdots, X_m\}$, $X_i \subseteq T$, and c_i be the cost associated with set X_i. We construct an instance G of the (bandwidth, total cost, Steiner tree) problem in polynomial time, as follows. Let the graph G have a vertex t_i for each corresponding t_i element of T, a vertex x_i for each set X_i, and two extra vertices, n and s. For each set X_i, we attach an edge between vertices n and x_i of cost c_i, and bandwidth $(B, 0)$. [3] Moreover, for each element t_i and set X_j such that $t_i \in X_j$, we attach an edge of cost 0 and bandwidth $(B, 0)$. We also attach a similar edge between s and n. All other edges are assumed to have infinity cost and $(0, 0)$ bandwidth available. Finally, let $S = \{s\}$ be the source set, $D = \{t_1, t_2, \cdots, t_k\}$ the destination set, and $B_s = B$. The arrangement for a particular instance of the MIN SET COVER ($T = \{t_1, t_2, \cdots, t_7\}$, $X = \{X_1, X_2, X_3, X_4\}$, $X_1 = \{t_1, t_2, t_3\}$, $X_2 = \{t_3, t_4, t_5\}$, $X_3 = \{t_5\}$, $X_4 = \{t_6, t_7\}$) is shown in Figure 8.

It is easy to see that any feasible mtree must contain a path from t_i to n, using an edge (x_j, n) for some X_j such that $t_i \in X_j$, for all i. Therefore, any feasible mtree provides a feasible solution of equivalent cost to the original set cover instance. The proof now follows from Theorem 4.2.

Based on Theorem 4.3, we advocate the use of a cluster type of polynomial time algorithm [19] to solve Problem 4.2, because such strategy is likely to provide a ln factor bound on the mtree cost.

4.1.3 A Cluster Algorithm for Low-Cost Multicast Trees

The basic idea is to design an algorithm that works in phases. In each phase, the algorithm connects a given number n of clusters that were previously disconnected. The algorithm stops after $\log_n K$ phases, for a multicast group of size K. Let OPT be the minimum cost feasible tree

[3] B in the up-down direction, 0 in the opposite direction.

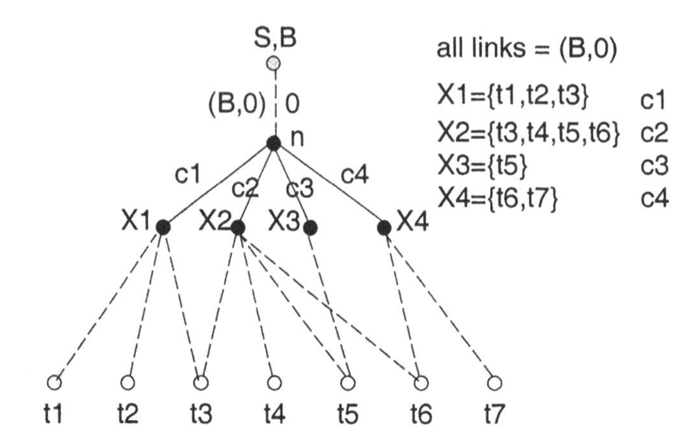

Figure 8: Reduction from MIN SET COVER to feasible minimum cost multicast problem.

with respect to cost function c, and let OPT_c be its cost. In order to guarantee a tree whose cost is no greater than $\log_n K \times OPT_c$, the cost of the partial connectivity we add at each phase must be linearly dependent with OPT_c. This being true, Theorem 4.3 states that no better polynomial time algorithm exists. Thus, the cluster strategy is an excelent candidate to produce low-cost mtrees in polynomial time, at least in the worst case sense. We now present a family of algorithms, called Bandwidth Constrained Steiner Tree algorithms, or BCST-algorithms.

BCST-Algorithms

1 Initialize the set of clusters in phase 1, Φ_1, to contain $\mid S \cup D \mid$ singleton sets, one for each member of the multicast connection. Define the single vertex as the "center" or head of its respective cluster. $i = 1$.

2 Repeat until there remains a single cluster in Φ_i:

2.1 Construct a complete graph $G_i(E_i, V_i)$ as follows. $v \in V_i$ iff v is a center of a cluster in Φ_i. Between every pair of vertices $(x, y), x, v \in V_i$, include an edge in E_i of cost defined as the minimum path cost between these two vertices in which the bidirectional bandwidth constraint defined by the clusters' bandwidth is required. The algorithm to find such path, BD-Dijkstra, is polynomial ([3]).

2.2 Find a minimum cost matching of largest cardinality in G_i.

2.3 For each edge (x, y) selected in the matching, merge their respective clusters C_x, C_y, by adding the path in the original graph corresponding to that edge to form cluster C_{xy}. The vertex (edge) set of the newly formed cluster is defined to be the union of the vertices (edges) of its cluster components, plus the vertices (edges) of the connecting path.

2.4 Elect a new vertex center for each newly formed cluster. Increment i $(i = i + 1)$.

BCST-algorithms maintain a set of clusters Φ_i in each phase i. Each cluster is formed by multicast members that are already connected among themselves. Notice that we have referred to BCST as a *family* of algorithms because two steps were not completely specified. They are basically: the specification of a cluster bandwidth, in step (2.1); and the election of the cluster head, in step (2.4). Note that, when considering minimum paths in step (2.1), any pair of cluster heads should be considered, regardless of whether they are source–source, destination–destination, or source–destination type. The following lemma is needed to bound the cost of the resulting mtree.

Lemma 4.1 *[20] Let T be a tree with an even number of marked vertices. Then, there is a pairing $(v_1, w_1), \cdots, (v_k, w_k)$ of the marked vertices such that the paths $p(v_i, w_i)$ in T are edge-disjoint.*

In particular, if the tree is an edgeweighted tree, a pairing of marked vertices that minimizes the sum of the lengths (cost) of the tree-path between these vertices are edge-disjoint. This is why we need to find a minimum cost matching in step (2.2). Notice that connecting source–source or destination–destination pairs does not pose any major problem to the construction of the mtree. In fact, it is likely that two sources or destinations are close to one another in the mtree. Moreover, multicast members are likely to be both sources and destinations.

In order to study hard bounds on both cost and overload, let there be a solution for Problem 4.2. Namely, let there be at least one feasible tree connecting the multicast members. Figure 9 gives an example of such a tree.

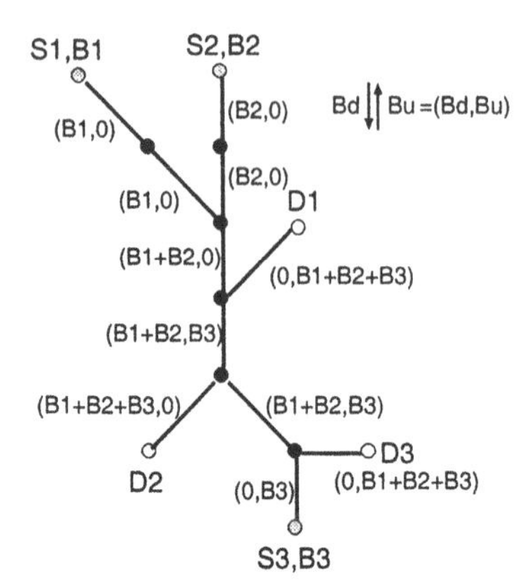

Figure 9: Feasible mtree. Only link bandwidths are shown

It is not difficult to see that bounds on cost and overload are related. Thus we devise two strategies for the amount of bandwidth B_p (bandwidth of path p) required by paths connecting clusters. (i) $B_p = B_{ch}$, where B_{ch} is the cluster head bandwidth requirement. We call this MINBCST-algorithm; (ii) $B_p = \sum_{s \in S} B_s$, where S is the set of all sources. We call this MAXBCST-algorithm. In the first one, we use Lemma 4.1 to argue that any tree found by such an algorithm has a total cost MT_c no greater than $\log_2 k \times OPT_c$.

Lemma 4.2 *A MINBCST-algorithm outputs an mtree* $MT_c \leq \log_2 k \times OPT_c$.

The following remarks are due.

Remark 4.1 *Any source–destination path connecting a given* (s, d) *pair in a feasible mtree must have at least* B_s *available in this path.*

Remark 4.2 *Any source–source or destination–destination pair in a feasible mtree does not necessarily require any available bandwidth in its connecting path.*

Remark 4.1 is obvious. Remark 4.2 comes from the fact that in case two sources (destinations) are placed next to each other on a "tip" of a feasible mtree, there is one flow direction in which no bandwidth is necessary.

Notice, however, that for the MINBCST-algorithm, the overload of a given edge belonging to the resulting mtree can be as large as the total bandwidth required by all sources. This is because a zero-bandwidth edge next to a destination, which would require a bandwidth of $\sum_{s \in S} B_s$ may be part of the mtree. Therefore, for this algorithm, if a feasible tree is found, the cost bound provided by Lemma 4.2 is guaranteed. However, for heavy loaded networks, it is likely that the mtree computed will be unfeasible.

In the MAXBCST-algorithm, trivially any mtree found is feasible. However, we cannot guarantee any cost bounds. This is so because edges belonging to the theoretical minimum-cost tree that do not necessarily have bandwidth of $\sum_{s \in S} B_s$ will be ignored in the tree computation. Therefore, the cost of the resulting mtree will have no relation to OPT_c. Notice that the BCST-algorithm does not require a feasibility check, which is necessary for any other algorithm with overload bounds greater than zero. We next claim that we cannot trade off performance bounds on cost for bounds on bandwidth overload any further.

Lemma 4.3 *Any BCST-algorithm with cost bound of* $\log_2 k \times OPT_c$ *cannot have bandwidth overload bound lower than* $\sum_{s \in S} B_s$.

Lemma 4.4 *Any BCST-algorithm that requires more bandwidth from paths connecting multicast members than what is prescribed by Remarks 4.1 and 4.2 cannot provide any bounds on the cost of its resulting mtree.*

The last two lemmas are somehow complementary, in the sense that if we try to improve cost bounds any further, we cannot claim any bounds on link overload, and vice versa. Therefore, the only algorithm over which we can claim hard bounds on cost is the MINBCST-algorithm. In [4], we compare simulation results of both the MAXBCST-algorithm and MINBCST-algorithm in solving identical bandwidth constrained Steiner tree problems.

5 Combinatorial Techniques in Wireless Ad Hoc Routing

In real communication networks, routing problems may present themselves in a multitude of ways. Sometimes, a routing problem requires several combinatorial approaches. In what follows, we describe a wireless routing problem in which the underlying network topology may change

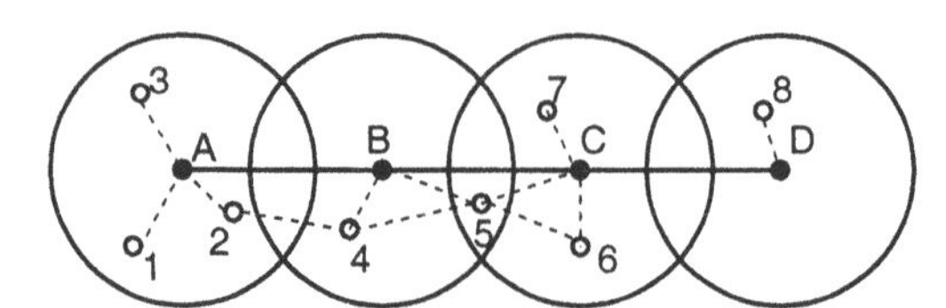

Figure 10: Conventional cellular networks.

with time, as a result of the inherent mobility of mobile radio nodes. A self-organizing clustering scheme is used as a necessary precondition to route packets across such a network.

5.1 Adaptive Clustering

In the network architecture of interest, nodes are organized into non-overlapping *clusters*. The clusters are independently controlled and are dynamically reconfigured as nodes move. This network architecture has three main advantages. First, it provides spatial reuse of the bandwidth due to node clustering. Secondly, bandwidth can be shared or reserved in a controlled fashion in each cluster. Finally, the bandwidth allocation is robust in the face of topological changes caused by node motion, node failure, and node insertion/removal because the cluster algorithm itself can efficiently adapt to such changes.

5.1.1 Network Architecture

An important wireless network feature addressed in this architecture is multihopping, i.e., the ability of the radios to relay packets from one to another without the use of base stations. Most of the nomadic computing applications today are based on a single-hop radio connection to the wired network. Figure 10 shows the cellular model commonly used in wireless networks. A, B, C, and D are fixed base stations connected by a wired backbone. Nodes 1 through 8 are mobile nodes. A mobile node is only one hop away from a base station. Communications between two mobile nodes must take place through fixed base stations and the wired backbone.

Motivated by radio packet multihopping, another wireless network model has emerged to serve a growing number of applications that rely on a quickly deployable wireless infrastructure. For instance, multihopping through wireless repeaters strategically located on campus permits

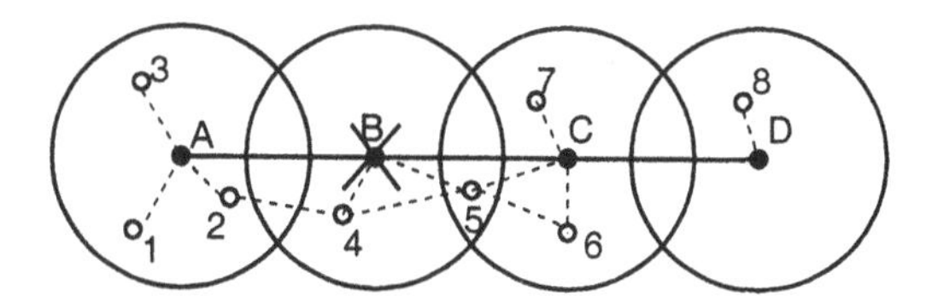

Figure 11: Multihop scenario when base station B fails.

us to reduce battery power and to increase network capacity. More precisely, by carefully limiting the power of radios, we conserve battery power. Furthermore, we also cause less interference to other transmissions farther away; this gives the additional benefit of "spatial reuse" of channel spectrum, thus increasing the capacity of the system. The multihop requirement may also arise in cellular networks. If a base station fails, a mobile node may not be able to access the wired network in a single hop. For example, in Figure 11, if base station B fails, node 4 must access base stations A or C through node 2 or node 5 which act as wireless multihop repeaters.

We consider a networking environment in which the users are mobile, the topology changes, interference occurs when multiple transmissions take place over (possibly different) links on the same or different codes, real-time multimedia traffic must be supported as well as datagram traffic, there is no stable communication infrastructure, and there is no central control. For this environment and scenarios, an architecture and networking algorithms that support a rapidly deployable radio communications infrastructure are developed. The network provides guaranteed Quality of Service (QoS) to real-time multimedia traffic among mobile users without requiring a fixed infrastructure (e.g., no base station).

5.1.2 Clustering in Multihop Wireless Networks

The objective of the proposed clustering algorithm is to find an interconnected set of clusters covering the entire node population. Namely, the system topology is divided into small partitions (*clusters*) with independent control. A good clustering scheme will tend to preserve its structure when a few nodes are moving and the topology is slowly changing. Otherwise, high processing and communication overheads will be paid to reconstruct clusters. Within a cluster, it should be easy to schedule packet transmissions and to allocate the bandwidth to real-time traffic.

Across clusters, the spatial reuse of codes must be exploited. Because there is no notion of clusterhead, each node within a cluster is treated equally. This permits us to avoid vulnerable centers and hot spots of packet traffic flow.

Optimal cluster size is dictated by the tradeoff between spatial reuse of the channel (which drives toward small sizes), and end-to-end delay minimization (which drives toward large sizes). Other constraints also apply, such as power consumption and geographical layout. Cluster size is controlled through the radio transmission power. For the cluster algorithm, we assume that transmission power is fixed and is uniform across the network. Within each cluster, nodes can communicate with each other in at most two hops. The clusters can be constructed based on the node ID. The following algorithm partitions the multihop network into some nonoverlapping clusters. We make the following operational assumptions underlying the construction of the algorithm in a radio network, which are common to most radio data link protocols [14].

- A1. Every node has a unique ID and knows the IDs of its 1-hop neighbors. This can be provided by a physical layer for mutual location and identification of radio nodes.
- A2. A message sent by a node is received correctly within a finite time by all its 1-hop neighbors.
- A3. Network topology does not change during the algorithm execution.

The clustering algorithm is depicted in Figure 12. We can see from the algorithm above that each node only broadcasts one *cluster* message before the algorithm stops, so the time complexity is $O(|V|)$ where V is the set of nodes. The clustering algorithm converges very rapidly. In the worst case, the convergence is linear in the total number of nodes. Consider the topology in Figure 13(a). After clustering (Figure 13(b)), six clusters result in the system: $\{1,2\}, \{3,4,11\}, \{5,6,7,8,9\}, \{10,12,13\}$, $\{14,15,16,17\}, \{18,19,20\}$. To prove the correctness of the algorithm we have to show that: (1) every node eventually determines its cluster; (2) in a cluster, any two nodes are at most two hops away; (3) the algorithm terminates.

Lemma 5.1 *Every node can determine its cluster and only one cluster.*

Proof of Lemma 5.1 The cluster ID of each node is either equal to its node ID or the lowest cluster ID among its neighbors. Every other

```
Distributed Clustering Algorithm ( Γ )
Γ : the set of ID's of my one-hop neighbors and myself
{
    if (my_id == min(Γ))
    {
        my_cid = my_id;
        broadcast cluster(my_id, my_cid);
        Γ = Γ - {my_id};
    }

    for(;;)
    {
        on receiving cluster(id, cid)
        {
            set the cluster ID of node id to cid;
            if (id == cid and (my_cid == UNKNOWN or my_cid > cid))
                my_cid = cid;
            Γ=Γ - {id};
            if (my_id == min( Γ))
            {
                if (my_cid == UNKNOWN) my_cid = my_id;
                broadcast cluster(my_id, my_cid);
                Γ  = Γ - (my_id};
            }
        }
        if ( Γ== ∅) stop;
    }
}
```

Figure 12: Distributed clustering algorithm.

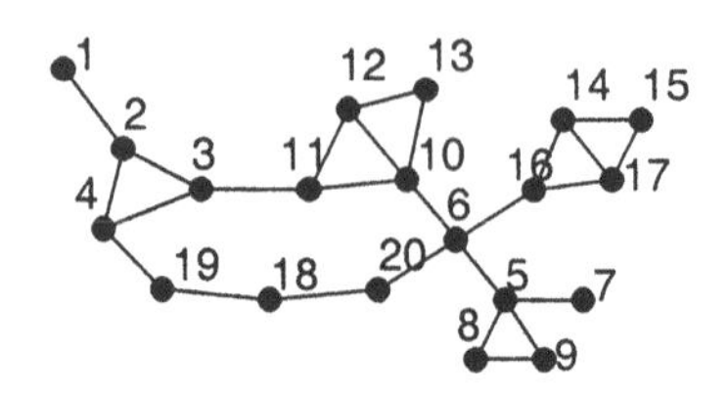

a) System Topology

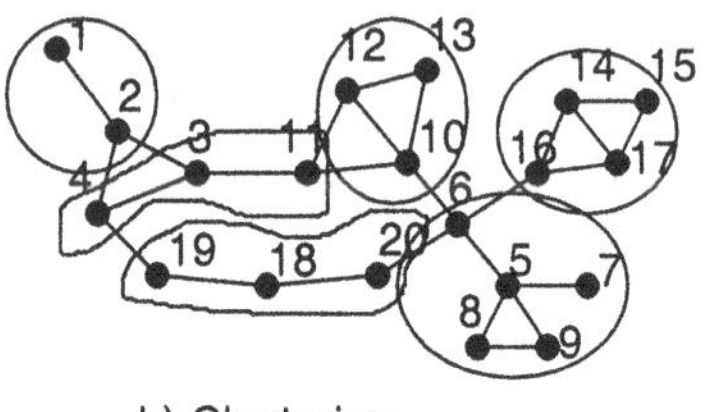

b) Clustering

Figure 13: Clustering example.

node must have its cluster ID once the said node becomes the lowest ID node in its locality. This cluster ID will be broadcast at this time, and will not be changed before the algorithm stops. Hence, every node can determine its cluster and only one cluster.

Lemma 5.2 *In a cluster, any two nodes are at most two hops away.*

Proof of Lemma 5.2 Let any node, say x, whose cluster ID ($cid(x)$) is not equal to node ID ($nid(x)$). From this algorithm, there must exist a neighbor of x, say y, $cid(y) = nid(y) = cid(x)$. Thus the hop distance between x and y is 1. So for any two nodes x and $x^{'}$, $cid(x) = cid(x^{'}) = nid(y)$, the hop distance between x and $x^{'}$ is no more than 2.

Theorem 5.1 *Eventually the algorithm terminates.*

Proof of Theorem 5.1 Because every node can determine its cluster (Lemma 5.1), the set Γ will eventually become empty. Thus, the algorithm will terminate.

Theorem 5.2 *Each node transmits only one message during the algorithm.*

Proof of Theorem 5.2 A node broadcasts messages only at the time it becomes the local minimal ID node ($min(\Gamma) == my\_id$ and the cluster

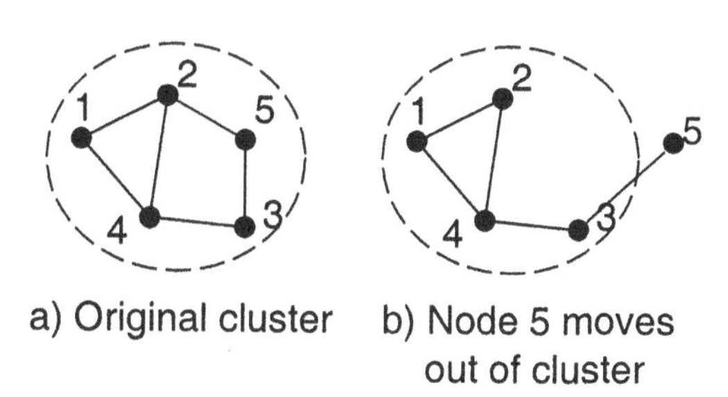

Figure 14: Re-clustering example.

ID that is broadcast will not be changed before the algorithm stops). Thus, only one message is sent out before the algorithm stops.

Theorem 5.3 *The time complexity of the algorithm is* $O(|V|)$.

Proof of Theorem 5.3 From the distributed clustering algorithm, each message is processed by a fixed number of computation steps. From Theorem 5.2, there are only $|V|$ messages in the system. Thus, the time complexity is $O(|V|)$.

5.1.3 Cluster Maintenance

In the dynamic radio network, (1) nodes can change location; (2) nodes can be removed; and (3) nodes can be added. A topological change occurs when a node disconnects and connects from/to all or part of its neighbors, thus altering the cluster structure. System performance is affected by frequent cluster changes. Therefore, it is important to design a cluster maintenance scheme to keep the cluster infrastructure as stable as possible. In this respect, the proposed cluster algorithm is more robust than the one reported in [14], because there are fewer restrictions on clusters. The cluster maintenance scheme was designed to minimize the number of node transitions from one cluster to another.

Consider the example shown in Figure 14(a). There are 5 nodes in the cluster and the hop distance is no more than 2. Because of mobility, the topology changes to the configuration shown in Figure 14(b). At this time, $d(1,5) = d(2,5) = 3 > 2$, where $d(i,j)$ is the hop distance between nodes i and j. So the cluster needs to be reconfigured. Namely, we should decide which node(s) should be removed from the current cluster. We let the highest connectivity node and its neighbors stay in the original cluster, and remove the other nodes. Recall that each node only keeps the information of its "locality", that is, one- and two-hop neighbors. Upon discovering that a member, say x, of its cluster is no longer in its locality, node y should check if the highest connectivity node is a one-hop neighbor. If so, y removes x from its cluster. Otherwise, y changes cluster. Two steps are required to maintain the cluster architecture.

Step 1: Check if there is any member of my cluster has moved out of my locality.

Step 2 If Step 1 is successful, decide whether I should change cluster or remove the nodes, not in my locality, from my cluster.

Consider the example in Figure 14(b). Node 4 is the highest connectivity node. Thus, node 4 and its neighbors 1, 2, 3 do not change cluster. However, node 5 should either join another cluster or form a new cluster. If a node intends to join a cluster, it has to check first if all members of this cluster are in its locality. Only in this case can it join the cluster.

5.2 QoS Routing in Multihop Wireless Networks

Multimedia applications such as digital audio and video have much more stringent QoS requirements than traditional datagram applications. For a network to deliver QoS guarantees, it must reserve and control resources. Routing is the first step in resource reservation. The routing protocol first finds a path with sufficient resources. Then, the resource setup protocol makes the reservations along the path.

5.2.1 Bandwidth in the Clustering Infrastructure

The key resource for multimedia QoS support is bandwidth. Thus, we first must define bandwidth in our cluster infrastructure. Assume that a real-time packet is transmitted every cycle time. For the purpose of real-time connection support, we can define "bandwidth" as the number of real-time connections that can pass through a node. Because in our scheme a node can at most transmit one packet per frame, the bandwidth of a node is given by:

$$Bandwidth = \left\lfloor \frac{cycle\ time}{frame\ time} \right\rfloor \tag{38}$$

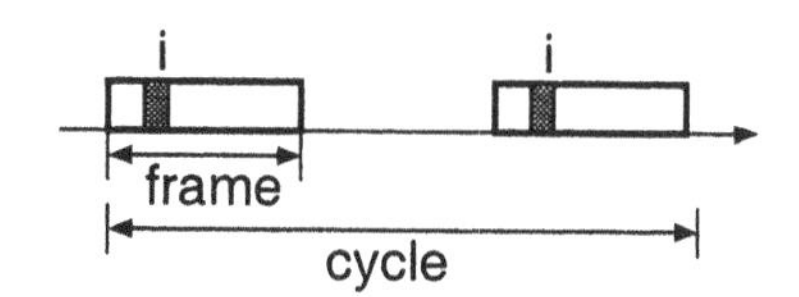

Figure 15: Node bandwidth.

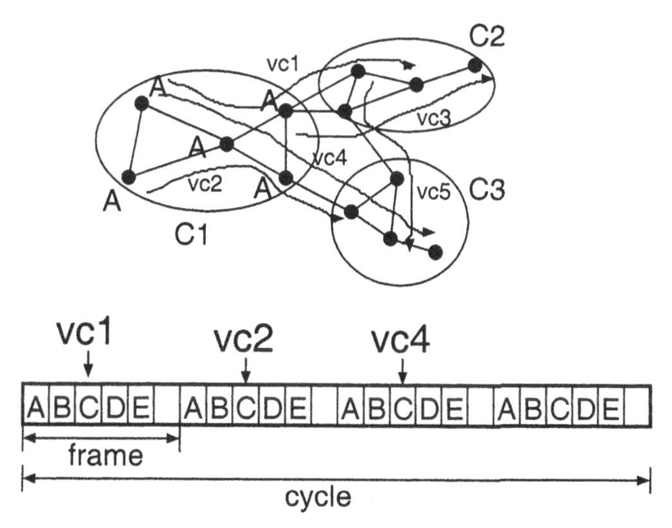

Figure 16: Bandwidth of node C in cluster C1.

The frame time of a cluster depends on how many nodes there are in the cluster. Figure 15 shows the slots dedicated to node i in the cycle, which correspond to node i bandwidth. For a numerical example, consider Figure 16, where the cycle time is 24. Consider cluster C_1, where frame size is equal to 6 slots. Thus, the node bandwidth in C_1 is $24/6 = 4$. Because there are 3 VCs passing through node C, the available bandwidth for node C is 1.

5.2.2 QoS Routing Scheme

The goal of the bandwidth routing algorithm is to find the shortest path such that the free bandwidth is above the minimum requirement. To compute the bandwidth-constrained shortest path, we use the DSDV (Destination Sequenced Distance Vector) routing algorithm [18] which was proven to be loop-free. Loop freedom follows from the fact that

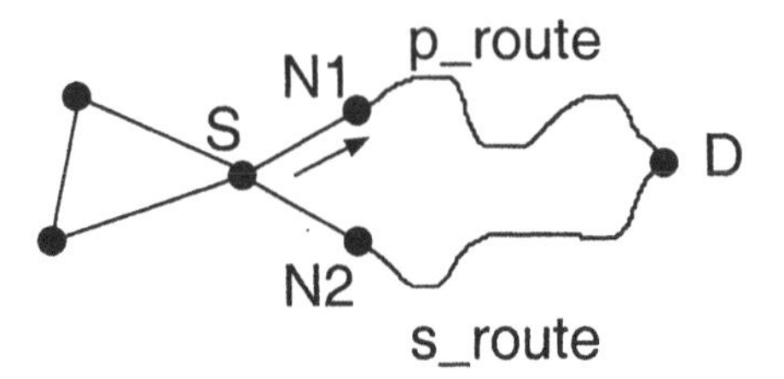

Figure 17: Standby routing.

the updates generated by a destination are sequentially numbered. In our shortest-path computation, the weight of each link is equal to 1 (i.e., minimal hop distance routing). The bandwidth constraint is simply accounted for by setting to infinity the weights of all the links to/from a node with zero bandwidth. An advantage of this scheme is to distribute real-time traffic evenly across the network. A cluster with small frame size will allow more connections to pass through it, because it has more bandwidth per node.

In addition to load balancing, our routing scheme also supports alternate paths. This is very important in a mobile environment, where links will fail because of mobility. In such an environment, routing optimality is of secondary importance. The routing protocol must be capable of finding new routes quickly when a topological change destroys existing routes. To this end, we propose to maintain secondary paths that can be used immediately where the primary path fails. In Figure 17, each node uses the primary route to route its packets. When the first link on the path (s,N1) fails, the secondary path (s,N2) becomes the primary path, and another standby path (s,N3) will be computed, as shown in Figure 18. It is worth emphasizing that these routes must use different immediate successors to avoid failing simultaneously.

The secondary (standby) route is easily computed using the DSDV algorithm. Referring to Figure 17, each neighbor of node S periodically informs S of its distance to destination D. The neighbor with the shortest distance yields the primary route. The runner-up yields the secondary route. Note that the secondary route is not the second shortest path, but the shortest path given that the first node is different for the two paths. Thus, this scheme guarantees that the first link is different for the two paths, enhancing redundancy. Furthermore, the standby route computation requires no extra table, message exchange, or computation

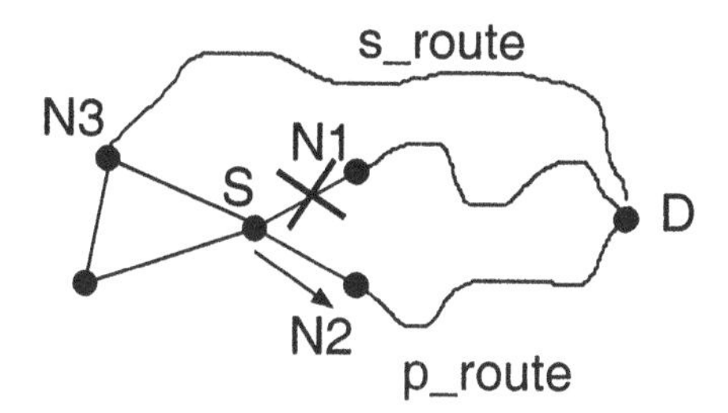

Figure 18: Routing failure scenario with standby route computation.

overhead. Also, the standby route is loop-free, as is the primary route. Evaluation of the clustering and routing algorithms can be found in [17].

6 Final Remarks

This Chapter has attempted to illustrate popular combinatorial techniques used to solve routing problems in various communication networks. The combinatorial techniques showcased include dynamic programming, integer linear programming, Steiner tree construction, and dynamic cluster formation. We have drawn application examples from packet networks to optical WDM networks to wireless networks.

Additional routing problems, not addressed in this Chapter, include:

- Routing with redundancy (disjoint path routing): routing techniques are similar to single path routing;
- Ad hoc routing: typically used in wireless networks. Because ad hoc networks tend to change rapidly in time, combinatorial optimization is of lesser importance than robustness. The routing techniques used are generally heuristics with little combinatorial flavor.

Moreover, complexity analysis (NP-completeness [13]), illustrated in the reductions from partition and min-set cover to multicast routing problems, is an extremely important tool to evaluate how complex a given routing problem is. Moreover, routing algorithms can be constructed based on complexity analysis, as is the case of the Steiner tree algorithms described earlier.

References

[1] R.E. Bellman, "Dynamic Programming," *Pinceton University Press*, Princeton, NJ, 1957.

[2] D. Bertsekas, R. Gallager, "Data Networks," *Prentice-Hall, Englewood Cliffs, NJ,* 1992.

[3] D. Cavendish, A. Fei, M. Gerla, R. Rom "On the Construction of Low Cost Multicast Trees with Bandwidth Reservation," *UCLA Technical Report Number 970043*, December, 1997.

[4] D. Cavendish, A. Fei, M. Gerla, R. Rom "On the Construction of Low Cost Multicast Trees with Bandwidth Reservation," *LNCS Conference on High Performance Computer Networks - HPNC98*, Springer-Verlag, New York, 1998.

[5] D. Cavendish and M. Gerla, "Internet QoS Routing Using the Bellman-Ford Algorithm," *IFIP Conference on High Performance Networking - HPN98*, Austria, 1998.

[6] D. Cavendish, "On the Complexity of Routing and Wavelength Assignment in WDM Rings with Limited Wavelength Conversion," *In Proc. of Asian-Pacific Optical and Wireless Communications Conference and Exhibit*, November 2001, Beijing, China.

[7] D. Cavendish and B. Sengupta, "Routing and Wavelength Assignment in WDM Ring Networks with Heterogeneous Wavelength Conversion Capabilities," *In Proc. of Infocom 2002*, Vol. 3, pp. 1415–1424, 2002.

[8] D. Cavendish, A. Kolarov, and B. Sengupta, "Routing and Wavelength Assignment in WDM Mesh Networks", *Proc. of Globecom 2004*, Vol. 2, pp. 1016–1022, December 2004.

[9] Cormen, Leiserson and Rivest, "Introduction to Algorithms," McGraw-Hill, New York, 1990.

[10] S. Deering et al., "The PIM Architecture for Wide-area Multicast Routing," *IEEE/ACM Transactions on Networking*, vol. 4, pp.153–162, April 1996.

[11] S. E. Dreyfus and A. M. Law, "The Art and Theory of Dynamic Programming," *Academic Press*, New York, 1977.

[12] U. Feigne, "A Threshold of $ln\ n$ for Approximating Set Cover," *In Proceedings of the 28th Annual ACM Symposium on the Theory of Computation*, 1996.

[13] M. R. Garey and D. S. Johnson, "Computers and Intractability," *Freeman*, San Francisco, 1979.

[14] M. Gerla and J. T.-C. Tsai, "Multicluster, Mobile, Multimedia Radio Network," *ACM-Baltzer J. Wireless Networks*, Vol. 1, no. 3, pp. 255–265, 1995.

[15] M. Henig, "Efficient Interactive Method for a Class of Multiattribute Shortest Path Problems," *In Proceedings of Management Science*, Vol. 40, # 7, pp. 891-897, July 1994.

[16] J. F. Kurose and K. W. Ross, "Computer Networking: A Top-Down Approach Featuring the Internet," *Addison-Wesley*, New York, 1999.

[17] C. R. Lin and M. Gerla, "Adaptive Clustering for Mobile Wireless Networks," *IEEE Journal on Selected Area in Communications*, Vol. 15, No. 7, Sept. 1997.

[18] C. E. Perkins, and P. Bhagwat, "Highly Dynamic Destination-Sequenced Distance-Vector Routing (DSDV) for Mobile Computers," *In Proceedings of ACM SIGCOMM'94*, pp. 234–244.

[19] M. V. Marathe, R. Ravi, R. Sundaram, S. S. Ravi, D. J. Rosenkrantz, H. B. Hunt, "Bicriteria Network Design Problems," *In Proceedings of 22nd International Colloquium on Automata Languages and Programming*, LNCS 944, pp. 487–498, 1995.

[20] R. Ravi, M. Marathe, S. S. Ravi, D. J. Rosenkrantz, and H. B. Hunt III, "Many Birds with One Stone: Multi-objective Approximation Algorithms," *In Proceedings of the 25th Annual ACM Symposium on the Theory of Computing*, pp. 438–447, 1993.

[21] Z. Wang and J. Crowcroft, "Bandwidth-Delay Based Routing Algorithms," *In Proceedings of GLOBECOM95*, Vol. 3, pp. 2129–2133, 1995.

[22] G. Wilfong and P. Winkler, "Ring Routing and Wavelength Assignment," *In Proc. of the Ninth Annual ACM-SIAM Symposium on Discrete Algorithms*, pp. 333–341, 1998.

Chapter 18

Stretch-Optimal Scheduling for On-Demand Data Broadcasts

Yiqiong Wu, Jing Zhao, Min Shao, and Guohong Cao
Department of Computer Science & Engineering
The Pennsylvania State University, PA 16802
E-mail: {ywu,jizhao,mshao,gcao}@cse.psu.edu

1 Introduction

For dissemination-based applications, unicast data delivery as used by the current Web servers may not be scalable. With unicast, a data item must be transmitted for each client that has made the request, and hence the load on the server and the network increases with the number of clients. Broadcasting techniques have good scalability because a single broadcast response can potentially satisfy many clients. Recent technology advances in satellite networks, cable networks, wireless LANs, and cellular networks make it possible to disseminate information through broadcasting.

A key consideration in designing broadcast systems is the algorithm used to schedule the broadcast. Scheduling algorithms have been extensively studied in the context of operating systems [11]. Traditional scheduling algorithms such as first-come-first-served (FCFS), round-robin, and shortest job first, are used in processor scheduling and disk scheduling. These scheduling algorithms are used to optimize the response time, throughput, or fairness. Because most of these scheduling algorithms are designed for point-to-point communication environments, they may not be applicable to broadcasting environments.

A number of researchers addressed issues on broadcast schedules. For example, Wong [14] studied several scheduling algorithms such as first-come-first-served, longest wait time (LWT), and most requests first (MRF) in broadcasting environments. Researchers in [5, 8] investigated techniques to index broadcast data to save power in mobile environments. Su and Tassiulas [12] formulated broadcast scheduling as a dynamic optimization problem and proposed efficient suboptimal solutions that can achieve a mean access latency close to the lower bound. Hameed and Vaidya [7] applied existing fair queueing algorithms to broadcast scheduling. Most of the research [2, 6, 12] on broadcast scheduling focuses on *push-based* broadcasting; that is, the server delivers data using a periodic broadcast program based on precompiled access profiles and typically, without any active user intervention. The application of such schemes is limited because they are not capable of adapting to dynamic user requirements.

There is another kind of broadcasting: *on-demand (pull-based)* broadcasting, where a large and dynamic client population requests data from an information server that broadcasts data to the clients based on these requests. Aksoy and Franklin [3] initiated the study of on-demand broadcast scheduling. They proposed an $R \times W$ scheduling algorithm that provides balanced treatment of both hot and cold data resulting in a good overall performance. The algorithm combines MRF and FCFS, and uses a novel pruning technique to reduce the computation overhead. However, the $R \times W$ algorithm assumes that each data item has the same data size. Hence, it is not suitable for requests with variable data size, because response time alone is not a fair measurement for requests with different data size. Acharya and Muthukrishnan [1] addressed the broadcast scheduling problem in heterogeneous environments, where the data have different sizes. The solution is based on a new metric called $Stretch$, defined as the ratio of the response time of a request to its service time. Based on stretch, they proposed a scheduling algorithm, called longest total stretch first (LTSF) to optimize the stretch and achieve a performance balance between worst cases and average cases. A straightforward implementation of LTSF is not practical for a large system, because the server has to recalculate the total stretch of each data item with pending requests at each broadcast interval to decide which data to broadcast next.

In this chapter, we propose a stretch optimal scheduling algorithm for on-demand broadcast systems based on the observation that LTSF is not optimal in terms of the overall stretch. One nice property of the proposed algorithm is that it is extremely simple and the computation overhead is significantly lower compared to LTSF. Analytical studies as well as detailed simulation experiments are used to verify these claims. Simulation results show that our algorithm can significantly reduce the overall stretch and still have a comparative worst-case system response time compared to existing scheduling algorithms.

The rest of the chapter is organized as follows. Section 2 develops the necessary background. In Section 3, we present the stretch optimal scheduling algorithm, develop analytical models to demonstrate its optimality in terms of stretch, and address some implementation issues. Section 4 evaluates the performance of the proposed algorithm. Section 5 concludes the chapter.

2 Preliminaries

In this section, we describe our system model, the performance metrics and the related scheduling algorithms.

2.1 The System Model

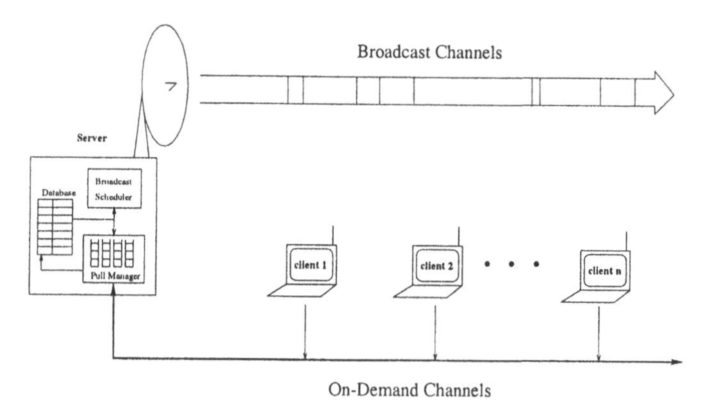

Figure 1: An on-demand broadcast system.

Figure 1 illustrates a broadcast environment that consists of a number of clients. These clients use uplink channels to send requests to the server, which uses a downlink (broadcast) channel to broadcast the responses. In the context of Hughes DirecPC architecture [13], the uplink channel would be phone lines from the clients to the DirecPC office and the downlink would be the satellite channel. In cellular networks that provide general packet radio service (GPRS) [9], the uplink would be the packet random access channel, and the downlink would be the packet broadcast channel.

Similar to previous work on broadcast scheduling [3, 12, 14], we make several assumptions. There is a single broadcast channel monitored by all clients and the channel is fully dedicated to data broadcast. When a client needs a data item that it cannot find locally, it sends a request for the data to the server. These requests are queued up at the server upon arrival. Based on the scheduling algorithm, the server repeatedly chooses an item based on these requests, broadcasts it over the satellite link, and removes the associated requests form the queue. Clients continuously monitor the broadcast channel after having made the requests. We do not consider the effects of transmission errors. After an item is broadcast, it is received by all clients that are waiting for it. Unlike some previous work [3, 12, 14], we do not assume that the data items have equal sizes.

2.2 Performance Metrics

One popular performance metric is the response time of a request, i.e., the time between sending a request and receiving the reply by a client. In heterogeneous settings, the response time is not a fair measure given that individual requests significantly differ from one another in their service time. In this chapter, we adopt an alternate performance measure, namely the stretch [1, 16, 15] of a request, defined to be *the ratio of the response time of a request to its service time*. Note that the service time is the time to complete the request if it were the only job in the system. The rationale for this choice is based on our intuition; i.e., clients with larger jobs should be expected to be in the system longer than those with smaller requests. The drawback of minimizing response time for heterogeneous workloads is that it tends to improve the system performance of large jobs (because they contribute the most to the response time). Minimizing stretch, on the other hand, is more fair for all job sizes.

We use the average stretch to measure the overall system performance. Because it is important to ensure that the scheduling algorithm does not starve the requests of unpopular data, we also measure the worst-case wait time, which is the maximum amount of time that a client request waits in the service queue before being served. To make a scheduling decision, there is a *decision overhead*. When the database size is very large, the decision overhead may be significant. As a result, some broadcast bandwidth may be wasted because the server spends a large amount of time on making scheduling decisions. Thus, we also measure the decision overhead.

2.3 Related Scheduling Algorithms

Some of the scheduling algorithms related to our work are listed as follows and are used for comparison.

- First-Come-First-Served (FCFS): Broadcasts the data items in the order they are requested. To avoid redundant broadcasts, requests for data items that already have entries in the queue are ignored.

- Longest Waiting Time (LWT): Selects the data for which the total waiting time of pending requests is the largest, i.e., the sum of all requests waiting time.

- Most Request First (MRF): The data items are broadcast in the order of the number of pending requests. The item with the greatest number of pending requests is chosen for broadcast.

- Shortest Service Time First (SSTF): The data items are broadcast in the order of the data service time. The data service time is relative to the data size. The data item with the smallest size is chosen for broadcast.

- Requests times Wait ($R \times W$): The data items are broadcast in the order of the value of function $R \times W$, where R represents the number of pending requests and W represents the first unserved request waiting time. The item with the largest $R \times W$ is chosen for broadcast. The $R \times W$ algorithm is a combination of MRF and FCFS in a way that ensures good scalability and low decision overhead.

- Largest Total Stretch First (LTSF): The data items are broadcast in the order of the total current stretch, i.e., the sum of the current stretches of all pending requests for the data. The data item with the longest total current stretch is chosen for broadcast.

3 The Stretch Optimal Scheduling Algorithm

In this section, we first use an example to show that the LTSF algorithm is not optimal in terms of the overall stretch. Then, we propose a stretch optimal scheduling algorithm and discuss some implementation issues.

3.1 An Example

In LTSF, the server maintains a queue Q_i for each data item i. Because there may be multiple requests for the same item i, let $t_{k,i}$ denote the arrival time of the kth request of item i. s_i denotes the data size of item i, and B denotes the broadcast bandwidth. Suppose the current time is t. Broadcasting item i at time t has the following stretch.

$$Stretch(i) = \sum_{k \in Q_i} \frac{t + \frac{s_i}{B} - t_{k,i}}{\frac{s_i}{B}} \quad (1)$$

The LTSF algorithm selects the data item with the maximum stretch based on Equation (1) to broadcast. Although it is intuitively correct in a point-to-point communication environment, it may not be optimal in an on-demand broadcasting environment. For example, suppose a database has two data items such that $s_1 = 1$, $s_2 = 2$. Assume that the request queues of these items are: $Q_1 = \{t_{1,1} = 8\}$, $Q_2 = \{t_{1,2} = 5, t_{2,2} = 9\}$. For simplicity, let $B = 1$. At time $t = 10$, the server needs to decide which item to broadcast first. If the LTSF algorithm is used:

$$Stretch(1) = \frac{10 + 1 - 8}{1} = 3$$

$$Stretch(2) = \frac{10 + 2 - 5}{2} + \frac{10 + 2 - 9}{2} = 5$$

As a result, the second item is broadcast first, and then the first item is broadcast at $t = 12$ (broadcasting the second item needs 2 time units). Then, the overall stretch is:

$$5 + \frac{12 + 1 - 8}{1} = 10$$

However, if the first item is broadcast first, and the second item is broadcast at $t = 11$. Then, the overall stretch is:

$$3 + \frac{11 + 2 - 5}{2} + \frac{11 + 2 - 9}{2} = 9$$

Thus, the LTSF algorithm cannot minimize the overall stretch.

3.2 The Stretch Optimal Scheduling Algorithm

Before presenting our stretch optimal scheduling algorithm, we first introduce a broadcast scheduling function for item i:

$$S(i) = \sum_{k \in Q_i} \frac{1}{s_i^2} \quad (2)$$

Theorem 3.1 *The scheduling algorithm, which selects the data item with maximum $S(i)$ (Equation (2)) to broadcast, can minimize the overall stretch.*

Proof. We give the sketch of the proof. Based on Equation (1), for any two data items i and j, the scheduling algorithm should determine which one to broadcast first. If the server broadcasts item i first, the overall stretch is:

$$Stretch_{i \to j} = \sum_{k \in Q_i} \frac{t + \frac{s_i}{B} - t_{k,i}}{\frac{s_i}{B}} + \sum_{k \in Q_j} \frac{t + \frac{s_i}{B} + \frac{s_j}{B} - t_{k,j}}{\frac{s_j}{B}}$$

Similarly, if item j is broadcast first, the overall stretch is:

$$Stretch_{j \to i} = \sum_{k \in Q_j} \frac{t + \frac{s_j}{B} - t_{k,j}}{\frac{s_j}{B}} + \sum_{k \in Q_i} \frac{t + \frac{s_j}{B} + \frac{s_i}{B} - t_{k,i}}{\frac{s_i}{B}}$$

We should minimize the overall stretch. Let

$$Stretch_{i \to j} - Stretch_{j \to i} = \sum_{k \in Q_j} \frac{s_i}{s_j} - \sum_{k \in Q_i} \frac{s_j}{s_i}$$

To minimize the overall stretch, if $Stretch_{i \to j} - Stretch_{j \to i} < 0$, item i should be broadcast first; otherwise, item j should be broadcast first. Thus, in order to broadcast item i first,

$$Stretch_{i \to j} - Stretch_{j \to i} < 0 \Longrightarrow \sum_{k \in Q_j} \frac{s_i}{s_j} < \sum_{k \in Q_i} \frac{s_j}{s_i} \Longrightarrow \sum_{k \in Q_j} \frac{1}{s_j^2} < \sum_{k \in Q_i} \frac{1}{s_i^2} \tag{3}$$

To minimize the overall stretch, item i is broadcast when $\sum_{k \in Q_i} \frac{1}{s_i^2}$ has the maximum value. □

An example Suppose a database has three data items such that $s_1 = 1$, $s_2 = 2$, and $s_3 = 3$. Assume that the request queues of these items are: $Q_1 = \{t_{1,1} = 8\}$, $Q_2 = \{t_{1,2} = 9, t_{2,2} = 9, t_{3,2} = 9, t_{4,2} = 9, t_{5,2} = 9\}$, $Q_3 = \{t_{1,3} = 2, t_{2,3} = 9\}$. At time $t = 10$, the server needs to decide which item to broadcast first. If Theorem 3.1 is followed, because $S(1) = 1$, $S(2) = \frac{5}{4}$, and $S(3) = \frac{2}{9}$, the data broadcasting order is 2-1-3, which has an average stretch of 1.81. The scheduling order of other algorithms and their average stretch costs are listed in Table 1.

The algorithm Based on Equation (2), the overall stretch can be optimized. However, due to the removal of the time parameter in Equation (2), some cold

Table 1: An example

Scheduling Algorithm	Broadcasting Order	Average Stretch
SSTF	1-2-3	1.875
	1-3-2	2.62
Our	2-1-3	1.81
MRF	2-3-1	2.1
FCFS	3-1-2	2.94
LWT, RxW, LTSF	3-2-1	2.88

data items may never be served, and the algorithm may suffer from the starvation problem. To address this problem, we add a scheduling deadline rule. When the first request for a data item i arrives, a scheduling deadline is assigned to this pending request. When the deadline passes, the data item has the highest priority to be broadcast. The stretch optimal algorithm is as follows.

1. Scan the request queue of each data item to check if any request has reached the scheduling deadline. If the request for item i has reached the scheduling deadline, let $v = i$ and go to Step 3 directly. If multiple entries have passed their scheduling deadline, use their $S(i)$ to break the tie.

2. Scan the request queue of each data item to find the data item with maximum $S(i)$, and set $v = i$. If multiple entries have the same $S(i)$, use their request arrival time to break the tie.

3. Broadcast item v.

Even with the tie-breaking rules, it is possible that multiple candidates have passed their scheduling deadline, and have the same S value. In this case, the server just randomly picks one.

We observe some interesting properties of the stretch optimal scheduling algorithm.

- For two requested data items with the same number of pending requests, the one with smaller data size should be chosen for broadcast. Hence, if there are a similar number of pending requests for each data item, the proposed algorithm has similar performance to SSTF.

- When the requested data items have the same size, the one with more pending requests should be chosen for broadcast. Hence, when the data sizes are similar, the proposed algorithm has similar performance to MRF.

3.3 Implementation Issues

To implement the stretch optimal scheduling algorithm, the server maintains two service queue data structures: an S-Heap, and a D-Link, which are shown in Figure 2. The service queue structure contains a single entry for each data item that has outstanding requests. In addition to an ID, each entry contains S, $S(i)$, and $1stARV$ which is the arrival time of the oldest outstanding request for the data item. When a request arrives at the server, a hash lookup is performed on the ID of the requested data item. If an entry already exists, the S value of that entry is simply increased by $\frac{1}{s_i^2}$.[1] The server also updates the S-Heap, which is constructed based on the value of function $S(i)$. If no entry is found (i.e., there is currently no outstanding request for the data item), a new entry is created with the S value initialized as $\frac{1}{s_i^2}$ and $1stARV$ initialized to the current time. Then, the server inserts the new S value to S-Heap and adds the new $1stARV$ value to the end of D-Link, which is sorted based on $1stARV$ of each data item.

To make a broadcast scheduling decision, the server first checks the head of the D-Link to see whether the scheduling deadline has been passed. If the deadline has been passed, it deletes the head of the D-Link, removes the entry from the service queue, and removes the corresponding node from the S-Heap. Otherwise, it removes the root of the S-Heap, removes the corresponding entry from the service queue, and removes the corresponding node from the D-Link. The S-Heap or D-Link operation (remove or insert) takes $O(\log N)$, where N is number of nodes in the heap (at most equal to the number of data items in the database). Thus, the scheduling decision overhead is $O(\log N)$. Compared to the decision overhead of LTSF, which is $O(N)$, our scheduling algorithm can significantly reduce the decision overhead.

4 Performance Evaluation

4.1 Simulation Model

We developed a simulation model written in CSIM [10]. The model represents an environment similar to that described in Section 2.1. Each client is simulated

[1]To reduce the computation overhead, a variable can be used to represent $\frac{1}{s_i^2}$, and $\frac{1}{s_i^2}$ only needs to be calculated once.

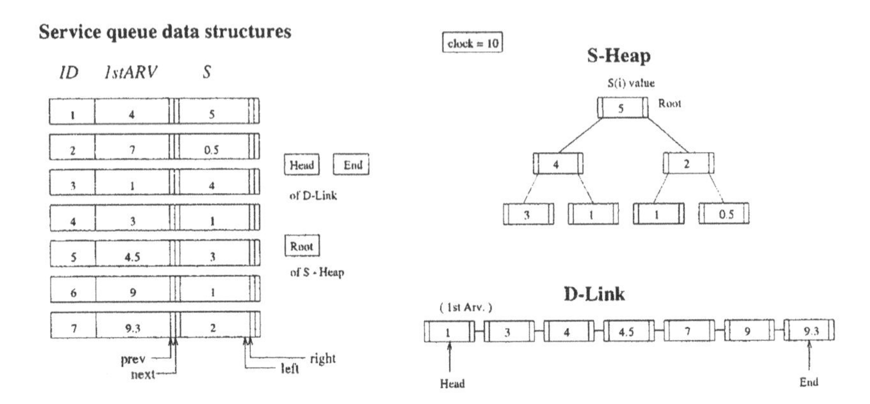

Figure 2: The service queue data structure.

by a process that generates a stream of queries. Clients send requests to the server whenever the query is generated. The aggregated stream of requests, generated by the clients, follows a Poisson process with mean λ. The density function for the interarrival times of accessing item i is:

$$f(t) = \lambda_i e^{-\lambda_i t}$$

where $\lambda_i = p_i * \lambda$, p_i denotes the probability of clients requesting data item i. The access pattern follows the $Zipf$ distribution (similar assumptions are made by other researchers as well [4, 7, 17]). In the $Zipf$ distribution, the access probability of the ith data item is represented as follows.

$$P_i = \frac{1}{i^\theta \sum_{j=1}^{n} \frac{1}{j^\theta}}$$

where $0 \leq \theta \leq 1$, and n is the database size. When $\theta = 1$, it is the strict Zipf distribution. When $\theta = 0$, it becomes the uniform distribution. Large θ results in more "skewed" access distribution. The data item size varies from s_{min} to s_{max}, and has the following two types of distributions.

- **Random:** The distribution of data sizes falls randomly between s_{min} and s_{max}. Zipf distribution favors data items with small sequence numbers. Because the data size is random, the data size and the access distribution are not correlated.

- **Increase:** The size (s_i) of the data item (i) grows linearly as i increases; i.e., $s_i = s_{min} + (i-1) * \frac{s_{max}-s_{min}}{n-1}$. In other words, the data item with smaller size will be accessed more frequently than bigger ones.

Most system parameters and their default values are listed in Table 2.

Table 2: Default system parameters

Database items	2000 items
Number of clients	200
Broadcast bandwidth (B)	115Kbps
The minimum data item size (S_{min})	1KB
The data item size ratio $R = \frac{S_{max}}{S_{min}}$	100
Mean query generate rate λ	$\frac{1}{100s}$
Zipf distribution parameter θ	0.6

4.2 The Effects of the Access Pattern

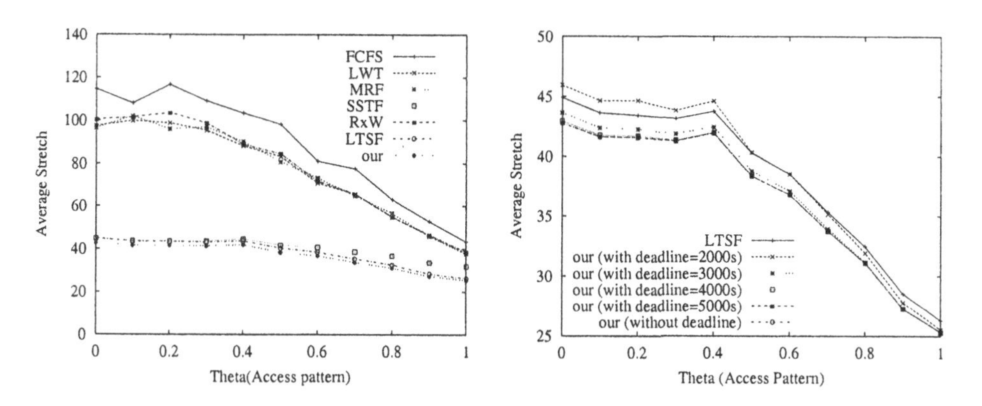

Figure 3: The average stretch under different access patterns (random distribution).

Figure 3 shows the average stretch as a function of the access skew coefficient θ. As can be seen, our algorithm has the lowest average stretch. Because LTSF also considers stretch when making scheduling decisions, it has similar performance to ours. Because FCFS, LWT, MRF, and $R \times W$ do not consider stretch in broadcast scheduling, therefore their average stretches are much higher than our algorithm and the LTSF algorithm. Note that SSTF has similar average stretch to our algorithm, because it broadcasts data in the order of data service time, which reflects the essence of the stretch-based algorithms.

Generally speaking, in order to reduce the average stretch, the server should

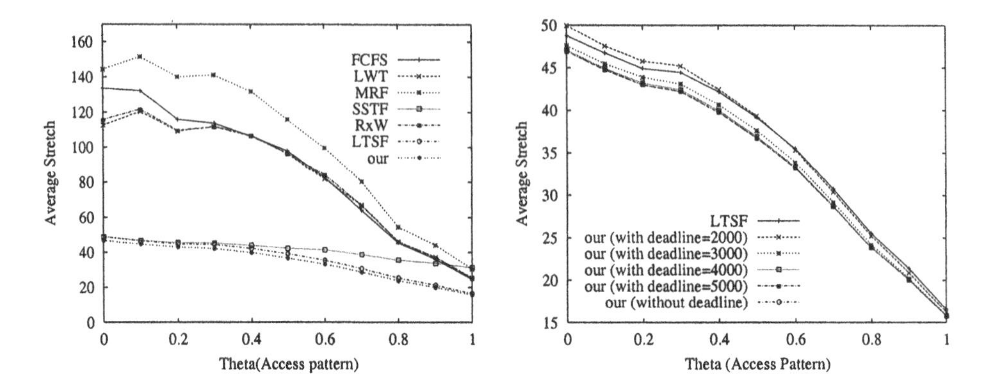

Figure 4: The average stretch under different access patterns (increase distribution).

try to schedule data items with small data size and high popularity. When $\theta = 0$, the access pattern is uniformly distributed, and data items have similar popularity. As a result, the data size is the major factor. Because FCFS, LWT, MRF, and $R \times W$ do not consider data size, they have much worse performance compared to our algorithm. The SSTF, however, considers the data size as a major factor when making scheduling decisions, and it has a similar average stretch to our algorithm when $\theta = 0$. When $\theta = 1$, the access pattern is much more skewed, and data popularity becomes a major performance factor. Because FCFS, LWT, MRF, and $R \times W$ consider data popularity when making scheduling decisions, the performance difference between these algorithms and our algorithm becomes much smaller as $\theta = 1$. The SSTF algorithm, however, does not consider popularity when making scheduling decisions, and its average stretch becomes much worse than our algorithm when $\theta = 1$.

The right graph of Figure 3 also shows the performance of our algorithm with different deadlines. It is easy to see that the average stretch reduces as the scheduling deadline increases. As expected, the worst-case waiting time decreases as the scheduling deadline increases (as shown in Figure 5). As shown in Figure 5, our algorithm has a comparative worst-case waiting time even without a scheduling deadline. When the scheduling deadline is 2000, it has the best worst-case waiting time compared to other scheduling algorithms.

From Figure 3 and Figure 4, we can see that the difference between random distribution and the increase distribution is not that significant. Due to space limits, we only show results of the random distribution in the following sections.

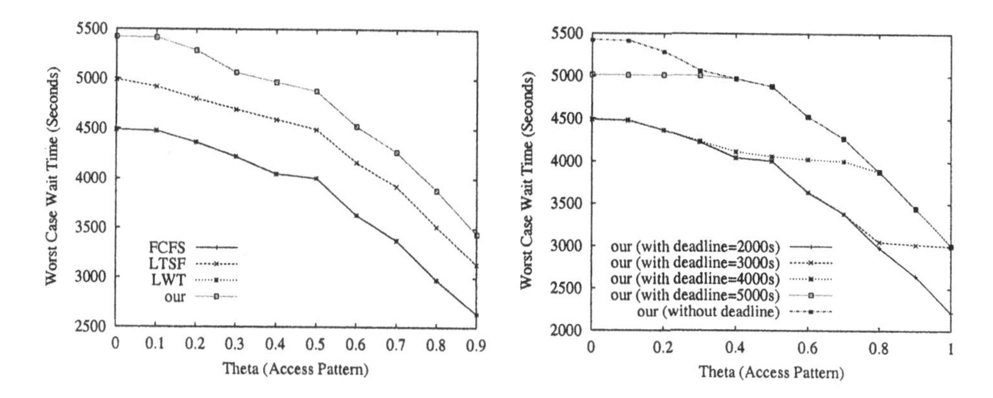

Figure 5: The worst waiting time under different access patterns.

4.3 The Effects of the System Work Load

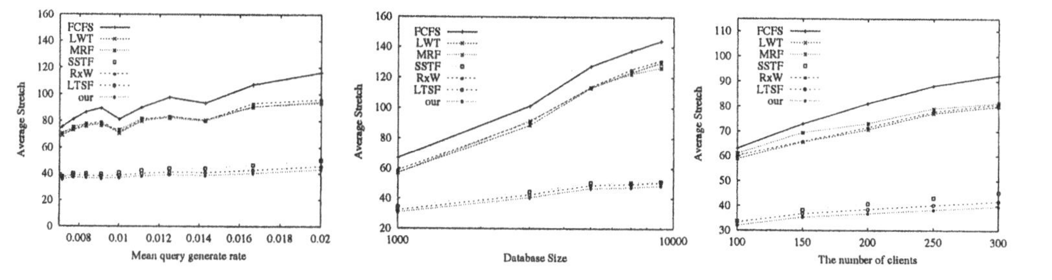

Figure 6: The average stretch under different system workload.

Figure 6 shows the average stretch as a function of the mean query generate rate, the database size, and the number of clients. Three figures have similar trends; that is, as the work load (in terms of query generate rate, database size, or number of clients) increases, the average stretch also increases. From the figure, we can see that our algorithm outperforms other scheduling algorithms in various scenarios.

4.4 The Scheduling Overhead

Figure 7 shows the scheduling overhead of the LTSF algorithm and our algorithm. In LTSF, the server has to recalculate the total stretch for every data item with pending requests in order to decide which data item to broadcast next, and hence the scheduling algorithm becomes a bottleneck due to its high computational overhead. As explained in Section 3.3, by using heaps, the computation complexity of our algorithm can be reduced to $O(\log N)$. Results from Figure 7 also verify that our algorithm can significantly reduce the computation overhead compared to the LTSF algorithm. Moreover, in the LTSF algorithm, the computation overhead increases linearly when the database size or the number of clients increases, whereas the computation overhead in our algorithm increases much more slowly, and then our algorithm is more scalable compared to the LTSF algorithm.

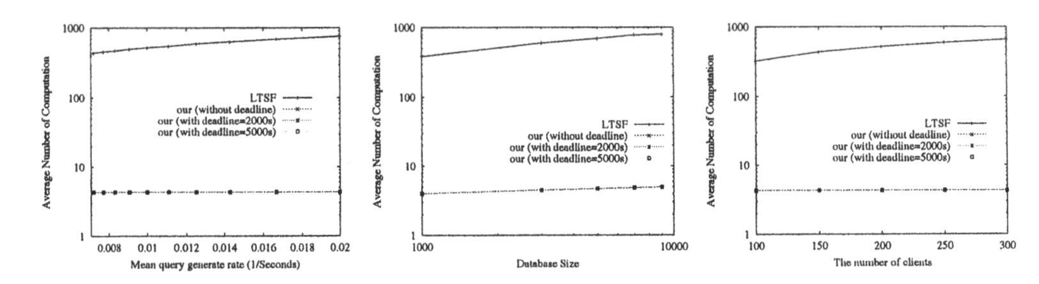

Figure 7: The scheduling overhead under different system workload.

5 Conclusions

This chapter has addressed the challenges of large-scale on-demand data broadcasts introduced by broadcast media such as satellite networks, cable networks, wireless LANs, and cellular networks. In such environments, the scheduling problem is different from that in a point-to-point communication environment or a push-based broadcast environment. Moreover, when variable-sized heterogeneous requests are considered, most of the previous scheduling algorithms fail to perform well. As stretch is widely adopted as a performance metric for variable-size data requests, we proposed a broadcast scheduling algorithm to optimize the system performance in terms of stretch. One nice property of the proposed algorithm is that it is extremely simple and the computation overhead is very low. Analytical results described the intrinsic behavior of the algorithm. Simulation results demonstrated that our algorithm significantly outperforms existing scheduling algorithms under various scenarios.

References

[1] S. Acharya and S. Muthukrishnan. Scheduling On-Demand Broadcasts: New Metrics and Algorithms. *ACM MobiCom'98*, pages 43–54, Oct. 1998.

[2] S. Acharya, R. Alonso, M. Franklin, and S. Zdonik. Broadcast disks: Data Management for Asymmetric Communication Environments. *Proc. ACM SIGMOD*, pages 199–210, May 1995.

[3] D. Aksoy and M. Franklin. RxW: A Scheduling Approach for Large-Scale On-Demand Data Broadcast. *IEEE/ACM Transactions on Networking*, 7(6), Dec. 1999.

[4] M. Ammar and J. Wong. On the Optimality of Cyclic Transmission in Teletext Systems. *IEEE Transactions on Communications*, pages 8–87, Jan. 1987.

[5] A. Datta, D. Vandermeer, A. Celik, and V. Kumar. Broadcast Protocols to Support Efficient Retrieval from Databases by Mobile Users. *ACM Transactions on Database Systems*, 24(1):1–79, March 1999.

[6] H. Dykeman, M. H. Ammar, and J. Wong. Scheduling Algorithms for Videotext Systems under Broadcast Delivery. *In Proc. International Conference of Communications*, pages 1847–1851, 1996.

[7] S. Hameed and N. Vaidya. Efficient Algorithms for Scheduling Data Broadcast. *ACM/Baltzer Wireless Networks (WINET)*, pages 183–193, May 1999.

[8] T. Imielinski, S. Viswanathan, and B. Badrinath. Data on Air: Organization and Access. *IEEE Transactions on Knowledge and Data Engineering*, 9(3):353–372, May/June 1997.

[9] R. Kalden, I. Meirick, and M. Meyer. Wireless Internet Access Based on GPRS. *IEEE Personal Communications*, pages 8–18, Apr. 2000.

[10] H. D. Schwetman. DSIM: A C-Based Process Oriented Simulation Language. *Proc. Winter Simulation Conf.*, pages 387–396, 1986.

[11] A. Silberschatz and P. Galvin. Operating System Concepts. *Addison-Wesley*, 1997.

[12] C. Su and L. Tassiulas. Broadcast Scheduling for Information Distribution. *IEEE INFOCOM*, pages 107–117, 1997.

[13] Hughes Network Systems. DirecPC Home Page. *http://www.direcpc.com*, 2000.

[14] J.W. Wong. Broadcast Delivery. *Proceedings of the IEEE*, 76(12), Dec. 1999.

[15] X. Wu and V. Lee. Preemptive Maximum Stretch Optimization Scheduling for Wireless On-Demand Data Broadcast. *Proc. of Int'l Database Engineering and Applications Symposium (IDEAS)*, 2004.

[16] Y. Wu and G. Cao. Stretch-Optimal Scheduling for On-Demand Data Broadcasts. *IEEE International Conference on Computer Communications and Networks*, pages 500–504, Oct. 2001.

[17] L. Yin and G. Cao. Adaptive Power-Aware Prefetch in Wireless Networks. *IEEE Transactions on Wireless Communication*, 3(5):1648–1658, Sept. 2004.

Chapter 19

Dynamic Simulcasting: Design and Optimization

Jiangchuan Liu
School of Computing Science
Simon Fraser University, Burnaby, BC, Canada
E-mail: jcliu@cs.sfu.ca

Bo Li
Department of Computer Science
The Hong Kong University of Science and Technology, Clear Water Bay, Kowloon, Hong Kong
E-mail: bli@ust.hk

Alan T.S. Ip
Department of Computer Science and Engineering
The Chinese University of Hong Kong, Shatin, N.T., Hong Kong
E-mail: tsip@cse.cuhk.edu.hk

Ya-Qin Zhang
Microsoft Corporation, One Microsoft Way, Redmond, WA 98052-6399, USA
E-mail: yzhang@microsoft.com

1 Introduction

With the rapid development and deployment of broadband networks, real-time video distribution is emerging as one of the most important network applications. The multiuser nature of video programs makes broadcast or multicast an efficient method for delivering video content to a large population of receivers. Such systems have also been effectively supported by existing network infrastructures, *e.g.*, by wireless networks and the Internet using IP multicast or application-layer multicast [27]. A key challenge, however, is how to handle user heterogeneity, as receivers with different platforms, such as PDAs, laptops, and PCs, or with different connection speeds, such as 1.5 Mbps ADSL or 10 Mbps Ethernet, all expect to access video services nowadays. Clearly, a single-rate transmission, though simple, is difficult to satisfy these receivers given their diverse bandwidth requirements and only one choice of streaming rate.

Simulcasting has been introduced as a viable vehicle to mitigate such mismatches and hence improve user satisfaction [3, 8, 31]. A simulcast server maintains some replicated streams for the same video content but with different rates, and delivers each stream to a specific set of receivers via a multicast channel. Two questions arise for video simulcasting naturally: how to choose an appropriate number of streams, and how to allocate bandwidth to the streams to reduce bandwidth mismatches.

In this chapter, we for the first time present a formal study of the above issues, to which we refer as the *stream replication problem*. The key objective here is to strike a balance between bandwidth economy and user satisfaction; in other words, given some bandwidth constraint, a configuration of replication should minimize the expected bandwidth mismatch for all the receivers. Clearly, the optimal replication must be adaptively configured in a nonstatic network environment, leading to a *dynamic simulcasting* paradigm.

We first formulate the stream replication problem for a given number of streams, and derive an optimal algorithm. This optimization follows the design nature of most existing simulcasting systems, i.e., the number of streams is predetermined by system operators or service providers [8]. Through theoretical analysis and experimental results, we then demonstrate that the number of replicated streams is a critical factor in the overall system optimization. It is necessary to choose an optimal, not an ad hoc, number of streams based on the available resources as well as receivers' requirements. In addition, recent advances in video coding have shown that a stream can be replicated or encoded in real-time by fast compression domain transcoding [16]. The fast and low-cost operations for stream setup and termination have also been supported in advanced video streaming standards, such as the MPEG-4 Delivery Multimedia Integration Framework (DIMF) [1]. Given such flexibility, it is possible to adaptively regulate the number of streams to accommodate the receivers' requirements. Therefore, we further consider the use of a flexible number of video streams, and derive an efficient algorithm to jointly optimize the number as well as the bandwidth allocated to each stream.

We also investigate the bandwidth allocation among different video programs (sessions). We note that, for optimal replication, the expected mismatch of the receivers is a stepwise function of the session bandwidth (the total bandwidth of the replicated streams). As a result, an equal bandwidth allocation often leads to a waste of session bandwidth. We therefore introduce a novel mismatch-aware allocation scheme. It intelligently distributes the unused bandwidth to other sessions, yet preserves general fairness properties.

The performance of dynamic simulcasting is examined under a variety of settings. The results conclusively demonstrate that it significantly outperforms static nonoptimal schemes, and the use of joint optimization for stream number and bandwidth further reduces the bandwidth mismatch. We also study the impact from a number of key factors, including the available bandwidth and the receivers' bandwidth distribution. The results offer some new insights into video simulcasting, and quantitatively demonstrate the various tradeoffs between bandwidth efficiency and user satisfaction under different conditions; thus they provide a general guideline for capacity planning and bandwidth allocation.

The rest of the chapter is organized as follows. Section 2 presents the background and some related work. The system model for dynamic simulcasting is described in Section 3. Section 4 formulates the problem of optimal stream replication for a single session, and presents efficient algorithms. The intersession bandwidth allocation is studied in Section 5. Some implementation and computation issues are discussed in Section 6. We then present the performance results in Section 7, and conclude the chapter in Section 8.

2 Related Work

2.1 Simulcasting Protocols

There has been a significant amount of work on simulcasting in the literature. A representative is the Destination Set Grouping (DSG) protocol [3], which targets the best-effort Internet with IP multicast. In DSG, a source maintains a small number of video streams (say 3) at different rates. A receiver subscribes to a stream that best matches its bandwidth. It also estimates the network status according to the packet loss ratios, and reports its estimation to the sender through a scalable feedback protocol. If the percentile of the congested receivers for a stream is above a certain threshold, the bandwidth of the stream is reduced by the sender. If all the receivers experience no packet loss, the stream bandwidth is increased. The choice of the threshold is experience-based, which is not necessarily optimal. In [26], some heuristics are proposed for stream bandwidth adjustment. The objective is to reduce the overall bandwidth originated from the server as well as the aggregate bandwidth in local regions, while not minimizing bandwidth mismatches. An optimal bandwidth allocation algorithm that minimizes bandwidth mismatches for a 2-stream case is proposed in [7]. The algorithm performs an exhaustive search on the receivers' expected bandwidths, based on an observation that the optimal value of the source bandwidth must be one of them. Due to the high complexity of exhaustive search, it is not easy to directly extend this algorithm with more layers.

Simulcasting has also been introduced in many commercial video streaming systems. For example, RelaNetworks' RealSystem G2 supports simulcasting under the name of *SureStream* [8], which generates a fixed number of streams at prescribed rates, and a receiver can dynamically choose a stream commensurate with its bandwidth. Nevertheless, the use of dynamical allocation on the sender's side has not been addressed in these commercial systems.

2.2 Simulcasting Versus Layered Transmission

Simulcasting is one of the representatives of multirate multicasting for heterogeneous receivers. Another is *layered transmission* [9, 10, 12], in which a sender generates multiple streams (called *layers*) that can progressively refine the video quality. A receiver thus can subscribe to a subset of layers commensurate with its bandwidth.

Layered transmission also suffers from bandwidth mismatches, because adaptation on the receiver's side is at the coarse-grained layer level. To minimize this mismatch, protocols using dynamic layer rate allocation on the sender's side have been proposed for cumulative layering, where the layers are subscribed cumulatively starting from a base layer [36, 10, 11, 12]. However, the constraints and hence the optimization strategies for layer rate allocation and stream replication are quite different. For illustration, consider a single video session (program) with a given total bandwidth N and total number of replicated streams (or layers) K. As explained later, we assume bandwidth allocation is discrete. For simulcasting, the problem of stream rate allocation is thus to find an optimal K-partition for integer N; for cumulative layering, it is to find an optimal enumeration of K numbers with the maximum one being less than or equal to N. In addition, it can be proved that the more layers a layered transmission protocol uses, the smaller the mismatch that a receiver would experience (ignoring the layering overheads) [12]. Thus, the use of "thin layers" has been advocated in some existing protocols [6]. For simulcasting, this is not true, because stream replication would introduce very high redundancy with a large number of streams. As such, it is necessary to find an optimal number of streams.

Different from cumulative layering, noncumulative layering allows a receiver to subscribe any subset of layers that is commensurate with its bandwidth, and hence is more flexible. In this scenario, Byers et al. [21] suggest the use of a Fibonacci-like layer bandwidth allocation, which not only enables a fine-grained receiver adaptation, but also minimizes the layer join or leave actions for receivers in a dynamic environment. Their work does not specifically target video distribution; the constraints of session bandwidth and layer number for existing non-cumulative layered video coders, such as the multiple description coders [1], have yet to be addressed.

The above adaptation algorithms generally rely on end-to-end services. There are also distributed rate adaptation algorithms involving operations at both end-nodes and intermediate switches. Fei et al. [14] study the construction of multicast trees for rate-adaptive replicated servers. Kar et al. [13] study the problem of maximizing the total utility for layered multicast. They assume that some junction nodes are deployed inside the network, and propose two distributed algorithms that converge to the optimal allocation using coordination among the sender, receivers, and junction nodes. In this chapter, we focus on the end-to-end adaptation only, which does not rely on special assistance from intermediate nodes, offering solutions readily applicable to the current best-effort Internet.

Finally, it is worth noting that simulcasting and layering have been compared in many different contexts, such as Internet multicasting [4], TCP-friendly streaming [29], and proxy cache assisted streaming [30]. Though it is often believed that layering achieves higher bandwidth utilization as there is no overlapping among the layers, it suffers from high complexities as well as the structure constraints for both encoding and decoding [18, 28]. As a result, a layered video stream usually has a lower quality than a single-layer stream of the same rate, and, more importantly, it is not compatible with many existing video formats and decoding algorithms. To the contrary, simulcasting produces independent streams, which can serve receivers with simpler and even heterogeneous decoding algorithms. Therefore, simulcasting remains a promising technique to address user heterogeneity and, as mentioned before, has been supported in many commercial streaming systems.

3 System Model and Definitions

3.1 System Model

In our system, a video server distributes a set of video programs using the simulcasting technique: each video program has several replicated streams of different rates. The descriptions of the video programs are advertised to the receivers via a dedicated multicast channel. A receiver interested in a particular video program can thus subscribe to one of the streams to receive the video. We refer to a program and its receivers as a *simulcast session* (or *session*), and the bandwidth allocated to the session as *session bandwidth*, which imposes an upper bound for the total bandwidth of the replicated streams.

The status of the system can be characterized by a 3-tuple $(C, P, M_{s,t})$, where C is the total bandwidth of the server; P is the total number of sessions, and each session has an index in $[1, \ldots, P]$; $M_{s,t}$ is the ratio that a receiver in session s has expected bandwidth t; that is, assume in session s, the total number of receivers is n and the number of receivers with expected bandwidth t is n_t, we have $M_{s,t} = n_t/n$. We stress that this model captures the essentials of many existing video multicasting or broadcasting systems, in which the expected bandwidths of the receivers are heterogeneous and limited by their processing capabilities or access links; the video server, though having a higher output bandwidth, has to accommodate many simultaneous sessions (each with several replicated streams). Therefore, the bandwidth to the sessions as well as to the streams should be carefully allocated. To this end, an end-to-end adaptation framework is adopted in our system, in which both the sender and the receivers perform adaptation to maximize the overall user satisfaction. Specifically, a receiver always tries to subscribe to a stream that best matches its expected bandwidth. It also periodically reports its expectation to the server. The server thus dynamically allocates bandwidth among the sessions as well as the streams within a session according to these expectations. In a system that allows a flexible setting for the number of streams, the server also adaptively estimates the optimal number for each session and then regulates the setting accordingly.

In this chapter, we mainly focus on developing the optimization framework for dynamic simulcasting. The mechanisms for receiver bandwidth estimation and report are out of its scope. These two issues have been extensively studied in the literature, and many of the algorithms can be applied in our system, e.g., [10] (receiver-based bandwidth estimation) and [2, 10] (scalable reporting).

3.2 Measurement of Bandwidth Mismatch

Note that a receiver cannot subscribe to a fraction of a video stream. Assume a receiver's expected bandwidth is t, and the bandwidth of the stream is r, we measure the bandwidth mismatch as:

$$RM(t,r) = \begin{cases} (t-r)/t, & 0 < r \leq t \\ 1, & r > t \;\; or \;\; r = 0 \end{cases} \tag{1}$$

We use such a relative measure (RM) instead of an absolute mismatch measure because RM will not enlarge the impact of the mismatches perceived by wideband receivers. For example, consider a 1 Mbps receiver that subscribes to a stream of 768 Kbps. The absolute mismatch is 256 Kbps, which is even larger than the maximum absolute mismatch that a 128 Kbps receiver could experience. However, for the 1 Mbps receiver, the degradation of its satisfaction is actually not that severe, and, obviously, the RM measure of 0.25 more fairly reflects this degradation. Furthermore, the RM measure has the following properties: (1) it is monotonically decreasing with the increase of r, provided the receiver successfully subscribes to the video stream, i.e., in the case of $r \leq t$; (2) by assigning a very large output, it penalizes the practically undesirable cases, e.g., $r = 0$, the receiver does not receive any stream, or $r > t$, the receiver cannot subscribe to the stream completely.

For a simulcast session, our objective for both sender and receiver adaptations is to minimize the expected RM for all the receivers. There could be other mismatch or user satisfaction measures, as well as mappings from the mismatch to some application-level performance degradation; for example, different fairness measures, such as the Inter Receiver Fairness (IRF) function [7], or subjective/objective video quality measures, such as the Peak Signal-to-Noise Ratio (PSNR) [1]. Nevertheless, the optimization algorithms presented in this chapter are general enough, which does not impose strict constraints on the measurement function, and can accommodate other mismatch or fairness measures as well.

For convenience, we list the major notations used in this chapter as follows.

C :	The total bandwidth of the server;
P :	The total number of video programs (also the number of sessions);
$M_{s,t}$:	The ratio of the receivers that have expected bandwidth t in session s; $\sum_{t>0}^{\infty} M_{s,t} = 1$; $\sum_{t>0}^{\infty} M_{s,t} = 1$ for each $s \in [1, \ldots, P]$;
l_s :	The total number of the replicated streams for session s;
$r_{s,i}$:	The bandwidth of stream i for session s;
$\vec{R}_s$:	The bandwidth allocation vector for the streams in session s, $\vec{R}_s = r_{s,1}, r_{s,2}, \ldots, r_{s,l_s}$;
$\phi(t, \vec{R}_s)$:	The rate of the best-matching stream for a receiver of bandwidth t, $\phi(t, \vec{R}_s) = \max_{r \leq t, r \in \vec{R}_s} r$;
N_s :	The session bandwidth for session s;
T_s :	The maximum bandwidth of the receivers in session s; $T_s = \max_{M_{s,t}>0} t$;
$ERM(s, N_s)$:	Expected Relative Mismatch (ERM) for session s with session bandwidth N_s;
K :	The predetermined number of streams for the fixed stream number case (OptFN or ExpFN).

4 Intra Session Optimization

In this section, we study the problem of intra session optimization at the sender's end; that is, given the bandwidth of a session, what is the optimal setting of the replicated streams for this session (video program)? This includes optimally allocating the bandwidth of the session to its streams and, if needed, determining the optimal number of streams. The method for inter session allocation and some practical issues are discussed in the next two sections.

4.1 Problem Formulation

Let $\vec{R}_s = (r_{s,1}, r_{s,2}, \ldots, r_{s,l_s})$ denote the *bandwidth allocation vector* for session s, where l_s is the total number of the replicated streams for session s, and $r_{s,i}$ is the rate of stream i. Without loss of generality, we assume that $r_{s,1} < r_{s,2} <, ..., < r_{s,l_s}$. In practice, such rates take only discrete values, for two reasons: First, given a finite number of quantizers, the output rate of a video compressor is always discrete [1]. Second, bandwidth allocation is channelized in many multicast or broadcast networks. Therefore, for convenience, we assume the bandwidths (rates) discussed in this chapter are all integer multiples of a basic unit.

For a given $\vec{R}_s$, a receiver with bandwidth t should subscribe to the stream with the best-matching bandwidth $\phi(t, \vec{R}_s) = \max_{r \le t, r \in \vec{R}_s} r$. This is a relatively simple operation. The challenging problem is how to determine $\vec{R}_s$ on the server's side, to which we refer as *intra session allocation.*

For session s, the input for intra session allocation includes the session bandwidth N_s and the receivers' bandwidth distribution $M_{s,t}$. The output is the minimum Expected Relative Mismatch (ERM) for all the receivers in the session, together with the corresponding bandwidth allocation vector $\vec{R}_s$. Assume T_s is the maximum bandwidth of the receivers in session s. Clearly, an optimal allocation should satisfy that $r_{s,1} > 0$ and $r_{s,l_s} \le T_s$. The optimization problem thus can be formally described as follows:

$$\begin{aligned} &Minimize \quad ERM(s, N_s) = \textstyle\sum_{t=1}^{T_s} M_{s,t} RM[t, \phi(t, \vec{R}_s)], \\ &Subject\ to \quad 0 < r_{s,1} < r_{s,2} <, \ldots, < r_{s,l_s} \le T_s, \ \textstyle\sum_{i=1}^{l_s} r_{s,i} \le N_s \end{aligned} \tag{2}$$

As discussed before, we consider two versions of the above optimization problem: (1) Optimal bandwidth allocation for a given (fixed) number of streams (OptFN), and (2) joint optimization for the number of streams and their respective bandwidths (OptNB). The latter not only provides a general tool for the choice of stream number in designing a simulcast system, but also serves as the foundation for the system employing a flexible setting of number.

4.2 Optimization for Fixed Number of Streams (OptFN)

In this scenario, we assume the total number of streams is fixed to a given K, or $l_s = K$. Hence, only the bandwidth of each stream ($r_{s,k}$ for $k = 1, 2, \ldots, K$) is to be determined.

Lemma 4.1 . *There exists an optimal bandwidth allocation vector for problem OptFN.*

Proof. The number of valid allocations is finite because $r_{s,k} \in Z^+$ and $r_{s,K} \leq T_s$, $k = 1, 2, \ldots, K$. Moreover, the RM measure is well defined for each valid allocation. Hence, there exists an optimal vector. □

We now show an efficient algorithm to solve this problem. Define $\alpha(n, m, k)$ as

$$\min_{l_s=k, r_{s,k}=m, \sum_{i=1}^{k} r_{s,i}=n} \sum_{t=1}^{T_s} M_{s,t} RM[t, \phi(t, \vec{R}_s)],$$

that is, the minimum ERM when a total number of k streams are generated with a total bandwidth n, and the bandwidth of stream k is m. The solution to problem OptFN is clearly given by $\min\limits_{1 \leq n \leq N_s, 1 \leq m \leq T_s} \alpha(n, m, K)$. We have the following recurrence relation for $\alpha(n, m, k)$,

$$\alpha(n, m, k) = \begin{cases} \sum\limits_{t=0}^{m-1} M_{s,t} RM(t, 0) + \sum\limits_{t=m}^{T_s} M_{s,t} RM(t, m), \; if \; m = n > 0, k = 1 \\ \min\limits_{1 \leq j < m} \{\alpha(n - m, j, k - 1) - DIFF(m, j)\}, \\ \qquad\qquad if \; m \leq n \leq N_s, 1 < k \leq K, k \leq m \leq min\{n, T_s\} \\ \infty, \; otherwise \end{cases} \quad (3)$$

where $DIFF(m, j) = \sum_{t=m}^{T_s} M_{s,t}[RM(t, j) - RM(t, m)]$. The first equation in (3) stands for a boundary case with only one stream ($k = 1$), which occupies all the session bandwidth. For $k > 1$, one more stream is to be added based on a case of $k - 1$. Without loss of generality, assume this stream is stream k, the highest stream. The difference of ERM, when this stream is added, depends only on the bandwidth of itself and that of stream $k - 1$, because only the receivers that originally subscribe to stream $k - 1$ have the potential of subscribing to stream k. Therefore, given bandwidth m of stream k, the difference of ERM is $\sum_{t=m}^{T_s} M_{s,t}[RM(t, j) - RM(t, m)]$, or $DIFF(m, j)$. The minimum ERM for this k-stream case thus can be obtained by checking each possible bandwidth allocated to stream $k - 1$, i.e., bandwidth of $1, 2, \ldots, m - 1$, as shown in the second equation in (3).

As a result, the OptFN problem can be solved by dynamic programming. For the RM function, we have $DIFF(m, j) = \sum_{t=m}^{T_s} M_{s,t}$ $[RM(t, j) - RM(t, m)] = \sum_{t=m}^{T_s} M_{s,t}(m - j)/t = (m - j) \sum_{t=m}^{T_s} M_{s,t} t^{-1}$. For each given m, $\sum_{t=m}^{T_s} M_{s,t} t^{-1}$ does not change in the execution of the algorithm. Hence, the values of $\sum_{t=m}^{T_s} M_{s,t}t^{-1}$ for $m = 1, 2, \ldots, T_s$ can be precalculated and stored in space $O(T_s)$. Because the size of array $\alpha(n, m, k)$ is $N_s \cdot min\{N_s, T_s\} \cdot K$, and, for each entry of α, we need $O(N_s)$ iterations to find the bandwidth for stream $k-1$ (assume $T_s \leq N_s$), the complexity of the optimal allocation algorithm is bounded by $O(N_s^3 K)$. The corresponding allocation vector can be easily found by backtracking relation (3) [25].

Here, the optimization structure for $\alpha(n, m, k)$ depends on the intrinsic property of the adaptation scheme at the receiver's end; that is, the utility (degree of satisfactory) increases when the bandwidth mismatch decreases, and a receiver always subscribes to the best-matching stream. This holds not only with the specific RM function, but also with most other mismatch or fairness measures. Therefore, by using appropriate expressions for $DIFF(m, j)$, the above algorithm can accommodate these utility functions.

4.3 Joint Optimization for Stream Number and Bandwidths (OptNB)

In this scenario, both the number of streams (l_s) and their bandwidth ($r_{s,i}$) are to be optimized. For discrete bandwidth allocation, however, there is an upper bound of l_s, given by $l_s^{\max} = \lfloor \sqrt{1 + 8N_s}/2 - 1/2 \rfloor$. This corresponds to stream bandwidth allocation $(1, 2, 3, \ldots, l_s)$ subject to $\sum_{t=1}^{l_s} t \leq N_s$. Thus, from Lemma 4.1, there also exists an optimal solution to problem OptNB. A *naive* method to find the solution is to try l_s from 1 to $l_s^{\max}$, and call the algorithm for OptFN for each l_s. The complexity of this exhaustive search is $O(N_s^{3\frac{1}{2}})$. Nevertheless, a more efficient algorithm can be designed as follows.

Let $\beta(n, m) = \min_{r_{s,l_s}=m, \sum_{k=1}^{l_s} r_{s,k}=n} \sum_{t=1}^{T_s} M_{s,t} RM[t, \phi(t, \vec{R}_s)]$, that is, the minimum ERM when the session bandwidth is n, and the bandwidth of stream l_s is m. There is no constraint on l_s; therefore the solution to problem OptNB is simply given by $\min_{1 \leq n \leq N_s, 1 \leq m \leq T_s} \beta(n, m)$. We also have a recurrence relation for $\beta(n, m)$, as follows,

$\beta(n, m) =$

$$\begin{cases} \sum_{t=0}^{m-1} M_{s,t} RM(t,0) + \sum_{t=m}^{T_s} M_{s,t} RM(t,m), \quad if\, m = n > 0 \\ \min_{1 \le j < m} \{\beta(n-m,j) - DIFF(m,j)\}, \\ \qquad\qquad\qquad if\, m \le n \le N_s, 1 \le m \le min\{n, T_s\} \\ \infty, \quad otherwise \end{cases} \tag{4}$$

The explanation of this relation is similar to that of (3). The first equation represents the same boundary as in (3), except that the existence of only one stream is not explicitly stated, but implied by $m = n$. In the second equation of (4), because there is no limit on the number of streams, the index of k is omitted. As a result, calculating $\beta(n, m)$ and obtaining the optimal allocation for OptNB needs only $O(N_s^3)$ time, which is much lower than the exhaustive search algorithm and, interestingly, even lower than OptFN.

4.4 Remarks on the Lower Bound of ERM

Intuitively, ERM can be reduced if more session bandwidth is allocated. However, our observation from the solutions for OptFN (fixed number of streams) is that ERM cannot be further reduced after a certain N_s. A trivial bound is $N_s^0 = \sum_{k=1}^{K}(T_s - k + 1) = K(2T_s - K + 1)/2$, because the allocation of the highest total bandwidth is $(T_s - K + 1, T_s - K + 2, \ldots, T_s)$. The tight bound N_s^{bound} can thus be represented by $\min\{n' : \min_{1 \le m \le T_s} \alpha(n', m, K) = \min_{1 \le m \le T_s} \alpha(N_s^0, m, K)$. An effective method to find this N_s^{bound} is based on the optimal bandwidth allocation for cumulative layered multicast with K layers. If there is no constraint of session bandwidth, we can build a mapping from the cumulative layer bandwidth to the stream bandwidth: $r_{s,k} = \sum_{i=1}^{k} r'_{s,i}$, where $r'_{s,i}$ is the bandwidth of layer i [4]. This makes the two schemes achieve the same session ERM. Specifically, when the layer bandwidth allocation is optimal, this mapping gives the optimal stream bandwidth allocation with no bandwidth constraint. As a result, N_s^{bound} is given by $\sum_{k=1}^{K}\sum_{i=1}^{k} r'_{s,i}$, which can be calculated in time $O(T_s^2 K)$ [12]. This mapping is illustrated in Figure 1.

In a bandwidth-limited case, however, the optimization structure for stream replication is different from that for cumulative layering, and the choice of K becomes critical. As shown in our numerical results, the use of a flexible number of streams can remarkably reduce ERM for session bandwidths beyond N_s^{bound}.

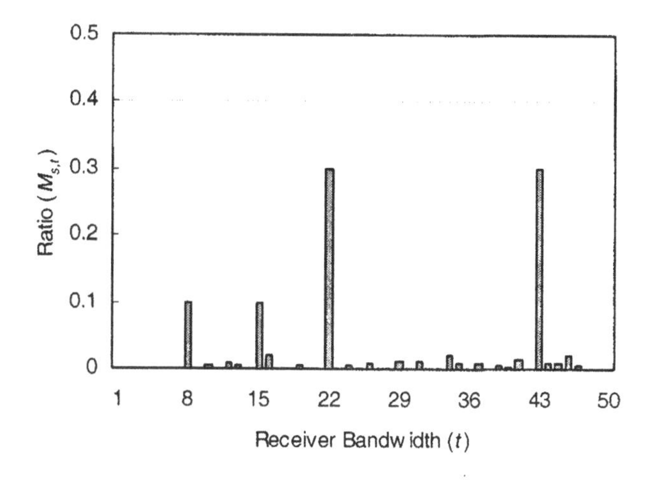

Figure 1: In this simple example, most receivers' bandwidths are distributed at four points: 8 (10%), 15 (10%), 22 (30%), and 43 (30%). Assume the number of streams is fixed to 3. In the case of cumulative layered multicasting, the cumulative layer bandwidth allocation (8, 22, 43) minimizes the expected mismatch, as long as the session bandwidth is no less than 43. Note that this session bandwidth is even lower than the maximum receiver bandwidth (about 47). If there is no constraint on session bandwidth, this also gives the optimal stream bandwidth allocation for simulcasting, but the total bandwidth of the streams now is 73 (=8+22+43). The mismatch cannot be further reduced by adjusting (either increasing or decreasing) the bandwidth of any stream, even if there is extra session bandwidth. On the other hand, for a limited session bandwidth, say 70 (< 73), the allocation (8,15,43) becomes the optimal choice for simulcasting. Actually, this is the optimal allocation for any session bandwidth between 66 (=8 + 15 + 43) and 72. In other words, the session ERM for OptFN is a stepwise function of the session bandwidth.

5 Inter Session Bandwidth Allocation

We now consider the issue of bandwidth allocation for different sessions, or *inter session bandwidth allocation.* Our basic goal is to achieve a fair yet efficient allocation. There are various notions of fairness for session bandwidth allocation, especially in a broadcast or multicast scenario

[19, 20]. In a centralized network, such as a cellular network, fairness is also related to administrative issues as well as charging policies. Hence, rather than define a new inter session allocation framework and claim its fairness, we try to identify the unique properties of our application, and enhance the system performance within existing frameworks. Specifically, we observe that the session ERM for our optimal replication algorithm is a stepwise function of session bandwidth, and the ERM of OptFN even becomes flat after a certain session bandwidth (see Figure 1 and the numerical results in Section 7 for illustration). Thus, some bandwidth allocated to a session can be wasted; it would be beneficial to distribute such bandwidth to other sessions.

We refer to such an enhancement as *ERM-Aware Allocation* (EAA). For illustration, we use an *Equal Share based Allocation* (ESA) as the basic allocation framework, which allocates the bandwidth uniformly among the sessions. Denote $mERM(s,n)$ as the minimum ERM when bandwidth n is allocated to session s, and $\tau(s,d)$ as $[mERM(s,n_s) - mERM(s,n_s+d)]/d$, i.e., the reduction of $mERM$ per bandwidth unit when d units are added to session s. The following heuristic algorithm provides a simple EAA implementation.

1 : $N_s \leftarrow \lfloor C/P \rfloor$, $s = 1, 2, \ldots, P$;
2 : While $mERM(s, N_s) = mERM(s, N_s - 1)$ do $N_s \leftarrow N_s - 1$; $s = 1, 2, \ldots, P$;
3 : $\sigma \leftarrow C - \sum_{s=1}^{P} N_s$;
4 : Repeat
5 : $\varphi \leftarrow \arg_d \max_{s=1,2,\ldots,P,\text{ and } d \le \sigma} \tau(s,d)$,
$s' \leftarrow \arg_s \max_{s=1,2,\ldots,P,\text{ and } d \le \sigma} \tau(s,d)$;
6 : if $\tau(s',\varphi) > 0$, then $N_{s'} \leftarrow N_{s'} + \varphi$, $\sigma \leftarrow \sigma - \varphi$;
7 : Until $\tau(s',\varphi) = 0$ or $\sigma = 0$.

Given an equal allocation (Step 1), the above algorithm first reduces the bandwidth of each session as much as possible without increasing the session's current $mERM$ (Step 2). It will then reallocate the unused bandwidth to the sessions (Steps 3 to 7), and each time a session that has the maximum ERM reduction per bandwidth unit is selected (Step 5).

Assume N'_s, $s = 1, 2, \ldots, P$, are the session bandwidths allocated by ESA. It can be proved that $\sum_{s=1}^{P} mERM(s, N_s) \le \sum_{s=1}^{P} mERM(s, N'_s)$ and $mERM(s, N_s) \le mERM(s, N'_s)$, $s = 1, 2, \ldots, P$. Hence, EAA can reduce not only the average ERM for all the sessions, but also the ERM of each session. In the worst case, EAA yields the same ERM as ESA.

6 Implementation Issues and Computation Overhead

The optimal stream replication algorithms can be implemented offline, assuming the distributions of the receivers' expectations are available and stationary for a large population [23]. On the other hand, an online implementation, or dynamic simulcasting, would achieve higher bandwidth utilization in a nonstatic environment. Nevertheless, there is a potential issue of computational overheads associated with online adaptation. The choice of a video codec that is compatible with such a dynamic allocation algorithm is also very important. In this section, we discuss these practical issues.

6.1 Computation Overhead and Algorithm Optimization

We have applied a set of techniques to speed up the optimization algorithms. First, when calculating $mERM(s,n)$ for $n = N_s$, the values of $\alpha(n,m,k)$ for $n < N_s$ are all available in intermediate stages; we can thus obtain $mERM(s,n)$for any $n \leq N_s$ in time $O(N_s^3 K)$ only. For EAA allocation, if necessary we can incrementally calculate the session ERM for each $n > N_s$ in time $O(N^2 K)$. Similar techniques can be used in OptNB as well. Second, our algorithms are based on the bandwidth distribution of all the receivers, and not the bandwidth of an individual receiver. In a typical multicast environment, the bandwidths of the receivers usually follow some clustered distribution. For instance, they often use standard access technologies, such as a 128 Kbps ISDN line, a 1.5 Mbps ADSL, or a 10 Mbps shared Ethernet; or those in a local region may share upstream links, and hence experience the same bottleneck bandwidth. Although individual receivers may experience short-term bandwidth fluctuations, or dynamically join or leave a session provided that the system supports these operations (e.g., in the IP multicast environment), such clustered distributions could persist for a relatively long time. As such, the adaptation algorithm can be executed infrequently, only when the distribution has substantially changed. This can be identified by statistical methods, such as the Pearson's χ^2-test or the Kolmogorov-Smirnov (K-S) test [24]. Finally, the curve of the minimum ERM for a particular session is in-

Setting (C, P, T_s)	Execution Time (ms)					
	OptFN*	EAA	Total	OptNB*	EAA	Total
(256,10,15)	1.5	0.5	15.5	1.1	0.5	11.6
(1024,20,30)	2.6	1.2	53.2	1.9	1.1	39.1
(1200,15,50)	4.2	1.4	64.4	3.9	1.3	59.8

* Execution time for one session.

Table 1: Execution times for the allocation algorithms

dependent of the receiver status of other sessions. When the status of a session changes, only its own $mERM$ curve needs to be recalculated, together with an execution of inter-session allocation.

We have implemented the optimization algorithms using C++ on an Intel Pentium III 900 MHz PC with 256 MB memory. The execution times of different settings are listed in Table 1. It can be seen that the computation overhead is not significant; the results can be obtained in a relatively short time that is suitable for real-time adaptation.

6.2 Video Stream Replicating

In practice, replicated video streams can be obtained through rate control at the source coding stage, or through media scaling mechanisms, e.g., *transcoding*, which converts an existing video stream to a stream with a different bit-rate or format [16]. Our optimization algorithm does not specify any particular video coding scheme in the application layer. It can cooperate with different media scaling schemes. Nevertheless, a scheme with a wide dynamic range, fast responsiveness, and fine granularity in terms of rate adaptation is of particular interest. This has been demonstrated by advanced video transcoders using motion vector replication or frequency-domain manipulation [16]. Furthermore, it is worth pointing out that the emerging MPEG-7 standard defines *transcoding hints*, a set of meta-data that effectively help the transcoding procedure meet the specific speed or bandwidth requirements yet preserve high video quality [17].

7 Performance Evaluation

In this section, we evaluate the performance of dynamic simulcasting, and try to identify the key factors that influence the performance. For the sake of comparison, we also implement a nonoptimal scheme that is often cited in the literature: the exponential (also called multiplicative

[5] allocation with a fixed number of streams (ExpFN) [9, 12]. In ExpFN, the stream bandwidths form a geometric progression, i.e., $r_i = \lfloor \rho^{i-1} r_1 \rfloor$, $i = 2, 3, \ldots, K$. Such an exponential setting can cover a broad dynamic range with a limited number of streams, so as to meet the diverse bandwidth demands from receivers. To achieve a fair comparison, we assume that both the minimum and the maximum receiver bandwidths are known, and r_1 is set to the minimum. Given constraints $\sum_{i=1}^{K} r_i \leq N_s$ and $r_K \leq \eta T_s$, the spanning factor ρ can be simply determined by a bisection search. Here, $\eta < 1$ is a damping factor, which ensures a reasonable portion of receivers can subscribe to stream K. Without this factor, the bandwidth of stream K will be set to T_s, and thus only the receivers of the maximum bandwidth can subscribe to stream K. Such receivers are very few, possibly only one, resulting in a waste for the high bandwidth setting of stream K. In our experiments, η is set to 0.85, the same as that in [10].

7.1 Numerical Results

7.1.1 Intra-session Bandwidth Distribution

We first study the performance of the stream replication schemes in a single session. To reflect the heterogeneous nature of the receivers, we model the expected bandwidths of the receivers in the session by a multimodal distribution. Specifically, we observe that most access and video decoding components on the receiver's side follow some specific standards, yet some use customized software or hardware [1]. Therefore, a mixture Gaussian model [15] is used to represent the bandwidth distribution for a large population. This model consists of w clusters, each following a Gaussian distribution. In our simulation, the minimum and maximum receiver bandwidths are 2 and 50, respectively. Assume each bandwidth unit is 32 Kbps; this range covers the bandwidths of many available network access techniques. The standard deviation of a cluster is set to 10% of the cluster mean. Thus most bandwidth differences are within ±10%, yet a few reach about ±40% or more, which reflects the flexibility in device design. By using different w, this model can be viewed as a generalization of those models used in previous studies [5, 7].

We assume the session has 500 receivers and draw 500 samples from the model to obtain a bandwidth distribution instance of the receivers. All the results presented are averages over ten instances.

In Figure 2, we show the impact of the bandwidth distribution for the single session. It can be seen that the session ERMs of the optimal

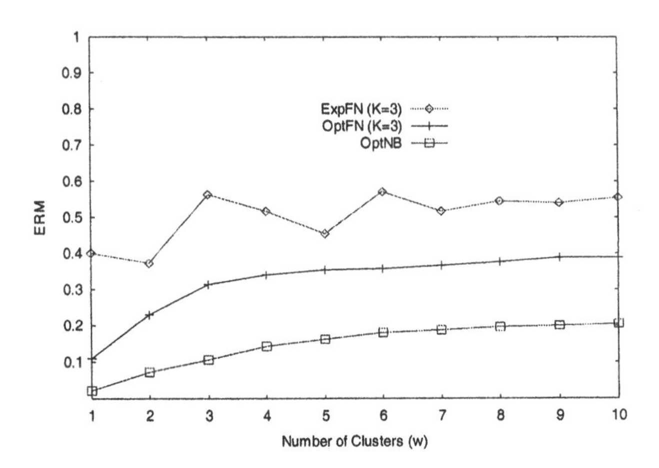

Figure 2: Session ERM for different bandwidth distributions. $N_s = 75$, $K = 3$ for both OptFN and ExpFN. Note that an ERM reflects the bandwidth mismatch, which is the lower the better.

replication schemes for dynamic simulcasting (OptFN and OptNB) are relatively small when there are only a small number of clusters (w). With the increase of w, their session ERMs also become larger. Note that w can be viewed as the *degree of heterogeneity* of the receivers: the higher the value of w, the more heterogeneous the session is, as the receivers' bandwidths are distributed in more clusters. Consequently, it is more difficult for the streams to match the demands of the receivers, especially when the number of streams is predetermined. On the contrary, because ExpFN is actually not aware of the distribution, its ERM does not evidently increase with w, and, for all w, the performance of this nonadaptive is much lower than the two optimal schemes.

In the following studies, we use two distributions of $w = 3$ and $w = 6$ as representatives (see Figure 3 for their instances).

7.1.2 Effect of Session Bandwidth

In this set of experiments, we study the effect of the bandwidth allocated to a session. Figure 4 shows the session ERM as a function of the session bandwidth under different settings. It is clear that both optimal allocation schemes significantly outperform ExpFN; at a medium to high bandwidth, the improvement of ERM is often over 0.2. For example, in Figure 4(a), the ERM of OptNB is reduced to 0.15 with a medium

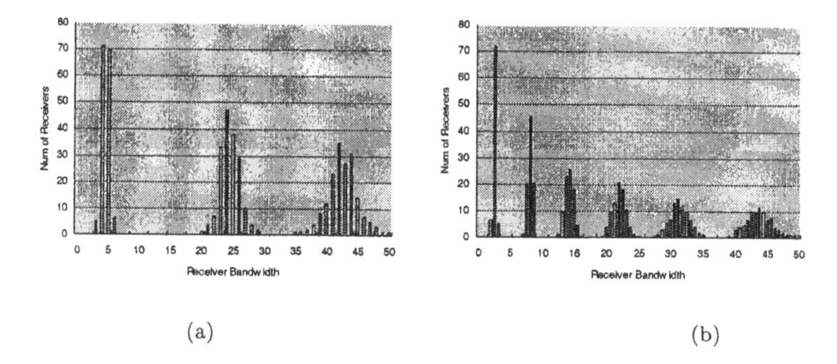

(a) (b)

Figure 3: Bandwidth distribution of the receivers in a session. The total number of receivers is 500, uniformly distributed in the clusters, i.e., each cluster has around $500/w$ reeivers. (a) $w = 3$; (b) $w = 6$.

session bandwidth (75). The ERM of ExpFN, however, remains higher than 0.5, which translates into an average bandwidth utility under 50%.

An interesting phenomenon of ExpFN is that its performance is not necessarily improved by allocating more bandwidth to the session. In Figure 4(b), though the ERM of ExpFN at the session bandwidth of 60 is lower than that at 40, it is noticeably higher than that at only 50. In Figures 4(a) and (c), the performance is even worse for all the bandwidths greater than 50. This is because the receivers' bandwidth distribution is not taken into account in this allocation scheme, and hence some unreasonable stream bandwidth settings could occur. To the contrary, as shown in Figure 4, the ERM of OptFN or OptNB is nonincreasing with the increase of the session bandwidth. This can also be formally proved from recurrence relations (3) and (4). Considering the performance gap and the unpredictable behavior of the nonoptimal scheme, we believe that the optimal replication effectively complements the expansion of session bandwidth.

The number of streams also influences the performance for the replication schemes. In Figure 4(b), the ERM of OptFN is close to that of the joint optimization scheme (OptBN) for bandwidths between 30 and 65, which basically means that 5-stream is the best choice in this interval. However, with lower or higher session bandwidths, it is no longer optimal, and the gaps are about 0.5 or more at some points. In Figure 4 (a), OptFN is close to OptNB only for session bandwidths between 20 and 25, and the gaps are usually more than 0.15 for other bandwidths. Moreover, the ERM of OptFN becomes flat for bandwidths over 52, be-

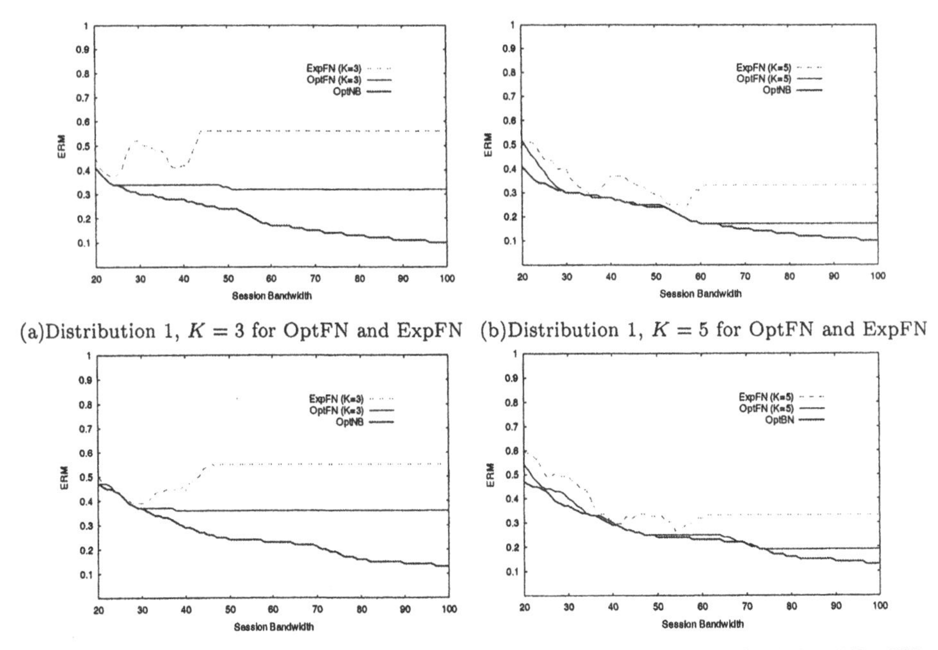

(a)Distribution 1, $K = 3$ for OptFN and ExpFN (b)Distribution 1, $K = 5$ for OptFN and ExpFN

(c)Distribution 2, K=3 for OptFN and ExpFN (d)Distribution 2, K=5 for OptFN and ExpFN

Figure 4: ERM as a function of session bandwidth for different allocation schemes.

cause the optimal allocation for this 3-stream setting has been reached. As illustrated in Figure 1, this allocation corresponds to the optimal allocation for cumulative layered multicast. However, OptNB can generate more streams to use the extra bandwidth and hence further reduce the bandwidth mismatch. As a result, with large session bandwidths (> 95), the ERM of OptNB is reduced to less than 0.1, which is much lower than that of OptFN. Similar observations can be made from Figures 4 (c) and (d) as well.

7.1.3 Impact of the Number of Streams

To further study the impact of the number of streams (K) used in OptFN and ExpFN, we let K vary from 1 to 9. Figure 5 shows the results when bandwidths 55 and 75 are allocated to the session, respectively. It can be seen that the variations of the ERMs with different numbers of streams are as large as 0.3 (excluding the single stream case, $K = 1$) for both OptFN and ExpFN. For OptFN, obviously there is an optimal setting

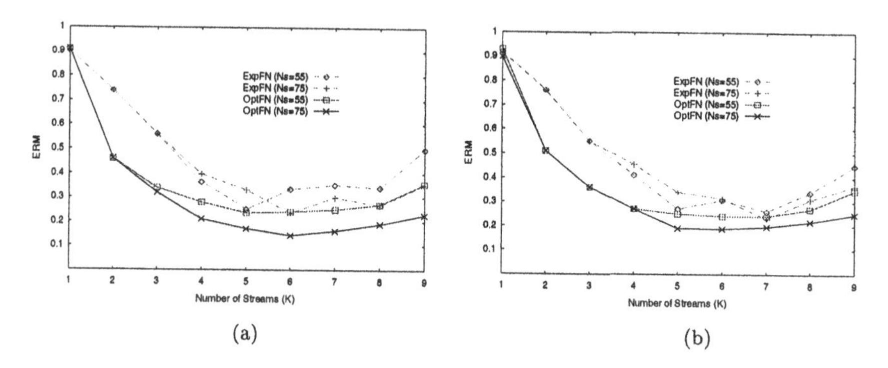

Figure 5: ERM as a function of the number of streams for OptFN and ExpFN. (a) Distribution 1. (b) Distribution 2.

for K, at which the session ERM is minimized. Intuitively speaking, if K is very small, the receivers' choice is limited and the adaptation is not flexible; an extreme case is $K = 1$, the single-rate transmission, where all the receivers have to subscribe to this single stream. On the other hand, if K is large, the redundancy of replication will contradict the improvements given by the flexible choices.

The exact optimal setting of K can be found by the algorithm for OptBN. Note that this optimal setting in Figure 5(a) (distribution 1) is different from that in Figure 5(b) (distribution 2) for the same session bandwidth. Moreover, as shown in Figure 4, when the session bandwidths are different, the optimal settings are also different, even for the same distribution. In other words, there is no universal choice for the optimal number of streams, suggesting a dynamic setting.

For the ExpFN allocation, the ERM is also reduced by using a proper number of streams, but remains much higher than OptFN for most settings.

7.1.4 Perceived Video Quality

Because our target application is video distribution, we also examine the video quality achieved by different replication schemes. We use the standard MPEG-4 video encoder with TM-5 rate control to generate replicated video streams at different rates. The average video quality of all the receivers for a standard test sequence Foreman (CIF) is presented in Figure 6, where the quality is measured by the Peak Signal-to-Noise Ratio (PSNR) of the Y channel [1]. It can be seen that the optimal

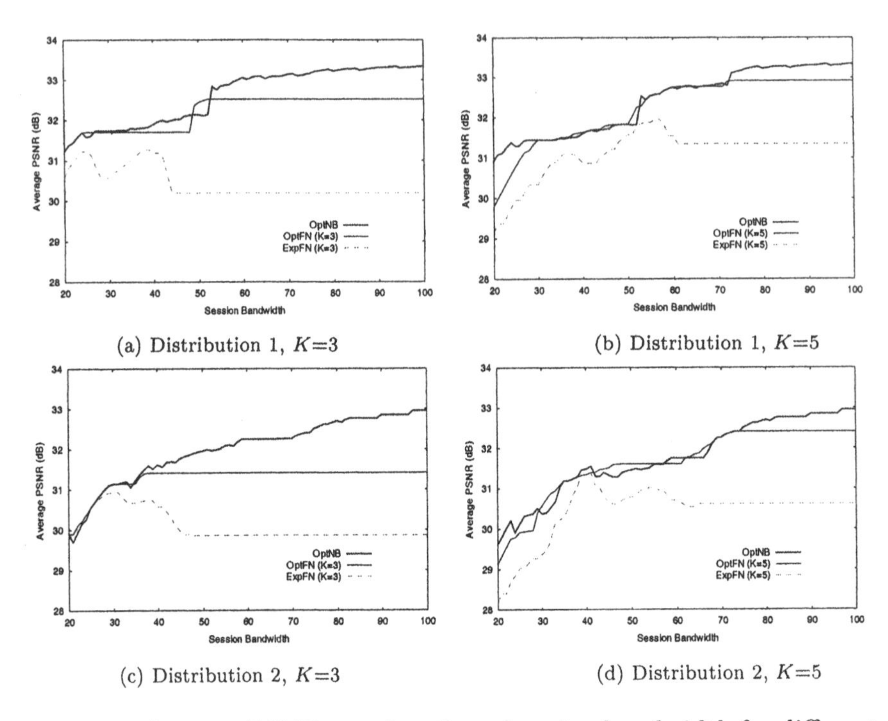

(a) Distribution 1, K=3

(b) Distribution 1, K=5

(c) Distribution 2, K=3

(d) Distribution 2, K=5

Figure 6: Average PSNR as a function of session bandwidth for different allocation schemes. Each point is calculated at a certain session bandwidth, and those for the same allocation scheme are then connected with a line.

replication algorithms for dynamic simulcasting generally improve the perceptual video quality. With medium and high session bandwidths, the gaps between OptNB and OptFN are about 0.5 to 1 dB, and the gaps between OptNB and ExpFN are usually larger than 2 dB; both are noticeable from the video coding point of view. This is consistent with our observations on the relationship between ERM and session bandwidth in Section 7. Because PSNR often has a linear relationship with the transmission bandwidth, Figure 6 is not simply an inverse and rescaled version of Figure 4. In particular, for OptNB, although its ERM is nonincreasing with session bandwidth, we find that PSNR is not necessarily nondecreasing; see for example, Figure 6(d), session bandwidth 20 through 50.

7.1.5 Effect of Inter Session Bandwidth Allocation

Finally, we study the effects of inter session bandwidth allocation. We assume that the demand probabilities for different video programs follow a Zipf distribution with a skew factor of 0.271, as suggested by movie rental statistics [22, 23]. The number of clusters for each session is uniformly distributed in between 2 and 7. In the experiments, we assume that there are 2500 receivers belonging to 15 sessions, and draw 2500 samples from the above model to obtain a receivers' status distribution for the whole system.

The performances of different combinations of the intra- and inter-session allocation schemes are compared in Figure 7. The results are consistent with our previous observations in intra session allocation; that is, OptNB generally outperforms OptFN if the same inter session allocation scheme is employed. However, the impact of different inter session allocation schemes is also non-negligible. It can be seen that, the ERM-Aware Allocation (EAA) consistently outperforms the Equal Share Allocation (ESA), both with OptNB and with OptFN. At low or medium bandwidths, the performance gaps can be larger than 0.1. Although the contribution of EAA is not so significant as that of intra session optimization, in view of its relatively low computation overhead, we believe that it is still worth consideration in practice. More interestingly, for bandwidth around 500, the ERM of OptFN plus EAA is quite close to that of OptNB plus EAA, and is better than that of OptNB plus ESA. This is because the preset number of streams (5 in our study) is likely to be the optimal choice for medium bandwidths (see Figure 4), and the choice of inter session allocation thus has more influence on the ERM. To conclude, EAA is particularly suitable for the cases where the stream number is fixed and the bandwidth resource is relatively scarce.

7.2 Simulation Results

In this set of experiments, we simulate the simulcasting algorithms using the LBNL network simulator *ns-2* [35]. To be compatible with the current Internet where TCP is the dominant traffic, we advocate the TCP-friendly bandwidth adaptation paradigm in this simulation [34]. We stress however that our optimal stream replication algorithms can be used with other adaptation paradigms. In the TCP-friendly paradigm, the expected bandwidth of a video receiver is estimated as the long-term throughput of a TCP connection as if the connection were running over

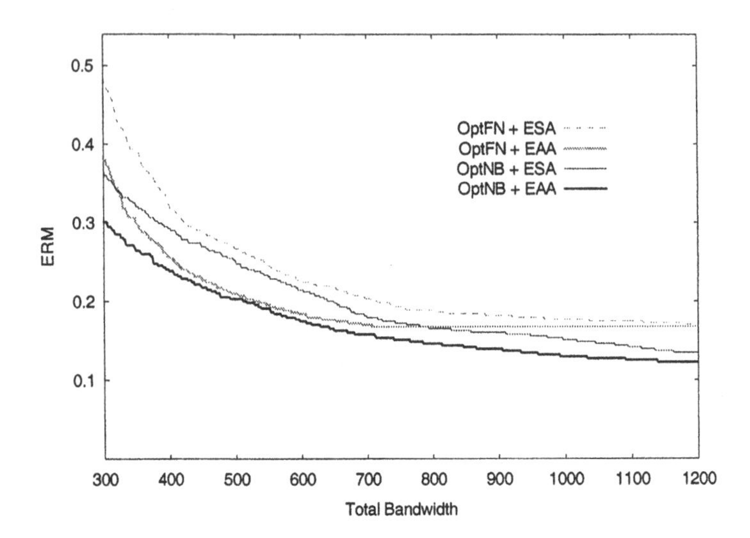

Figure 7: ERM as a function of the total bandwidth for all the sessions for different combinations of the intra session and inter session allocation schemes.

the same path. A receiver thus can control the subscription bandwidth to avoid starving the background TCP flows and, meanwhile, try to achieve a fair share with them. We assume that the sender performs reallocation every 10 seconds, which is also the adaptation period for each receiver. In addition, when severe congestion occurs (over 10% packet loss in our simulation), a receiver can instantly switch to a stream of lower rate.

Regarding the estimation of equivalent TCP throughput, there have been many efforts in the literature [34]; a general conclusion is that an estimation model relies on Round-Trip Time (RTT), packet size, and loss event rate. The latter two can be easily estimated on the receiver's side; the estimation of RTT, however, involves a feedback loop between the sender and the receiver. Because the topologies for our simulation are relatively small, we assume that each receiver reports its expected bandwidth to the sender every two seconds. Such report also serves a request for RTT estimation. In a large multicast network, to avoid the well-known feedback implosion problem, some feedback mergers can also be deployed inside the network [36].

Two typical network topologies are used in our simulation: the NSFnet and the China Telecom Network (CTnet), as depicted in Figure 8. In each topology, a node represents a FIFO drop-tail router with a queue

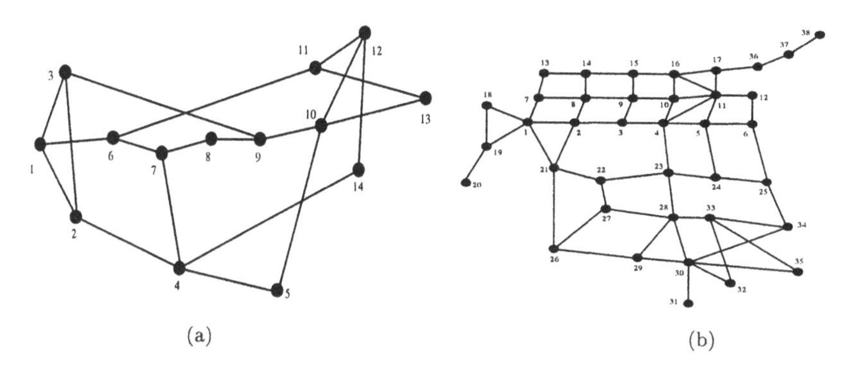

Figure 8: Simulation topologies: (a) 14-node NSFnet network; (b) 38-node CTnet network.

size of 25 packets, and each edge represents a link of bandwidth 2 Mbps. Given a topology, we use the following method to produce an instance for simulation:

Placement of end-nodes: A video server is attached to a randomly selected node, and an average of 5 video receivers is attached to each of the remaining nodes.

Cross-traffic: TCP Reno connections are randomly placed between node pairs (except for the node attached to the server), such that there are on average 3 TCP connections running over a link, and the minimum and maximum numbers are 0 and 9, respectively. As a result, the expected available bandwidth for a link is on average 500 Kbps; yet the minimum and maximum are 200 Kbps and 2 Mbps, respectively. The packet size is 500 bytes for both TCP and video traffic.

To mitigate the effect of randomness, we generated 10 instances for each topology using the above method. All instances were simulated for 1000 seconds, which is long enough for observing steady-state behaviors. We sample an RM (Relative Mismatch) value for each receiver every 5 seconds. Figure 9 shows the cumulative distributions of all sampled RM values for the three replication algorithms. We can see that for our optimal stream replication algorithms, more receivers have an RM value close to 0, the optimal value, and the probability of the RMs greater than 0.30 is relatively small (the probability is less than 0.2 for OptNB, and 0.25 for OptFN). On the contrary, for ExpNB, about 30% of the receivers would experience an RM that is even higher than 0.5, i.e., a bandwidth mismatch more than half of one's expected bandwidth. Consequently, the ERM values for the NSFnet are 0.21 (OptNB), 0.27

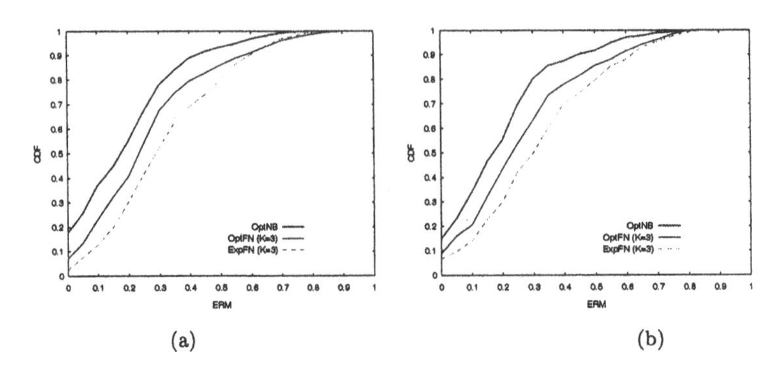

Figure 9: Cumulative distribution function (CDF) of the RM samples. (a) 14-node NSFnet; (b) 38-node CTnet.

(OptFN), and 0.34 (ExpFN), and that for the CTnet are 0.24 (OptNB), 0.29 (OptFN), and 0.37 (ExpFN). Such performance gaps are consistent with our observations in the numerical studies. They also imply that, under the TCP-friendly adaptation paradigm, a simulcast system employing OptFN is fairer in bandwidth sharing than that employing ExpFN, and the use of OptNB can further improve fairness.

8 Conclusions and Future Work

This chapter presented a formal study of the problem of adaptive stream replication for video simulcasting. Our main objective is to minimize the expected mismatch for all the receivers in a session. We formulated the optimization problems for stream bandwidth allocations, both with a flexible number of streams and with a fixed number of streams. We then presented efficient solutions, which led to the design of a novel simulcasting paradigm, called *dynamic simulcasting*. We also discussed some of the key implementation issues for dynamic simulcasting, including the computation overhead and the choice of video encoders. Its performance was studied under a variety of configurations. The results conclusively demonstrated that the optimal replication schemes in dynamic simulcasting can significantly improve the overall system performance in terms of both bandwidth utilization and perceptual video quality.

The algorithms we discussed and analyzed in this chapter involve end-to-end adaptation only, which is actually independent of the underlying physical medium. Therefore, it is applicable to diverse broadcast-

or multicast-capable networks. We are currently developing a comprehensive video simulcasting system using the optimal replication schemes. This is non-trivial from both theoretical and practical points of view, as it involves not only the optimization of individual components, but also the integration of the whole system. One of the important issues is stream switching. Given that a video stream generally has a complex syntax and its content is highly dependent, quality drift would occur after switching, and such drift could propagate to many following frames. There has been some preliminary work on seamless switching [32, 33]; however, it remains a difficult undertaking worth further investigation. Other issues include real-time video replication and dynamic stream creation or termination with low overheads. The impacts from channel errors and background traffic as well as the stability of the system are also possible avenues for further research.

References

[1] Y. Wang, J. Ostermann, and Y.-Q. Zhang, *Video Processing and Communications*, (Prentice-Hall, Upper Saddle River, NJ, September 2001).

[2] H. Schulzrinne, S. Casner, R. Frederick, and V. Jacobson, *RTP: A transport protocol for real-time applications*, (RFC 1889, January 1996).

[3] S. Cheung, M. H. Ammar, and X. Li, On the use of destination set grouping to improve fairness in multicast video distribution, in *Proceedings of IEEE INFOCOM'96.* (March 1996) pp. 553-560.

[4] T. Kim and M. H. Ammar, A comparison of layering and stream replication video multicast schemes, in *Proceedings of NOSSDAV'01.* (June 2001).

[5] S. Gorinsky and H. Vin, The utility of feedback in layered multicast congestion control, in *Proceedings of NOSSDAV'01.* (June 2001).

[6] L. Wu, R. Sharma, and B. Smith, Thinstreams: An architecture for multicast layered video, in *Proceedings of NOSSDAV'97.* (May 1997).

[7] T. Jiang, E. W. Zegura, and M. H. Ammar, Inter-receiver fair multicast communication over the Internet, in *Proceedings of NOSSDAV'99.* (June 1999).

[8] SureStream - Delivering superior quality and reliability, *RealNetworks White Paper.* (March 2002)

[9] S. McCanne, V. Jacobson, and M. Vetterli, Receiver-driven layered multicast, in *Proceedings of ACM SIGCOMM'96.*(August 1996) pp.117-130.

[10] D. Sisalem and A. Wolisz, MLDA: A TCP-friendly congestion control framework for heterogeneous multicast environments, in *Proceedings of IEEE/IFIP IWQoS'00.* (June 2000).

[11] Y. Yang, M. Kim, and S. Lam, Optimal partitioning of multicast receivers, in *Proceedings of IEEE ICNP'00.* (November 2000).

[12] J. Liu, B. Li, and Y.-Q. Zhang, A hybrid adaptation protocol for TCP-friendly layered multicast and its optimal rate allocation, in *Proceedings of IEEE INFOCOM'02.* (June 2002).

[13] K. Kar, S. Sarkar, and L. Tassiulas, Optimization based rate control for multirate multicast sessions, in *Proceedings of IEEE INFOCOM'01.* (April 2001).

[14] Z. Fei, M. H. Ammar, and E. Zegura, Multicast server selection: Problems, complexity and solutions, *IEEE Journal on Selected Areas in Communication.* (2002).

[15] D. Titterington, A. Smith, and U. Makov, *Statistical Analysis of Finite Mixture Distributions*, (Wiley, New York, 1985).

[16] J. Youn, J. Xin, and M.-T. Sun, Fast video transcoding architectures for networked multimedia applications, in *Proceedings of IEEE International Symposium of Circuits and Systems (ISCAS'00).* (May 2000).

[17] P. Kuhn, T. Suzuki, and A. Vetro, MPEG-7 transcoding hints for reduced complexity and improved quality, in *Proceeding of PacketVideo'01.* (April 2001).

[18] W. Li, Overview of the fine granularity scalability in MPEG-4 video standard, *IEEE Transactions on Circuits and Systems for Video Technology.* (March 2001) vol. 11, no. 3, pp. 301-317.

[19] H. Wang and M. Schwartz, Achieving bounded fairness for multicast and TCP traffic in the Internet, in *Proceedings of ACM SIGCOMM 98.* (September 1998).

[20] A. Legout, J. Nonnenmacher, and E. W. Biersack, Bandwidth allocation policies for unicast and multicast flows, *IEEE/ACM Transactions on Networking.* (August 2001) vol. 9, no. 4.

[21] J. Byers, M. Luby, and M. Mitzenmacher, Fine-grained layered multicast, in *Proceedings of IEEE INFOCOM'01.* (April 2001).

[22] G. Zipf, *Human Behavior and the Principle of Least Effort* , (Addison-Wesley, Reading, MA, 1949).

[23] A. Dan, D. Sitaram, and P. Shahabuddin, Scheduling policies for an on-demand video server with batching, in *Proceedings of ACM Multimedia'94.* (October 1994).

[24] A. Alan, *Introduction to Categorical Data Analysis*, (John Wiley and Sons, New York, 1996).

[25] T. H. Cormen, C. E. Leiserson, R. L. Rivest, and C. Stein, *Introduction to Algorithms*, (2nd Edition, MIT Press, Cambridge, MA, 2001).

[26] X. Li and M. H. Ammar, Bandwidth control for replicated-stream multicast video distribution, in *Proceedings of HPDC'96.* (August 1996).

[27] S. Banerjee and B. Bhattacharjee, A Comparative Study of Application Layer Multicast Protocols, *Technical Report*, Univeristy of Maryland, College Park. (2002).

[28] J. D. Lameillieure and S. Pallavicini, A comparative study of simulcast and hierarchical coding, *Technical Report*, (European RACE project, subgroup WG2, February 1996).

[29] P. de Cuetos, D. Saparilla, and K. W. Ross, Adaptive streaming of stored video in a TCP-friendly context: multiple versions or multiple layers, in *Proceedings of Packet Video Workshop.* (April 2001).

[30] F. Hartanto, J. Kangasharju, M. Reisslein, and K. W. Ross, Caching video objects: Layers vs versions?, in *Proceedings of IEEE International Conference on Multimedia and Expo (ICME'02).* (August 2002).

[31] J. Liu, B. Li, and Y.-Q. Zhang, Adaptive Video Multicast over the Internet, *IEEE Multimedia.* (January/February 2003) vol. 10, no. 1, pp. 22-31.

[32] J. Macher and G. Anderson, Multi-program transport stream switching, White Paper. *Thales Broadcast & Multimedia.*

[33] Y.-K. Chou, L.-C. Jian, and C.-W. Lin, MPEG-4 video streaming with drift-compensated bit-stream switching, in *Proceedings of IEEE Pacific-Rim Conference on Multimedia (PCM'02).* (December 2002).

[34] S. Floyd and K. Fall, Promoting the use of end-to-end congestion control in the Internet, *IEEE/ACM Transactions on Networking.* (August 1999) vol. 1, no. 4, pp. 458-471.

[35] S. McCanne and S. Floyd, *The LBNL Network Simulator, ns-2.* Available at http://www.isi.edu/nsnam/ns/.

[36] B. Vickers, C. Albuquerque, and T. Suda, Source adaptive multi-layered multicast algorithms for real-time video distribution, *IEEE/ACM Transaction on Networking.* (December 2000) vol. 8, no. 6, pp. 720-733.

Chapter 20

Optimization of Failure Recovery in High-Speed Networks

Terence T. Y. Kam
Electrical and Computer Engineering
Indiana Univ. Purdue Univ. Indianapolis, Indiana, IN 46202
E-mail: tkam@iupui.edu

Dongsoo S. Kim
Electrical and Computer Engineering
Indiana Univ. Purdue Univ. Indianapolis, Indiana, IN 46202
E-mail: dskim@iupui.edu

1 Introduction

In broadband digital networks, high reliability is imperative. A short service outage such as a link failure or a node failure can be significantly disruptive to network services. This is due to high traffic volume carried by a single link or node. The problem is compounded when end-users try to re-establish connection simultaneously within a short period of time, in that the retrials increase network traffic and further deplete network resources. The fast restoration of a network in the event of failure thus becomes a critical issue in deploying high-speed networks. Self-healing algorithms have been widely recognized as a key mechanism for implementing fast restoration, with the primary objective of self-healing networks being to optimize networks while anticipating reasonable failure coverage.

A shortest-path rerouting method has been used in many telecommunications and data networks for implementing self-healing capability. The shortest-path rerouting method may produce fast network restoration but the shortest path does not necessarily yield an optimal network in terms of flow or bandwidth assignments. This chapter explores an optimal source-based rerouting method to implement highly reliable self-healing virtual circuit networks. The topology of networks is assumed to be arbitrary and a pair of disjoint paths is provided between two termination nodes of each traffic demand. High reliability is achieved by having a pre-established protection path in addition to a working path between two termination nodes. The demand traffic is carried on the working path by default. In the event that failure is detected in any network component along the working path, one of the termination nodes switches to the protection path, enabling traffic to bypass the faulty working path.

One of two switching methods can be used to reroute traffic from working path to protection path in the event of failure, namely one-and-one switching or one-to-one switching. In the one-and-one switching, a source node of a demand transmits the traffic through its working and protection paths simultaneously. The one-to-one switching on the other hand routes traffic only through either the working or protection path, but not both. This method is deemed to consume less resources than the one-and-one switching. Both switching techniques utilize path restoration as opposed to span restoration. Path restoration is able to distribute the affected traffic all over the network and relieve congestion around the failed component.

The goal of network planning is to minimize the cost of network operation and to provide reliable services to users within a budget plan. Network cost between a pair of source and destination nodes can be minimized by selecting the shortest path. However, the shortest path does not necessarily ensure an optimal network when the capacity of physical links and other characteristics related to quality-of-service of network requests are taken into consideration. In this chapter, a single fault is assumed within the network at a given time. Faults are prioritized based on failure cost (FC), which is defined as the product of the total working flow of a particular link and its failure probability. This indicates the significance of the failure on this link. The network is then optimized by considering the largest FC. For a given network, the communication cost (T) is expressed as a function of the link capacity matrix (C) and the requirement vector (R) subjected to the multicommodity flow (F). The optimization problem then involves minimizing T which in turn minimizes the average source-to-destination delay, producing an optimized network.

A protection path has to be established for each traffic demand. This path must be node-disjoint from the working path except at the termination nodes. This is to ensure that a link failure will not affect both the working and protection paths at the same time. To search for protection paths, the k-successive disjoint path (KSP) algorithm obtains a set of disjoint protection paths for each demand. The shortcoming of the KSP method is that it fails to generate a reasonable number of candidate paths when a trap topology is encountered in the network traffic. Furthermore, the number of paths generated by this method is restricted by the degree of termination nodes. To overcome these problems, the m-ordered shortest-path (MSP) algorithm is proposed. This method enumerates all possible paths between a source-destination demand pair in nondecreasing order of network cost and takes the first m paths as candidate paths. This method generates more paths for the next optimization process and ensures that the number of possible paths is not restrained by the degree of the termination nodes. With the set of possible paths generated, a protection path is selected from the set by optimizing the network with a minimum T. Three optimization methods are examined. A global optimization checks every combination of protection paths for all traffic demands. Although this method guarantees a globally optimum network, the complexity of the algorithm makes it impractical for complicated networks. To work around this, a simplex method, which produces only a locally optimum network, can be used. This method, however, does not generate a near-optimal solution. The last optimization algorithm is a random simplex method. This method goes beyond local optimization in that it carries the search to a next region until a subsequent number of regions fail to produce a new optimal network. The random simplex method generates a solution closer to the global optimum in relatively smaller time complexity.

2 Switching Concepts and Methodology

To reroute traffic during a failure in the network, switching and failure detection systems are required. Network traffic between two end nodes is carried through a default working path. When a failure occurs in one of the components along the working path, the network has to be able to

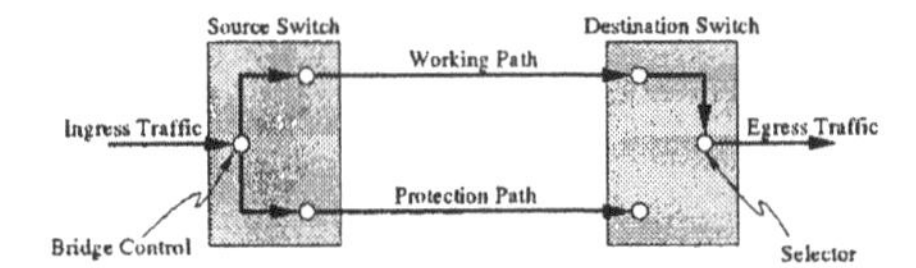

(a) One-and-one protection switching. In normal operation, the source switch duplicates ingress traffic through the working and protection paths. The destination selects traffic from the working path but ignores the protection path.

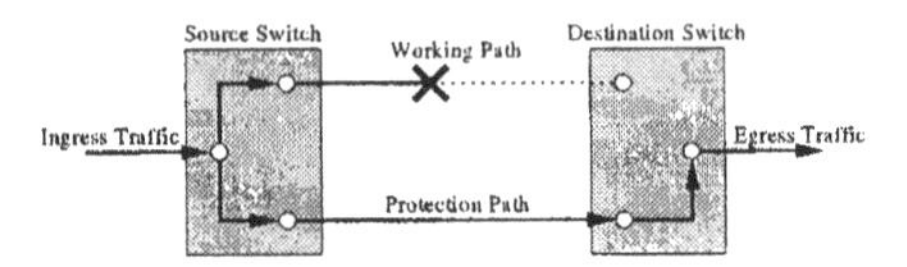

(b) One-and-one protection switching. In protection mode, the destination switch selects traffic from the protection path, ignoring the working path.

Figure 1: Operation of one-and-one protection switching.

detect the fault that is notified to the affected nodes of the fault occurrence. The affected nodes must then be able to switch the traffic flow, redirecting traffic to the predefined protection path. Efficient failure detection and notification as well as traffic switching are important in high-speed self-healing networks to minimize the ripple effect of network faults. A single link failure can send a ripple of failures through the network if there is a broadcast of failure messages from all nodes across the network and end users try to re-establish communication simultaneously. This will result in the failure of other nodes due to a surge in network load. Effective mechanisms ensure that the impact of rerouting is contained within a reasonable area around the affected nodes. This section expands the concepts of protection switching and explores the failure detection and notification methodologies for implementing self-healing networks based on the optimal rerouting scheme.

Switching Methods. Two methods of switching for self-healing networks are widely accepted: one-and-one protection switching and one-to-one protection switching. In switching, the source and the destination nodes, i.e., the end nodes, play pivotal roles in redirecting network traffic. The nodes can be assumed to be switches with controllers. Every node in the network must have the ability to signal other nodes in the event of failure.

In one-and-one protection switching, the source node duplicates the ingress traffic, i.e., data to be transmitted, and sends them to the destination through both the working and protection paths simultaneously.

The destination node selects data from the working path and ignores data from the protection path under normal operation. This can be seen in Figure 1(a). Note that the working and protection paths may consist of multiple intermediate nodes besides the source and destination nodes. When a fault occurs on one of the intermediate nodes or links, a node adjacent to the fault sends an alarm message to the destination node downstream using the portion of the working path that is not affected by the failure. Upon receiving the alarm message, the destination node switches to select the duplicated data from the protection path as shown in Figure 1(b). It is thus able to receive the data immediately without having to depend on the source node to detect failure and retransmit traffic. The destination node can subsequently send a management information message to the source node, notifying it of the failure. The one-and-one protection switching provides for fast rerouting but consumes more bandwidth because of the redundant network traffic sent through the protection path.

The one-to-one protection switching on the other hand is a switching mechanism that consumes fewer resources than the one-and-one protection switching. In the protection switching, the source node forwards ingress traffic only through the working path under normal operation as shown in Figure 2(a). The destination node is able to recognize a failure when the fault occurs in any intermediate nodes or links along the working path. This is due to the disruption of received data on the destination node, or an alarm message sent by an intermediate detecting node. The destination node then sends a switching command to the source node through the protection path, requesting data to be transmitted through the protection path. The one-to-one switching consumes fewer network resources because no duplicate traffic exists. However, rerouting may not be comparatively as fast as the one-and-one switching methods because of information exchange between source and destination nodes when failure occurs.

Failure Detection and Notification. Failure can occur on any nodes or links and can be detected by any node as discussed previously. Nodes should also be able to send alarm signals when a fault occurs. Recovery can take place over a span of a number of affected links or over the whole path between source and destination nodes.

A span restoration provides a logical link between the nodes directly adjacent to the failed link or node so that all traffic on the failed link

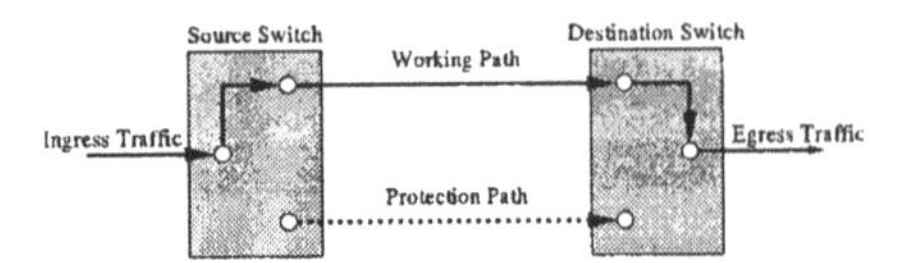

(a) One-to-one protection switching. In normal operation, the source switch forwards traffic only through the working path, but the protection path is ready.

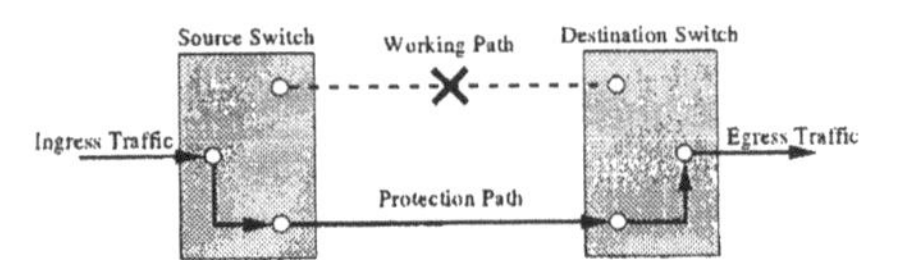

(b) One-to-one protection switching. In protection mode, the destination switch sends a signal to the source switch in order that the traffic be redirected through the protection path.

Figure 2: Operation of one-to-one protection switching.

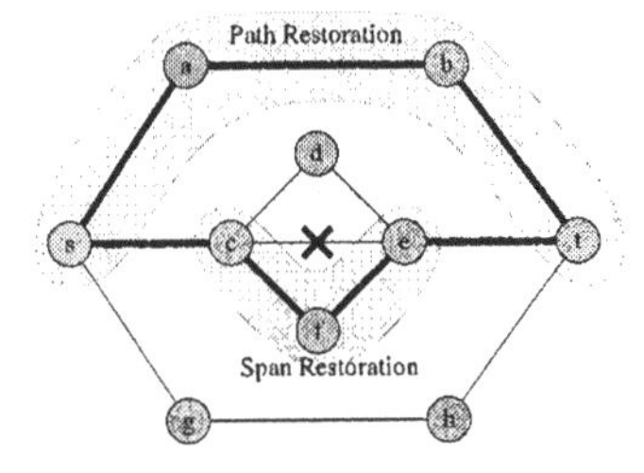

Figure 3: Comparison between span restoration and path restoration. Path restoration completely reroutes traffic from source to destination whereas span restoration only reroutes traffic to bypass the faulty link.

is rerouted on the logical link. In contrast, path restoration reroutes each traffic demand from its originating source to its final destination. Figure 3 illustrates the difference of two restoration mechanisms for a path between nodes s and t, where the working path before the fault on link (c-e) was the path (s-c-e-t). The span restoration is said to be simpler and faster than optimal path restoration, although it may result in network congestion on the restored logical link. A path restoration is able to distribute the affected demand traffic all over the network to avoid the concentration of traffic load around the failed component. Due to the nature of the path restoration, a node that detects the failure has to send an alarm signal only to the termination nodes affected by the failure. Otherwise, a failure can result in an explosion of broadcast messages, severely increasing the network load.

To avoid broadcast messages, the detecting node can send alarm messages only to the downstream nodes using subpaths that are not

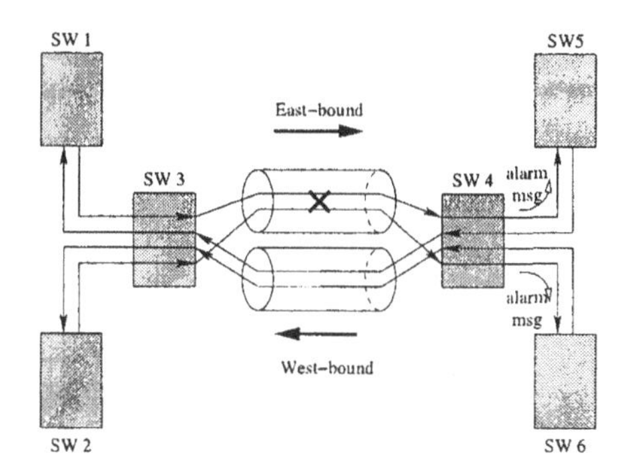

Figure 4: Failure detection and propagation.

affected by the failure. Figure 4 illustrates the failure notification method that eliminates the use of broadcast messages. Two virtual circuits are configured as (1-3-4-5) and (2-3-4-6). A failure on a directional link (3-4) is first detected by node 4. Upon detecting the failure location, node 4 sends individual alarm messages to the downstream termination nodes 5 and 6 using subpaths (4-5) and (4-6) that are not affected by the failure. The termination nodes can then initiate the recovery procedure by switching demand traffic from the current failed working path to their pre-established protection paths. A significant improvement and conservation of network resources, this scheme sends failure messages only to the termination nodes whose paths are known to pass the failed link.

The path restoration is used in the following optimal source-based rerouting scheme. The working path and the protection path in the scheme are node-disjoint to ensure that any given link failure will not affect both working and protection paths at the same time, and recovery occurs from the transmitting source to the terminating destination.

3 Fault Recovery

As was discussed previously, though traffic for a demand is carried on the working path under normal network operations, a protection path is established in addition to the working path. When a failure occurs in the network, traffic is immediately rerouted to the protection. The failure can appear in different forms: a link failure where connection between two switches breaks down, or a node failure where a switch malfunctions. Note that a node failure affects all links that are connected

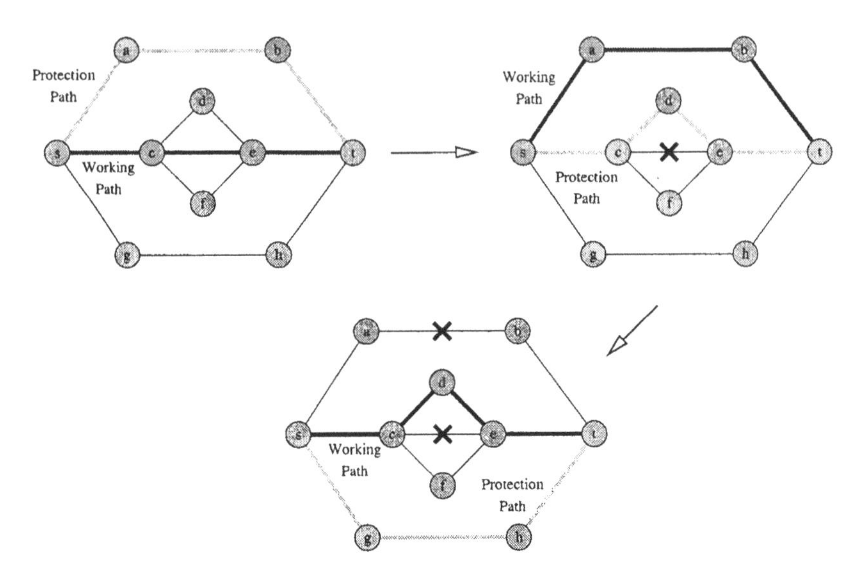

Figure 5: A look at fault recovery for two successive failures.

to the node. The choice of working and protection path for a particular demand plays an important role in determining the efficiency of fault recovery. Selection of working and protection path is not based only on a single traffic demand between two nodes, rather all demands in the network have to be considered to optimize network resources.

Multiple Fault Recovery. Up till this point, the discussion has assumed a single fault in the network, although very often multiple faults can occur consecutively. Fault recovery therefore has to take into account more than one failure in the network, affecting both the working and protection path in sequence. The recovery from multiple faults is accomplished by ensuring that a pair of working and protection paths exists at all times for a traffic demand using a regressive algorithm until all possible paths are exhausted between the source and destination nodes.

When a failure occurs in the working path, traffic is rerouted to the protection path; the protection path thus becomes the working path at this point. In order to have two disjoint paths, a new protection path that is fault-free has to be re-established for the demand. Thus, should a successive fault occur in the network and affect the working path, traffic can again be rerouted to the protection path. The process repeats for

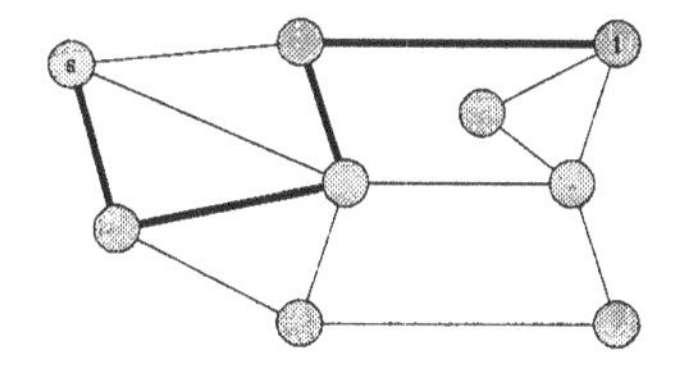

Figure 6: An example of a trap topology in a network.

subsequent failures until no disjoint paths can be further established. At this stage, the network is unable to reroute the demand and the traffic fails. Figure 5 depicts the network recovery for two successive failures.

Trap Topology. Because working and protection paths have to be completely disjoint except at the termination nodes, the selection of the paths is vital in determining the ability of the rerouting algorithm to generate more disjoint paths. This gives rise to a phenomenon known as a trap topology. The trap topology is essentially the selection of a path between two nodes that separates the part of the network between the nodes into two regions. An example of trap topology is shown in Figure 6. Trap topology is a significant factor in determining the effectiveness of fault recovery in a network.

The presence of a trap topology in a network results in fewer disjoint paths between two termination nodes that can be found. If the working path had been chosen as shown in Figure 6, it would be impossible to find a node-disjoint path between the source and destination nodes. A trap topology therefore will significantly deteriorate a network's ability to recover from faults. The choice of an initial working path is important in ensuring the network is not separated into two parts where a trap results.

When a trap topology occurs, the possible choices for protection paths are drastically reduced and in most cases, the path searching algorithm is unable to generate protection paths, causing an unrecoverable network failure. To avoid a trap, the rerouting scheme has to determine an initial working path selectively. The algorithm is discussed in a later section. By realizing the existence of trap topology and taking steps to eliminate it, the performance of the rerouting scheme can be improved and the network able to withstand more faults before breaking down.

4 Network Planning

The goal of network planning is to minimize a network cost, to increase the efficiency of network resources, and to provide reliable services to customers within reasonable scope of a budget plan. For example, in a virtual private network such as an ATM network, a network planner arranges permanent virtual paths (PVP) for satisfying source destination traffic demands. Network cost can often be minimized by finding the shortest path between every source destination pair, using a path search algorithm such as Dijsktra's algorithm. However, the shortest paths do not usually guarantee an optimal network when factors such as the maximum capacity of physical links and other network characteristics related to quality-of-service are considered. The entire traffic demands must be considered as a whole.

The problem of allocating optimal resources to network demands was extensively studied in [6, 12, 17, 19], and the optimization of self-healing techniques on connection-oriented networks have been exploited on dynamic reconfiguration in response to a change in traffic demands caused by a network failure [16]. Reconfiguration must take place within a reasonable amount of time. In high-speed networks operating in the giga-bps to tera-bps range such as the dense wavelength division multiplexing technology, a network failure, even in a brief time span, is not acceptable. Restoration has to be quick and transparent to end-users, in addition to keeping network cost at the minimum.

Failure Cost. To accomplish fault recovery within a reasonable scope, a single fault in the network components is assumed at one time, although multiple faults can occur consecutively. For any given network, each link in the network is described by the traffic it carries, its capacity, and its failure probability. To optimize network resources, such as residual capacity and the address translation space, in the event of network failure, the faults are prioritized based on failure cost (FC). For a given link, let f_i denote the total working flow on the link i, and p_i be its failure probability. The failure cost on the link i, ρ_i, is then defined as $\rho_i = p_i f_i$.

The failure cost reflects the effective significance of a link failure. By calculating the FC, the severity of failure in each link can be determined. Intuitively, the failure cost increases with increasing traffic flow or load of a link. This is because when a link carries more traffic, it becomes more utilized and failure in that link will become critical to the network. By determining failure costs, the network can be optimized by considering the largest failure cost as the highest priority.

Optimality of Network Traffic. For each network, the capacities of the links are described as a matrix known as the link capacity matrix $\mathcal{C}$. Similarly, the traffic demands in the network are defined as a requirement vector $\mathcal{R}$. For a given network topology with a link capacity matrix $\mathcal{C} = [C_{ab}]$ and requirement vector $\mathcal{R}$, the rerouting problem (minimizing network cost) is expressed to minimize the average source-to-destination delay by minimizing the communication cost T, defined as

$$\text{Minimize } T = \sum D(C,F) = \sum_{\text{all } a,b} \frac{f_{ab}}{C_{ab} - f_{ab}} + \Delta f_{ab},$$

subjected to F is a multicommodity flow
satisfying the requirement vector $\mathcal{R}$, and
$F \leq C$

where Δ is a very small constant. Many studies have considered $D(C, F)$ as a continuous function due to its computational complexity and the optimal flow algorithms assumed that the adaptation to a discrete function is applied in the refinement stage [6]. The restriction works for an initial optimal flow assignment or dynamic self-healing networks [16] that utilize a signaling protocol for switched virtual circuits. It is, however, difficult to apply the algorithm for incremental network optimization and permanent virtual circuit restoration. In such a case a discrete function is assumed. The problem then lies in developing an algorithm to compute the function in a practical amount of time.

5 Optimal Protection Rerouting Scheme

The pair of paths are node-disjoint except at the termination nodes as was previously discussed. Thus, the protection path can restore not only link failure but also node failure along the working path as long as the failed node is not either one of the termination nodes. The challenge then is to devise an algorithm that is able to return as many candidate paths as possible between two nodes and then selecting one that would produce an optimal network when all other traffic demands are considered. For path search, common path search algorithms such as Dijsktra's algorithm can be used. The path search algorithm can be run consecutively to return a pool of possible paths from which one is to be chosen.

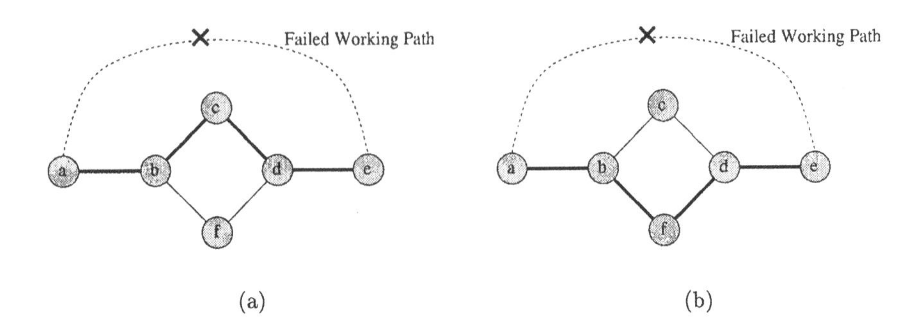

Figure 7: The first-order trap topology.

k-Successively Shortest-Paths Method (KSP). Traditionally, the protection path (as well as the working path) for a demand in the network can be found as one of k-successive disjoint shortest paths (KSP). The KSP method returns a set of paths, known as candidate paths, where any pair of paths in the set is link-disjoint from each other. From the candidate paths, the optimal rerouting scheme selects a path that will constitute an optimal network. Because all candidate paths have to be disjoint to one another, the KSP method only returns a few paths. This is especially the case in trap topology, discussed in an earlier section. In addition, the number of candidate paths generated by the KSP method cannot exceed the degree of the termination nodes. This poses a problem as the degree in most nodes in practical communication networks is small.

Figures 7 and 8 illustrate the problem with the KSP method in trap topology. In the first-order trap topology, there is no cycle containing two termination nodes so that the KSP method considers only one candidate path on which high flows can be assigned as in Figure 7(a). KSP only selects either path (a-b-c-d-e) or (a-b-f-d-e), not both because both paths are not link-disjoint. The second-order trap topology occurs when the first candidate path partitions the network into two separate networks. In Figure 8(a), there is no other candidate path if the KSP method chooses the path (a-b-c-d) as the shortest path, although two other paths, (a-e-c-d) and (a-b-f-d), can be candidate paths.

As can be seen, the KSP method presents a shortcoming. It is not able to return the appropriate number of candidate paths even though such paths exist in the network. KSP is also unable to generate the number of candidate paths that is more than the degree on the termination nodes. As such a more robust and flexible method has to be used.

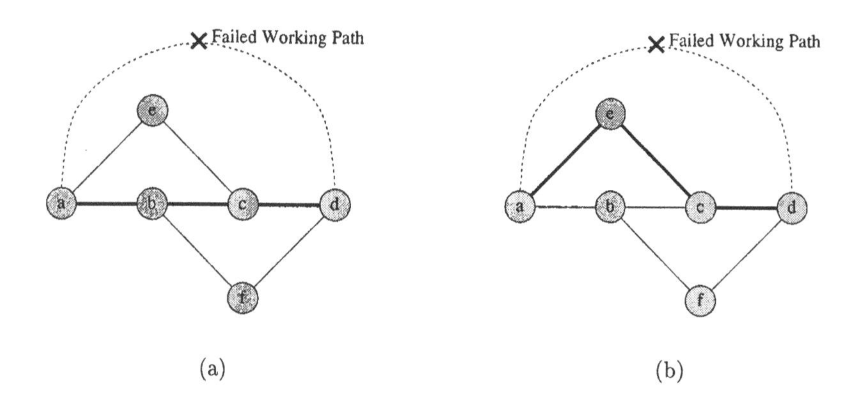

Figure 8: The second-order trap topology.

m-Ordered Shortest-Paths Method (MSP). To overcome the shortage of candidate paths generated by the KSP method, the m-ordered shortest-paths method (MSP) is proposed. The MSP method makes use of the observation that a candidate path does not necessarily have to be disjoint with other candidate paths. This is because the requirement is for the working path to be disjoint with the protection path. The set of candidate paths for the protection path does not have to be disjoint with the others as long as each candidate path is disjoint with the working path. By modifying the search algorithm to search consecutively and hiding a link in the network each time, the MSP scheme is able to list all possible paths between two nodes in nondecreasing order of network costs, with the paths not necessarily disjoint with each other, and take the first m paths as candidate paths.

Figures 7(b) and 8(b) illustrate how the MSP method is able to return more paths and work around the problem presented by trap topology on the same networks. By hiding a link on a candidate path and running the search algorithm again, a different path that is not disjoint can be returned. The process is repeated to generate all possible paths between two nodes. In Figure 7(b), the MSP method is able to return paths (a-b-f-d-e) in addition to path (a-b-c-d-e) which was the only path returned by KSP. Similarly, the MSP is able to find paths (a-b-c-d), (a-e-c-d), and (a-b-f-d) shown in Figure 8(b). The MSP method allows more candidate paths to be utilized in the optimization process and the number of candidate paths is not limited by the degree of termination nodes. The MSP algorithm is outlined in Algorithm 1.

Algorithm 1 List m shortest paths in nondecreasing order.

```
INPUT: source s, target t, integer m and digraph G
OUTPUT:path[m]
begin
  z is a tuple of cost c, path p and a set of excluded edges x
  Inverse priority queue Q of z
  Q := ∅;
  i := 0;
  [p, c] := G.find_min_path(s, t);
  Q.put(p, c, ∅);
  while i < m do
        z := Q.pop();
        path[i] = z.p;
        i := i + 1;
        Hide all edges in z.x from G;
        foreach e ∈ z.p do
          Hide e from G;
          [p, c] := G.find_min_path(s, t);
          Q.put(p, c, z.x ∪ e);
          Restore e ∈ G;
        od
  od
end
```

Optimal Rerouting. The selection of the protection path from a pool of candidate paths is based on minimizing the network costs, which signifies the end-to-end delay in the network. All traffic demands in the network at a particular time have to be considered as a whole.

Let R be a set of demands. For each demand $r = (s_r, d_r, \omega_r) \in R$, where s_r is a source node, d_r is a destination node, and ω_r is a bandwidth, the working path can be realized by using maximum flow optimization. The protection rerouting path, however, cannot be allocated by maximum flow optimization practically, because the bandwidth of a demand is fixed and all traffic flow of the demand should follow the same path in a real environment.

Let τ_r be an ordered set of candidate paths of demand r. x_r is a binary vector satisfying $\sum x_r(i) = 1$, meaning that the vector is a selector to choose one of the candidate paths. Then $\omega_r \tau_r x_r^T$ is the flow assignment of demand r. The optimal rerouting for protection switching on capacity matrix C and a set of demands R, ORP(C, R), is defined as the problem of finding a path for each demand, which minimizes the communication cost T formulated as follows

$$\text{Minimize } T = \sum_{\forall \text{links } (a,b)} \frac{f_{ab}}{C_{ab} - f_{ab}} + \Delta f_{ab},$$

$$\text{subject to } F = \sum_{\forall r \in R} \omega_r \tau_r x_r^T, \tag{1}$$

$$\sum x_r(i) = 1 \text{ for each demand } r, \text{ and} \tag{2}$$

$$F \leq C \tag{3}$$

where f_{ab} is the total flow on link (a, b). Constraint (1) addresses that the flow assignment for all demands be considered, or the flow assignment F satisfies the given demands R. Constraint (2) corresponds to one and only one protection path is assigned to each demand. This is the requirement of the rerouting scheme: only one protection path is required for each traffic demand in addition to the working path. Constraint (3) imposes that the flow assignment is not larger than the given capacity. This is intuitive and has to be set for real networks.

Algorithm 2 illustrates the global optimization of protection switching for a single fault at a time. While considering one link failure at a time, some demands may complete searching their protection paths on a previous link failure, so only demands that have not been reduced will be considered at the point, denoted by R^P. The algorithm also uses resources that are freed up, which are termed the residual capacity.

Suppose that a link failed; any demand whose working path passes the link will be rerouted on its disjoint protection path. The bandwidth allocated on the working path is no longer needed by the demand. However, in one-to-one protection switching, the bandwidth can be utilized for searching protection paths of other demands. The residual capacity of the network is augmented by the bandwidths of the demands if the failed link is on their working paths.

Algorithm 2 returns a protection path for each traffic demand using either the KSP or MSP method, and is optimized based on minimizing network cost. By reusing the bandwidth and resources that are freed up when traffic reroutes from working to protection, i.e., the residual capacity, network resources are fully utilized making the resulting network more efficient and cost-effective.

Algorithm 2 Rerouting algorithm for self-healing networks.

1. The residual capacity of the network after establishing all working paths in R is

$$C_R = C - \sum_{\forall r \in R} \omega_r \mu_r$$

where (μ_r is the working path of the demand $r = (s_r, d_r, \omega_r)$.

2. Let R^P be the subset of R that requires protection switching, and L be an ordered set of links on the failure cost.

3. Suppose that a link $l \in L$ with the largest failure cost failed and let $R_l \subseteq R$ be a subset of demands whose working paths μ_r contain the failed link l.

4. Revoke any demand $r \in R_l$ from the network as

$$C_l^R = C^R + \sum_{\forall r \in R_l} \omega_r \mu_r$$

5. Find a set of demands $R_l^P \in R^P$ if μ_r contains the failed link l for $r \in R_l^P$.

6. Search m candidate protection paths for each demand $r \in R_l^P$, say τ_r, and minimize the total cost T using ORP(C_l^R, R_l^P).

7. Update $R^P = R^P \setminus R_l^P$, and $L = L \setminus \{l\}$.

8. Repeat 3 unless R^P and L are empty.

Avoiding Trap Topology. In addition to ensuring an optimal network, the path search algorithm has to address the problem of trap topology, especially for selecting the working path. As was shown in Figure 6 previously, the choice of working path has to prevent the existence of trap topology. If the working path is selected in such a way that it creates a trap topology, no protection path can be established for the given traffic demand, even though such paths may exist in the network. This is due to the fact that the working and protection paths have to be disjoint from each other. The capabilities of a network to recover from a fault, especially numerous consecutive faults, therefore depend on the effectiveness of the path search algorithm to eliminate the occurrence of trap topology when choosing a path.

To prevent the selection of paths that will present the problem of trap topology, a check trap topology routine has to be added to the rerouting algorithm. When the rerouting algorithm returns a particular path, the check trap topology routine checks the path against the network. It hides all links in the path from the network and runs the path search algorithm again to check if another path can be found between the source and destination nodes, with the first path factored out. Failure to find a subsequent path will mean that the previous returned path causes a trap topology and is thus discarded. The link is then unhidden and the path search algorithm is run again to find other paths. The check trap topology routine examines all paths that are returned and only paths that do not result in trap topology in the network will be added to the set of candidate paths.

6 Optimization Algorithm

The path search algorithm produces a number of possible paths for each traffic demand; this set of paths is known as the candidate paths. Referring to Algorithm 2, when a link with the largest failure cost fails, the set of requests affected by the failure is obtained. The algorithm then uses either the MSP or KSP method to find a set of possible protection paths that are disjoint from the working path for each request.

Because each request has a set of possible protection paths, the algorithm needs to enlist optimization methods and find a total minimum cost for the whole network. Due to the fact that the number of requests can be very large for a practical network, such optimization algorithms can have huge time complexities and consume tremendous processing resources that make them impractical for implementation. Consider a group of n requests each with a maximum of k protection paths; the optimization algorithm will need to examine a total of k^n combinations of protection paths to obtain one that will result in minimum communication cost. An algorithm with this time complexity is almost impossible to implement for real-life networks where traffic demand is high.

It is valuable to examine the implementation of optimization algorithms that are used to find protection paths for a group of requests that will constitute a minimum communication cost T. In particular, this refers to the ORP(C_l^R, R_l^P) routine in Algorithm 2. For the purpose of discussion, the focus is on optimization of protection paths obtained from the MSP method because the MSP algorithm is free from the trap topologies and performs better than the KSP. Three optimization techniques are discussed, namely, the global optimization method, the simplex optimization method, and the random simplex optimization method.

Global Optimization. The first optimization method, which is the most straightforward, is global optimization. This algorithm examines every combination of protection paths to find a case yielding a global minimum total cost. Although this algorithm guarantees an optimal solution and works well for small networks, the algorithm runs in exponential time with the number of network requests, i.e., $O(k^n)$. For practical network deployments, the global optimization method is not feasible and to an extent impossible. For a network with a demand of more than 100 requests, for example, this optimization method is computationally impossible.

Simplex Optimization. A possible solution to the shortcomings of the global optimization method is to use local optimization, also known as the simplex method. As the name suggests, the simplex method finds a locally optimum solution. The simplex algorithm does not examine all possible combinations of protection paths but rather, considers only a cluster of combinations. The algorithm starts at one combination and examines the possible combinations in the neighborhood of the starting point. It goes on to find the combination of protection paths that will result in a total cost that is minimum in the cluster. The simplex algorithm runs relatively quickly with a time complexity of $O(kn)$ where n is the number of requests involved and k is the number of candidate protection paths for each request.

However, the shortcoming of the simplex method is that it may produce a nonoptimal solution. For small networks with scattered demands, the simplex method often provides good approximation of protection paths whose total cost is close to the global minimum. But as the network complexity and traffic volume increase, the difference between the simplex method and a global method can be very significant. This outcome will not be desirable if optimal networks are critical.

Random Simplex Optimization. To overcome the problem of simplex optimization, a random simplex method is discussed that works almost similar to the simplex method in that it searches the locally optimal combination of protection paths in a cluster of possible combinations. In addition to the local minimum, the random simplex algorithm generates a random seed with which it carries an additional search to another cluster and finds its local minimum. Comparing the new minimum with the previous one, the algorithm determines a global minimum at that time. It continues to search other local minima. The search for local minima in other regions continues until a stipulated number of seeds are used. A default number of seeds is set and the number is reset every time a new minimum is found. The algorithm quits only after no new minimum is found after the number of the seed count goes down to zero.

The algorithm does not incur the same overhead as global optimization because it does not test all combinations. In general, the running time of the random simplex outperforms that of the global method, but yields comparable results. However, because of the random nature of the random simplex algorithm, the algorithm may yield different results for each execution. This makes the random simplex method unpredictable and it may not find the best possible solution.

Clearly, the random simplex method does not take as much time to run as it takes the global method. At the same time, the method ensures a better result than the simplex method most of the time because it randomly searches other regions. Thus the random simplex method incorporates the advantages of both the simplex and global methods. The effectiveness of the random simplex method can be seen through simulation results presented in the next section.

Path Search. The first simulation looks at the comparison between the MSP and KSP path search algorithms. A small network that emulates a high-speed backbone network is shown in Figure 9(a). Nodes 0 to 3 represent routers or gateways, and nodes 4 to 7 are backbone switches that are interconnected with high-speed links indicated by thick lines. Demands are randomly generated between source and destination router pairs. There can be more than one demand between each destination and source router. The failure probability of each link in the network is also set.

For this experiment, the effectiveness of MSP and KSP in finding the appropriate protection path is examined. Therefore to ensure consistency, the working paths are set the same for both methods, meaning, the working paths are obtained using a similar algorithm. The results of

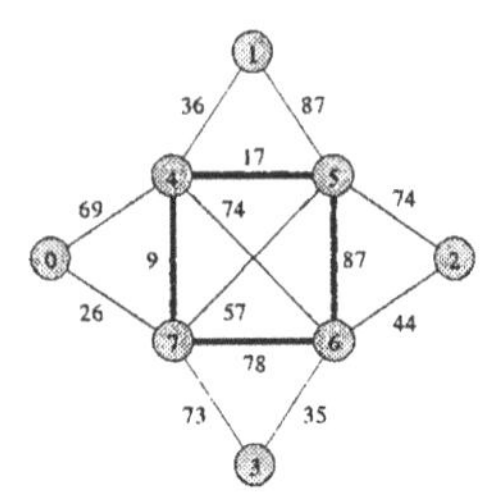

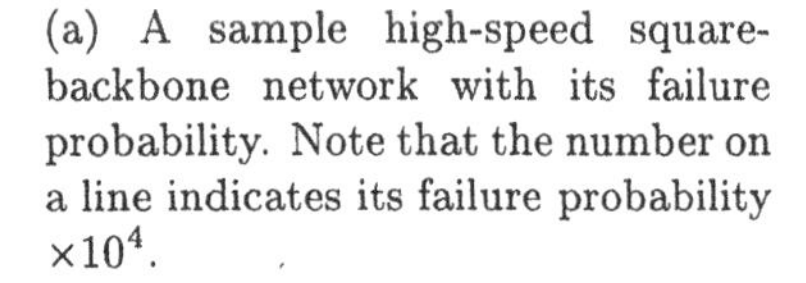

(a) A sample high-speed square-backbone network with its failure probability. Note that the number on a line indicates its failure probability $\times 10^4$.

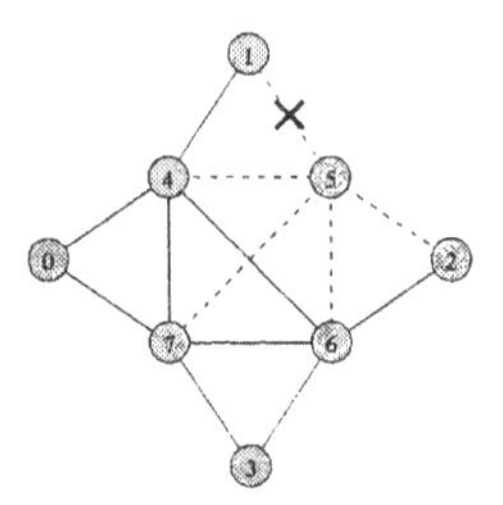

(b) Assuming link (1-5) fails and discarding the working path (1-5-2) as well as links incident to nodes on the working path, the resulting network has a first-order trap topology which KSP is unable to handle.

Figure 9: A square backbone network.

the experiment are shown in Table 1, where it is obvious that the MSP method produces a lower total network cost when compared to the KSP method. Upon examining the resulting protection paths more closely, it can be seen that the difference lies in the traffic demands from node 1 to node 2. With the working path set at (1-5-2), any link failure along the path will result in a first-order trap topology. This is shown in Figure 9(b). The KSP method, where candidate paths must be disjoint, can only return path (1-4-6-2), which is the shortest path, as the protection path and nothing else. On the other hand, the MSP method is able to return path (1-4-7-6-2) and possibly other paths as well. Choosing path (1-4-7-6-2) as the protection path for some of the demands between nodes 1 and 2 accounts for the lower total network cost.

This simulation shows that the MSP method is a more effective path search algorithm than the KSP method in that it is able to produce a resulting network with lower communication cost. In some networks with higher traffic load, MSP may even find protection paths when the KSP method cannot. As such the MSP method is a more robust way of searching paths between two nodes.

Multifault Recovery. The optimal rerouting scheme is tested for its ability to recover from multiple consecutive faults. To test multi-fault

Src.	Dst.	Flow	Working Path	Protection Path KSP	Cost	MSP	Cost
0	1	10	0-4-1	0-7-5-1	0.232	0-7-5-1	0.232
0	2	10	0-7-6-2	0-4-5-2	0.439	0-4-5-2	0.439
0	3	10	0-7-3	0-4-6-3	0.421	0-4-6-3	0.421
1	2	5	1-5-2	1-4-6-2	0.647	1-4-7-6-2	0.633
1	2	5	1-5-2	1-4-6-2	0.647	1-4-7-6-2	0.633
1	2	5	1-5-2	1-4-6-2	0.647	1-4-7-6-2	0.633
1	2	10	1-5-2	1-4-6-2	0.647	1-4-6-2	0.633
1	3	10	1-4-7-3	1-5-6-3	0.421	1-5-6-3	0.421
2	3	5	2-6-3	2-5-7-3	0.439	2-5-7-3	0.439
2	3	5	2-6-3	2-5-7-3	0.439	2-5-7-3	0.439
Total Cost					1.739		1.725

Table 1: Traffic demands and the returned protection paths and total cost for both MSP and KSP methods.

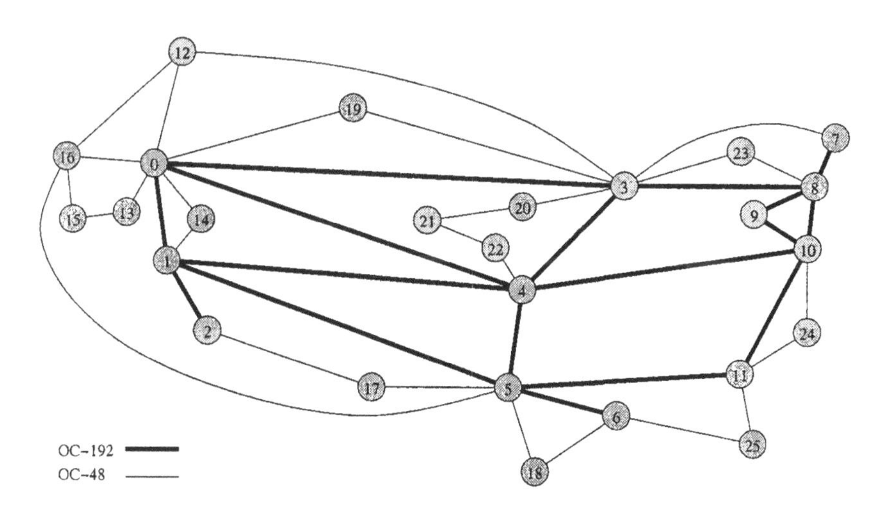

Figure 10: The AT&T backbone network.

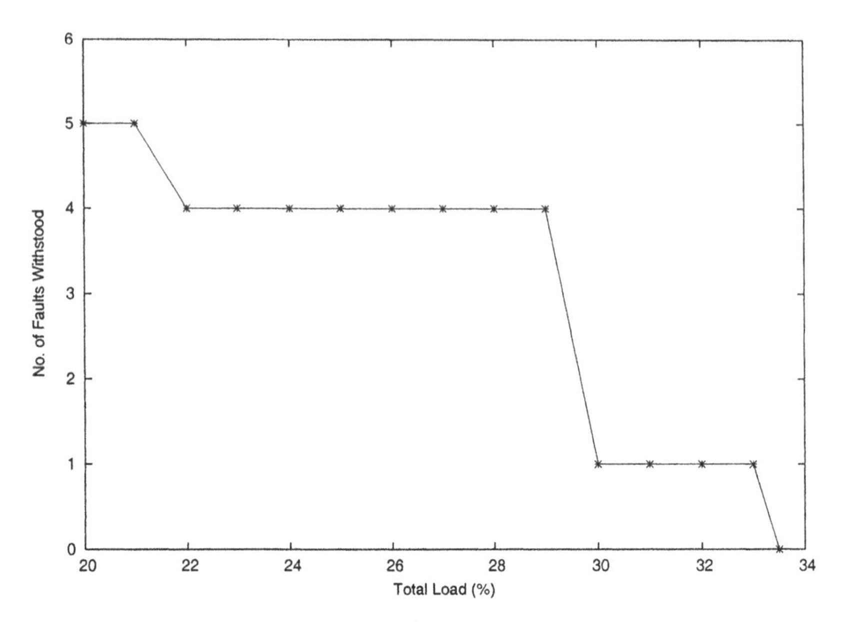

Figure 11: Maximum number of faults recovered varying with network load.

recovery, the simulation is executed regressively. After obtaining the working and protection paths for each demand, subsequent faults in the network are simulated by eliminating a particular link. The simulation finds a new working and protection paths pair. Of interest is how many faults the network is able to withstand for a particular set of demands before it fails to route the traffic. The network shown in Figure 10 is the AT&T backbone network across the continental United States. Using this network, a set of demands that is scattered among all the nodes in the network is created. Varying the total load of all demands and increasing the number of demands, one link is randomly removed from the network at a time and the regressive simulation observes how many faults the network is able to handle before it fails to find a protection path for one or more demand. To simplify the data, the total load is defined as a percentage of total flow in each link over the total bandwidth of the network. The result of the simulation is shown in Figure 11.

From the figures, it can be seen that the number of faults the network is able to withstand gradually decreases with increasing total load of the network. This is an expected observation as increasing the load will more likely overload links in the network, particularly links with

Src.	Dst.	Flow	Working Path	Protection Path		
				Simplex	Global	Rsimplex
0	1	10	0-4-1	0-7-6-5-1	0-7-6-5-1	0-7-5-1
0	1	10	0-4-1	0-7-6-5-1	0-7-6-5-1	0-7-6-5-1
0	1	10	0-4-1	0-7-6-5-1	0-7-5-1	0-7-5-1
0	1	5	0-4-1	0-7-5-1	0-7-6-5-1	0-7-6-5-1
0	2	10	0-7-6-2	0-4-5-2	0-4-5-2	0-4-5-2
0	2	10	0-4-5-2	0-7-6-2	0-7-6-2	0-7-6-2
0	3	5	0-7-3	0-4-6-3	0-4-5-6-3	0-4-5-6-3
0	3	10	0-7-3	0-4-5-6-3	0-4-6-3	0-4-6-3
0	4	5	0-4	0-7-4	0-7-4	0-7-4
0	5	5	0-7-5	0-4-5	0-4-5	0-4-5
0	5	10	0-4-5	0-7-5	0-7-5	0-7-6-5
0	6	5	0-7-6	0-4-5-6	0-4-5-6	0-4-5-6
0	6	10	0-7-6	0-4-6	0-4-6	0-4-6
0	7	10	0-7	0-4-7	0-4-7	0-4-7

Table 2: Demand flow assignment for the square backbone network.

smaller capacities, resulting in failure to find protection paths. A more important factor in determining the ability to handle multiple faults, however, is the link at which the fault occurs. In a large-scale network with numerous demands, some links carry more traffic than the others, and there are only two links connected to nodes with two degrees of termination. Links like these are said to be critical and cause vulnerable spots in the network. Should these links fail, the network will not be able to recover as there is no room for alternate paths. This will result in failure in the demands that depend solely on the faulty links.

Optimization Method	Total Cost	Process Time
Simplex	3.57644	0.06s
Random simplex	3.57313	1.27s
Global	3.57313	1.96s

Table 3: Cost and time comparison for the three different optimization methods on the square backbone network.

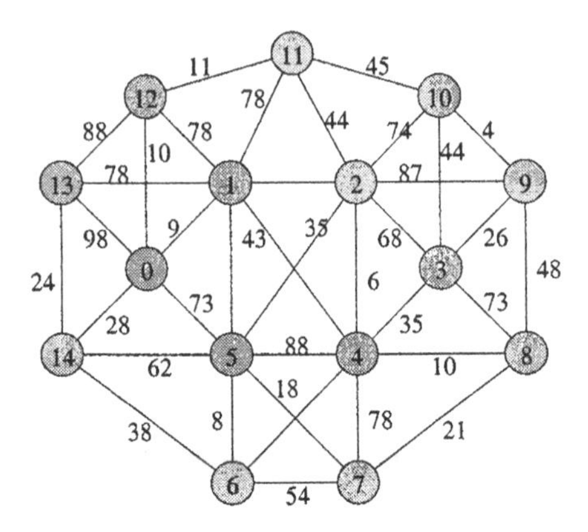

Figure 12: An arbitrary network.

Optimization Algorithm. The optimization process of the rerouting scheme can be accomplished using three different methods: global, simplex and random simplex. To illustrate the capabilities, and performance of the different optimization methods, simulation results using two different network demands and topologies are examined. Performance and capability are measured based on the returned protection paths, network cost, and the running time.

First, consider the square backbone network in Figure 9. A different set of requests is fed to the network and the working path for each demand is obtained. Then using the MSP method and either one of the three different optimization methods, the protection path is found for each working path. The requests and their resulting protection and working paths are shown in Table 2. Notice that the three different methods yield different combinations of protection paths.

The total cost and the process time for each method are shown in Table 3. We can observe that the total cost using the random simplex and the global optimization methods are the same whereas the simplex method yields a total cost that is not globally minimum. The result from the simplex method may be only slightly higher in this example, but in complex networks with high traffic volume, the difference can be very significant. In terms of processing time, we see that the simplex method runs the fastest. The process time is obtained from the program runtime, which may not reflect the actual runtime of the specific algorithm. But this process time data is a good enough approximation to examine the runtime of each method. In this small network with scattered demands,

Src.	Dst.	Flow	Working Path	Protection Path		
				Simplex	Global	Rsimplex
4	11	10	4-1-11	4-2-11	4-8-3-10-11	4-3-10-11
4	11	10	4-1-11	4-2-11	4-2-11	4-2-11
4	11	10	4-1-11	4-2-11	4-2-11	4-3-10-11
4	11	5	4-1-11	4-1-11	4-1-11	4-1-11
4	11	5	4-1-11	4-5-0-12-11	4-5-0-12-11	4-8-3-10-11
4	11	10	4-1-11	4-8-3-10-11	4-2-11	4-2-11
4	11	10	4-1-11	4-8-3-10-11	4-3-10-11	4-3-10-11
4	11	5	4-1-11	4-3-10-11	4-8-3-10-11	4-5-0-12-11
4	11	5	4-1-11	4-3-10-11	4-3-10-11	4-2-11
4	11	10	4-1-11	4-3-10-11	4-3-10-11	4-3-10-11
4	11	10	4-1-11	4-3-10-11	4-3-10-11	4-8-3-10-11
4	11	5	4-1-11	4-3-10-11	4-3-10-11	4-2-11

Table 4: Demand flow assignment for the arbitrary network.

Optimization Method	Total Cost	Process Time
Simplex	1.60957	0.19s
Random simplex	1.60593	9.07s
Global	1.60593	4965.35s

Table 5: Cost and time comparison for the three different optimization methods on the arbitrary network.

the global optimization method still runs in acceptable time, but for larger networks as seen in the next network, global optimization is not feasible.

A different and more complicated network is looked at in the next simulation. The network is an arbitrary network as shown in Figure 12. To increase the complexity of the demand, a single path is loaded with the same demand; i.e., multiple requests are sent between the same source and destination. Using the MSP method and either one of the three different optimization methods, the protection path is found for each working path. The result is shown in Table 4, where the protection paths combination using each method is different. The total cost and process time is shown in Table 5. The process time is obtained the same way as in the previous network example. Notice that similar to the previous example, the total cost from the protection path combination obtained from the global and random simplex optimization is the same, wherease the simplex method yields a higher network cost. The random simplex method produced a combination with a total cost that is globally minimum and its process time range is insignificantly more than the simplex method. But examining the process time for the global optimization method for this network, it can be seen that the process time increased tremendously and is about 500 times greater than the random simplex method. This example shows the impracticality of global optimization.

From the simulations, we can find that the simplex optimization method does not always yield an optimum solution for protection paths although it has shorter runtime. Using the global optimization method on the other hand guarantees an optimum solution. But the time complexity of global optimization renders the algorithm not feasible for practical implementation. The random simplex optimization combines the characteristics of both simplex and global optimization, yielding results that are in most cases optimum and with acceptable runtime. The random simplex algorithm, however, can still be improved by incorporating different optimization techniques.

References

[1] A.W. Brander and M.C. Sinclair. A comparative study of *k*-shortest path algorithms. In *Proc. 11th UK Performance Engineering Worksh. for Computer and Telecommunications Systems*, September 1995.

[2] O. Brun and J. Garcia. Resource allocation in communication networks. In *5th IEEE International Conference on High-Speed Networks and Multimedia Communication*, July 2002.

[3] G. Dahl and M. Stoer. A cutting plane algorithm for multicommodity survivable network design problems. *INFORMS Journal on Computing*, 10:1–11, 1998.

[4] H. Fujii and N. Yoshikai. Restoration message transfer mechanism and restoration characteristics of double-search self-healing ATM network. *IEEE J. on Sel. Areas in Comm.*, 12(1):149–158, January 1994.

[5] B. Gendron, T. Crainic, and A. Frangioni. Multicommodity capacitated network design In *B. Sans'o and P. Soriano (eds), Telecommunications Network Planning*, Kluwer, Norwell, MA., 1998.

[6] M. Gerla and L. Kleinrock. On the topological design of distributed computer networks. *IEEE Trans. Communications*, COM-25(1):48–60, January 1977.

[7] K. Gummadi, M. J. Pradeep, and C. S. R. Murthy. An efficient primary-segmented scheme for dependable real-time communications in multihop networks. *IEEE/ACM Trans. Networking*, 11(1):81–94, February 2003.

[8] A. Hines. *Planning for Survivable Networks.* Wiley, 2002.

[9] T. Kam and D.S. Kim. Optimal source-based rerouting for self-healing networks. In *5th IEEE International Conference on High-Speed Networks and Multimedia Communication*, July 2002.

[10] T. Kam and D.S. Kim. Self-healing networks with multi-fault recovery. In *IEEE Conference on Electro Information Technology*, June 2003.

[11] R. Kawamura, K. Sato, and I. Tokizawa. Self-healing ATM networks based on virtual path concept. *IEEE Trans. Communications*, 12(1):120–127, January 1994.

[12] A. Keshenbaum, P. Kermani, and G. Grover. Mentor: An algorithm for mesh network topological optimization and routing. *IEEE Transactions on Communications*, 39(4):503–513, April 1991.

[13] E. Kranakis, D. Krizanc, and A. Pelc. Fault-tolerant broadcasting in radio networks. *J. Algorithms*, 39(1):47–67, 2001.

[14] T. Van Landegem, P. Vankwikelgerge, and H. Vanderstraeten. A self-healing ATM network based on multilink principles. *IEEE J. on Sel. Areas in Comm.*, 12(1):139–148, January 1994.

[15] K. Murakami and H. Kim. Virtual path routing for surviable ATM networks. *IEEE/ACM Transactions on Networking*, 4(1):22–39, February 1996.

[16] K. Murakami and H. Kim. Optimal capacity and flow assignment for self-healing ATM networks based on line and end-to-end restoration. *IEEE/ACM Transactions on Networking*, 6(2):207–221, April 1998.

[17] Z. Qin, T. Lo, and F. Wu. The optimal virtual path design of ATM networks. In *22nd Conference on Local Computer Networks*, pages 159–167, November 1997.

[18] M. Stoer and G. Dahl. A polyhedral approach to multicommodity survivable network design. *Numerische Mathematik*, 68(1):149–167, 1994.

[19] Z. Wang and J. Crowcroft. Quality-of-service routing for supporting multimedia applications. 14(7):1228–1234, September 1996.

[20] E.M. Wong, A.M. Chan, and T.P. Yum. Analysis of rerouting in circuit-switching networks. *IEEE/ACM Transactions on Networks*, 8(3):419–427, June 2000.

Chapter 21

An Approximation Algorithm for the Dynamic Facility Location Problem

Yinyu Ye
Department of Management Science and Engineering
Stanford University
Stanford, CA, 94305
E-mail: yinyu-ye@stanford.edu

Jiawei Zhang
IOMS-Operations Management Group
Stern School of Business
New York University
New York, NY 10012
E-mail: jzhang@stern.nyu.edu

1 Introduction

Facility location is one of the most important aspects of logistics and resource allocations. The goal of research in this area is to support decisions regarding building or opening facilities or servers, e.g., plants, warehouses, and wireless antenna towers, among a set of possibilities such that all the demands can be served. Usually the objective of minimizing cost or maximizing profit is also involved. A good location decision could save a company or an organization millions of dollars; see, e.g., [3]. The importance of facility location decisions in supply chain design has been outlined in a recent paper by Daskin, Snyder, and Berger [9].

Although facility location models were originally proposed for solving logistic problems, they have a lot of modern applications in designing communication networks and routing information flows; see e.g. [4, 11, 1]. Other applications of facility location models include: clustering in a very large database, vehicle routing, tool selection, and installing caches through the Internet to minimize latency of access and to reduce network congestion.

In this chapter, we address discrete facility location problems, where the set of potential locations for building facilities is finite. These problems are all NP-hard; i.e., unless $P = NP$ they do not have polynomial time algorithms to find an optimal solution. There is a vast literature on these NP-hard facility location problems and many solution approaches have been developed in the last four decades. We refer the reader to the excellent book [21] and a more recent survey [8] on discrete facility location. In this chapter, we focus on approximation algorithms for solving these NP-hard problems. Approximation algorithms possess a tradeoff between solution speed and solution quality. They usually run extremely faster, even in real-time, with a provable solution quality. This feature is especially valuable for resource allocation problems that are highly dynamic and unpredictable.

Definition. Given a minimization problem, an algorithm is said to be a (polynomial) *ρ-approximation algorithm*, if for any instance of the problem, the algorithm runs in polynomial time and outputs a solution that has a cost at most $\rho \geq 1$ times the minimum cost, where ρ is called the *performance guarantee* or the *approximation ratio* of the algorithm.

Because the minimum cost is unknown in general, in order to establish a performance guarantee of an algorithm, we need to compare the cost of the solution produced by the algorithm to a lower bound of the minimum cost. Therefore, it is important to have a good lower bound. For most of the discrete facility location problems, linear programming relaxations provide good lower bounds for the minimum cost.

One of the most-studied facility location models is the so-called *uncapacitated facility location problem (UFLP)*. In the UFLP, we are given a finite set $\mathcal{F}$ of *facilities*, a finite set $\mathcal{D}$ of *clients*, a cost f_i for opening facility $i \in \mathcal{F}$, and a connection cost c_{ij} for connecting i to j for $i, j \in \mathcal{D} \cup \mathcal{F}$. The objective is to open a subset of the facilities in $\mathcal{F}$ and connect each client to an open facility so that the total cost is minimized. We assume that the connection costs are metric, meaning that they are nonnegative, symmetric, and satisfy the triangle inequality; i.e., for any

$i, j, k \in \mathcal{D} \cup \mathcal{F}$, $c_{ij} \geq 0$, $c_{ij} = c_{ji}$, and $c_{ij} \leq c_{ik} + c_{kj}$. For brevity we refer to the uncapacitated facility location problem as the uncapacitated problem.

We may use several terms interchangeably. For example, a client may also be called a *city*, *end-user*, or simply a *demand point*. If client j is connected to facility i, we may say that client j is *assigned to* facility i or is served by facility i. Similarly, the connection cost c_{ij} may also be called the *service cost* or *assignment cost*.

The UFLP is a static problem; it concerns how many facilities to build and where to locate them. In this chapter, we consider the *dynamic facility location problem (DFLP)*, which is a generalization of the UFLP when the time factor is present. The DFLP also addresses the issue of when to build a facility. More precisely, in the DFLP, we are given a set of facilities $\mathcal{F}$, a set of clients $\mathcal{D}$, and the time periods numbered from 1 to T. At each time period t, a client $j \in \mathcal{D}$ is specified by a demand d_j^t that can be served by facilities open at the beginning of the time period t, i.e., at time period t or earlier. A cost f_i^t is incurred when the facility $i \in \mathcal{F}$ needs to be open at time period t, where f_i^t may be equal to ∞ if facility i is not available at time period t. A cost c_{ij}^{st} is incurred for supplying one unit demand of client j in time period t from facility i at the beginning of time period s ($c_{ij}^{st} = \infty$ for $t < s$). The objective is to choose a subset of facilities $\mathcal{F}$ to open at each time period, such that all demands of the clients are satisfied and the total cost is minimized. Here we assume that $c_{i_1 j_1}^{st} \leq c_{i_2 j_1}^{s't} + c_{i_2 j_2}^{s't'} + c_{i_1 j_2}^{st'}$ for any $i_1, i_2 \in \mathcal{F}$, $j_1, j_2 \in \mathcal{D}$ and time periods $1 \leq s \leq s' \leq t, t' \leq T$. We refer to the dynamic facility location problem as the dynamic problem. It is clear that when $T = 1$, the DFLP becomes the UFLP.

More precisely, the DFLP can also be formulated as an integer program (Van Roy and Erlenkotter [27]):

$$\text{minimize} \quad \sum_{t=1}^{T} \sum_{i \in \mathcal{F}} \sum_{s=1}^{T} \sum_{j \in \mathcal{D}} d_j^t c_{ij}^{st} x_{ij}^{st} + \sum_{s=1}^{T} \sum_{i \in \mathcal{F}} f_i^s y_i^s \tag{1}$$

$$\text{subject to} \quad \sum_{i \in \mathcal{F}} \sum_{s=1}^{t} x_{ij}^{st} = 1 \quad \text{for all} \quad j \in \mathcal{D}, 1 \leq t \leq T \tag{2}$$

$$x_{ij}^{st} \leq y_i^s \quad \text{for all} \quad i \in \mathcal{F}, j \in \mathcal{D}, 1 \leq s, t \leq T \tag{3}$$

$$x_{ij}^{st}, y_i^s \in \{0, 1\} \quad \text{for all} \quad i \in \mathcal{F}, j \in \mathcal{D}, 1 \leq s, t \leq T, \tag{4}$$

where $y_i^s = 1$ if facility i is open at time period s, $y_i^s = 0$ otherwise; $x_{ij}^{st} = 1$ if the demand of client j at time period t is supplied from facility

i at time period s. The constraint (2) indicates that the demand of each client j at any period t must be served by a facility i at a time period s earlier than t. The constraint (3) indicates that if a facility i at time period t supplies the demand of a client, then i must be operated at time period t.

Our main result is a 1.86-approximation algorithm for the DFLP. To the best of our knowledge, this is the first approximation algorithm for the dynamic problem. As a corollary of our result, an improved approximation algorithm is obtained for a two-stage stochastic facility location problem studied by Ravi and Sinha [22], by showing that it can be formulated as a special case of the DFLP. This improves the previously best known performance guarantee 3.16 for the two-stage stochastic facility location problem.

This chapter is organized as follows. Because the DFLP generalizes the UFLP, in Section 2, we summarize the known approximation algorithms for the UFLP. We present the linear programming relaxation of the DFLP and its dual in Section 3. In Section 4, we present our primal–dual algorithm for the DFLP that generalizes the one by Jain and Vazirani [15]. The algorithm is analyzed in Section 5. Finally, we show that the two-stage stochastic facility location problem is a special case of the DFLP and thus admits a 1.85-approximation algorithm.

2 Approximation Algorithms for the UFLP

First of all, we note that a result of Guha and Khuller [10], combined with an observation of Sviridenko [24], implies that no polynomial time algorithm for the uncapacitated problem can have a performance guarantee better than 1.463 unless $P = NP$. Therefore, under the assumption that $P \neq NP$, we cannot expect a ρ-approximation algorithm with $\rho < 1.463$ for any of the problems that generalize the UFLP, in particular, the DFLP.

The first approximation algorithm for the uncapacitated problem was developed 20 years ago by Hochbaum [12] with performance guarantee $O(\ln |\mathcal{D}|)$. The first constant-factor approximation algorithm was given by Shmoys, Tardos, and Aardal [23], who presented a 3.16-approximation algorithm based on rounding a (fractional) optimal solution of the linear programming relaxation of an integer programming formulation of the UFLP. Such an approach is called *linear programming rounding*. Subsequently, many approximation algorithms with better performance guarantees have been proposed. These algorithms fall into three classes.

The first class contains approximation algorithms that are based on linear programming rounding, including Guha and Khuller [10], Chudak and Shmoys [6, 7], and Sviridenko [25]. The main idea is to solve the linear programming relaxation of the UFLP, and then carefully round the optimal fractional solution into a feasible integer solution. Both deterministic and randomized rounding techniques have been developed for solving the UFLP.

The second class contains local search algorithms including Korupolu, Plaxton, and Rajaraman [16], Charikar and Guha [5], and Arya et al. [2]. In such an algorithm, one starts with a feasible solution, and then improves the solution by some simple operations such as open one facility, close one facility, or swap two facilities. Such algorithms, although extremely simple, have constant performance guarantees.

The last class of algorithms is primal–dual in flavor. Such an algorithm simultaneously increases dual variables, and tries to find a feasible primal solution according to the information provided by the dual variables. The first such algorithm was proposed by Jain and Vazirani [15]. The algorithms of Mettu and Plaxton [20], Thorup [26], Mahdian et al. [18], and Jain, Mahdian, and Saberi [14] also fall into this class.

The approximation ratios and running times of these algorithms are summarized in Table 1, where $n = |\mathcal{D}| + |\mathcal{F}|$.

One might observe that the DFLP can actually be formulated as a special case of the UFLP, where we regard each demand-period pair as a single client. However, in such a formulation, the triangle inequalities do not hold any more, which is essential in establishing the performance guarantees of the algorithms for the UFLP. Therefore, we need to tackle the DFLP directly by taking advantage of its special structures. Nevertheless, our algorithm is inspired by the algorithm of Jain and Vazirani [15] for the UFLP.

3 Linear Programming Relaxation for the DFLP

Consider the linear programming relaxation of the integer program (1) to (4):

$$\text{minimize} \quad \sum_{t=1}^{T} \sum_{i \in \mathcal{F}} \sum_{s=1}^{T} \sum_{j \in \mathcal{D}} d_j^t c_{ij}^{st} x_{ij}^{st} + \sum_{s=1}^{T} \sum_{i \in \mathcal{F}} f_i^s y_i^s \qquad (5)$$

Approximation ratio	Reference	Running time
$O(\ln \lvert\mathcal{D}\rvert)$	Hochbaum [12]	$O(n^3)$
3.16	Shmoys et al. [23]	linear programming
2.41	Guha and Khuller [10]	linear programming
1.736	Chudak [6]	linear programming
$5+\epsilon$	Korupolu et al. [16]	$O(n^6 \log(n/\epsilon))$
3	Jain and Vazirani [15]	$O(n^2 \log n)$
1.853	Charikar and Guha [5]	$O(n^3)$
1.728	Charikar and Guha [5]	linear programming
1.861	Mahdian et al. [18]	$O(n^2 \log n)$
1.61	Jain et al. [14]	$O(n^3)$
1.582	Sviridenko [25]	linear programming
1.517	Mahdian et al. [19]	$O(n^2 \log n)$

Table 1: Approximation algorithms for the uncapacitated problem

$$\begin{aligned} \text{subject to} \quad & \sum_{i\in\mathcal{F}} \sum_{s=1}^{t} x_{ij}^{st} = 1 \quad \forall j \in \mathcal{D}, 1 \le t \le T \\ & x_{ij}^{st} \le y_i^s \quad \forall i \in \mathcal{F}, j \in \mathcal{D}, 1 \le s,t \le T \\ & x_{ij}^{st}, y_i^s \ge 0 \quad \forall i \in \mathcal{F}, j \in \mathcal{D}, 1 \le s,t \le T. \end{aligned}$$

The dual of the linear programming relaxation is

$$\begin{aligned} \text{maximize} \quad & \sum_{t=1}^{T} \sum_{j\in\mathcal{D}} \alpha_j^t \\ \text{subject to} \quad & \alpha_j^t - d_j^t c_{ij}^{st} \le \beta_{ij}^{st} \quad \forall i \in \mathcal{F}, j \in \mathcal{D}, 1 \le s,t \le T \\ & \sum_{t=1}^{T} \sum_{j\in\mathcal{D}} \beta_{ij}^{st} \le f_i^s \quad \forall i \in \mathcal{F}, 1 \le s \le T \\ & \beta_{ij}^{st} \ge 0, \quad \forall i \in \mathcal{F}, j \in \mathcal{D}, 1 \le s,t \le T. \end{aligned}$$

It is straightforward to show that the above dual is equivalent to the following linear program.

$$\begin{aligned} \text{Maximize} \quad & \sum_{t=1}^{T} \sum_{j\in\mathcal{D}} d_j^t \alpha_j^t \\ \text{subject to} \quad & \alpha_j^t - c_{ij}^{st} \le \beta_{ij}^{st} \quad \forall i \in \mathcal{F}, j \in \mathcal{D}, 1 \le s,t \le T \qquad (6) \end{aligned}$$

$$\sum_{t=1}^{T}\sum_{j\in\mathcal{D}} d_j^t \beta_{ij}^{st} \le f_i^s \quad \forall i \in \mathcal{F}, 1 \le s \le T \tag{7}$$
$$\beta_{ij}^{st} \ge 0, \quad \forall i \in \mathcal{F}, j \in \mathcal{D}, 1 \le s, t \le T.$$

Intuitively, the dual variable α_j^t can be interpreted as the amount of money that client j is willing to spend to get one unit of its demand of period t served, and the dual variable β_{ij}^{st} can be interpreted as the part of this that is contributed to pay for opening facility i in period s.

4 Approximation Algorithm

Our algorithm builds upon the primal–dual algorithm of Jain and Vazirani [15] for the uncapacitated problem. The algorithm first constructs a feasible dual solution and then finds a feasible integer primal solution based on the dual solution.

Phase 1: Finding a Feasible Dual Solution

The algorithm is a dual ascent algorithm. We introduce a notion of time θ in the algorithm. Initially $\theta = 0$, all dual variables are set to be 0, all facility–period pairs (i, s) are closed, and all demand–period pairs (j, t) are said to be *unfrozen*. We start increasing dual variables α_j^t for all (j, t) at the same rate as long as they are unfrozen, i.e., at time θ, $\alpha_j^t = \theta$ for all unfrozen demand–period pairs. Three types of events occur during the execution of the algorithm.

(1) The first type of event occurs when $\sum_{t=1}^{T}\sum_{j\in\mathcal{D}} d_j^t \beta_{ij}^{st} = f_i^s$ for some (i, s). At this point, (i, s) is said to be *tentatively open*, and we stop increasing α_j^t and $\beta_{i'j}^{s't}$ for all (j, t) such that $\beta_{ij}^{st} > 0$. We call (i, s) the *connecting witness* of (j, t).

(2) The second type of event occurs when $\alpha_j^t = c_{ij}^{st}$ for some (j, t) and (i, s) that is not tentatively open. This happens at time $\theta = c_{ij}^{st}$. At this point, we say that (j, t) is *tight* with (i, s). Beyond this point, we start increasing the dual variable β_{ij}^{st} if (j, t) is unfrozen.

(3) The third type of event occurs when $\alpha_j^t = c_{ij}^{st}$ for some (j, t) and (i, s) that is tentatively open. In this case, we stop increasing α_j^t and call (i, s) the connecting witness of (j, t).

Notice that only unfrozen demand–period pairs participate in these events, and we only raise the dual variables α_j^t and β_{ij}^{ts} for those unfrozen pairs. The algorithm will stop when all demand–period pairs are frozen.

From the statement of the dual ascent algorithm, we know that $(\alpha_j^t, \beta_{ij}^{st})$ is a dual feasible solution.

Phase 2: Finding a Feasible Primal Solution

In this phase, we specify which facilities to open in each period, and how the demands should be served. Let G be the set of tentatively open facility–period pairs in the first phase. We say that $(i,s),(i',s') \in G$ are dependent if there is a pair (j,t) such that $\beta_{ij}^{st} > 0$ and $\beta_{i'j}^{s't} > 0$. We pick a maximal independent set of facility-period pairs $G' \subset G$ such that for any $(i,s) \in G \setminus G'$, there exists $(i',s') \in G'$ where $s' \leq s$, and (i,s) and (i',s') are dependent. Then we open the facility–period pairs in G' and assign the demands–period pairs to the closest open facility–period pairs.

5 Analysis of Algorithm

In order to analyze the algorithm, we let θ_i^s be the time that (i,s) get tentatively open. Moreover, denote

$$D' = \{(j,t) : j \in \mathcal{D}, 1 \leq t \leq T, \exists (i,S) \in G', s.t.\ \beta_{ij}^{st} > 0\}.$$

Because G' is an independent subset of G, for each $(j,t) \in D'$, there exists exactly one pair $(i,s) \in G'$ such that $\beta_{ij}^{st} > 0$. We denote such $i = \delta_{jt}$ and $s = \gamma_{jt}$. It then follows that for each $(j,t) \in D'$, $\beta_{\delta_{jt}j}^{\gamma_{jt}t} > 0$.

Theorem 5.1 *The two-phase algorithm presented in Section 4 is a 1.86-approximation algorithm for the dynamic facility location problem.*

Proof. We first show that the cost of opening facility–period pairs in G' is at most $\sum_{t=1}^{T}\sum_{j\in\mathcal{D}'} d_j^t \beta_{\delta_{jt}j}^{\gamma_{jt}t}$. For each facility–period pair (i,s) that is tentatively opened, we must have that

$$f_i^s = \sum_{t=1}^{T}\sum_{j\in\mathcal{D}} d_j^t \beta_{ij}^{st}.$$

Therefore,

$$\begin{aligned}
\sum_{(i,s)\in G'} f_i^s &= \sum_{(i,s)\in G'}\sum_{(j,t)\in D} d_j^t \beta_{ij}^{st} \\
&= \sum_{(i,s)\in G'}\sum_{(j,t)\in D'} d_j^t \beta_{ij}^{st}
\end{aligned}$$

$$= \sum_{(j,t)\in D'} \sum_{(i,s)\in G'} d_j^t \beta_{ij}^{st}$$
$$= \sum_{(j,t)\in D'} d_j^t \beta_{\delta_{jt}j}^{\gamma_{jt}t}.$$

Next we bound the service cost. We show that for $(j,t) \in D'$, there exists a pair $(i,s) \in G'$ such that $c_{ij}^{st} \le \alpha_j^t - \beta_{\delta_{jt}j}^{\gamma_{jt}t}$; for $(j,t) \notin D'$, there exists a pair $(i,s) \in G'$ such that $c_{ij}^{st} \le 3\alpha_j^t$.

If $(j,t) \in D'$, then $(\delta_{jt}, \gamma_{jt}) \in G'$. Because $\beta_{\delta_{jt}j}^{\gamma_{jt}t} > 0$, it must be the case that

$$\alpha_j^t = c_{\delta_{jt}j}^{\gamma_{jt}t} + \beta_{\delta_{jt}j}^{\gamma_{jt}t}.$$

If $(j,t) \notin D'$, then we consider one of its connecting witnesses, say (i,s). If $(i,s) \in G'$, then we know that $c_{ij}^{st} = \alpha_j^t \le 3\alpha_j^t$. Otherwise, there must exist $(i',s') \in G'$ such that (i,s) and (i',s') are dependent, and $s' \le s$. That is, there exists (j',t') such that both $\beta_{ij'}^{st'}$ and $\beta_{i'j'}^{s't'}$ are positive. Therefore,

$$c_{ij'}^{st'} \le \alpha_{j'}^{t'} \quad \text{and} \quad c_{i'j'}^{s't'} \le \alpha_{j'}^{t'}.$$

Moreover, $\alpha_{j'}^{t'} \le \theta_i^s$. Therefore, by the triangle inequality,

$$c_{i'j}^{s't} \le c_{i'j'}^{s't'} + c_{ij'}^{st'} + c_{ij}^{st} \le 2\theta_i^s + c_{ij}^{st} \le 3\alpha_j^t,$$

where the last inequality follows from $t_i^s \le \alpha_j^t$ and $c_{ij}^{st} \le \alpha_j^t$, where both are true because (i,s) is the connecting witness of (j,t).

Then we know the total facility cost F and total service cost C satisfying

$$3F + C \le \sum_{t=1}^{T} \sum_{j\in\mathcal{D}} 3d_j^t \alpha_j^t \le 3OPT,$$

where OPT is the total of an optimal solution. It is known that such an algorithm would imply a 1.86-approximation algorithm; see Guha and Khuller [10], and Mahdian, Ye, and Zhang [19]. This completes the proof of the theorem. ■

6 Two-Stage Stochastic Facility Location

In this section, we consider a version of the two-period (or two-stage) stochastic uncapacitated facility location problem proposed by Ravi and

Sinha [22]. (We mention that a similar formulation for the maximization version of the facility location problem was proposed by Louveaux and Peeters [17].) We denote this problem as the stochastic problem. In this problem, we are given a set of facilities $\mathcal{F}$ and a set of clients $\mathcal{D}$, and the cost c_{ij} of serving one unit demand of client $j \in \mathcal{D}$ from facility $i \in \mathcal{F}$. It is assumed that the c_{ij} satisfies the triangle inequality. However, the demand of each client is unknown at the first stage. There are m possible scenarios of the demands, and each of them happens with probability p_k, $k = 1, 2, \ldots, m$ such that $\sum_{i=1}^{m} p_k = 1$. In scenario k, client j has demand d_j^k. The cost of opening a facility i depends on when to open it: the cost is f_i^0 if we open facility i at the first stage, i.e., before any scenario realized; the cost becomes f_i^k if facility i is open after scenario k is realized. Again, we want to minimize the total cost. This problem can be formulated as the following integer program [22].

$$\begin{array}{lll}
\text{Minimize} & \sum_{i\in F} f_i^0 y_i^0 + \sum_{k=1}^{m} p_k \left(\sum_{i\in F} f_i^k y_i^k + \sum_{i\in F}\sum_{j\in D} d_j^k c_{ij} x_{ij}^k \right) & \\
\text{subject to} & \sum_{i\in F} x_{ij}^k = 1 & \forall j \in \mathcal{D}, 1 \le k \le m \quad \text{such that} \quad d_j^k > 0 \\
& x_{ij}^k \le y_i^0 + y_i^k & \forall i \in \mathcal{F}, j \in \mathcal{D}, 1 \le k \le m \\
& x_{ij}^k, y_i^k \in \{0,1\} & \forall i \in \mathcal{F}, j \in \mathcal{D}, 1 \le k \le m,
\end{array}$$

where y_i^0 is the first stage variable and y_i^k is the second stage variable for scenario $k = 1, 2, \ldots, m$.

Theorem 6.1 *The two-stage stochastic facility location problem can be approximated by a factor of* 1.86 *in polynomial time.*

Proof. We first show that the two-period stochastic problem can be formulated as a special case of the metric dynamic problem with $T = 2$. Then the theorem follows from the result of the previous section.

Given any instance of the stochastic problem, we construct an instance of the metric dynamic problem with $T = 2$ as follows. The set of clients is

$$\{(j,k) | j \in \mathcal{D}, k = 1, 2, \ldots, m\}.$$

The demand of each client (j,k) is 0 at time period 1 and $p_j d_j^k$ at time period 2. The set of facilities is

$$\mathcal{F}^0 \cup \mathcal{F}^1 \cup \ldots \cup F^m,$$

where

$$\mathcal{F}^k = \{(i,k)|i \in \mathcal{F}, k = 0,1,2,\ldots,m\}.$$

The opening cost of each facility $(i,0) \in \mathcal{F}^0$ is f_i^0 at time period 1 and ∞ at time period 2. For $k \geq 1$, the opening cost of each facility $(i,k) \in \mathcal{F}^k$ is ∞ at time period 1 and $p_k f_i^k$ at time period 2. The cost of serving one unit demand of the client (j,k) at time period 2 from a facility (i,l) is c_{ij} if $k = l$ and (i,l) is from time period 2, or $l = 0$ and (i,l) is from time period 1 and ∞ otherwise.

It is left to show that the triangle inequality holds. Consider any two clients (j_1,k_1) and (j_2,k_2) from time period 2, and any two facilities (i_1,l_1) and (i_2,l_2).

Case 1. Both facilities (i_1,l_1) and (i_2,l_2) are from the same time period. If none of the service costs between these two clients and the two facilities is ∞, then they must satisfy the triangle inequality. If one of them is ∞, say the service cost between client (j_1,k_1) and facility (i_1,l_1), then by construction, $k_1 \neq l_1$ if both (i_1,l_1) and (i_2,l_2) are from time period 2 and $l_1 \neq 0$ if both (i_1,l_1) and (i_2,l_2) are from time period 1. In the first case, one of the following must be true: $k_1 \neq l_2$, $k_2 \neq l_2$, or $k_2 \neq l_1$. Therefore, one of the service costs between (j_1,k_1) and (i_2,l_2), between (j_2,k_2) and (i_1,l_1) and (i_2,l_2) must be ∞. The triangle inequality holds. In the latter case, the service cost between (j_2,k_2) and (i_1,l_1) must also be ∞. The triangle inequality holds too.

Case 2. Facility (i_1,l_1) is from period one and (i_2,l_2) is from period two. Again, if none of the service costs between these two clients and two facilities is ∞, then we are done. Otherwise, recall our definition of the triangle inequality for the metric dynamic problem. We only need to check the case where the service cost from (i_1,l_1) to one of the clients, say (j_1,k_1), is ∞. In this case, $l_1 \neq 0$, thus the service cost between (j_2,k_2) and (i_1,l_1) must also be ∞. Again, the triangle inequality holds. This completes the proof of the lemma. ■

7 Conclusions

In this chapter, we presented some preliminary results on the dynamic facility location problem. The model was introduced by Van Roy and Erlenkotter [27]. There are many other multiperiod facility location models in the literature [13]. It is important to develop approximation algorithms for these models as well.

We also showed that the scenario-based stochastic facility location problem is a special case of the dynamic facility location problem. This leads to a 1.86-approximation algorithm for solving the stochastic problem.

We remark that the service cost c_{ij}^{st} could accommodate transportation cost, production cost, and inventory cost in the context of logistic problems. For example, c_{ij}^{st} is the sum of the unit transportation cost c_{ij}, the unit production cost q_i^t at facility i at time period s, and $\sum_{k=s}^{t-1} h_i^k$ where h_i^k is the cost for holding one unit at facility i at time period k. If c_{ij} satisfies triangle inequality, and h_i^k depends only on k, then we can verify that c_{ij}^{st} also satisfies triangle inequality. However, we remark that, in case h_i^k depends on both i and k, we can show that c_{ij}^{st} satisfies a relaxed inequality $c_{i_1j_1}^{st} \leq c_{i_2j_1}^{s't} + 2(c_{i_2j_2}^{s't'} + c_{i_1j_2}^{st'})$. Under the relaxed assumption, our algorithm has a performance guarantee of 2.21. Of course, more challenging research directions are in solving dynamic facility location problems with capacity constraints and nonmetric service costs.

Acknowledgement. Research supported in part by the Boeing project on *Dynamic Resources Allocation.*

References

[1] K. Andreev, B. M. Maggs, A. Meyerson, and R. K. Sitaraman, "Designing overlay multicast networks for streaming," *SPAA*, 149–158, 2003.

[2] V. Arya, N. Garg, R. Khandekar, A. Meyerson, K. Munagala, and V. Pandit, "Local search heuristic for k-median and facility location problems," *Proceedings of the ACM Symposium on Theory of Computing*, 21–29, 2001.

[3] J. D. Camm, T. E. Chorman, F. A. Dill, J. R. Evans, D. J. Sweeney, and G. W. Wegryn, "Blending OR/MS, judgement, and GIS: Restructuring P&G's supply chain," *Interfaces*, 27, 128–142, 1997.

[4] P. Chardaire, *Facility Location Optimization and Cooperative Games*, Ph.D. Thesis, School of Information Systems, University of East Anglia, Norwich NR4 7TJ, UK, 1998.

[5] M. Charikar and S. Guha, "Improved combinatorial algorithms for facility location and k-median problems," *Proceedings of the 40th IEEE Foundations of Computer Science (FOCS)*, 378–388, 1999.

[6] F. A. Chudak, "Improved approximation algorithms for uncapacited facility location," in R.E. Bixby, E.A. Boyd, and R.Z. Ríos-Mercado (Eds.), *Integer Programming and Combinatorial Optimization*, LNCS 1412, Springer-Verlag, New York, 180–194, 1998.

[7] F. A. Chudak and D. B. Shmoys, "Improved approximation algorithms for the uncapacitated facility location problem," *SIAM Journal. on Computing*, 33(1), 1–25, 2003.

[8] J. Current, M. Daskin, and D. Schilling, *Discrete Network Location Models*, in Z. Drezner and H. Hamacher (Eds.), Facility Location Theory: Applications and Methods, Springer-Verlag, Berlin, 81–118, 2002.

[9] M. S. Daskin, L. V. Snyder, and R. T. Berger, "Facility location in supply chain design," *Working Paper*, No. 03-010, Department of Industrial Engineering and Management Sciences, Northwestern University, Evanston, Illinois, 60208-3119, U.S.A.

[10] S. Guha and S. Khuller, "Greedy strikes back: Improved facility location algorithms," *Journal of Algorithms,* 31, 228–248, 1999.

[11] S. Guha, A. Meyerson, and K. Munagala, "Hierarchical placement and network design problems," *IEEE Symposium on Foundations of Computer Science (FOCS)*, 603–612, 2000.

[12] D. S. Hochbaum, "Heuristics for the fixed cost median problem," *Mathematical Programming,* 22, 148–162, 1982.

[13] S. K. Jacobsen, "Multiperiod capacitated location models," in P. Mirchandani, R. Francis (Eds.), *Discrete Location Theory,* Wiley, New York, 119-171, 1990.

[14] K. Jain, M. Mahdian and A. Saberi, "A new greedy approach for facility location problems," *Proceedings of the 34th ACM Symposium on Theory of Computing (STOC)*, 731–740, 2002.

[15] K. Jain and V.V. Vazirani, "Approximation algorithms for metric facility location and k-median problems using the primal-dual schema and lagrangian relaxation," *Journal of the ACM*, 48, 274–296, 2001.

[16] M. R. Korupolu, C. G. Plaxton, and R. Rajaraman, "Analysis of a local search heuristic for facility location problems," *Proceedings of the 9th Annual ACM-SIAM Symposium on Discrete Algorithms (SODA)*, 1–10, 1998.

[17] R. V. Louveaux and D. Peeters. "A dual-based procedure for stochastic facility location," *Operations Research*, 40, 564–573, 1992.

[18] M. Mahdian, E. Markakis, A. Saberi, and V.V. Vazirani, "A greedy facility location algorithm analyzed using dual fitting," *Proceedings of 5th International Workshop on Randomization and Approximation Techniques in Computer Science*, LNCS 2129, Springer-Verlag, New York, 127–137, 2001.

[19] M. Mahdian, Y. Ye, and J. Zhang, "Improved approximation algorithms for metric facility location problems," *5th International Workshop on Approximation Algorithms for Combinatorial Optimization (APPROX)*, LNCS 2462, Springer-Verlag, New York, 229–242, 2002.

[20] R. R. Mettu and C. G. Plaxton, "The online median problem," *SIAM Journal on Computing*, 32, 816–832, 2003.

[21] P. Mirchandani and R. Francis (Eds.), *Discrete Location Theory*, Wiley, New York, 119–171, 1990.

[22] R. Ravi and A. Sinha. "Hedging uncertainty: Approximation algorithms for stochastic optimization problems," in G. L. Nemhauser and D. Bienstock (Eds.), *Integer Programming and Combinatorial Optimization*, LNCS 3064, Springer-Verlag, New York, 101–115, 2004.

[23] D. B. Shmoys, E. Tardos, and K. I. Aardal, "Approximation algorithms for facility location problems," *Proceedings of the 29th Annual ACM Symposium on Theory of Computing (STOC)*, 265–274, 1997.

[24] M. Sviridenko, cited as personal communication in [7], July, 1998.

[25] M. Sviridenko, "An 1.582-approximation algorithm for the metric uncapacitated facility location problem," *Proceedings of the 9th Conference on Integer Programming and Combinatorial Optimization (IPCO)*, 240–257, 2002.

[26] M. Thorup, "Quick k-median, k-center, and facility location for sparse graphs," In *Automata, Languages and Programming, 28th International Colloquium*, LNCS 2076, Springer-Verlag, New York, 249–260, 2001.

[27] T. J. Van Roy and D. Erlenkotter, "A dual-based procedure for dynamic facility location," *Management Science*, 28, 1091–1105, 1982.

Chapter 22

Genetic Code-Based DNA Computation for the Hamiltonian Path Problem

Mingjun Zhang
Life Sciences and Chemical Analysis Division
Agilent Technologies, Palo Alto, CA 94304
E-mail: mingjunzhang@ieee.org

Maggie X. Cheng
Computer Science Department
University of Missouri, Rolla, MO 65401
E-mail: chengm@umr.edu

Tzyh-Jong Tarn
Department of Electrical and Systems Engineering
Washington University, St. Louis, MO 63103
E-mail: tarn@wuauto.wustl.edu

1 Introduction

In the late 1950s, the Nobel laureate Richard Feynman first introduced the idea of computation at a molecular level. In 1994, the concept of DNA computation was demonstrated using experiments to solve a directed Hamiltonian Path Problem (HPP) by Adleman [1]. Since then, the possibility of DNA computation has attracted many researchers' attention. Lipton [8] showed that Adleman's technique for finding a solution to the HPP could be generalized to solve the problem of finding a satisfying assignment of an arbitrary directed network. An optimal solution may be obtained for solving an NP-complete problem using DNA computation. This advantage has many implications for applications in networks, such as data packet routing and scheduling, network resource management, etc.

The central idea of DNA computation is the Watson–Crick model of DNA structure, which specifies complementary binding properties of DNA molecules. Within cells of any organism, there is a substance called DeoxyriboNucleic Acid (DNA), which is a double-stranded helix of nucleotides that carry the genetic information of a cell. This information is the code used within cells to form proteins and is the building block upon which life is formed. A single-stranded DNA consists of a chain of simpler molecules called bases, which protrude from a sugar–phosphate backbone. The four bases are adenine (A), thymine (T), guanine (G), and cytosine (C). Any single-stranded DNA will adhere tightly to its complementary strand, in which C always pairs with G and T always pairs with A, and vice versa. For example, a single-stranded DNA segment consisting of the base sequence $CTGCA$ will only stick with a complementary sequence of $GACGT$. The idea of DNA computation is to use single-stranded DNAs to code the problem, and let the DNA strands react in test tubes or substrate surface, which is a parallel binding process of single-stranded DNAs. Then, find binding results and interpret answers by applying biomolecular techniques.

DNA computation is attractive mainly for three reasons: (1) fast parallel information processing, (2) remarkable energy efficiency, and (3) high storing capacity. A number of theoretical models for creating DNA computers have been developed. Mathematical proofs have shown that some models are at least equivalent to a classical Turing machine. Among these models, the most popular model comes directly from Adleman's paper [1]. The technique can be refined into a process consisting of two phases. First, randomly generate all possible solutions to the problem. Second, isolate correct solutions through repeated separations of the DNA from incorrect solutions and potentially good solutions.

However, DNA computation currently is too error-prone to achieve its great potential. Many ideas of DNA computation assume zero error rate. In reality, errors appear at every stage of DNA computation. In [1, 8], the problem of high error rates has been identified as the most challenging problem for the success of DNA computation. An error-resistant method is strongly desired for DNA computation. One open question is whether the error rates in DNA manipulations can be adequately controlled [2, 9]. Some algorithms have been proposed to handle a few of the apparently crippling error rates. Eng and Serridge [5] proposed a surface-based DNA computation algorithm to solve the minimal set cover problem. The technique decreases errors caused by a potential DNA strand lost by affixing the DNA onto a glass or silicon surface. In [3], a model of DNA computation has been developed to use dynamic programming, and the large size of memory available to DNA computers. The goal is to reduce error rates by increasing DNA strand storage for each computation problem. A more thorough study of decreasing error rate can be found in [2], where methods for making volume-decreasing algorithms (the number of strands decreases as the algorithm executes) more resistant to certain types of errors are proposed. One effort in the paper is to convert the decreasing volume problem to a constant volume problem (the number of strands remains the same throughout the computation). The basic idea is to add DNA strand redundancy by increasing the volume of solutions. The technique requires increasing the steps of operations and cannot be applied to an algorithm that is constant volume to begin with. Another effort proposed in the paper is to reduce the false negative error rate in the bead separation procedure by double encoding DNA bases. The idea is to have each DNA encoded base appear twice in separate locations in the strand to increase possibilities of being extracted. However, it is still not clear at the present stage whether the error rate can be reduced sufficiently to allow for a general-purpose DNA computation. Much work remains to be done.

In this chapter, a genetic code-based method for DNA computation is proposed to solve NP-complete problems in networks. The method is error resistant and can efficiently reduce the error rates. The goal of the method is to understand what the meaning of an encoded sentence is, even if some characters are mutated. The new method is based on the genetic coding theory of biological systems, which has a significant low error rate for coding genetic information. To illustrate the idea, cities from the Hamiltonian path problem have been coded. In addition, formal mathematical formulation of DNA computation has been presented. This chapter is organized as follows. In Section 2, a new method of genetic code-based DNA computation for NP-complete problems in networks is presented, and applied to solve the Hamiltonian Path problem. Following an abstract model of DNA computation, related theorems of DNA computation are presented in Section 3. The discussions and conclusions are given in Section 4.

2 Genetic Code-Based DNA Computation for NP-Complete Problems

As discussed earlier, two steps are usually involved in solving a DNA computation problem. They are (1) problem coding and (2) DNA strand operations performed by molecular techniques. The errors mainly come from the following three sources.

- DNA strand extraction operations. The operations are to remove strands from a test tube containing a given pattern. In reality, only about 95% of the strands matching the pattern can be removed. Sometimes, strands that do not match may be accidentally removed. Even with a 99% successful extraction rate, the chance of getting a good strand after multiple steps of extraction exponentially decreases. If only one "solution" is in the test tube, it is almost impossible to exactly extract the strand through a couple of biological reactions and operations. The biological process of DNA molecular techniques is messy.

- Random errors. Random nucleotide errors, such as substitutions, insertions, and deletions, may occur in DNA strands, which may lead to wrong binding results. The most common types of substitution errors are transitions (C by T, T by C, A by G, G by A) and transversion (C by A, A by C, C by G, G by C, T by A, A by T, T by G, G by T).

- PCR mistakes. The polymerase enzymes do make mistakes when they are synthesizing copies of the DNA strands.

To avoid difficulty in reducing errors at later stages, our strategy is to address the problem at the early stage of DNA computation. The high error-resistant coding approach may be used to reduce the error rate at the problem coding stage. Next, the idea of a genetic code-based algorithm is proposed to reduce the error rates.

2.1 The Genetic Code is Highly Error-Resistant

Researchers broke the genetic code in the early 1960s. The code was a triplet code based on three-letter codons. The complete genetic code is shown in Figure 1, where the 64 triplets stand for one or another of the 20 amino acids, and the stop codons. AUG coded for methionine is the start codon, and the initial signal for translation. Three of the codons, UAA, UAG, and UGA, are stop codons, or chain terminators; when the translation machinery reaches one of these codons, translation stops, and the polypeptide is released from the translation complex. Clearly, a given amino acid may be encoded by more than one codon, but a codon can code for only one amino acid. It is called redundancy and overlapping. The redundancy and overlapping are not evenly divided among the amino acids. For example, methionine and tryptophan are

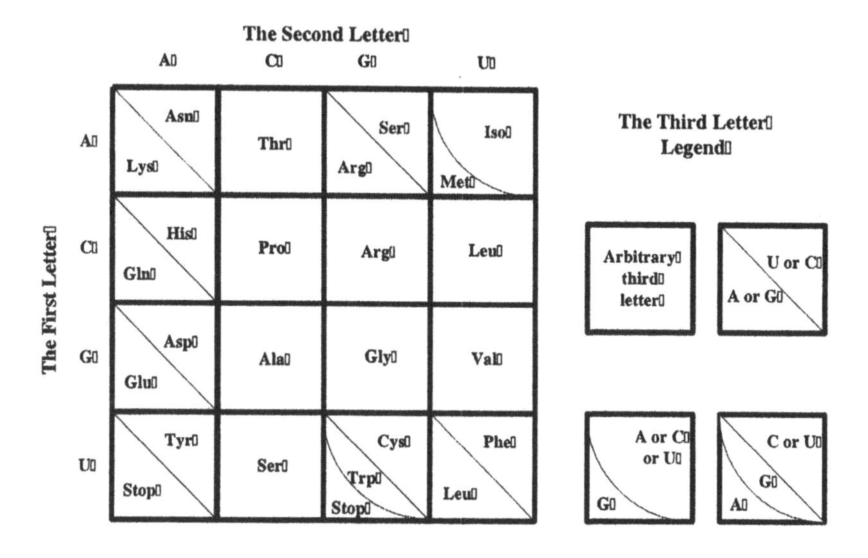

Figure 1: Table of the genetic code.

represented by only one codon each, whereas leucine, arginine and serine are represented by six different codons.

The genetic code provides the specificity for protein synthesis. The genetic information in an mRNA molecule can be thought of as a series of nonoverlapping three-letter "words." Each sequence of the three nucleotides along the chain specifies a particular amino acid. Each codon is complementary to the corresponding triplet in the DNA molecule from which it is transcribed [10].

The fact that there are 20 amino acids means that a remarkably large number of proteins can conceivably be constructed, far more in fact than ever have actually existed. Consider, for example, that an average protein contains about 200 amino acids. The number of possible 200 amino acid proteins is 20 raised to the 200th power, or 2×10^{200}. This number is astonishingly large, and the size of the number is emphasized by the fact that the number of distinct proteins present in a cell is probably below 10^5.

The genetic code has great quality assurance. One reason for this great quality assurance is the redundancy and overlapping of the genetic code. Redundancy and overlapping are good for error resistance. Because of the redundancy and overlapping, some misbinding results in no change in the coding word. These misbinding bases are called silent or synonymous errors, which are not expressed in protein expression. Similar to silence mutations in cells, the silent or synonymous errors are quite common, and account for genetic diversity that is not expressed as phenotype differences.

From a mathematical point of view, the genetic coding system seems to be an optimal solution. Because there are only four base letters (A,G,C,U), a one-letter code clearly could not unambiguously encode 20 amino acids. A two-letter code could contain only $4 \times 4 = 16$ codons—still not enough. But a triplet code could contain up to $4 \times 4 \times 4 = 64$ codons. It seems that it is enough and would have enough redundancy for errors during transcriptions and translation. There is no necessity for a four-letter code, which could contain $4 \times 4 \times 4 = 215$ codons for 20 amino acids, and seems to be too inefficient. If a DNA computation problem is coded using redundant genetic codes, the code should be highly error-resistant and the error rates low. This is our idea of problem coding for DNA computation.

2.2 Genetic Code-Based DNA Computation

Different from biological systems that are highly diverging and require a large set of DNA codons for coding genetic information, the set of "codons" for DNA computation may be small. To take advantage of chemical properties of the molecular structures and obtain a reduced set of "codons" (called a code set for DNA computation), we reduce the regular codons of biological systems to obtain a small coding set. First, all coding sets should be closed. i.e., elements of the coding set and their complements (anticodon) are within the same coding set. This is based on a concern that DNA computation does not use the same mechanism as biological systems for recognizing DNA strands. The slack on distinguishing a code and its complement will make the DNA computation easier and robust. In the meantime, it is preferred to still use the triple code mechanism. There may be biochemistry and stability reasons for the triple codes, though it is not clear currently. It is better to still keep this feature. Start by reducing sets of codons from biological systems as follows.

- Methionine (start codon): AUG. To avoid confusion, the complement of the codon, UAC (tyrosine), is also used as a start code for DNA computation.

- Serine: use UCU, UCC, UCA, UCG, AGU, and AGC, where the complements of the last four codons fall nicely into the set itself. So, AGA, AGG, AGU, AGC, UCA and UCG are all used to code serine.

- Proline: use CCU, CCC, CCA, and CCG. Then, GGA, GGG, GGU, and GGC coded for glycine are used to code proline.

- Alanine: use GCU, GCC, GCA, and GCG. Then, CGA, CGG, CGU, and CGC are also used to code alanine.

- Leucine: use CUU, CUC, CUA, CUG, UUA, and UUG. Then, GAA, GAG, GAU, GAC, AAU, and AAC are also used to code leucine.

- Valine: use GUU, GUC, GUA, and GUG. Then, CAA, CAG, CAU, and CAC are used to code valine.

- Threonine: use ACC, ACA, and ACG. Then, UGG, UGU, and UGC are also used to code threonine.

- Phenylalanine: use UUU and UUC. Then, AAA and AAG can also be used to code phenylalanine.

- Tyrosine: use UAU. Then, AUA is used as well.

- Stop codons: use UAA, UAG, and UGA. Then, AUU, AUC, and ACU are used to code the stop code.

The above coding reductions result in seven coding sets, which are enough to code significantly large problems by varying the length of wording. For the convenience of DNA computation, the above coding sets can be further simplified and renamed as follows and shown in Figure 2.

- Start coding set: AUG, AUA, UAC, or UAU.

- Coding set 1: GAA, GAG, GAU, GAC, CUU, CUC, CUA, or CUG.

- Coding set 2: CAA, CAG, CAU, CAC, GUU, GUC, GUA, or GUG.

- Coding set 3: AAA, AAG, AAC, AAU, UUU, UUC, UUG, or UUA.

- Coding set 4: ACC, ACA, ACG, ACU, UGG, UGU, UGC, or UGA.

- Coding set 5: CCU, CCC, CCA, CCG, GGA, GGG, GGU, or GGC.

- Coding set 6: GCU, GCC, GCA, GCG, CGA, CGG, CGU, or CGC.

- Coding set 7: UCU, UCC, UCA, UCG, AGU, AGC, AGA, or AGG.

- Stop coding set: UAA, UAG, AUU, or AUC.

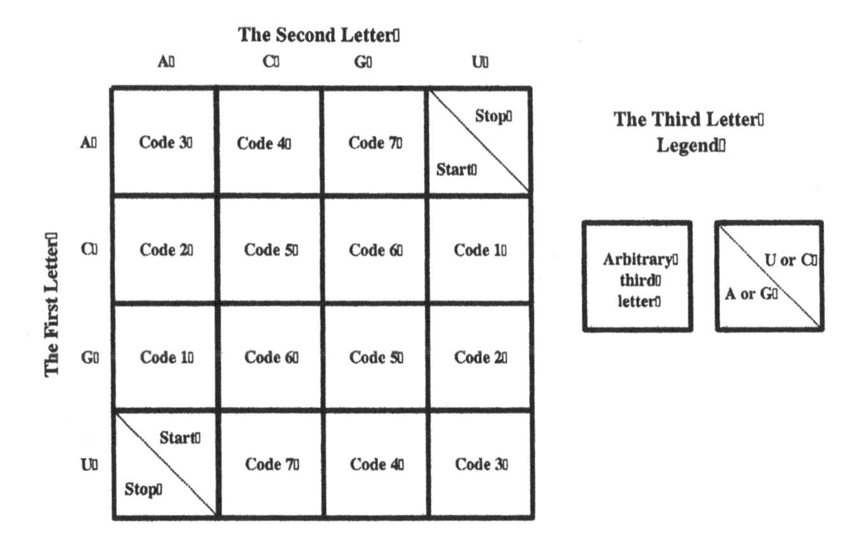

Figure 2: New code set.

It can be easily checked that all the above coding sets are closed with respect to complementary operations. i.e., the complementary base of each member of the set falls into the set itself. This can be proved by construction. In addition, the mappings from a code to the code set are multiple-to-one and most of them are symmetric. This will allow a significant amount of overlapping, which leads to high error resistance. By coding this way, many DNA base mutations remain silent.

For DNA computation, the next step is to apply DNA molecular techniques to obtain biological solutions including extraction and interpretation of biological meanings from the solutions. The above coding approach does not affect any of these operations. The above idea has been applied to solve an NP-complete problem, the Hamiltonian Path Problem (HPP).

2.3 DNA Computation for Solving the Hamiltonian Path Problem

The HPP problem is to find (if there is) a Hamiltonian path for a given graph. A Hamiltonian path is a sequence of compatible one-way edges of a directed graph that begins and ends at a specified vertex and enters every other vertex exactly once. Known algorithms for this problem have exponential worst-case complexity. The problem has been proved to be NP-complete.

To apply the above ideas, the first step is to code the problem. Assume that the graph has $n > 0$ vertices (cities) and i is the index of a vertex. The following steps are usually followed.

1. Associate the start vertex $i = 1$ with one code (three bases) from the start coding set. Associate the end vertex $i = n$ with a code from the stop coding set. It is preferred to have all edges coded at the same length for the convenience of later sequence extraction. For the start and end cities, only the first three DNA bases are needed. To match sequence length requirements for each vertex, fill in the rest of the sequence with random DNA bases.

2. Associate each other vertex $i(1 < i < n)$ with a $3m$-mer sequence of DNA generated by m codes (usually an even number to keep left and right edges of a city the same length) from Figure 2, and denote it by O_i, whose complementary is denoted by $\bar{O}_i$. For each edge $i \rightarrow j$, an oligonucleotide $O_{i \rightarrow j}$ is created, which is the 3' $3m/2$-mer of O_i followed by the 5' $3m/2$-mer of O_j. This construction preserves edge orientation. The $\bar{O}_i$ serve as splints to bring oligonucleotides associated with compatible edges together for ligation.

3. Keep only those paths that begin with codes from the start coding set, and end with codes from the stop coding set. This can be done by PCR amplifying products of Step 1 using primers starting with codes from the start coding set or the stop coding set. Thus only those molecules encoding paths, which begin with vertex 1 and end with vertex n, are amplified.

4. Keep only those paths that enter exactly N vertices. N is the total number of vertices of the graph. The product of Step 2 was run on an agarose gel and the $3m$ base pair band (corresponding to dsDNA encoding paths entering exactly n vertices) was excised and soaked to extract DNA.

5. Keep only those paths that enter all vertices of the graph exactly once. To achieve this by affinity, purify the product of Step 4 using a biotin–avidin magnetic beads system. This can be done by first generating single-stranded DNA from the dsDNA product of Step 4 and then incubating the ssDNA with $\bar{O}_2$ conjugated to magnetic beads. Only those ssDNA molecules containing O_2 (and hence encoded paths that entered vertex 2 at least once) annealed to the bound $\bar{O}_2$ and were retained. The process is repeated successively with $\bar{O}_3, \bar{O}_4, \ldots, \bar{O}_{n-1}$ and $\bar{O}_n$.

6. The remaining DNA sequences (paths) in the test tube represent solutions. To implement them, the product of Step 5 is PCR amplified and

run on a gel.

To further illustrate the idea, consider an $n = 7$ vertex HPP graph as given in [1]. We use 12 DNA bases (4 codes from the coding sets) to uniquely code each of the 7 cities. The following is one possibility to code the cities.

- City 1: code the start city with AUG plus three additional DNA bases to match the length requirements. Only the first three bases are used to identify the city. We choose AUGUAC here.
- City 2: one edge is coded by GAACAG. Theoretically, any pair of codes from the coding set 1 and coding set 2 can be used. The other edge is coded by any pair from coding set 3 and coding set 4. AAAUGC is used here.
- City 3: one edge is coded by any pair of codes from coding set 2 and coding set 3. We choose CAAUUC. The other edge is coded by any pair from coding set 4 and coding set 5, such as ACAGGA.
- City 4: one edge is coded by any pair of codes from a pair of coding set 5 and coding set 6. CCUGCU is used here. The other edge is coded by any pair from coding set 7 and coding set 1, such as UCUCUG.
- City 5: one edge is coded by any pair of codes from coding set 6 and coding set 7. GCAUCG is used here. The other edge is coded by any pair of codes from coding set 1 and coding set 3, such as CUAUUC.
- City 6: one edge is coded by any pair of codes from coding set 2 and coding set 7. GUCAGC is chosen here. The other edge is coded by any pair from coding set 1 and coding set 4, such as CUAACG.
- City 7: code the end city with UAG, plus three random DNA bases to match the length requirement. Only the first three DNA bases will be used to identify the city. We choose UAGAUC.

Figure 3 shows a graph of codes for each edge and vertex. Upon coding all cities, the rest of the work is the same as Adleman has done in his experiment [1]. The same conclusions should be drawn from the experiment. The difference is that wide various instances of DNA strand solutions will be left in the final test tube. By doing this, more errors for sequence binding and extraction are allowed. The process is more error resistant.

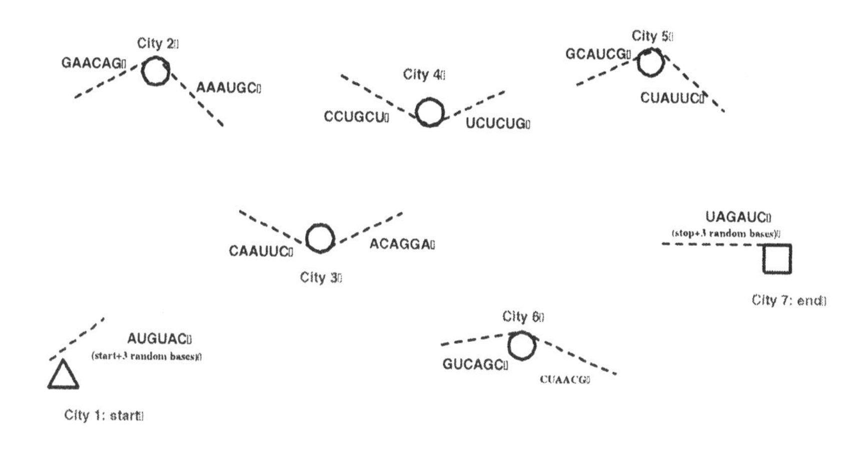

Figure 3: Coding for the seven cities of the HPP problem.

3 An Abstract Mathematical Model of DNA Computation

An abstract mathematical model of DNA computation is necessary for building relations between physical problems and the DNA computational mechanisms. Define the following.

- $X = x_i x_{i+1} ... x_j, Y = y_i y_{i+1} ... y_j$ as single-stranded DNA strings, where $i \in N$, $j \in N$, and $i < j$. N is the natural number. $x_i, y_i \in A, T, G, C$.
- The complementary sequence of X is expressed as $\bar{X}$, which follows the exact binding properties of Watson–Crick complementary principles.
- T_i is defined as a test tube, where $i \in N$. $T_i(+X)$ means a test tube containing a string X. $T_i(-X)$ represents a test tube that does not contain the string X.

As discussed earlier, DNA computation involves biomolecular techniques including hybridization, separation, cutting, and pasting at desired locations. These operations can be generalized and classified at DNA strand or test tube levels as follows, where "→" is used to point to "the results of" reactions from the left side operations.

- Ligation: plus operation "+". The ligation operation concatenates strands of DNA. It is often invoked after an annealing operation. Although it is possible to use some ligase enzymes to concatenate free-floating double-stranded DNA, it is more efficient to allow single strands to anneal together, connecting a series of single-strand fragments, and then use ligase to seal the covalent bonds between adjacent fragments. For two single-stranded DNAs X and Y, a ligation process can be expressed as "$X + Y \rightarrow XY$", where XY represents a new single-stranded DNA.

- Cut: minus operation "−". Restriction enzymes can cut a strand of DNA at a specific address of a subsequence, which is usually on the order of 4 to 8 nucleotides. Some restriction enzymes will only cleave single-stranded DNA, whereas others only cleave double-stranded DNA. If a single-stranded DNA X is cut at the position n from the $3'$ end, the process can be expressed as "$-X(n) \rightarrow Y + Z$", where Y and Z are two newly generated single-stranded DNAs and Y has a length of $n - 1$.

- Hybridization: multiplication operation "•". It is also called annealing, which is a process when single-stranded complementary DNA sequences spontaneously form a double-strand DNA in solution through hydrogen bonding. For a single-stranded DNA X, the DNA binding process can be described as "$X \bullet \bar{X} \rightarrow (X\bar{X})$", where $(X\bar{X})$ is a new double-stranded DNA and is stable. The strand cannot be bonded with other strands except after a further melting operation.

  Similarly, the operation can be used to anneal a short-stranded DNA to a longer-stranded DNA so that the shorter DNA strand extends from the end of the longer strand. Then the complement of the subsequence that is extended from the end can be added to the longer strand either through the use of a polymerase enzyme, or through introducing and annealing the sequence, cemented by the ligase.

- Melting: division operation "\". It is an inverse of the annealing process, i.e., the separation of double-stranded DNA into single-stranded DNA by raising the temperature. Heating can be selectively used to melt apart short double-stranded sequences while leaving longer double-stranded sequences intact. For example, "$\backslash(X\bar{X}) \rightarrow X + \bar{X}$" means melting the double-stranded DNA as two complementary single-stranded DNA segments.

In addition to strand-level operations, DNA computation also includes the following operation at the test tube level, which is performed using sets of DNA strands.

- Mergence: union operation $\cup$. It means two test tubes can be combined, usually by pouring one test tube into the other. "$T_1 \cup T_2 \rightarrow T$" represents adding two test tubes T_1 and T_2 together to produce a new test tube T.

- Separation: difference operation $-$. "$-(T_i, X) \rightarrow T_1(+X) \cup T_2(-X)$" represents a separation operation applied to test tube T_i. The operation produces two test tubes, where T_1 contains a string X and T_2 does not contain the string X. Either T_1 and T_2 could be an empty set ϕ. This step is done with gel electrophoresis and requires the DNA strands to be extracted from the gel once the strands of different length have been identified by some form of staining or radioactive tagging.

  Separation paths go through all vertices by (1) constructing a set of magnetic beads attached to the complementary sequence of the vertex, and (2) magnetically extracting beads binding to the complementary sequences.

- Amplification: product operation $\times$. Given a test tube of DNA strands, make multiple copies of a subset of the strands presented. Copies are made with PCR. PCR requires a beginning and an ending subsequence, called "primers", to identify the sequence to be replicated. Copies of these substances anneal to the single strands and polymerase enzymes build the complementary strands, and the process repeats, doubling the number of strands in the test tube each cycle. For example, "$\times T \rightarrow T_1 \cup T_2$" means two test tubes T_1 and T_2 are produced from one test tube T.

- Detection: question operation ?. After PCR amplification, gel electrophoresis is applied to see if anything of the appropriate length is left. For example, "$T?X \rightarrow$ True" means test tube T contains at least one string X. Otherwise, "$T?X \rightarrow$ False".

- Destroy: empty set intersection operation $\cap$. Subsets of strands can be systematically destroyed or "digested" by enzymes that preferentially break apart nucleotides in either single- or double-stranded DNA. The process can be expressed as "$T \cap \phi \rightarrow \phi$".

Assume two test tubes T_1 and T_2. Then some steps of the HPP can be programmed and described at the test tube level as:

- Amplification: $\times T_i (i \in N)$ generates new large sets of sequences.

- Mergence: $T_1 \cup T_2$ generates new test tube T.

- Separation: $-(T, X)$ separating X from test tube T, where X is DNA strands of length 84 and pass all cities.
- Detection: $T?X$ to find an answer.

The problem can be programmed and described at DNA strand level as:

- For a city coded by $X_iY_i, i \in N$, hybridization may lead to $(X_i\bar{X}_i)(Y_i\bar{Y}_i)$. The process can be described as $X_iY_i \bullet \bar{X}_i\bar{Y}_i \rightarrow (X_i\bar{X}_i)(Y_i\bar{Y}_i)$.
- Ligation also happens during the process. $\bar{Y}_{i-1}(X_i\bar{X}_i)(Y_i\bar{Y}_i)\bar{Y}_{i+1} + X_{i-1}Y_{i-1} \rightarrow X_{i-1}(Y_{i-1}\bar{Y}_{i-1})(X_i\bar{X}_i)(Y_i\bar{Y}_i)\bar{Y}_{i+1}$.

It is interesting to note that most operations for conventional computers require combinations of a number of basic operations. However, they may be completed by a single operation of DNA computation. The above symbolic operation may be used to directly program DNA algorithms.

3.1 Convert Character-Based DNA Sequence to Numerical Domain

To further convert the problem of DNA computation to the mathematics domain, the following mathematical functions and operations are defined. The goal is to connect the physical operations of DNA computation to the mathematical operations.

Define a function $f(x)$ of DNA bases as

$$f(x) = \begin{cases} 1, & x = A; \\ -1, & x = T; \\ i, & x = G; \\ -i, & x = C, \end{cases} \tag{1}$$

where A valued as 1 only pairs with C valued as -1, and G valued as i only pairs with C valued as $-i$. The mapped numerical values are of opposite sign to each other. The above definition can be expressed in a complex number domain as in Figure 4.

Then, the complementary base of each DNA base x can be determined by the following inverse function

$$y = f^{-1}(-f(x)) = \begin{cases} T, & x = A; \\ G, & x = C; \\ C, & x = G; \\ A, & x = T. \end{cases} \tag{2}$$

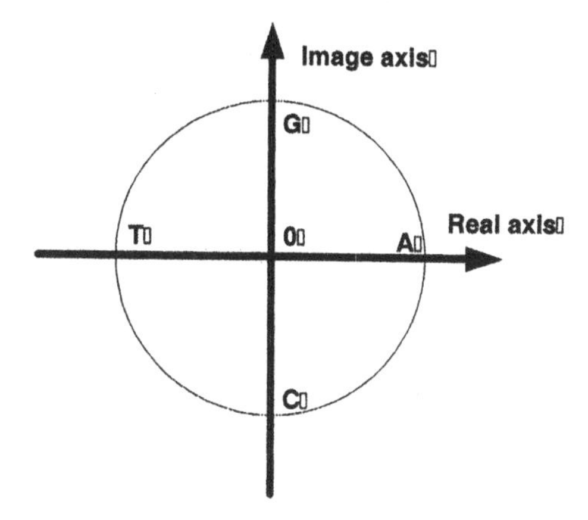

Figure 4: Convert DNA bases to complex numbers.

By applying equations (1) and (2), the complementary sequence (both numerical or character values) of any DNA strand can be easily obtained. This also means only single-stranded DNA needs to be worked out during the problem formulation stage of DNA computation. The complementary strands can be automatically generated.

Instead of mapping DNA bases as complex numbers, DNA bases may be mapped as integer numbers only. Define a function $f(x)$ of DNA bases as

$$f(x) = \begin{cases} 0, & x = A; \\ 1, & x = C; \\ 2, & x = G; \\ 3, & x = T. \end{cases} \tag{3}$$

Similarly, the complementary base of x can be determined by the following inverse function.

$$y = f^{-1}(\{3\} - f(x)) = \begin{cases} T, & x = A; \\ G, & x = C; \\ C, & x = G; \\ A, & x = T. \end{cases} \tag{4}$$

For example, the numerical sequence of a DNA segment $X = AGGCAT$ is $f(X) = f(AGGCAT) = 022103$ by the above definition. The complementary segment of X can be easily obtained as $Y = f^{-1}(\{3\} - f(X)) =$

$f^{-1}(311230) = TCCGTA$, where $\{3\}$ represents a sequence of 3 with length equal to the DNA segment. The mathematical operation is conducted base by base. The above function space expression converts a DNA segment to a numerical sequence, which has nice properties for mathematical sequence analysis.

3.2 A Vector Space Expression of DNA Sequences

Consider the four DNA bases $[A, T, G, C]^T$ as a basis of a vector space, for any DNA strand $X = x_1 x_2 ... x_i ... x_n$, it can be expressed as a vector by a transfer matrix Π as

$$\begin{aligned} X &= \Pi \begin{bmatrix} A \\ T \\ G \\ C \end{bmatrix} \\ &= \begin{bmatrix} p_{11} & p_{12} & p_{13} & p_{14} \\ p_{21} & p_{22} & p_{23} & p_{24} \\ \vdots & \vdots & \vdots & \vdots \\ p_{n1} & p_{n2} & p_{n3} & p_{n4} \end{bmatrix} \begin{bmatrix} A \\ T \\ G \\ C \end{bmatrix}, \end{aligned} \tag{5}$$

where $\sum_{j=1}^{4} p_{ij} = 1, \forall i \in N$, and

$$\begin{aligned} p_{i1} &= \begin{cases} 1, & x_i = A. \\ 0, & otherwise. \end{cases} \\ p_{i2} &= \begin{cases} 1, & x_i = T. \\ 0, & otherwise. \end{cases} \\ p_{i3} &= \begin{cases} 1, & x_i = G. \\ 0, & otherwise. \end{cases} \\ p_{i4} &= \begin{cases} 1, & x_i = C. \\ 0, & otherwise. \end{cases} \end{aligned} \tag{6}$$

The complementary sequence of $\bar{X}$ can be easily obtained by swapping column one with column two, and column three with column four. It is

$$\bar{X} = \begin{bmatrix} p_{12} & p_{11} & p_{14} & p_{13} \\ p_{22} & p_{21} & p_{24} & p_{23} \\ \vdots & \vdots & \vdots & \vdots \\ p_{n2} & p_{n1} & p_{n4} & p_{n3} \end{bmatrix} \begin{bmatrix} A \\ T \\ G \\ C \end{bmatrix}. \tag{7}$$

For example, for a single-stranded DNA $X = ACGTGGATCT$, its transfer matrix is

$$\Pi_1 = \begin{bmatrix} 1 & 0 & 0 & 0 \\ 0 & 0 & 0 & 1 \\ 0 & 0 & 1 & 0 \\ 0 & 1 & 0 & 0 \\ 0 & 0 & 1 & 0 \\ 0 & 0 & 1 & 0 \\ 1 & 0 & 0 & 0 \\ 0 & 1 & 0 & 0 \\ 0 & 0 & 0 & 1 \\ 0 & 1 & 0 & 0 \end{bmatrix}. \quad (8)$$

The complementary sequence of X is $\bar{X} = TGCACCTAGA$, whose transfer matrix is

$$\Pi_2 = \begin{bmatrix} 0 & 1 & 0 & 0 \\ 0 & 0 & 1 & 0 \\ 0 & 0 & 0 & 1 \\ 1 & 0 & 0 & 0 \\ 0 & 0 & 0 & 1 \\ 0 & 0 & 0 & 1 \\ 0 & 1 & 0 & 0 \\ 1 & 0 & 0 & 0 \\ 0 & 0 & 1 & 0 \\ 1 & 0 & 0 & 0 \end{bmatrix}. \quad (9)$$

The above definitions make it possible to define each DNA strands as vectors. To illustrate some nice theorems for this formulation, the following definition and theorems are proposed.

Definition: Equivalent Transfer Matrix. Because each strand of a double-stranded DNA uniquely determines the other strand, each single DNA strand can be alternatively used to describe the same DNA double strand. Transfer matrices of a single-stranded DNA and its complement are called equivalent to each other. For example, the above Π_1 is an equivalent transfer matrix of Π_2 and vice versa.

Theorem 3.1 *Under the formulation of (5), two DNA transfer matrices are equivalent, if and only if one matrix is the result of swapping column one with column two, and column three with column four of the other matrix.*

Proof Suppose the above statement is not true. By mathematical contradiction, the Watson–Crick complementary principles of DNA strands are violated.

■

Theorem 3.2 *Under the formulation of (5), two DNA sequences are complementary to each other, if and only if their transfer matrices are equivalent.* ■

The proof is straightforward.

Theorem 3.3 *Necessary condition for similar sequences: if two DNA sequences are similar, then sums of each column of the transfer matrices are similar.*

Proof Because the two sequences are similar, they at least have similar numbers of A, T, G, C bases. The value of each column of the transfer matrix represents numbers of DNA bases. As a result, sums of each column have to be similar. ■

To code a problem for DNA computation, it is critical to make sure that all codes are not similar to each other and they are not complementary to each other. The above theorems may be used to check similarities and complementary properties of the DNA sequences.

The above mathematical formulations may also inspire an idea of hybrid DNA computation. Because the DNA computation is remarkable for its ability of parallel computation, and conventional computers are efficient for routine computations, such as matrix comparison and algebraic operations, a computational problem may be coded as a DNA computation at core, and peripheral conventional computing devices used for routine calculations. The issue deserves to be further investigated.

4 Discussion and Conclusions

Even though DNA computation is still an open problem for practical implementation, numerous studies have suggested that it is a great field for fast computation to solve large NP-complete problems. Currently, error resistance is a major concern for DNA computation. This chapter proposes a genetic code-based DNA computation method to solve the problem. The new method has high error resistance because of its advantage of redundant and overlapping coding. The idea is inspired by the genetic code of biological systems, except that the codon sets have been reduced for the purpose of DNA computation, because DNA computation may not need significant coding sets to code biological diversities. In addition, a framework of mathematical formulation of the DNA computation has been presented for the first time. Some theorems of the formulation have also been presented based on the abstract mathematical model.

DNAs are miraculous little machines that store energy and information, they cut, paste, and copy. Similar to the way that electronic computers are used to control electrical and mechanical systems, a DNA computation mechanism may be used to control chemical and biological systems by being engineered into living cells and allowing them to run precise digital programs that interact with their natural biochemical processes. This has been believed to be the most attractive method for DNA computation applications.

The biological systems are good examples of communication networks, even though many of those communication mechanisms have not been fully understood. One example is the nerve cells, which are often organized into complicated intercommunicating networks. Typically each neuron is physically connected to tens of thousands of others. Using these connections (links), neurons (nodes) can pass electrical signals to each other. The communication between two arbitrary cells are not necessary to be directly connected. Some cells may need to go through alternative nodes (cells) to communicate with each other. Finding an optimal path for communication is equivalent to the Hamiltonian path problem. DNA computation may help to find a possible optimal solution in this case.

From the above discussion, it is clear that knowledge about the combinatorial optimization problem is beneficial for DNA computation; the advances in DNA computation may be applied to solve the complex problems that are hard to tackle using traditional combinatorial optimization methods.

References

[1] L. Adleman, Molecular computation of solutions to combinatorial problems, *Science*, v.266, pp. 1021–1024, Nov. 1994.

[2] D. Boneh and R. Lipton, Making DNA computers error resistant. *Technical report, Princetion University*, CS-TR-491-495, 1996.

[3] E. Baum and D. Boneh, Running dynamic programming algorithms on a DNA computer. *the 2nd DIMACS workshop on DNA based computers*, pp. 141–147, 1996.

[4] F. H. C. Crick, The origin of the genetic code. *Journal of Molecular Biology*, vol. 38, pp. 367–379, 1968.

[5] T. Eng and B. Serridge, A surface-based DNA algorithm for minimal set cover. *the 3rd DIMACS workshop on DNA based computers*, pp. 74–82, June 1997.

[6] L. R. Jaisingh. Theory and Problems of Abstract algebra. *McGraw-Hill*, second edition, 2004.

[7] L. Landweber and R. Lipton. DNA2DNA computations: A Potential "Killer App"? *the 3rd DIMACS workshop on DNA based computers*, pp. 59–68, June 1997.

[8] R. Lipton. Using DNA to solve NP-complete problem. *Science*, vol. 268, pp. 542–545, April 1995.

[9] C. C. Maley. DNA computation: Theory, practice, and prospects, *Evolutionary Computation*, vol. 6, no. 3, pp. 201–229, 1998.

[10] W. K. Purves, D. Sadava, G. H. Orians, and H. C. Heller. Life: The Science of Biology, *Sinauer*, 6th edition, 1994.